CHEMISTRY OF THE COVALENT BOND

Other *Freeman* Books in Chemistry

Chemistry of
the Covalent Bond

by LEALLYN B. CLAPP Brown University

Drawings by Evan Gillespie

W. H. Freeman WITHDRAWN **and Company**

SAN FRANCISCO AND LONDON

The Brown Plan
for Teaching Chemistry

For some time before Brown University adopted its new plan for teaching chemistry in 1948, the Department of Chemistry—especially Professor Paul C. Cross—had been dissatisfied with the course of study that it was offering.

Physical chemistry had gradually permeated general chemistry, qualitative analysis, and quantitative analysis to such a degree that the course in physical chemistry could offer little that was new. Professor Cross, in teaching physical chemistry, found that he was unable to select subject matter from which the cream had not already been skimmed in previous courses. He could, to be sure, present the subject matter with greater thoroughness and rigor, could reinforce the concepts and principles that had been only loosely learned; but the course suffered because he had been left with nothing excitingly new and fresh to offer. He felt that a course in physical chemistry, if it was to offer new material, would have to be given earlier in the program of study.

A sound development of the principles of physical chemistry in the freshman year would have been possible if the subject matter of freshman chemistry could have been properly treated by the methods of physical chemistry; but it could not be, for the freshman did not usually have the background in mathematics and physics that was requisite for the best presentation of physical chemistry. The solution adopted at Brown was to present physical chemistry in the second year. Subsequent courses in inorganic and analytical chemistry benefit from the rigorous prerequisite in physical principles. This required that the course offered in the first year, while the student was getting his background in mathematics (including calculus) and physics, should be descriptive.

The chemistry that is still primarily in the descriptive stage, that has been most difficult to treat mathematically, is what we have called the "chemistry of the covalent bond," which is largely the

v

chemistry of carbon, hydrogen, nitrogen, oxygen, sulfur, phosphorus, and the halogens. (Although the elements that form compounds containing covalent bonds are not confined to the upper right-hand corner of the Periodic Table, the elements named are those that furnish the best examples and the most generalizations.)

The first-year course at Brown University, then, is this Chemistry of the Covalent Bond. The second year of chemistry is a study of the Physical Principles of Chemistry, which takes advantage of the freshman year of physics and calculus. In the third year, with a good knowledge of atomic and molecular structure and the covalent bond, and with the gas laws and such concepts as electrode potential, ionization potential, electrical conductivity, and phase transition available as tools, the student undertakes a thorough study of inorganic chemistry. Then, with a solid foundation in organic, physical, and inorganic chemistry, he takes up analytical chemistry.

Conspicuously absent from this program is the systematic scheme of analysis. Volumetric techniques are taught in the laboratory of the second course, however, and some ion separations are carried out in the inorganic chemistry laboratory (fifth semester). The chemistry of sulfides is not considered essential.

These fundamental courses lead to advanced courses in organic, inorganic, physical, industrial, and biological chemistry. A year of quantitative analysis is required in the last semester of the junior year (gravimetric analysis) and the first semester of the senior year (instrumental analysis).

This program has had the benefit of eight years of experimentation at Brown and has withstood the criticism of nine new members of the department. As authors of the two introductory textbooks we are grateful to Brown University and to Professor Charles A. Kraus and the late Provost Samuel T. Arnold for maintaining a challenging atmosphere in which to work. We are especially indebted to Professor Paul C. Cross, who encouraged us to try a new approach to the teaching of undergraduate chemistry.

March 30, 1957   Leallyn B. Clapp
                                                     Robert H. Cole
                                                     James S. Coles

Preface

This book was written to fill the need for a text for the first-year course in chemistry at Brown University. Admission to this course presupposes the advantage of a year of high-school chemistry, and the concepts learned there are taken for granted.

Though I rejected the title "The Influence of Geometry on Chemistry," that phrase does describe the emphasis that is laid on structure. The implications of directed valence are not grasped in one application but gradually become part of the chemist through persistent cultivation. A favorable soil for this growth is provided by frequent return to the structural concepts of the first chapter.

The introduction of oxidation-reduction by means of examples chosen mostly from organic compounds has been eminently more satisfactory than its introduction by means of inorganic compounds in the other beginning courses that I have taught. The pitfalls of variable valence cannot be avoided entirely, but the students are protected by being held accountable for only a limited number of oxidizing agents (notably permanganate ion in acidic or basic solution and dichromate in acidic solution), for which they are asked to learn the changes in oxidation number. Nearly all the reducing agents are organic and contain the common functional groups. A hybrid method of determining oxidation numbers is presented, and the ion-electron method is used for balancing equations. Enough exercises are given to instill confidence in this procedure.

Mechanisms of reaction have not been emphasized, but neither have they been avoided. It has seemed desirable, at this stage, to dwell at some length on a few mechanisms—base-catalyzed aldol-type condensations, differences in olefin and carbonyl additions, and substitution in the aromatic ring (Friedel-Crafts, etc.)—rather than to give detailed accounts of a larger number of mechanisms.

A mechanism is used largely as a tool for learning and relating reactions rather than as an end in itself.

Chapter 18, "Evolution of the Covalent Bond," is difficult reading for freshmen, for by this time they know too much chemistry. Once the concepts of atoms, molecules, and electrons have been used for a year in high-school and a semester in college chemistry, it is difficult to separate oneself from these ideas and become an unprejudiced observer of 1850. But the attempt is rewarding if it leads the student into the stacks of a library to see what an original source is and if it gives him a glimpse of scientific ideas in the process of evolution. Incentive to do this is offered by a selected number of original references and by a gentle push in the direction of the library. Many students have been delighted to learn that a foreign language is useful at this level.

Part Five, "Inside the Tetrahedron," appears to me to cover a segment of chemistry that is neglected because no teacher has felt responsible for it. The teacher of organic chemistry has felt that his loyalty was to carbon compounds and, of course, has always had too much to talk about without seeking other territory. Consequently, the effect of changing the central atom in a tetrahedron to an element other than carbon is not often taken up in the first organic course. The teacher of general chemistry does not have the benefit of the student's acquaintance with the organic method of studying the chemistry of homologous series. Hence each compound would have to be treated separately, if at all, and this is not done in such courses. But the chemistry of this group of substances, especially of the covalent halides and oxyhalides (Chap. 26), is subject to some happy generalizations when treated as the chemistry of a functional group.

Since I have felt the desirability of a textbook that would not be a ponderous volume, I have included only the subject matter to which Brown freshmen students are actually exposed in a one-year course. For students who are not chemistry majors (at Brown mostly pre-meds and pre-nurses) the teacher can give a vigorous one-semester course at the junior level from the first twenty chapters by omitting Chapters 1, 9, and 13 and part of Chapter 18.

I am pleased to have the permission of Professor Gene B. Carpenter to include as part of Chapter 1 a presentation that he wrote for another purpose. I am grateful to my colleagues for criticism both barbed and gentle, especially to Professor (as he was then) James S. Coles and Professor Harold R. Nace, who sat through my lectures one year, and to Mr. Robert M. Sherman, who taught the course one year and created many of the exercises. My thanks go to Mrs. F. N. Tompkins and Miss Helen E. Lutes, who survived several mimeographed editions in the making, to Miss Eleanor J. Policelli, who did tedious work on the index, and to Mr. Gordon L. Willette, who gave the text a critical reading as a student.

March 30, 1957   Leallyn B. Clapp

Foreword to the Student

It has been assumed that a number of ideas and definitions that you learned in high-school chemistry are still at your command. If this is not so, you should review the following topics in a textbook of general chemistry:

the metric system of measurement

the meanings of a chemical equation and the quantitative use of equations in weight and volume problems

why reactions go to an end (three reasons)

how to write formulas for simple inorganic compounds such as $NaHSO_3$, $KMnO_4$, Na_2SO_4, $CaCO_3$, and H_2S

the importance and some uses of the periodic system

chemical symbols (how many of the first twenty elements can you name?)

the gas laws

the elementary chemistry of O, N, Cl, H, S, H_2O, Na, K, Ca, and Al

Dalton's atomic theory

Le Châtelier's principle

the calculation of percentage composition from a formula

The following terms are used in this text without definition:

acid	compound	molecule
alkali	concentrated	molecular weight
allotropic form	solution	neutron
amalgam	dilute solution	nucleus
amphoterism	electron	orbit
anhydrous	element	physical change
atom	exothermic reaction	proton
atomic weight	formula weight	saturated solution
boiling point	freezing point	solute
calorie (and	ion	solvent
kilocalorie)	mole	substance
chemical change		

Since you will be learning structural chemistry, you will be well advised to open this text with paper and pencil at hand. As an added incentive to do this, a number of exercises are included within the text of the chapters. The best procedure is to work these internal exercises immediately after reading the preceding paragraphs. A backward look at the forest from a point beyond the trees is given by more exercises at the ends of the chapters.

Contents

PART ONE

A FRAMEWORK OF CONCEPTS

---

1 Introduction to Science and Chemistry

"The research man may often think and work like an artist, but he has to talk like a bookkeeper, in terms of facts, figures, and logical sequence of thought."—H. D. Smyth

1-1. The Organization of Science

One aspect of science is the vast body of organized observations and theories preserved in books and journals. Such organized knowledge is called science only if it contains a fairly high proportion of theory. A biography, if tolerably well put together, is certainly organized knowledge; but it is not science unless it exhibits a serious attempt to explain actions and motives by psychological or psychiatric theories; thereupon it becomes a case history and may be considered as science.

As the understanding of various fields of study increases, the fields become increasingly scientific; so the area of knowledge included in science is continually expanding. The best that can be done, therefore, in classifying fields of learning is to list them in order from the more scientific to the less scientific, and to let the student decide where he wishes to draw the line marking off science from nonscience. The accompanying diagram (p. 4) indicates this scheme roughly and shows, by means of connecting lines, some of the relations between various fields. This diagram is not complete; historically it has proved unsound to concoct supposed limits to the fields that science may include. The boundaries between the fields are largely arbitrary, accidental, and unimportant.

What may be called the "inherent complexity" of the fields increases from top to bottom in the diagram. First, each field depends to some extent on those lying above it; to understand biology thoroughly, one first has to understand large parts of

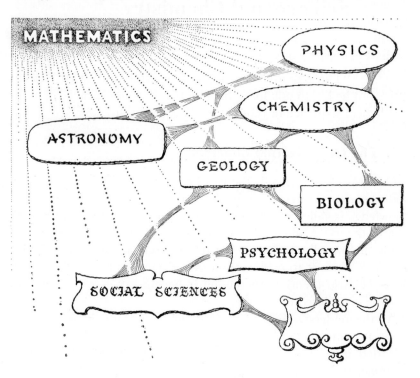

physics and chemistry. Second, the observations to be understood depend on more variables toward the bottom of the diagram. In physics, for example, the motion of a bullet depends on only a few variables, such as initial velocity and air resistance; in economics, however, stock-market fluctuations depend on so many factors that we cannot be sure that even the important ones have all been considered.

It is natural that the greatest progress should have been made in organizing the subjects of least inherent complexity; physics is therefore the most "scientific" science. But there is no special glory in working at one end of the diagram rather than at the other; the choice depends on whether one prefers to study simpler phenomena very intensively or more complicated phenomena more extensively. As the student is no doubt aware, the "practical complexity" of a subject does not follow the "inherent complexity"; physicists compensate for the simplicity of their subject by asking embarrassingly detailed questions about physical phenomena, and at the other end of the list workers have often given up trying to understand until more data are gathered.

Mathematics occupies a unique position. It is not itself a science; rather, it is the language of science. (This statement will not satisfy a mathematician, but it at least states one important aspect of mathematics.) Ordinary language is, of course, used also, but the use of mathematics to express scientific statements increases steadily. Some students may not entirely approve of this trend, but the fact is that it comes about because of the power and precision of mathematical symbolism rather than because in this way teachers can make a subject seem more difficult. For example, as long as the concept of force remained a vague philosophical idea, little progress could be made in the study of mechanics; but when it was defined in such a way that it could be measured and so subjected to mathematical manipulation, rapid progress resulted. Consider how much easier to use and to remember is the formula $F = ma$ than is the verbal form of Newton's laws of motion.

Chemistry, in particular, occupies a fundamental position toward the more scientific end of the list. It is a typical science in that it has both a good proportion of guiding theory and also much important knowledge that is only imperfectly understood and organized. Chemistry is thus a science in the making, and the study of chemistry therefore offers a good opportunity to develop an understanding of the nature of science.

Such a diagram as that given above offers to the science major a guide to the selection of courses.

1-2. The Nature of Science

"Science" is now a popular word; yet a satisfactory definition is difficult to compose. First, as we have seen, the scope of science is steadily broadening. Second, "science" means different things to different people: it means one thing to the writer of advertisements in such a sentence as "Science proves Hempo cigarettes less irritating," another to the practicing scientist. In view of such difficulties we shall be satisfied to define science by describing some of its characteristic features.

Mention of the importance of mathematical symbols suggests a useful way of regarding the relation of science to the physical world. The concepts and theories of science provide a *map* of the physical world. To most features of the physical world there are corresponding scientific concepts (and associated symbols), and to most relations discernible among these features there are corresponding laws or theories.

Since the universe is enormously complicated, it is useless to expect an absolutely accurate map. But the map that science provides is being constantly improved. And just as a rough map of a geographic region is usually better than no map at all, so approximate or incomplete scientific theories are usually better than no theory at all. In this course, as in any science course, you will encounter phenomena that are not yet adequately described by theory; and you will learn some rough, approximate theories in order to get a general view of the terrain more quickly than you could by the study of the corresponding more accurate theory. Science does not find "absolute" truth, but it approaches always closer to the truth.

Thus far science has been considered primarily as "organized knowledge," as one textbook definition has it. This is incomplete as a definition because it leaves out the scientist and the nature of his work, and it is the work that is most characteristic of science.

The professional activities of a scientist are supposed to be described by the term "scientific method." The great success of scientists in problem-solving has suggested the hope that their activities fit into a simple, consistent pattern, such as the standard sequence: hypothesis, experimental verification or rejection of hypothesis, new or improved hypothesis, and around again. In fact, however, scientific work is not so simple, and these steps are likely to be thoroughly tangled together. Rare indeed is the critical experiment that answers definitely "yes" or "no"; most experiments answer "maybe." So a description of the method of science is not easy; the following paragraphs merely emphasize certain important features.

As science grows, new concepts are invented and become important. Without the concept of molecules, for example, most of modern chemistry would disappear. But there is no systematic procedure, no orderly part of the scientific method, that leads the scientist to create fruitful concepts or to ask the right questions. It usually requires a genius to invent a truly new and useful idea, and the process is very similar to the creative process in an artist. Inspiration cannot be explained or taught. But, once a new idea has been created, we can make a rough estimate of its value by testing it against the following requirements of a good theory.

First, a scientific idea must be communicable. If a solitary worker asserts that he has a wonderful theory, and even if he can demonstrate scientific marvels, his idea does the world little good unless he can explain it to others. A new idea must be stated with clarity and precision if it is to become useful. The best way to avoid confusion is to define a new concept in such language that a prescription for its observation or measurement is evident. Scientific concepts not susceptible to measurement have rarely survived. Similarly, the statement of an idea must indicate a method by which its consequences can be observed.

Second, a good idea must be fruitful. Unless it suggests new experiments or other new ideas, it is not very useful and is generally forgotten. Similarly, it must be flexible enough to be capable of growth and modification.

Third, a theory should be economical. It should make as few and as simple assumptions as possible. An extensive theory following from a small number of key assumptions is considered to be "elegant." Note that there can be no theory without assumptions. A theory attempts to explain complicated things in terms of more fundamental things. But the most fundamental things cannot be explained in this way; instead, they must be described and inter-related by assumptions.

Fourth, a new idea should fit in with existing knowledge. Not only must it be consistent with known facts; it should also be a reasonable extension of accepted ideas. This is only a tentative requirement, for it would rule out the really revolutionary advances in science; but it is usually a good rule. This leads to the suggestion that the scientific climate must be just right for a certain theory to appear and be accepted. It is very common to find scientific discoveries made simultaneously and independently by different workers when the proper foundation has been laid. Theories have been promulgated and then have lain dormant until they were later "rediscovered" when the time was right.

A reasonably complete discussion of the nature of science in the making would require volumes of analysis of the interplay of new observations and experiments, new ideas, and new tools. This will not be attempted now, but in the body of the course attention will be directed to illustrations of this interplay.

1-3. The Study of Science

The extension of science into formerly distant fields, and the dependence of modern civilization on scientific discoveries, have progressed so far that today one cannot be a well-educated or a completely useful citizen without some knowledge both of the nature and of the facts of science.

There is no point in dwelling on the obvious dependence of civilization on science. But it seems worthwhile to emphasize the necessity of a basic knowledge of science for all citizens. In support of this statement, attention is directed to the numerous political questions of recent history that depend for their answers, to some extent, on scientific knowledge: (1) How much secrecy is desirable in the atomic energy program? The answer must depend on what scientific facts are already available to potential enemies and on whether we are likely, in the absence of secrecy, to import more ideas than we export. (2) Was the Secretary of Commerce justified in firing, on partly scientific grounds, the director of the Bureau of Standards? (3) Should clock manufacturers be given protection by high tariffs in the interests of national defense? (Or can precision instruments be made equally well by other industries?) (4) How much money should Congress allot to the National Science Foundation?

Such questions, bearing directly on national policy, are sure to increase in number. Yet a high government official recently revealed his opinion of scientific research when he said that "the scientists talk a lot about pure research" but that one definition of pure research is research that could not possibly be of value to the people who put up the money for it. It is a necessity, in a modern democracy, that some science and some understanding of science be taught to all.

The practical usefulness of scientific knowledge to the individual is fairly obvious. Chemistry, in particular, is a necessity or a help in nearly every profession: medicine, engineering, geology, physics, biology, running a home. Even if one plans to go into a nonscientific business, the advance of science is likely to engulf it in his lifetime.

It has been argued here that it is important to learn something both of the results and of the methods of science. But it is not easy to include both in the same course. We shall make the attempt in this course by organizing it about a few guiding ideas. Some of these are arrived at by a quasi-historical approach; others are arrived at by bold guesses that facts are shown to substantiate. Problems involving the derivation and application of important laws are considered, both for the sake of illustrating the types of reasoning involved and for the facts themselves.

These brief comments on the organization and nature of science suggest certain techniques required in the study of science.

The more knowledge of mathematics and physics the student can acquire, the easier is the study of chemistry.

Chemistry is by no means completely explained on the basis of a few simple rules. There are many important facts that fit no simple generalization. The only way to learn such facts is by the tedious process of memorization. This is one reason for being interested in chemistry: it is an article of scientific faith that ultimately it will be possible to bring order out of chaos by new theories. Naturally, every effort is made to hold memorization to a minimum; but when it is inevitable, the student will be advised to memorize.

The importance of the clarity and precision of concepts has been noted. It is clear from this that the science student must spend much of his study time learning carefully the specialized vocabulary of his science. Whenever a new term is introduced, it must be learned thoroughly enough so that the student can give a good definition in his own words. Note here that clear, precise written English is indispensable. Since communication is essential to science, an incorrect statement is just as wrong when the error arises from language or punctuation as when it arises from trouble with the facts.

Sciences are so constructed that the idea the student learns one day is part of the foundation of what he learns the next day. It is therefore important to spread out the study time over the semester, rather than to let it pile up for pre-examination cramming. Only if the student is reasonably up to date in his studying can he follow the lectures at all.

From what has been said so far it is apparent that the study of science will be easy only for the student having good study habits. It is for this reason, and not because they are intellectually more difficult than any other subjects, that sciences have a certain reputation for difficulty.

1-4. The Place of Chemistry

Chemistry is the study of the fine structure of matter and of the changes of this structure. That is to say, it is concerned with how atoms are joined to form molecules, how molecules are changed and modified, and how the properties of matter depend on this fine structure. It is not easy to mark it off from physics, for physicists also study the structure of atoms and molecules, or from biology, for biologists study the chemical changes intimately associated with living organisms. What is studied in the chemistry department of one university may be studied in the physics department of another. The easiest way to describe chemistry is to note some of the questions it tries to answer.

How are substances formed from the smallest chemical building blocks, the atoms?

How do the mechanical, electrical, thermal, optical, chemical, and physiological properties of a substance depend on its structure?

How can substances be transformed into others? How can particular substances be made?

How can an unknown substance be analyzed, and how can its structure be determined?

How can a chemical reaction be speeded up or slowed down?

How can energy be obtained from chemical reactions?

Not all of these questions will be answered in this text, and probably none will be answered completely to your satisfaction. That is well.

1-5. Introduction to Chemistry: Atoms

Atoms, and indeed all matter, are electrical in nature. An electric charge may perhaps best be described as a cloud of positive or negative electricity, with no sharply defined surface. Hence atoms, too, lack a definite boundary in space. It is possible to describe the behavior of electric charges by mathematical expressions without ever being able to see, hear, or feel them. The same can be done for atoms. Such mathematical expressions describing atoms may be presented in the form of a graphic, physical concept, which, though not defining precisely the actual structure of an atom, will be extremely useful. Only in advanced research is a more exact treatment of the properties of the atom warranted. Although the evidence for the validity of the atomic theory is now so overwhelming that it can no longer be doubted, the description of the atom is still changing.

Every atom consists of a nucleus surrounded by one or more electrons. The radius of the nucleus is small compared with that of the entire atom, but almost the entire mass of the atom is concentrated in the nucleus. The nucleus of the hydrogen atom contains about 99.95% of the mass of the entire atom. The nucleus of the atom is made up of neutrons (which are electrically neutral) and protons (each with a positive electric charge of one unit) and hence bears a positive charge. Only in the hydrogen atom are there no neutrons in the nucleus. Surrounding this compact nucleus is a cloud of negative electricity (composed of electrons) of total charge equal to the number of protons in the nucleus. Even though the electron may be pictured as a cloud of negative electricity, it may also be thought of as traveling in an orbit about the nucleus somewhat in the way the earth travels in an orbit about the sun. The electronic orbit is spherical or egg-shaped rather than planar. The electron in the hydrogen atom is most likely to be found within about 0.5 A (1 A = 1 angstrom = 10^{-8} cm) from the nucleus; it seldom strays more than 1 A from the nucleus.

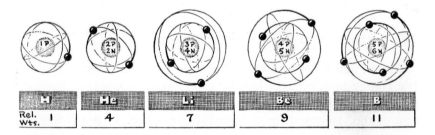

If the atomic weight of oxygen is taken as 16, that of hydrogen is 1, almost all of which is due to the proton. The mass of a neutron is also 1. The above diagrams indicate the relative weights of the nuclei of the first five elements of the Periodic Table. In each case the number of protons in the nucleus is exactly matched by the number of electrons outside. In helium, the two electrons move at the same distance from the nucleus and are said to occupy the same shell. In lithium, however, one of the three electrons is somewhat farther removed from the nucleus than the other two; this electron occupies a second shell, somewhat larger than the first. The first is referred to as the K shell and the second as the L shell. Atoms heavier than neon all have two such K electrons very close to the nucleus and eight electrons in the L shell, plus additional electrons in shells farther from the nucleus.

TABLE 1-1 | *Content of the Nuclei and Electronic Orbits of a Few of the Elements*

MASS NUM- BER*	ATOMIC NUM- BER	SYM- BOL	Nucleus		Outside Shells					
			NEU- TRONS	PRO- TONS	K	L	M	N	O	P
1	1	H	0	1	1					
4	2	He	2	2	2					
7	3	Li	4	3	2	1				
9	4	Be	5	4	2	2				
11	5	B	6	5	2	3				
12	6	C	6	6	2	4				
14	7	N	7	7	2	5				
16	8	O	8	8	2	6				
19	9	F	10	9	2	7				
20	10	Ne	10	10	2	8				
23	11	Na	12	11	2	8	1			
39	19	K	20	19	2	8	8	1		
85	37	Rb	48	37	2	8	18	8	1	
133	55	Cs	78	55	2	8	18	18	8	1

* Mass number is defined as the integral number closest to the atomic weight.

Succeeding shells are able to accommodate the following maximum numbers of electrons: K, 2; L, 8; M, 18; N, 32. The maximum may not always be attained, however, before the next succeeding shell starts to fill.

It is the number of electrons in the outermost shell that has the greatest influence in determining the chemical properties of the element.

In Table 1-1, H, Li, Na, K, Rb, and Cs each have one electron in the outside shell. That means that their chemical properties are similar. This is true for other elements that have the same number of electrons in the outer shells. The periodic arrangement of the elements is a recognition of this repeated occurrence of atoms having the same properties.

In writing an electronic picture of an atom, we usually show only the outside shell of electrons, the nucleus and all other planetary electrons (the "kernel") being represented by the symbol for the element.

Group 0	I	II	III	IV	V	VI	VII	0
	H ·							
He	Li ·	· Be	· B ·	· C ·	· N:	· O:	: F ·	: Ne :
Ne	Na ·	· Mg	· Al ·	· Si ·	· P:	· S:	: Cl ·	: A :
A	K ·	· Ca						

1-6. Electron Theory of Valence

1. Covalence.   When atoms form molecules, the electron atmosphere about each atom interpenetrates that of the other, and the shape of the electron smear may change. In the hydrogen molecule the electron atmospheres interact, with the result that the probability of finding the electrons belonging to the two atoms will be greatest between the two nuclei. This is the justification for writing a simple structure for the hydrogen molecule as H : H. The distance between the nuclei is approximately 0.60 A. It may be said that the electrons are shared by both atoms, each of which has furnished one electron. A pair of electrons shared by two atoms, each of which has furnished one, is called a covalent bond.

A qualitative and perhaps naive justification for writing a covalent bond (the electron pair) between the two atoms forming the bond is given in the accompanying diagram. Suppose that electron e_1 belongs to atom A, the kernel of which is shown bearing one positive charge; electron e_2 belongs to atom B. At the instant

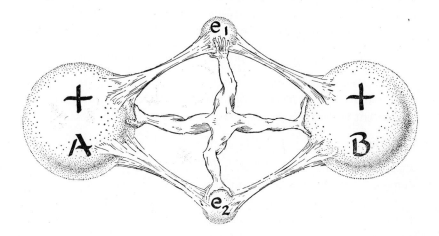

of bond formation there will be electrostatic attraction between A and e_1 and e_2 and between B and e_1 and e_2 (the elastic bands). Balanced against these attractions will be the repulsions between e_1 and e_2 and between A and B (indicated by arms). From this static picture one might conclude that the electron pair (e_1 and e_2) could be located at some position of compromise between the two nuclei. In reality the picture must be a dynamic one, and for that circumstance perhaps the only safe thing to say is that the probability of finding the electrons is greatest between the two nuclei.

For a large number of atoms that form covalent bonds a general rule may be stated as follows: In forming a covalent bond, each atom strives to achieve the structure of the next heavier rare gas in the periodic system. This general rule is often called the Rule of Eight because each rare gas (except helium) has eight electrons in its outside shell. The shared electrons are considered as belonging to the outside shell of each atom. Accordingly, each hydrogen atom in the molecule H_2 has the electronic structure of helium, the next rare gas. Halogen atoms are also bound by covalent bonds to form molecules, as : F̈ : F̈ : . Each fluorine atom alone has seven electrons in its outside shell, and by sharing a pair with a second fluorine atom, each one attains the electronic configuration of the rare gas neon, with eight electrons in the outside shell.

Hydrogen and iodine atoms are joined together by a covalent bond to form a covalent compound, H : Ï : , in which the hydrogen atom shares its single electron with the iodine, and, in turn, the iodine shares one of its electrons with the hydrogen, each achieving a different rare gas structure. Water is also a covalent compound, with the structure H : Ö : H, in which each of the two hydrogen atoms shares one electron with the oxygen atom and the latter shares one electron with each of the two hydrogens.

An example with a larger number of atoms is the compound methane, in which carbon and hydrogen form a covalent compound, CH_4, by sharing four pairs of electrons between four H's and one C. The hydrogens here have the helium configuration while carbon has the neon structure. Carbon and fluorine form a covalent compound, CF_4, in which each of the five atoms in the molecule has the neon electronic configuration. Again one pair of electrons is shared between the carbon and each fluorine, each of the atoms furnishing one of the pair.

$$
\cdot \overset{\cdot}{\underset{\cdot}{C}} \cdot \qquad H \cdot \qquad
\begin{matrix} H \\ H : \overset{\cdot\cdot}{\underset{\cdot\cdot}{C}} : H \\ H \end{matrix}
\qquad \cdot \overset{\cdot}{\underset{\cdot}{C}} \cdot \qquad : \overset{\cdot\cdot}{\underset{\cdot\cdot}{F}} \cdot
\qquad
\begin{matrix} : \overset{\cdot\cdot}{F} : \\ : \overset{\cdot\cdot}{F} : \overset{\cdot\cdot}{\underset{\cdot\cdot}{C}} : \overset{\cdot\cdot}{F} : \\ : \overset{\cdot\cdot}{F} : \end{matrix}
$$

Two pairs of electrons may be shared between atoms. Oxygen exhibits covalence in some compounds—for example, with carbon in carbon dioxide. In this picture the four electrons drawn between the atoms are shared in two pairs. The oxygens have the neon structure, and so does the carbon. In CS_2, carbon assumes the structure of neon and sulfur that of argon.

$\cdot \overset{..}{\underset{..}{C}} \cdot$ $\overset{..}{\underset{.}{O}} \cdot$ $: \overset{..}{O} : : C : : \overset{..}{O} :$ $\cdot \overset{.}{\underset{.}{C}} \cdot$ $: \overset{..}{S} \cdot$ $: \overset{..}{S} : : C : : \overset{..}{S} :$

Three pairs of electrons may be shared, as in N_2 or HCN:

$\cdot \overset{..}{N} \cdot$ $: N : : : N :$ $H : C : : : N :$

In order to abbreviate structures, the shared pairs of bonds are frequently represented by dashes, thus: the hydrogen molecule, H–H; fluorine, $: \overset{..}{\underset{..}{F}} – \overset{..}{\underset{..}{F}} :$ or, still more briefly, F–F, in which the unshared electrons are not shown at all; the oxygen molecule, O=O; nitrogen, N≡N.

When chemical combination is due to the sharing of pairs of electrons by atoms, the number of shared pairs is called the covalence. The covalence of oxygen is 2 in water, the covalence of carbon is 4 in CH_4, CF_4, CO_2, CS_2, and HCN, and the covalence of nitrogen is 3 in N_2 and HCN.

Substances composed of molecules formed by covalent bonds are characterized by low melting points, low boiling points, and low solubility in water, and very frequently they do not dissociate into ions even if they are soluble in water. Most substances of this type are nonconductors of electricity. [There are many important exceptions to this generalization, in which the character of the covalent bonds appears to be of secondary consideration—for example, diamond, graphite, tungsten carbide, silicon dioxide (sand), and aluminum nitride, all of which have very high melting points.]

2. Electrovalence. Electrovalent bonds are also called ionic valences or polar valences. An electrovalent bond is formed by a complete transfer of an electron (or electrons) from one atom to another. The atom releasing the electron is said to be the donor, and the other atom is called the acceptor. This type of bond is usually formed only when the donor has but one or two electrons in the outermost shell and the acceptor lacks but one or two electrons of the next heavier rare-gas structure. The donor will, in general, be a metal of Group I or II of the Periodic Table, and the acceptor of Group VIa or VIIa.

By the transfer of electrons, the donor becomes positively charged and the acceptor negatively charged. The electrostatic attraction between the opposite charges is the origin of the binding force in the chemical bond between the two elements.

Below are shown examples of atoms that form electrovalent bonds and the usual representation of the compounds formed. In chemical combinations of this kind the electrovalence of sodium and chlorine is 1, and the electrovalence of magnesium is 2. The electrovalence is the number of electrons transferred to or from an atom in the formation of this type of chemical bond.

$$\text{Na} \cdot \qquad \cdot \ddot{\underset{..}{\text{Cl}}} : \qquad \text{Na}^+ : \ddot{\underset{..}{\text{Cl}}} : ^-$$

$$\cdot \text{Mg} \cdot \qquad \cdot \ddot{\underset{..}{\text{Cl}}} : \qquad : \ddot{\underset{..}{\text{Cl}}} : ^- \quad \text{Mg}^{++} \quad : \ddot{\underset{..}{\text{Cl}}} : ^-$$

These ionic compounds formed by transfer of electrons from the metal atom to the nonmetal atom are held together by electrostatic attraction between the positive ions and the negative ions. In the solid state there are no simple molecules as such, but each positively charged ion is bound electrostatically to a number of negatively charged ions near it, and vice versa. Owing to these strong binding forces between atoms in the crystal, it requires much energy (heat) to separate the atoms from one another, and therefore the melting point of the solid is high. Likewise, even though the ions move about in the liquid formed by melting of the solid, the electrostatic forces between the ions of opposite charge are still sufficiently great to render difficult the separation of particles from the liquid to form a gas, and the boiling point is consequently high.

Because of the attraction between water molecules (which are "dipoles," with a positive charge on one end and a negative charge on the other) and the charged ions of electrovalent compounds, such compounds will dissociate into their ions in aqueous solutions. The ions are free to move about independently within the solution and hence are capable of conducting an electric current through the solution. Similarly, electrovalent compounds in the liquid state, where the ions are free to move about independently, are also good conductors of electricity.

Not all the cases mentioned so far are clear-cut in character. Sodium chloride in the solid state is definitely an electrovalent compound. The hydrogen iodide molecule is definitely covalent. Perhaps a large majority of compounds may best be described as intermediate in character; that is, they possess some properties of the covalent bond and some properties of the electrovalent bond. In Table 1-2 are some estimated percentages of the character of the bonds in various compounds.

TABLE
1-2
| *Character of the Bonds in Certain Compounds (Percentages Estimated)* *

	COVALENT	ELECTROVALENT		COVALENT	ELECTROVALENT
HI	95	5	BeF_2	21	79
HBr	89	11	BeO	37	63
HCl	83	17	BF_3	37	63
HF	40	60	BCl_3	78	22
LiI	57	43	CF_4	56	44
NaCl	5	95	CS_2	78	22

* From Pauling, *The Nature of the Chemical Bond.*

3. Coordinate Covalence.   A third way in which a chemical bond can be formed is by the sharing of a pair of electrons *both* furnished by one of the two atoms involved. Qualitatively, the properties of this coordinate covalent bond (also called semipolar) and the covalent bond are the same. They differ in the mode of

$$
\begin{matrix}
\text{H} & & \text{H} \\
\ddot{} & + & \ddot{} \, + \\
\text{H}\!:\!\ddot{\text{N}}\!: + \text{H} \longrightarrow \text{H}\!:\!\ddot{\text{N}}\!:\!\text{H} \\
\text{H} & & \text{H}
\end{matrix}
\qquad [1\text{-}1]
$$

formation. Ammonia has three covalent bonds between nitrogen and hydrogen, as shown in Expression 1-1; and, besides, nitrogen has in its octet one pair of unshared electrons. The simplest chemical substance lacking a pair of electrons in its outer shell is a proton (a hydrogen atom that has lost its lone electron). A source of protons is the hydrogen chloride molecule, which has 17% ionic and 83% covalent character in the gaseous state and is almost completely dissociated (100% ionic) in water solution. When the ammonium ion is formed by the reaction of ammonia with a proton, the unshared pair of electrons on the nitrogen becomes one of the four shared pairs of the ion, and subsequently it belongs to both the nitrogen and a hydrogen. The two electrons of that pair can no longer be distinguished from the other six electrons after the bond is formed.

In general, the coordinate covalent bond results in properties intermediate between those of electrovalent bonds and those of covalent bonds. For example, such bonds are weaker than electrovalent bonds, and therefore coordinate covalent compounds are much less stable toward heat than electrovalent compounds but are more stable than many covalent compounds. Solubility in water is also intermediate in value for many coordinate covalent compounds.

In the complex hydrated copper II ion shown below an unshared pair from each oxygen is shared with the copper atom (both elec-

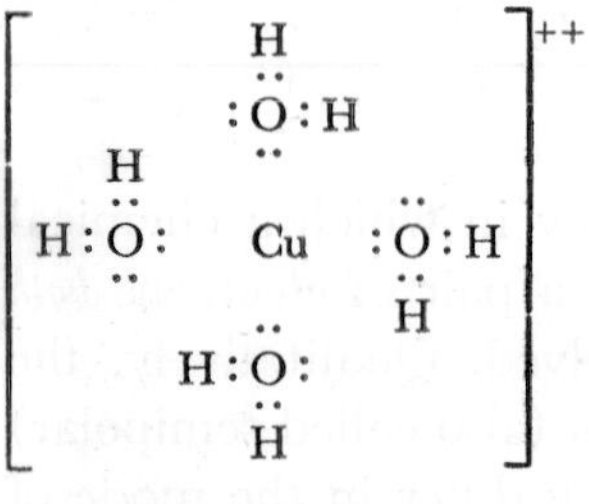

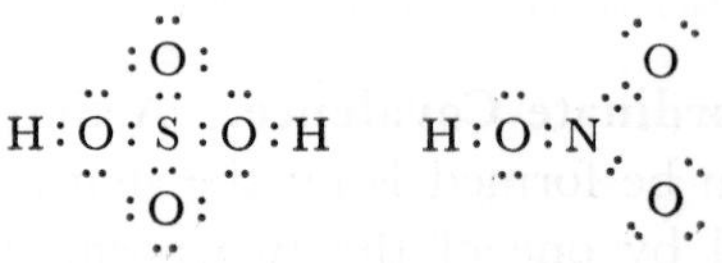

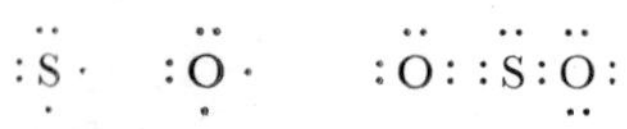

trons are furnished by the oxygen). In sulfur dioxide two types of bonds are present between sulfur and oxygen. One sulfur-to-oxygen bond is a coordinate covalent bond, and the pair is furnished by sulfur. The other sulfur-to-oxygen bond is a double bond of the covalent type. Sulfuric acid has two covalent bonds from sulfur to oxygen and two coordinate covalent bonds, as shown. One of the nitrogen-to-oxygen bonds in nitric acid is a coordinate covalent bond.

1-7. Mechanisms of Reaction of the Covalent Bond

In the most fundamental view of the electron theory, covalent bonds can be broken in only two ways. Either the two atoms take one electron each in splitting (Exp. 1-2), or one atom takes both

$$A:B \longrightarrow A \cdot + B \cdot \qquad [1\text{-}2]$$

$$A:B \overset{\nearrow A:^- + B^+}{\underset{\searrow A^+ + B:^-}{}} \qquad \begin{array}{c}[1\text{-}3]\\[1em][1\text{-}4]\end{array}$$

electrons (Exp. 1-3 or 1-4). The first type (Exp. 1-2) is called a *free-radical* reaction (A · and B · are free radicals), and the second (Exp. 1-3 or 1-4) may be called a *polar reaction.*

Free-radical reactions occur at high temperatures, under the influence of sunlight, or by the inducement of other free radicals or atoms. Although two of the first reactions to be studied (Exp. 2-3–2-6 and 3-21) are of the free-radical type, most of the reactions presented in this text are of the polar type.

Polar reactions occur at ordinary temperatures under the influence of a catalyst (frequently an acid or a base), are often affected by the solvent used, and are frequently influenced by the nature of the reagent itself. In a polar reaction the possibility of splitting as in Expression 1-3 (or 1-4) will seldom be equal to the possibility of splitting as in Expression 1-4 (or 1-3). This will be evident from your present knowledge of the Periodic Table. Only when A and B are like atoms is it probable that the electron pair is equally shared

$$: \overset{..}{\underset{..}{A}} \quad : \quad \overset{..}{\underset{..}{B}}: \qquad [1\text{-}5]$$

$$: \overset{..}{\underset{..}{A}} \quad : \overset{..}{\underset{..}{B}} : \qquad\qquad\qquad [1\text{-}6]$$

$$: \overset{..}{\underset{..}{A}} : \quad \overset{..}{\underset{..}{B}} : \qquad\qquad\qquad [1\text{-}7]$$

by the two atoms (Exp. 1-5). When A and B are unlike atoms, a more probable disposal of the shared electron pair is that shown in Expression 1-6, where B is electron-attracting (electronegative), or that shown in Expression 1-7, where B is electron-releasing (electropositive) in character. In Expression 1-6, A is electropositive with respect to B; in Expression 1-7, A is electronegative with respect to B.

The displacement reaction is one of the most important polar reactions in which a covalent bond is broken and a new one formed. In general terms, the reaction is shown in Expression 1-8, where

$$: \overset{..}{\underset{..}{A}} : \overset{..}{\underset{..}{B}} : + : \overset{..}{\underset{..}{Y}} : \longrightarrow : \overset{..}{\underset{..}{A}} : \overset{..}{\underset{..}{Y}} : + : \overset{..}{\underset{..}{B}} : \qquad [1\text{-}8]$$

$: \overset{..}{\underset{..}{Y}} :$ displaces $: \overset{..}{\underset{..}{B}} :$ from the compound AB. A part of the task of learning chemistry is to find out when, why, and how such reactions take place in particular cases.

1-8. Spatial Configuration of Atoms and Directed Valence

In electrovalent compounds, in which there is a definite separation of charged particles (ions), the charge on each ion may be considered as spread out or smeared over the entire surface of the ion. In other words, the charge (positive or negative) cannot be located at a particular point on the ion. In covalent molecules, in which electron pairs are shared, the bond has a definite location on the atoms involved. A limited number of molecular shapes are known. Why molecules have the particular shape they do have will not be discussed here. Suffice it to say that the quantum levels of the electrons involved in the formation of the bonds are principally responsible, but the character of the atoms themselves (sizes, for example) makes some modifications.

Diatomic molecules are necessarily linear, since two points determine a straight line. A few triatomic molecules are also linear. In carbon dioxide, the directed valences of carbon and oxygen are linear, O=C=O. Carbon disulfide has this same structure. Water, on the other hand, is a planar molecule in which the directed valences take on an angle of 105°:

H–O
    H

In some other planar structures the directed valences are symmetrical. In boron trichloride, for example, the bond angles are 120°:

$$120°→ Cl$$
Cl —— B
        Cl

The carbonate ion, $CO_3^=$, is also a trigonal plane, with carbon at the center.

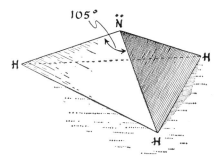

Four-atom molecules may have other shapes. Ammonia and phosphine, PH_3, have a tetrahedral structure in which the nitrogen (or phosphorus) is at the apex of a pyramid with angles of about 105° between the N-H (or P-H) bonds. This is not a regular tetrahedron and is better labeled a trigonal pyramid.

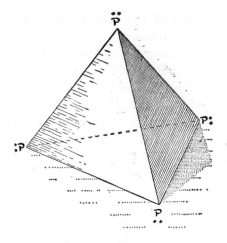

Phosphorus, in the gaseous state, exists as the P_4 molecule. The four P atoms are located at the vertices of a regular tetrahedron. Since the faces are equilateral triangles, the angles between bonds are 60°.

Sulfur vapor (as well as some liquid forms and the crystalline forms) occurs as S_8 molecules. The eight atoms are bound together in a ring in which the angle between bonds is about 106°.

Carbon (and other elements in Group IV) forms compounds in which the structure is a regular tetrahedron with the carbon atom

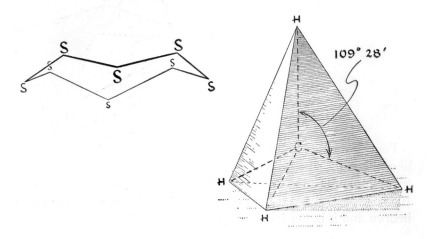

at the center. CCl_4 and CH_4 are examples of such compounds, with angles of 109°28′ between the C-Cl or C-H bonds, those angles following of necessity from the geometry of the figure. If one or more of the atoms attached to the carbon are different, there may be a slight distortion of the bond angles. If a chlorine atom is substituted for a hydrogen in the above structure, the angle is probably not distorted more than 2° by this strongly electro-negative atom.

The environment of an atom may have the most profound effect on the structure it assumes. This is evident in the oxygen-contain-

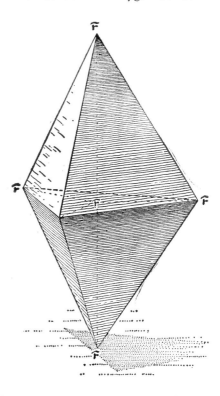

ing compounds just described (H_2O and CO_3^-, for example) and still more in the ion $Ni(NH_3)_4^{++}$, the structure of which, depending on the anion attached, may be either a square plane [see $Cu(H_2O)_4^{++}$, p. 20] or a tetrahedron with nickel at the center.

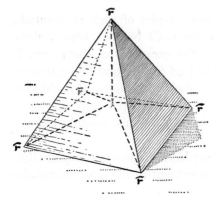

The square complex is red, and the tetrahedral nickel cation is blue-green.

Still other geometric configurations are possible in various molecules. PF_5, for example, is a bipyramid with the phosphorus atom at the center of the triangular base. IF_5, on the other hand, is a square pyramid with iodine equidistant from the five fluorine atoms. The SF_6 molecule has an octahedral configuration, as do a number of complex compounds of Co, Ni, Fe, Cr, and Pt. The last, in various compounds, may have either a square (planar) or an octahedral configuration.

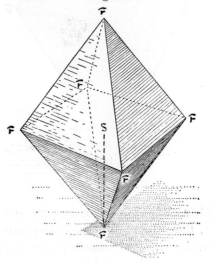

Most of this text will be devoted to the chemistry of covalent compounds with a tetrahedral configuration, chiefly with carbon, silicon, nitrogen, sulfur, and phosphorus as the central atoms. Carbon will be the immediate center of attention.

1-9. Effect of Atomic Size on Geometry of Molecules

The geometry of molecules and ions provides a simple answer to the question of permissible formulas for various substances. Why may phosphate be written in various ways—PO_3^{-3}, $P_2O_7^{-4}$, PO_4^{-3}—whereas nitrate is always represented as NO_3^-? Why is silicate ion written either as $SiO_3^=$ or as SiO_4^{-4} but carbonate invariably as $CO_3^=$? The key to this apparent puzzle is in the relative sizes of the nitrogen and phosphorus atoms and of the carbon and silicon atoms. The larger of the two atoms in each case will accommodate four oxygen atoms around it, but the smaller atom is not large enough to allow four oxygen atoms to touch it (that is, to form bonds). The formula for phosphate or silicate thus depends on other conditions also, but the formula for carbonate or nitrate is controlled primarily by geometry. The "Chart on Atomic Sizes" by J. A. Campbell [*Journal of Chemical Education*, 23, 526 (1946)], which shows the relative sizes of the atoms (see p. 28), reveals the possibility that these statements may be true. An analogy may help here. It is impossible to place four tennis balls around a half-inch marble so that all four touch the marble. There is no difficulty, on the other hand, in placing four tennis balls around a golf ball so that they all touch it.

The size and shape of molecules will also account, in part at least, for the sharp contrast in properties between carbon dioxide and silicon dioxide. Carbon dioxide is a gas ordinarily and forms a solid (Dry Ice) that sublimes readily at $-78°C$. This must mean that there is very loose bonding between molecules. In sand, SiO_2, on the other hand, with a melting point of 1,700°, the bonding between molecules must be very strong. Since both compounds are formed by covalent bonds, the type of bond does not explain the difference in properties. Carbon is such a small atom that the two

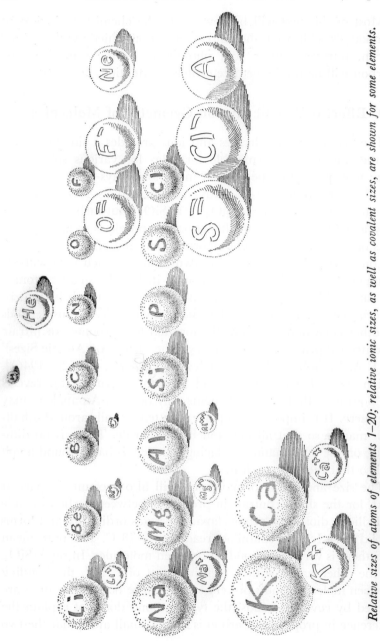

Relative sizes of atoms of elements 1–20; relative ionic sizes, as well as covalent sizes, are shown for some elements. (Redrawn from J. A. Campbell, loc. cit.)

strongly electronegative oxygen atoms repel each other and the molecule is forced into a linear structure. Silicon, larger than carbon, keeps the electronegative oxygens far enough apart so that four of them are compatible in the allotted space. "Silicon dioxide," then, is really a misnomer, for each particle of sand is one giant molecule in which each silicon has four single covalent bonds to oxygen. Each oxygen is attached to two silicons, and the empirical formula is SiO_2.

TABLE
1-3 | *Relative Sizes of Atoms and Ions*

	COVALENT RADIUS (ANGSTROMS)	IONIC RADIUS (ANGSTROMS)		COVALENT RADIUS (ANGSTROMS)	IONIC RADIUS (ANGSTROMS)
H	0.30	——	Na	1.57	0.95
			Mg	1.36	0.65
Li	1.23	0.60	Al	1.25	0.50
Be	0.89	0.31	Si	1.17	
B	0.80	0.20	P	1.10	
C	0.77		S	1.04	1.84
N	0.70		Cl	0.99	1.81
O	0.66	1.40			
F	0.64	1.36	K	2.03	1.33
			Ca	1.74	0.99

1-10. The Element Carbon

Carbon occurs in pure form in nature as an element in two allotropic forms: diamond and graphite. In diamond, the entire structure is really one giant molecule, in which the various atoms are bonded to one another in continuous tetrahedra by four equivalent linkages. The result is the hardest known substance. In graphite the carbon atoms are linked together in six-membered rings in

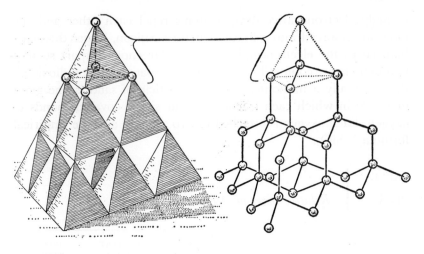

which each atom has one double and two single bonds. The forces binding successive planes together are not valence forces and are very weak. Hence the various planes can slide easily over one another. This accounts for the excellent lubricating properties of graphite.

"Amorphous" is the descriptive term commonly given to coal,

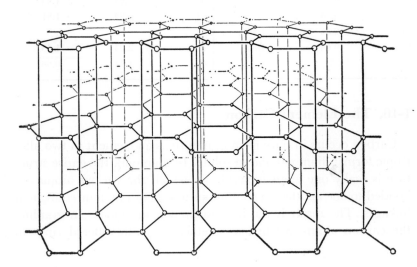

charcoal, coke, and carbon black. A better term is "microcrystal-line," since these substances are composed of thousands of micro-scopic crystals.

1-11. The Carbon Atom

The most important property of the carbon atom is that, with only a few exceptions, it is joined to other atoms by four covalent bonds. These four valences, as we have seen, are directed toward

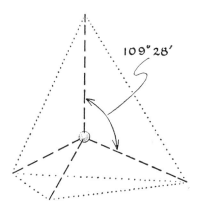

the vertices of a regular tetrahedron at an angle of 109°28' to one another. On a flat surface (such as the blackboard or a sheet of paper) the four bonds are written $-\overset{|}{\underset{|}{C}}-$, but it must always be borne in mind that they are not in a single plane and must be projected into three dimensions.

The property that makes carbon unique among all the elements is its propensity to build long chains by forming covalent bonds with other carbon atoms. Boron and silicon atoms may join atoms of their own kind in chains up to six or perhaps ten atoms in length, but carbon chains may be hundreds or even thousands of atoms in length. More important than mere length of chain is the much greater stability of the carbon compounds, which makes the larger molecules possible.

The bonds joining carbons may be single (one pair of electrons shared by the two atoms), double (two shared pairs), or triple (three shared pairs). The three types of bonds are pictured in the

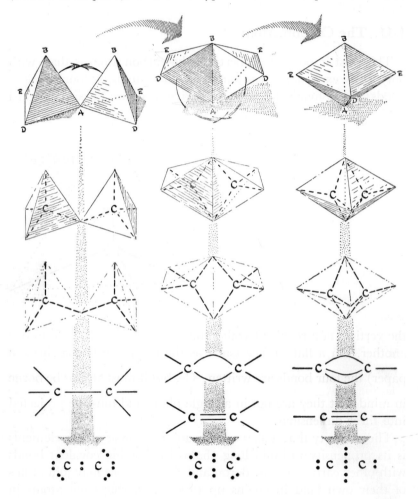

accompanying diagrams, where one may imagine a carbon atom at the center of each tetrahedron and a pair of electrons at each lettered point. As would be expected from purely geometric considerations, it is impossible for two tetrahedra to meet in four points. (Three points determine a plane.) The single bond allows free rotation of the two tetrahedra about the point where they are joined. The double bond restricts the rotation and requires that points D, D, E, and E lie in the same plane. The triple bond adds a further restriction. A line joining the two E's passes through the center of each tetrahedron; so, in a carbon compound containing a triple bond, the two atoms in the bond and the adjacent two atoms are necessarily linear.

Much of your future reasoning in this course will be based on these structural concepts:

1. There are always *four* bonds to carbon.
2. The bonds are arranged in a tetrahedral structure.
3. Carbons may be joined to other carbons in strong bonds.

The evidence for them will be reviewed in Chapter 18.

1-12. Aliphatic and Cyclic Compounds

The fact that carbon atoms may be joined in either chains or rings gives a convenient method of classification of carbon compounds. The open-chain compounds are called aliphatic compounds.

$$-\overset{|}{\underset{|}{C}}-\overset{|}{\underset{|}{C}}-\overset{|}{\underset{|}{C}}-\overset{|}{\underset{|}{C}}-\overset{|}{\underset{|}{C}}-\overset{|}{\underset{|}{C}}-\overset{|}{\underset{|}{C}}-$$

The ring, or cyclic, compounds are further divided into alicyclic compounds (only carbon atoms in the ring), heterocyclic compounds (one or more other atoms—such as oxygen, nitrogen, or sulfur—besides carbon in the ring), and aromatic compounds (a class of cyclic compounds having alternate double and single bonds joining the carbons).

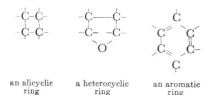

| an alicyclic ring | a heterocyclic ring | an aromatic ring |

1-13. Principle of Homology

Although there are more than a half million known carbon compounds, the study of them is greatly simplified by segregation into a few classes, the members of each class having nearly the same chemical properties. Each class is called a homologous series.

A homologous series is a family of carbon compounds that have (1) similar chemical properties, (2) regularly varying physical properties, and (3) molecular formulas differing by CH_2 or a multiple of CH_2. Before proceeding to the first homologous series, we shall define the various types of formulas to be employed in the study of carbon compounds.

1-14. Formulas

An *empirical formula* tells the kinds of atoms in a molecule and the simplest ratio that exists among them. Formaldehyde and grape sugar, for example, have the same empirical formula, CH_2O; that is, the ratio C:H:O is 1:2:1.

The *molecular formula* of a compound tells the kinds of atoms in the molecule and the actual number of each kind in one molecule. The molecular formula of formaldehyde is CH_2O, but that of grape sugar is $C_6H_{12}O_6$.

The *structural formula* shows the kinds of atoms, the number of each kind, and their arrangement in the molecule. Ammonium cyanate and urea both have the same molecular formula, CH_4N_2O, but different structural formulas:

ammonium cyanate   urea

Isobutane and n-butane have the same molecular formula, C_4H_{10}, but the atoms are arranged differently in space:

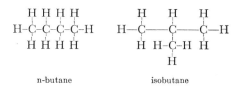

n-butane   isobutane

These structures may be written more compactly in *condensed*

$$\overset{\displaystyle H}{\underset{\displaystyle H}{\overset{|}{\underset{|}{H-C-}}}}$$

structural formulas, in which H–C– is written as CH_3– or –CH_3 and

$$\overset{\displaystyle H}{\underset{\displaystyle H}{\overset{|}{\underset{|}{-C-}}}}$$

is written as –CH_2–. In this condensed notation, n-butane is

CH_3–CH_2–CH_2–CH_3 and isobutane is CH_3–CH–CH_3.

$$\underset{CH_3}{\overset{|}{}}$$

Formulas written on a line do not, of course, show the proper

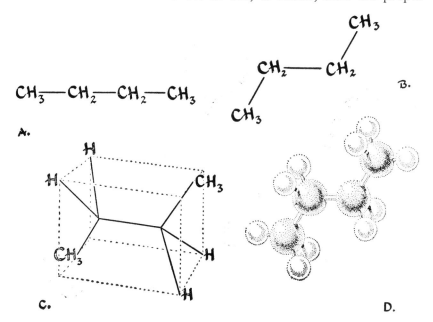

spatial relations. The open chain in n-butane, for example, is not C–C–C–C, as shown above, in A, but rather the chain B, in which the angle between carbons is 109°28'. This also is deceptive, of course, since it is a projection in only two dimensions and molecules occupy three dimensions. The three-dimensional molecule is shown in the box, C, and almost takes on an animal character in the ball-and-stick model, D. For other examples of structural formulas, see page 49.

When two different compounds have the same molecular formula but different structural formulas (as do ammonium cyanate and urea or n-butane and isobutane, the two pairs mentioned above), they are called *isomers* and are examples of *isomerism*. Isomers usually differ from each other in their physical and chemical properties. Isomerism is the most important implication of the idea of directed valence.

SUMMARY

The organization of science, its nature, and our reasons for studying science are given a positive treatment. An introduction to chemistry requires some elementary picture of the atom. The present-day electronic theory of covalence, electrovalence, and coordinate covalence sets the pattern for a structure of concepts on which to build the geometry of molecules. The tetrahedral geometry of carbon and the postulate that carbon atoms can be joined together by strong bonds complete the setting for the implications that follow in the next fifteen chapters.

REFERENCES

1. Standen, "On the Horns of the Sacred Cow," J. Chem. Education, 28, 608 (1951).
2. Campbell, "Relative Sizes of Atoms," J. Chem. Education, 23, 525 (1946).

3. Ehret, "The Role of Electrons in Interatomic Relations," J. Chem. Education, 25, 291 (1948).

4. Foster, "Effects of Molecular Shapes," J. Chem. Education, 29, 156 (1952).

5. Gillis, "Isolectronic Molecules," J. Chem. Education, 35, 66 (1958).

EXERCISES

1. Write electronic structures for the following atoms: Ca, Ti, I, Si.

In Exercises 2, 3, and 4 assume (a) that the first atom is the central atom unless otherwise stated; (b) that the Rule of Eight is sacred most of the time.

2. Write electronic structures for the following linear molecules: CO_2, HCl, NO, CS_2. Use the symbol to represent the "kernel" of the atom (p. 14).

3. Write electronic structures for the following planar molecules or ions: BCl_3, NO_3^-, $CO_3^=$, NOCl, SCl_2, H_2O, H_2S. (O and S are the central atoms in H_2O and H_2S, respectively.)

4. Write electronic structures for the following tetrahedral molecules: PCl_2, NI_3, N_2O_5 (two central nitrogens, O_2NONO_2), $SOCl_2$, SO_2Cl_2, $POCl_3$, $CO(NH_2)_2$, $ClSO_3H$ (central atom S), $HClO_4$ (central atom Cl), $HClO_3$ (central atom Cl), H_3PO_3 (central atom P), H_3PO_4 (central atom P), $SnCl_4$.

5. Write an electronic structure for PCl_5.

6. In Exercises 2, 3, and 4, which structures contain coordinate covalent bonds?

7. In Exercises 2, 3, and 4, which electronic structures, of necessity, violate the Rule of Eight?

8. Reconcile the two statements that "carbon is tetrahedral" and that "carbon dioxide is a linear molecule."

9. What are the salient features of the types of bonds that occur in molecules?

10. Why is the simplest carbonate ion $CO_3^=$ but the simplest silicate ion SiO_4^{-4}? (Silicate ion is sometimes written as $SiO_3^=$, but this has not been shown to exist.)

11. Why is the formula for nitrogen gas N_2 but the formula for phosphorus (in the gaseous state) P_4?

Properties of the Covalent Bond of Carbon with Hydrogen and Halogens

2 Alkanes, or Paraffins

In the next fifteen chapters we shall thoroughly explore the properties of the covalent bond in a tetrahedral carbon in order to see what effects particular atoms attached to carbon in a particular group have on those properties. We shall examine these bonds in approximately the following order (though overlapping will occur): C–H, C–X (halogen), C–O, and C–N.

The first family of compounds to be studied is certainly, in quantity, the most important, for it includes gasoline, kerosene, fuel oil, and lubricating oil, and it furnishes the raw materials for a large fraction of the entire chemical industry.

2-1. Saturated Hydrocarbons

The first homologous series to be studied is the saturated hydrocarbon series, also referred to as the alkanes, or paraffins. A

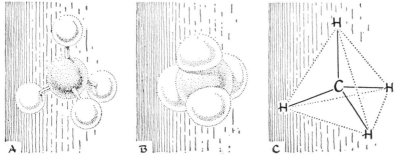

hydrocarbon is a compound in which only carbon and hydrogen are present, and it is saturated if only single covalent bonds link

TABLE	*Normal Alkanes and*
2-1	*Their Boiling Points*

MOLECULAR FORMULA	STRUCTURAL FORMULA	COMMON NAME	BOILING POINT (°C)	DIFFERENCE IN BOILING POINT (°C)
CH_4	H H : C : H H	methane	−161	
				72
C_2H_6	H H H : C : C : H H H	ethane	−89	
				47
C_3H_8	H H H H : C : C : C : H H H H	propane	−42	
				42
C_4H_{10}	H H H H H : C : C : C : C : H H H H H	n-butane	0	
				36
C_5H_{12}	$CH_3-CH_2-CH_2-CH_2-CH_3$	n-pentane	36	
				33
C_6H_{14}	$CH_3-CH_2-CH_2-CH_2-CH_2-CH_3$	n-hexane	69	
				29
C_7H_{16}	$CH_3-CH_2-CH_2-CH_2-CH_2-CH_2-CH_3$	n-heptane	98	
				28
C_8H_{18}	$CH_3-CH_2-CH_2-CH_2-CH_2-CH_2-CH_2-CH_3$	n-octane	126	
				25
C_9H_{20}	$CH_3-(CH_2)_7-CH_3$	n-nonane	159	
				23
$C_{10}H_{22}$	$CH_3-(CH_2)_8-CH_3$	n-decane	174	

the carbons in the molecule. The simplest alkane, containing one carbon and four hydrogens, is called methane (p. 41). Since its boiling point is $-161°C$, it is a gas at ordinary temperatures. Other members of this homologous series having only open chains are given in Table 2-1. (An open-chain compound is one in which every carbon atom is bonded to one or two, but not more than two, other carbon atoms.) The student should learn the names of the first ten compounds in this series.

The alkane series fits the definition of homologous series given in Chapter 1. Each succeeding member differs from the previous one in its molecular formula by CH_2. As indicated in the table, the boiling point rises with increasing chain length. The last column in the table shows that the amount of this rise becomes smaller with increasing chain length. Note that the change from one to two carbons (almost doubling the molecular weight) has a greater effect on the boiling point than the increase from eight to nine carbons, where the change in molecular weight is only about 12%. The chemical properties of the different alkanes (the reactions they undergo) are similar to one another, as will be shown later in this chapter.

2-2. Nomenclature

The common names of the first ten members of the series, all having open chains, were given in Table 2-1. There are alkanes with branched chains, however, in which at least one carbon atom is bonded to either three or four other carbons. The compound shown is called isobutane. It has the same molecular formula, C_4H_{10}, as normal butane (abbreviated as n-butane). Isobutane and n-butane are isomers. (See Chap. 1.) It is possible to write three pentanes of formula C_5H_{12}, as shown in the accompanying

```
    H  H  H
    ··  ··  ··
H : C : C : C : H   or   CH₃–CH–CH₃
    ··  ··  ··               |
    H      H                CH₃
    H : C : H
        ··
        H
              isobutane
```

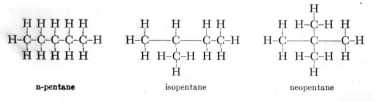

n-pentane

isopentane

neopentane

diagram. A chain without branches is given the prefix *normal* (abbreviated as *n*). A single branch of a CH_3 group on the second carbon in a chain gives rise to the prefix *iso*. The prefix *neo* means new and was given to neopentane because this was the last pentane discovered.

The number of possible isomers of other alkanes, as shown in Table 2-2, indicates that using the common names for alkanes of more than six or seven carbons would be a burden.

TABLE 2-2 *Numbers of Isomers*

NO. OF CARBONS IN ALKANE	NO. OF ISOMERS	NO. OF CARBONS IN ALKANE	NO. OF ISOMERS
C_1	1	C_6	5
C_2	1	C_7	9
C_3	1	C_8	18
C_4	2	C_{20}	366,319
C_5	3	C_{30}	4,111,846,763

2-3. Organic Groups

It is often convenient to speak of the various groups of atoms that may form a part of a molecule. For example, if one hydrogen atom were removed from the methane molecule, CH_4, the result would be the structure

$$\begin{matrix} & H & \\ H: & \overset{\cdot\cdot}{\underset{\cdot\cdot}{C}} & \cdot \\ & H & \end{matrix}, \text{ or } \begin{matrix} & H & \\ & | & \\ H- & C & - \\ & | & \\ & H & \end{matrix}$$

which is called the methyl

group. This group will be recognized as part of some of the molecules diagramed previously. It is not a compound and can have no independent existence for any appreciable length of time. Similarly, the ethyl group, C_2H_5-, may be obtained from the formula for ethane, CH_3-CH_3. We name such groups by changing the ending of the parent hydrocarbon from -*ane* to -*yl*. Two isomeric

Parent Hydrocarbon		*Group*	
NAME	FORMULA	NAME	FORMULA
methane	CH_4	methyl	CH_3-
ethane	CH_3-CH_3	ethyl	CH_3-CH_2-
propane	$CH_3-CH_2-CH_3$ {	n-propyl	$CH_3-CH_2-CH_2-$
		isopropyl	$CH_3-CH-CH_3$ $|$

groups are derived from propane as the hydrogen atom may be removed from an end carbon or from the middle carbon. The two groups are named n-propyl and isopropyl, respectively.

2-4. The IUC System

In order that we may identify any particular organic compound, a systematic scheme of naming compounds has been devised. An international commission of chemists met at Geneva (Switzerland) in 1892 for the first conference to devise a set of simple rules. Revisions of nomenclature and extensions of the system were made by the International Union of Chemists at Liége (Belgium) in 1930. The system is referred to either as the Geneva system or as the IUC system of nomenclature.

1. The first rule may be stated thus: select the longest chain of carbons and number it in the direction which will give the smallest digits to any substituents on the chain. The longest chain is given the common name mentioned on page 42. For example, the three different ways of numbering isopentane are as follows:

$$\overset{1}{C}H_3-\overset{2}{\underset{\underset{CH_3}{|}}{C}H}-\overset{3}{C}H_2-\overset{4}{C}H_3 \qquad \overset{4}{C}H_3-\overset{3}{\underset{\underset{CH_3}{|}}{C}H}-\overset{2}{C}H_2-\overset{1}{C}H_3 \qquad \overset{2}{C}H_3-\overset{3}{\underset{\underset{{}_1CH_3}{|}}{C}H}-\overset{4}{C}H_2-CH_3$$

(1)   (2)   (3)

The numbering systems used in examples 1 and 3 are equivalent* and correct, but example 2 is wrong, for its side chain would not have the smallest number possible.

2. The position of a substituent group is denoted by a number followed by a hyphen and a prefix giving the name of the group. The compound given above is named 2-methylbutane.

3. Each homologous series has its own characteristic ending, as *-ane* for alkanes.

Exercise 2-1. Using these rules and the examples in Table 2-4 (p. 48) as a guide, name by the IUC system the following compounds:

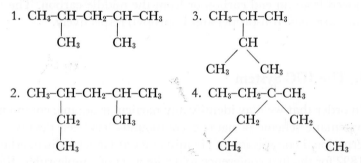

Exercise 2-2. Write the structural formula for each of the following: 2-methyl-3-ethylpentane, 2,4-dimethylhexane, 3,5-diethylheptane, 3-methyl-3-ethylhexane.

Ways of writing the structures and numbering the carbon chains of the five isomeric hexanes are shown in Table 2-4. Skeletons of the hexanes in various poses are shown in diagrams A–E on page 49.

---

* Anyone not convinced that examples 1 and 3 are equivalent structures should make ball-and-stick models of the two. If a chemist can superimpose the two models by turning groups in any desired way but without breaking a bond, the models, for his purposes, represent one substance. He is justified on the assumption that there is free rotation about single bonds in a molecule.

2-5. Occurrence of Alkanes

Alkanes occur in natural gas and petroleum. The natural gas used by many cities for home fuel is largely methane, with some ethane, propane, and the two butanes. One sample of natural gas contained 78% methane, 13% ethane, 6% propane, and 2% butanes. Ordinarily, natural gas contains 80–87% methane, 6–10% ethane, some higher hydrocarbons, and 1–8% nitrogen. Petroleum is made up very largely of saturated hydrocarbons containing from one to twenty carbon atoms per molecule, the smaller molecules being natural gas dissolved in petroleum.

The process of petroleum refining is mainly the separation of these alkanes into fractions, approximately according to chain length. The process is accomplished by fractional distillation. One petroleum company collects fractions according to the boiling points given in Table 2-3. When the wax oil is cooled, paraffin

TABLE

2-3 | *Petroleum Fractions*

APPROX. NO. OF C'S IN CHAIN	BOILING POINT	NAME OF FRACTION	USE
C_1–C_4	——	natural gas	fuel
C_5–C_6	20–60	petroleum ether	solvent
C_6–C_7	60–100	ligroin, naphtha	solvent, cleaning fluid
C_5–C_{11}	40–205	gasoline	motor fuel
C_{11}–C_{16}	205–300	kerosene, fuel oil	illumination, fuel
C_{15}–C_{18}	250–300	gas oil	cracking stock, Diesel fuel
C_{16}–C_{20}	above 275	wax oil	lubricating oil, mineral oil

wax, of melting point 40–55°, crystallizes out, and petroleum jelly (Vaseline) also separates. Petroleum coke is the final residue left from such an oil. Asphalt-base petroleums furnish heavy greases and leave asphalt as the final residue.

TABLE

2-4 | *Isomeric Hexanes*

ISOMER OF HEXANE	IUC NAME
$CH_3-CH_2-CH_2-CH_2-CH_2-CH_3$	hexane

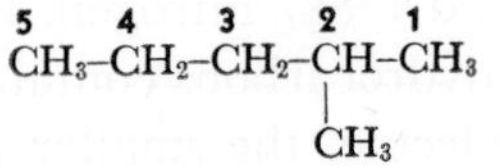

$$\overset{5}{CH_3}-\overset{4}{CH_2}-\overset{3}{CH_2}-\overset{2}{CH}-\overset{1}{CH_3}$$
$$\underset{CH_3}{|}$$

2-methylpentane

$$\overset{1}{CH_3}-\overset{2}{CH_2}-\overset{3}{CH}-\overset{4}{CH_2}-\overset{5}{CH_3}$$
$$\underset{CH_3}{|}$$

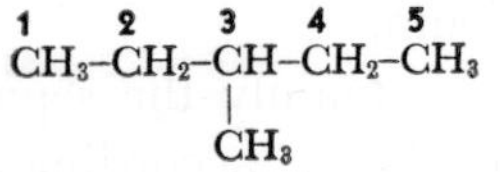

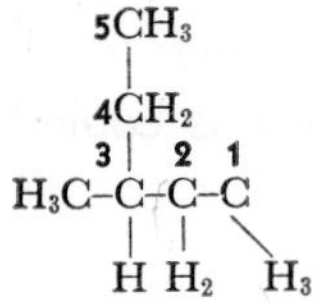

or

$$\underset{5CH_3}{|}$$
$$\underset{4CH_2}{|}$$
$$\overset{3}{H_3C}-\overset{|}{C}\overset{2}{-}C-\overset{1}{C}$$
$$\underset{H \ \ H_2 \ \ \ H_3}{}$$

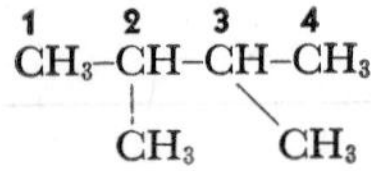

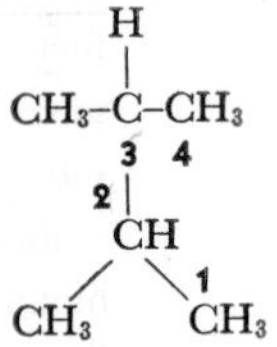

3-methylpentane

$$\overset{1}{CH_3}-\overset{2}{CH}-\overset{3}{CH}-\overset{4}{CH_3}$$
$$\underset{CH_3 \ \ \ \ CH_3}{|}$$

or

$$\underset{H}{|}$$
$$CH_3-\overset{|}{C}-CH_3$$
$$\overset{3 \ \ 4}{\underset{2}{|}}$$
$$CH$$
$$\underset{CH_3 \ \ \ \ \ CH_3}{}$$

2,3-dimethylbutane

$$\underset{CH_3}{|}$$
$$CH_3-\overset{|}{C}-CH_2-CH_3$$
$$\underset{CH_3}{|}$$

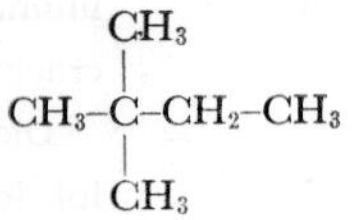

2,2-dimethylbutane

2-6. Gasoline

About 20% of the crude oil coming from an oil well is distilled into the gasoline fraction. This is called straight-run gasoline. Since the advent of the automobile, petroleum chemists have been

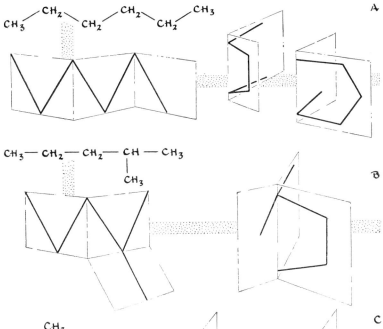

$CH_3 - CH_2 - CH_2 - CH - CH_3$
CH_3

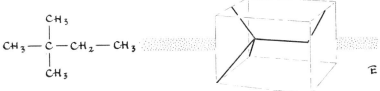

seeking new and better ways of increasing the gasoline fraction at the expense of fractions that are in lesser demand and cheaper. By various methods they have succeeded in more than doubling the gasoline fraction, chiefly at the expense of kerosene, gas oil, and lubricating oils.

The process called cracking takes long-chain hydrocarbons and clips them into smaller pieces. The means for doing this are more subtle than the word "clips" suggests, but the result is just that. The cracking stock is mainly the $C_{15}-C_{18}$ hydrocarbons, which are used also for Diesel fuel. By subjecting a hydrocarbon such as $C_{18}H_{38}$ to a temperature of 500–800°C, we may cleave it into pieces.

$$C_{16}H_{34} \longrightarrow \begin{array}{l} C_8H_{18} + C_8H_{16} \\ C_9H_{18} + C_7H_{16} \\ \text{or smaller pieces} \end{array}$$

The reverse process of building up larger molecules from smaller ones has also been effectively used for increasing the gasoline fraction. One method is frequently called polymerization, but "dimerization" would be a more specific term (that is, the chain length is ordinarily doubled). Sulfuric acid may be used to bring about the dimerization of a compound called isobutylene, C_4H_8, to give the dimer C_8H_{16}, which in turn can be made to add a mole-

$$\begin{array}{cc} CH_3 & CH_3 \\ | & | \\ CH_3-C-CH_2-CH-CH_3 \\ | \\ CH_3 \end{array}$$

"isooctane"

cule of hydrogen to yield "isooctane." (For equations, see pp. 86–87.) The name "isooctane" is familiar to you if you read newspaper advertisements.

Exercise 2-3. Supply the IUC name for "isooctane."

A second method of building larger molecules for the gasoline fraction is called alkylation. In this process, a saturated hydrocarbon such as isobutane (IUC name?) is added to an unsaturated compound called an alkene in the presence of a catalyst (for example, HF) to yield a member of the alkane series.

If a small amount (up to 3 ml per gallon) of tetraethyl lead (see the accompanying structural formula) is added to ordinary gasoline, a better gasoline for high-compression motors is obtained. To prevent the deposition of lead on the piston heads and spark plugs of the motor, a small amount of ethylene dichloride, $Cl-CH_2-CH_2-Cl$, and ethylene dibromide, $Br-CH_2-CH_2-Br$,

$$
\begin{array}{c}
CH_3 \\
| \\
CH_2 \\
| \\
CH_3-CH_2-Pb-CH_2-CH_3 \\
| \\
CH_2 \\
| \\
CH_3
\end{array}
$$

is also added to the gasoline. The lead chloride and lead bromide formed during combustion accompany the exhaust gases. Nevertheless, aircraft operators have been plagued with spark-plug fouling by lead deposits for many years. New gasoline additives recently found now promise some relief from this trouble. One of these is tricresyl phosphate (p. 582), which replaces some of the bromide and is apparently a better scavenger for lead.

Finally, a dye is added to gasoline for high-compression motors to distinguish it from gasoline for stoves and also because colored gasoline sells better. At present most "regular" gasolines contain some tetraethyl lead.

Straight-run gasoline in high-compression motors "knocks" badly. The constituents chiefly responsible are open-chain hydrocarbons such as n-hexane and n-heptane; n-heptane, in fact, knocks so badly in a standard motor that it has been assigned the octane number 0. Branch-chain hydrocarbons do not knock so severely. "Isooctane," mentioned above, was once thought to be the best gasoline that could be made, and hence it was given the octane number 100. A commercial gasoline that has an octane rating of 72 is, with regard to antiknock properties, as good as a gasoline containing 72% "isooctane" and 28% n-heptane.

Ratings above 100 are now easily reached; 2,2,3-trimethylbutane, for example, has an octane rating of 116.

Exercise 2-4. Write the structural formula for 2,2,3-trimethylbutane.

For a wider outlook on petroleum refining and the synthesis of gasoline, see the references at the ends of Chapters 2, 3, and 4.

2-7. Chemical Properties of Methane

1. Combustion. The most important property of any alkane is its complete oxidation by oxygen (burning in air) to carbon dioxide and water. This reaction, the principal one taking place in a Bunsen flame using natural gas, produces much heat. It is the

$$CH_4 + 2O_2 \longrightarrow CO_2 + 2H_2O + \text{heat} \tag{2-1}$$

basis of the entire gas industry, and it supplies a large fraction of the energy consumed in heating houses, cooking, and furnishing industrial power.

Incomplete combustion occurs if not enough oxygen is present to give the reaction written above. Carbon monoxide, a noxious gas, may result, along with deposition of carbon. The result may be a sooty flame.

$$2CH_4 + \tfrac{5}{2}O_2 \longrightarrow CO + C + 4H_2O \tag{2-2}$$

Exercise 2-5. Write an equation for the complete combustion of n-hexane.

2. Halogenation. Chlorine gas will react with methane gas only when subjected to sunlight or ultraviolet radiation. The reaction in sunlight cannot be controlled well and may reach explosive violence. It goes by a free-radical mechanism (p. 21) for which the sunlight supplies the initial energy. Theoretically, only one

$$: \overset{\cdot\cdot}{\underset{\cdot\cdot}{Cl}} : \overset{\cdot\cdot}{\underset{\cdot\cdot}{Cl}} : + \text{ sunlight} \longrightarrow 2 : \overset{\cdot\cdot}{\underset{\cdot\cdot}{Cl}} \cdot \qquad \text{[2-3a]}$$

$$: \overset{\cdot\cdot}{\underset{\cdot\cdot}{Cl}} \cdot + H : \overset{\overset{\textstyle H}{\cdot\cdot}}{\underset{\underset{\textstyle H}{\cdot\cdot}}{C}} : H \longrightarrow H : \overset{\cdot\cdot}{\underset{\cdot\cdot}{Cl}} : + H : \overset{\overset{\textstyle H}{\cdot\cdot}}{\underset{\underset{\textstyle H}{}}{C}} \cdot \qquad \text{[2-3b]}$$

$$H : \overset{\overset{\textstyle H}{\cdot\cdot}}{\underset{\underset{\textstyle H}{}}{C}} \cdot + : \overset{\cdot\cdot}{\underset{\cdot\cdot}{Cl}} : \overset{\cdot\cdot}{\underset{\cdot\cdot}{Cl}} : \longrightarrow CH_3 : \overset{\cdot\cdot}{\underset{\cdot\cdot}{Cl}} : + : \overset{\cdot\cdot}{\underset{\cdot\cdot}{Cl}} \cdot \qquad \text{[2-3c]}$$

packet of sunlight is necessary to set off the entire reaction, for
each chlorine atom initiating the removal of one hydrogen atom
from methane (Exp. 2-3b) leads to the generation of a new chlorine
atom (Exp. 2-3c). After the reaction has proceeded a little way,
both CH_4 and CH_3Cl are present and both subject to attack by
chlorine atoms. Eventually, then, the reaction continues through
four steps. It is commercially feasible because the products boil at

$$CH_4 + Cl_2 \xrightarrow{\text{sunlight}} \underset{\substack{\text{methyl} \\ \text{chloride}}}{CH_3Cl} + HCl \qquad \text{[2-4]}$$

$$CH_3Cl + Cl_2 \longrightarrow \underset{\substack{\text{methylene} \\ \text{chloride}}}{CH_2Cl_2} + HCl \qquad \text{[2-5]}$$

$$CH_2Cl_2 + Cl_2 \longrightarrow \underset{\text{chloroform}}{CHCl_3} + HCl \qquad \text{[2-6]}$$

$$CHCl_3 + Cl_2 \longrightarrow \underset{\substack{\text{carbon} \\ \text{tetrachloride}}}{CCl_4} + HCl \qquad \text{[2-7]}$$

temperatures far enough apart so that they can be fairly easily
separated by fractional distillation.

Exercise 2-6. Show how CH_2Cl_2, $CHCl_3$, and CCl_4 are obtained by free-
radical reactions from CH_3Cl and Cl_2 in the presence of sunlight.

As might be expected, the various halogens differ greatly in
their behavior toward alkanes. The order of decreasing activity is

F, Cl, Br, I. Fluorine attacks methane with explosive violence, bromine reacts very slowly, and iodine does not react at all, in the presence of sunlight.

3. Nitration. Nitric acid, as well as many other chemicals, can act in two ways on other substances. The way it acts in a particular case may depend on the other substance as well as on the conditions under which the reaction is carried out. It may dissociate as shown on the left. This action is favored when the nitric acid is dilute—that is, when a large amount of water is present.

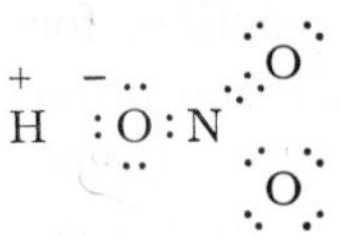

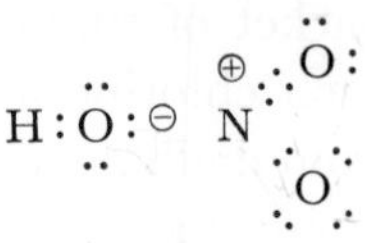

Here it is behaving like an acid; that is, the ionized proton is the active part. In concentrated solution it is not dissociated to so large an extent. In the presence of many organic substances it acts in another way, as if it were exhibiting a separation of electric charges as shown on the right (above). When it acts as a nitrating agent (see also p. 123), the attack is by the electron-deficient nitrogen. In concentrated sulfuric acid, evidence indicates, the following reaction takes place:

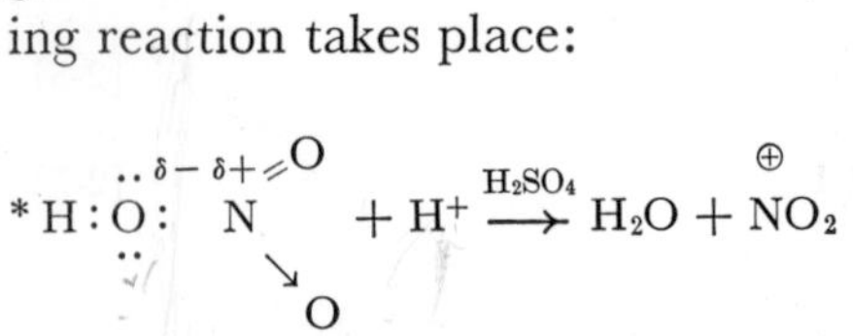

$$* H : \overset{..}{\underset{..}{O}} : \overset{\delta-}{N} \overset{\delta+}{\nearrow} \overset{O}{\underset{\searrow}{}} \quad + H^+ \overset{H_2SO_4}{\longrightarrow} H_2O + \overset{\oplus}{NO_2} \qquad [2\text{-}8]$$

Though it is untenable to assume that OH^- ever exists as such in the acid solution, nitric acid is nevertheless made to behave as a base in the presence of concentrated sulfuric acid. Rather more

---

* The symbols $\delta+$ and $\delta-$ (Greek letter, delta) will be used hereafter to mean a partial separation of charges as opposed to the complete separation indicated by + or −. A circled symbol, $\oplus$ or $\ominus$, will be used to designate a transient separation of charges, postulated generally for the course of the reaction. Its existence may or may not have been amply demonstrated. In general the partials will be used in the static case and circled charges in the dynamic case.

plausible is the assumption that the proton from the sulfuric acid first coordinates itself with the oxygen of the nitric acid and that the splitting out of water then follows, leaving the $\overset{\oplus}{NO_2}$ ion. As evidence one may cite the fact that salts in which $\overset{\oplus}{NO_2}$ is the cation have now been isolated (1950).

Nitric acid in the gas phase above 400° will react with methane to yield nitromethane and water. Whether the charge separation just described obtains under these conditions is not known with certainty. If it does, the energy with which $\overset{\oplus}{NO_2}$ seeks a pair of electrons on carbon probably makes the reaction go. At this high

$$CH_4 + HONO_2 \xrightarrow[13\%\ (1\ pass)]{475°} CH_3\text{-}N\overset{\displaystyle O}{\underset{\displaystyle O}{\diagdown}} + H_2O \qquad\qquad [2\text{-}9]$$

temperature, however, the reaction may well go by a free-radical mechanism.

2-8. The Higher Alkanes

Once the reactions of one member of a homologous series have been learned, the reactions of other members of the series follow with very few modifications. This observation simplifies tremendously the study of carbon compounds.

1. Combustion. All members of the alkane series burn in air or oxygen to give carbon dioxide and water. For example:

$$2\ \begin{matrix} CH_3 \\ \diagdown \\ \diagup \\ CH_3 \end{matrix} CH\text{-}CH_3 + 13O_2 \longrightarrow 8CO_2 + 10H_2O \qquad\qquad [2\text{-}10]$$

There is no modification in property here; only the coefficients in the new equation differ from those for the oxidation of methane.

Combustion of hydrocarbons is what makes the wheels go round in a large fraction of industry as well as in the automobile. About 60% of the energy used industrially is supplied by liquid and gaseous alkanes.

2. Halogenation.   Halogens react with higher members of the alkane series less readily as the carbon chain is lengthened. Only fluorine and chlorine will attack the long-chain alkanes.

One slight modification must be taken into account with any compound other than methane. There is a choice of carbons to attack, and the products are different. Actually, in the chlorina-

$$CH_3\text{--}CH_2\text{--}CH_3 + Cl_2 \xrightarrow{\text{light}} \begin{cases} CH_3\text{--}CH_2\text{--}CH_2\text{--}Cl + HCl \\ \text{or} \\ CH_3\text{--}\underset{\underset{Cl}{|}}{C}H\text{--}CH_3 + HCl \end{cases} \qquad [2\text{-}11]$$

tion of propane (Exp. 2-11), about equal quantities of the two products are obtained even though the two end carbons carry six hydrogens and the central carbon two hydrogens. One might think that all eight hydrogens would be replaced at the same rate, yielding three times as much of the first compound as of the second.

In isopentane, chlorine attacks the three types of carbon atoms at quite different rates at 300°C. The carbon carrying one hydrogen

$$\begin{matrix} CH_3 \\ \quad \diagdown \\ \qquad CH\text{--}CH_2\text{--}CH_3 \\ \quad \diagup \\ CH_3 \end{matrix}$$

is attacked more than four times as fast as the carbons carrying three hydrogens, and the $-CH_2-$ group is attacked more than three times as fast. The ratio of these rates is 4.43:3.25:1 at 300° but approach 1:1:1 as the temperature is raised still higher.

Exercise 2-7.   Write the structure for the monosubstituted chloro compound formed at the highest rate in chlorinating isopentane at 300°.

3. Nitration. Ethane, propane, and the higher alkanes will react with nitric acid in the gaseous state at slightly lower temperature than methane. This reaction is complicated to some extent by the breaking of carbon-carbon bonds along with nitration.

$$
\underset{\overset{|}{H}}{\overset{\overset{H}{|}}{CH_3{-}\underset{}{C}}}{:} \quad \overset{\delta-}{H} + \overset{\delta+}{HO}{:} \quad \overset{\delta-}{N}\overset{\overset{\delta+O}{\diagup}}{\diagdown O} \xrightarrow[\substack{33\% \\ (1\ pass)}]{450°} \underset{\substack{73\ parts \\ nitroethane}}{CH_3{-}CH_2{-}N}\overset{\overset{O}{\diagup}}{\diagdown O} + H_2O \qquad [2\text{-}12]
$$

$$
CH_3{-}CH_3 + 3HON\overset{\overset{O}{\diagup}}{\diagdown O} \longrightarrow \underset{\substack{27\ parts \\ nitromethane}}{CH_3{-}NO_2} + NO + CO_2 + 2H_2O + NO_2 \qquad [2\text{-}13]
$$

When propane is the starting alkane, 1-nitropropane, 2-nitropropane, nitroethane, and nitromethane are all obtained. This reaction is run on a commercial scale, and all four of these products are now available in carload lots from this one alkane.

2-9. The Cycloalkanes

The alkanes have the general formula C_nH_{2n+2}. Another homologous series of saturated hydrocarbons is obtained if the end carbons in a chain are joined in a ring. Since the two end carbons are joined to each other, the molecule will have two fewer hydrogens, and hence the general formula for the cycloalkanes, or cycloparaffins, is C_nH_{2n}. The simplest cyclic saturated hydrocarbon contains three carbons and is called cyclopropane. Others are shown in Table 2-5.

2-10. Chemical Properties of Cycloalkanes

There are great differences in chemical activity among these various cycloalkanes, as we shall see when we examine specific reactions.

1. Catalytic Hydrogenation. The gradation in the chemical

TABLE

2-5 | *Cycloalkanes*

FORMULA		IUC AND COMMON NAME	BOILING POINT (°C)
C_3H_6	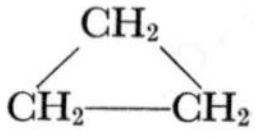 CH$_2$ CH$_2$——CH$_2$	cyclopropane	−34
C_4H_8	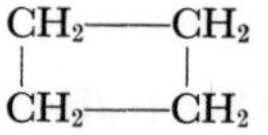 CH$_2$——CH$_2$ CH$_2$——CH$_2$	cyclobutane	13
C_5H_{10}	CH$_2$——CH$_2$ | CH$_2$ CH$_2$——CH$_2$	cyclopentane	51
C_6H_{12}	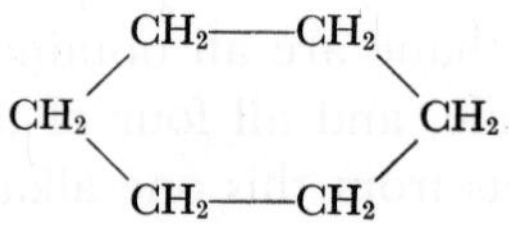 CH$_2$——CH$_2$ CH$_2$ CH$_2$ CH$_2$——CH$_2$	cyclohexane	81
C_7H_{14}	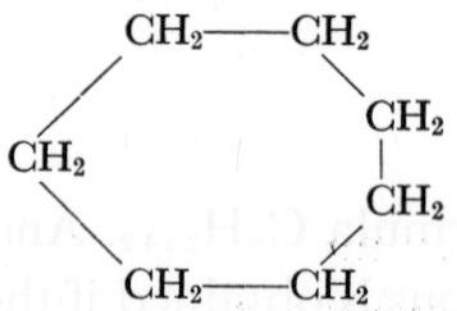 CH$_2$——CH$_2$ CH$_2$ CH$_2$ CH$_2$ CH$_2$ CH$_2$——CH$_2$	cycloheptane	118
C_8H_{16}	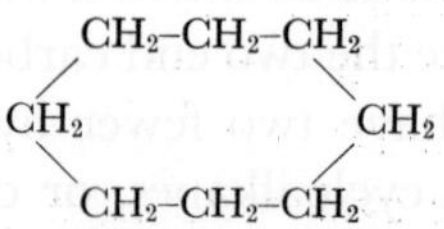 CH$_2$-CH$_2$-CH$_2$ CH$_2$ CH$_2$ CH$_2$-CH$_2$-CH$_2$	cyclooctane	151

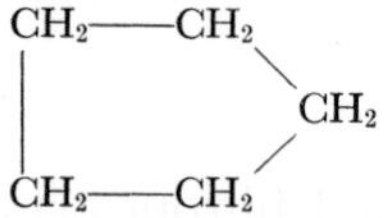

activity of cycloalkanes is shown by the temperature that is required to break the rings in the presence of hydrogen and a nickel
catalyst. Cyclopropane at 120°, cyclobutane at 200°, and cyclo-

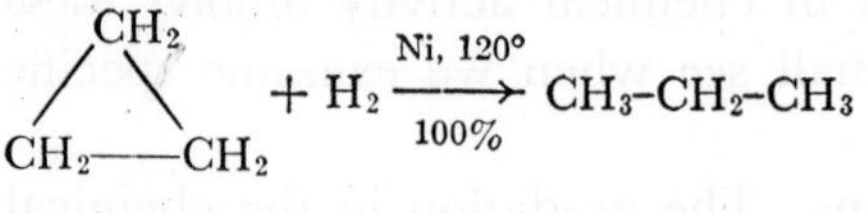

$$
\begin{array}{c}
\text{CH}_2 \\
\diagup \diagdown \\
\text{CH}_2\text{——CH}_2
\end{array}
+ \text{H}_2 \xrightarrow[100\%]{\text{Ni, 120°}} \text{CH}_3\text{-CH}_2\text{-CH}_3
\qquad [2\text{-}14]
$$

pentane at 325° are hydrogenated to the respective open-chain compounds. The cyclohexane ring cannot be broken by this method.

Exercise 2-8. Write the corresponding reaction of cyclobutane and hydrogen in the presence of Ni at 200°.

2. Reaction with Bromine. Cyclopropane will open in the presence of bromine at 25° to add a bromine atom at each carbon where the bond is broken. Cyclobutane will not react at $-10°$ and apparently has not been tried at higher temperatures. The

$$
\begin{array}{c}
CH_2\text{–}CH_2 \\
\diagdown \diagup \\
CH_2
\end{array}
+ Br_2 \longrightarrow Br\text{–}CH_2\text{–}CH_2\text{–}CH_2\text{–}Br \qquad [2\text{-}15]
$$

$$
\begin{array}{c}
\diagup CH_2\text{–}CH_2 \\
CH_2 \qquad \diagdown CH_2 \\
\diagdown CH_2
\end{array}
+ Br_2 \xrightarrow[\text{light}]{\text{ultraviolet}}
\begin{array}{c}
\diagup CH_2 \\
Br\text{–}CH \qquad \diagdown CH_2 \\
\diagdown CH_2\text{–}CH_2
\end{array}
+ HBr \qquad [2\text{-}16]
$$

rings of cyclohexane and cyclopentane (Exp. 2-16) are too stable to break. When ultraviolet radiation is used as a catalyst, substitution rather than addition takes place. It may be said, then, that the activity of cyclohexane and cyclopentane toward halogens is similar to that of the open-chain alkanes.

2-11. Baeyer's Strain Theory

These properties of the cycloalkanes are best accounted for by the geometry of the compounds. The three-membered ring in cyclopropane is necessarily a triangle, a planar figure, in which the angles are each 60°. Since the normal angle between bonds in a tetrahedral carbon atom is 109°28′, the bonds between the carbons of cyclopropane are distorted and under a strain. An estimate of the strain can be derived from mathematical considerations. The strain on each bond in cyclopropane is $\frac{1}{2}(109°28′ - 60°) = 24°44′$.

cyclobutane: $\frac{1}{2}(109°28' - 90°) = 9°44'$
cyclopentane: $\frac{1}{2}(109°28' - 108°) = 44'$
cyclohexane: $\frac{1}{2}(109°28' - 120°) = -5°16'$

Cyclopropane is under the greatest strain and hence should be the most reactive of these four compounds. Cyclopentane is under very slight strain, and cyclobutane is under moderate strain. The two reactions discussed above bear out this gradation in chemical activity from cyclopropane to cyclopentane.

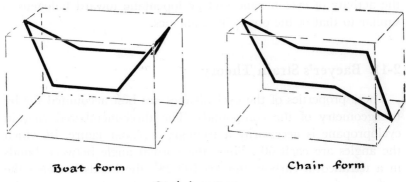

Boat form   Chair form

Cyclohexane

Cyclohexane is actually under even less strain than might be assumed from the calculation. The extraneous nature of the negative sign will be immediately clear if a model of the compound is made. The six carbon atoms of the ring do not lie in a single plane. This allows each carbon-carbon bond to assume its normal angle, and the ring is strainless. Rings containing up to thirty carbon atoms have been synthesized.

Baeyer's strain theory accounts for the gross difference in the chemistry of the three-, four-, and five-membered rings but does not apply to larger ring structures.

A quantitative measure of the strain is available in the heat of combustion of these cyclic compounds. If energy is needed to form the strained ring, this energy should be liberated when the ring is broken. If the compounds are burned, the heat evolved should be a measure of the Baeyer strain. The results are shown in Table 2-6. For comparison, a value is also given for ethylene (Chap. 4) and for a $-CH_2-$ group in an open-chain compound. The value for cyclohexane bears out the contention that the six-membered ring is under no strain.

TABLE 2-6 | *Heats of Combustion of Hydrocarbons*

COMPOUND	KILOCALORIES PER $-CH_2-$ GROUP (14 GRAMS)
Ethylene, $CH_2=CH_2$	170
Cyclopropane	168.5
Cyclobutane	165.5
Cyclopentane	159
Cyclohexane	158
Other alkanes (continuous chain)	158

2-12. Uses of Cycloalkanes

Cyclopropane is used as a general anesthetic. It does not have some of the dangerous properties associated with ether in the same use. Cyclohexane is used to some extent in aviation gasoline because it has good antiknock properties.

SUMMARY

The device of studying the vast array of carbon compounds by families of substances having closely related properties at once reduces the apparently impossible task of learning chemistry to a systematic scheme. The first homologous series, saturated hydrocarbons, is treated with respect to its nomenclature, physical and chemical properties, and uses. Petroleum refining is given a cursory glance, and the student is urged to read Reference 1 below for a more complete outlook.

The chemistry of the cycloalkanes, a second homologous series, is governed in part by geometry, as described in Baeyer's strain theory.

REFERENCES

1. Shoemaker, d'Ouville, and Marschner, "Recent Advances in Petroleum Refining," J. Chem. Education, 32, 30 (1955).
2. Gilbert, "The Reactive Paraffins," J. Chem. Education, 18, 435 (1941).
3. Anderson, "What a Chemistry Teacher Should Know about Oil," J. Chem. Education, 32, 563 (1955).

EXERCISES

(Exercises 1–8 will be found within the text.)

9. Name each of the following by the IUC system:

1. 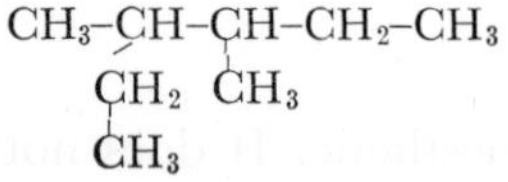 $CH_3-CH-CH-CH_2-CH_3$
       CH_2 CH_3
       CH_3

2. $CH_3-CH_2-C\!-\!-\!-\!-CH-CH_2-CH_2-CH_3$
                CH_2 $CH-CH_3$
                CH_3 CH_3

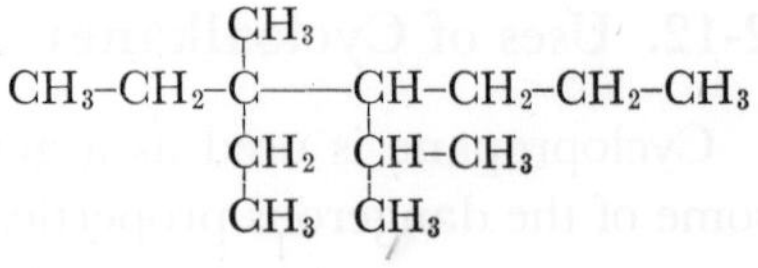

$$CH_3$$
3. $CH_3-C-CH_2-CH-CH_3$
   $\quad\ \ CH_3 \quad\ \ CH_3$

4. $(CH_3)_2-CH-CH_2-CH_2-CH_2-CH_3$

5. $CH_3-CH-(CH_2-CH_2-CH_3)_2$

10. Write the condensed structural formula for

1. 2,2,4-trimethylpentane
2. 2,2-dimethyl-3-ethyl-4-isopropyloctane
3. 2,4-diethylhexane
4. 3,4-diethylheptane

Criticize the name given to No. 3, and suggest a correct name.

11. Write condensed structural formulas for all nine isomeric forms of heptane. Name each by the IUC system.

12. Balance the equation for the complete combustion of heptane.

13. Write equations for the monosubstitution reaction of chlorine with (1) cyclohexane, (2) 2-methylbutane, and (3) methylcyclopentane. Indicate all expected isomers.

14. Define or identify by a suitable illustration each of the following terms:

1. chain isomers
2. substitution
3. cracking
4. polymerization
5. octane number
6. hydrogenation

15. Calculate the percentage composition of PCl_5; of H_2SO_4; of "isooctane."

16. How many kilocalories of energy are released when one gram of gasoline is burned? Assume that gasoline has the formula C_8H_{18} and that the heat of formation of CO_2 is 94.38 kcal per mole and that of H_2O is 68.38 kcal per mole.

17. The composition of a volatile hydrocarbon is found by analysis to be 83.3% carbon and 16.7% hydrogen. What is the probable empirical formula? the molecular formula? Write a structure for one possible compound fitting this analysis.

18. Write electronic structures for chloroform, cyclobutane, and nitroethane.

3 Alkyl Halides

The alkyl halides have the general formula $C_nH_{2n+1}X$, in which X is any one of the four halogens (fluorine, chlorine, bromine, or iodine). The compounds are completely described by the phrase "monohalogenated saturated hydrocarbons." If the letter R represents the group C_nH_{2n+1}, the general symbol for any alkyl halide is RX.

TABLE
3-1 | *Alkyl Halides*

BOILING POINT	RX	COMMON NAME	IUC NAME
$-24°C$	CH_3-Cl	methyl chloride	chloromethane
12	CH_3-CH_2-Cl	ethyl chloride	chloroethane
47	$CH_3-CH_2-CH_2-Cl$	n-propyl chloride	1-chloropropane
	$CH_3-CH-CH_3$ | Cl	isopropyl chloride	2-chloropropane
78	$CH_3-CH_2-CH_2-CH_2-Cl$	n-butyl chloride	1-chlorobutane
	$CH_3-CH_2-CH-CH_3$ | Cl	sec-butyl chloride	2-chlorobutane
	CH_3 \ $CH-CH_2-Cl$ / CH_3	isobutyl chloride	1-chloro-2-methyl-propane
	CH_3 \ $C-Cl$ / | CH_3 CH_3	tert-butyl chloride	2-chloro-2-methyl-propane

3-1. Nomenclature of Alkyl Halides

We form the common name of an alkyl halide (Table 3-1) by substituting *yl* for *ane* in the name of the corresponding hydrocarbon and adding *chloride, fluoride, bromide,* or *iodide.* We can classify alkyl halides by specifying the number of alkyl groups on the carbon that carries the halogen. A primary alkyl halide is one that carries only one alkyl group on the carbon bearing the halogen (or three H's in the case of methyl halides); a secondary alkyl halide has two alkyl groups on this carbon; and a tertiary alkyl halide carries three alkyl groups on this carbon. Two of the four butyl chlorides bear common names derived from this classification: secondary butyl chloride and tertiary butyl chloride. The name "primary butyl chloride" cannot be used since it is not unique; both n-butyl chloride and isobutyl chloride have primary carbon atoms carrying the chlorine.

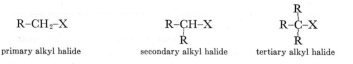

The IUC names for the alkyl halides are derived according to the rules for the alkanes given in Chapter 2, with one addition: the prefix *chloro* (or *bromo* or *iodo*) is used with the hydrocarbon stem, and a number with a hyphen designates its position on the chain.

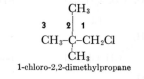

$$CH_3-CH_2-CH_2-CH_2-Cl$$
1-chlorobutane

3-2. Preparation of the Alkyl Halides

1. Direct Substitution. The action of halogens (especially chlorine) has already been mentioned as a reaction of the alkanes (pp. 52, 56). As a method of preparation of alkyl halides it is useful only for the lower members of the series (1–3 carbons). The longer alkanes not only are sluggish in their reactivity but always yield iso-meric mixtures that are not easily separated by fractional distilla-

$$CH_3-CH_3 + Cl_2 \xrightarrow[\text{light}]{\text{ultraviolet}} CH_3-CH_2-Cl + HCl \qquad\qquad [3\text{-}1]$$

$$2CH_3-CH_2-CH_3 + 2Cl_2 \xrightarrow[\text{light}]{\text{ultraviolet}} CH_3-CH_2-CH_2-Cl + CH_3-\underset{\underset{\text{Cl}}{|}}{CH}-CH_3 + 2HCl$$

$$[3\text{-}2]$$

tion. Isopropyl and n-propyl chlorides can be separated since there is a difference of 10° in the boiling points, which are 37° and 47°, respectively.

Exercise 3-1.   Write all the possible monochloro substitution products obtainable from isopentane by direct chlorination. Name them by the IUC system.

2. From Alcohols (see Chap. 7).   The general formula for an alcohol is ROH. Alcohols react with hydrohalogen acids to give alkyl halides and water. In this equation HX may be HCl, HBr,

$$R-OH + HX \longrightarrow R-X + H_2O \qquad\qquad [3\text{-}3]$$

or HI. In reactions with the same alcohol HI reacts faster than HBr, and HBr faster than HCl. The class of the alcohol also deter-mines the rate of reaction. Tertiary alkyl halides are frequently formed within seconds after the reacting substances are mixed. Secondary alkyl halides form less rapidly, and primary alkyl halides frequently require long refluxing. Competing reactions take place in the formation of secondary and tertiary alkyl halides; the yields of these two classes are generally lower, as a consequence, even though the reactions are faster than those in which primary halides are formed.

The reaction of an alcohol and a hydrohalogen acid involves, first, the formation of a coordinate covalent bond between the oxygen of the alcohol and the proton of the acid, and then, simul-

taneously or consecutively loss of water and the formation of a
new bond between the R group and the halogen.

$$R:\ddot{O}:H + H:\ddot{X}: \longrightarrow \left[R:\overset{..}{\underset{H}{O}}:H \right]^{+} :\ddot{X}:^{-} \longrightarrow [R^{+}]\ [X^{-}] + HOH \qquad [3\text{-}4]$$
$$\qquad\qquad\qquad\qquad\qquad\qquad\qquad\qquad \llcorner_{\rightarrow R\text{-}X}$$

(In this text brackets are used in three ways. The first use, as here, is to designate
structures whose existence may be transitional and may never have been demonstrated. For the second use see pages 108–112 and Chapters 19 and 20; for the
third use see page 183.)

Other reagents that are used to convert alcohols to alkyl halides
are PCl_3, PBr_3, PCl_5, and PI_3. All of these substances fume in moist
air because of their reactions with the water vapor. This reaction
between a phosphorus halide and water is closely analogous to the
reaction between the phosphorus halide and an alcohol:

$$PCl_3 + 3H\text{-}OH \longrightarrow P(OH)_3 + 3HCl \qquad\qquad\qquad [3\text{-}5]$$

$$PCl_3 + 3R\text{-}OH \longrightarrow P(OH)_3 + 3R\text{-}Cl \qquad\qquad\qquad [3\text{-}6]$$

As a rationalization of the mechanism for the hydrolysis (and
alcoholysis) of PCl_3, one may assume that the first reaction is a
coordination of a proton from water by the unshared pair of electrons on phosphorus. (See p. 565 also.) The resulting tetrahedral
cation undergoes a double decomposition with water in successive
steps to $H_4PO_3^{\oplus}$. Loss of a proton at this stage yields phosphorous
acid; three molecules of HCl are liberated in the reaction. The

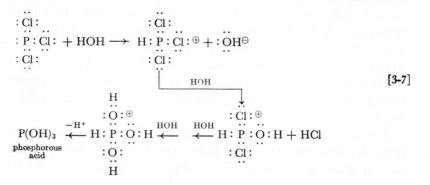

exact mechanism is not known, and this theory is still in the argumentative stage.

An alcohol may react in an analogous manner with the same intermediates, the difference being that RCl rather than HCl is the second product.

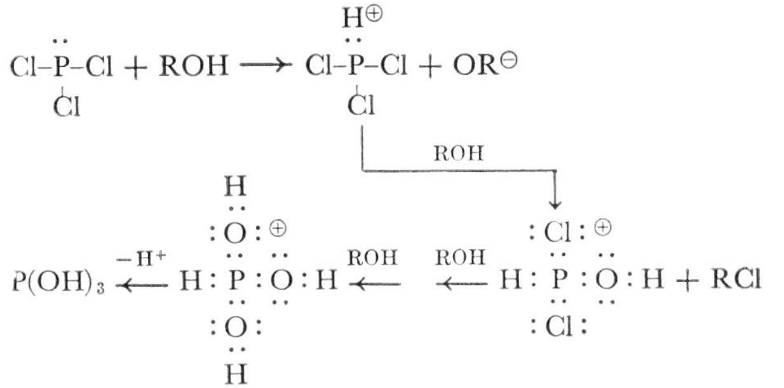

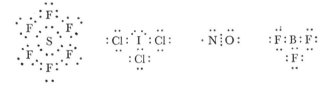

[3-8]

Phosphorus pentachloride is interesting because it apparently violates the Rule of Eight. Other exceptions to the rule are SF_6, ICl_3, NO, and BF_3:

Phosphorus pentachloride is a yellow solid that reacts violently with water and with other compounds in which an OH group is attached to C or S.

$$PCl_5 + H\text{-}OH \longrightarrow POCl_3 + 2HCl \qquad\qquad [3\text{-}9]$$

$$PCl_5 + R\text{-}OH \longrightarrow POCl_3 + HCl + RCl \qquad\qquad [3\text{-}10]$$

$$PCl_5 + CH_3CH_2\text{-}OH \longrightarrow POCl_3 + HCl + CH_3\text{-}CH_2\text{-}Cl \qquad [3\text{-}11]$$
ethyl chloride

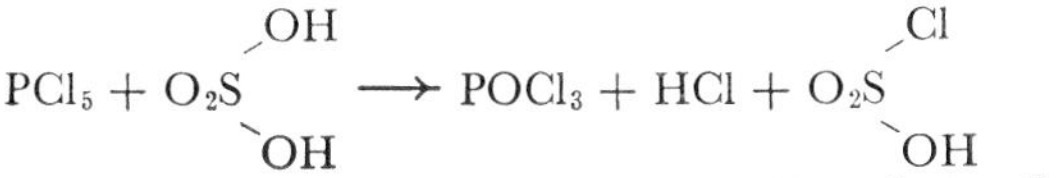

$$PCl_5 + O_2S\diagdown^{OH}_{OH} \longrightarrow POCl_3 + HCl + O_2S\diagdown^{Cl}_{OH} \qquad [3\text{-}12]$$
chlorosulfonic acid

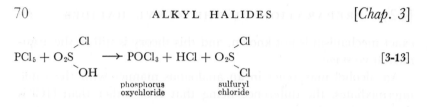

$$PCl_5 + O_2S \begin{smallmatrix} Cl \\ \diagup \\ \diagdown \\ OH \end{smallmatrix} \longrightarrow POCl_3 + HCl + O_2S \begin{smallmatrix} Cl \\ \diagup \\ \diagdown \\ Cl \end{smallmatrix}$$ [3-13]

phosphorus   sulfuryl
oxychloride   chloride

The crystalline structure of PCl_5 is probably the following ionic pair, in which one phosphorus follows the Rule of Eight and the other surrounds itself with six pairs of electrons:

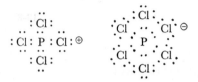

Phosphorus tribromide and phosphorus tri-iodide are ordinarily not isolated when they are to be used in making alkyl halides. Instead, red phosphorus and the alcohol are placed in a flask, and then bromine or iodine is added to the mixture. The following reactions take place:

$2P + 3Br_2 \longrightarrow 2PBr_3$ [3-14]

$PBr_3 + ROH \longrightarrow RBr + P(OH)_3$ [3-15]

$$PBr_3 + CH_3\text{–}\underset{\underset{\overset{|}{OH}}{\overset{|}{CH_2}}}{CH}\text{–}CH_3 \xrightarrow[90\%]{} CH_3\text{–}\underset{\underset{\overset{|}{Br}}{\overset{|}{CH_2}}}{CH}\text{–}CH_3 + P(OH)_3$$ [3-16]

Exercise 3-2. Choose a reaction to show how to make sec-butyl bromide.

3-3. Reactions of the Alkyl Halides

A. *Hydrolysis*

Hydrolysis may be defined as any reaction in which water is one of the reactants; it does not include reactions in which water acts only as a solvent. Many times water will serve both purposes at the same time. Alkyl halides react upon being boiled with water to give an alcohol and the hydrohalogen acid. This is the reverse of one reaction mentioned as a method of preparing an alkyl halide. It has been pointed out already that HX will react with ROH to give the two starting compounds shown in Expression 3-17.

$$RX + HOH \rightleftharpoons ROH + HX \qquad\qquad\qquad [3\text{-}17]$$

It seems reasonable to suppose that after a while the reaction must reach an equilibrium—that is, that the products on the right will be formed at the same rate at which they are consumed by the reverse reaction.

All chemical reactions are reversible. A number of the reactions already given were not said to be reversible. Chemists have a habit of saying that a reaction goes in a particular direction if the equilibrium is highly favorable to one set of products over the other. It should be kept in mind, however, that all reactions are reversible to some extent.

The question whether an equilibrium can be shifted to favor one set of products over the other may arise in your mind. One of the products formed in the hydrolysis just mentioned is an acid. If this acid is removed from the scene of action, the reaction can be made to continue to the right until practically all of the reactants are consumed. If HX is a gas, it is removed from contact with the alcohol.

The formation of the alcohol is favored by the use of a very large excess of water. If a large volume of water is used, the two products are separated by many water molecules and hence have less chance of colliding with each other. The probability that the reverse reaction will take place is therefore smaller.

Water reacts with an alkyl halide because water is a base. Since it is a very weak base, one might expect the stronger base OH⁻ to be a better reagent than water for the hydrolysis. Indeed, use of a strongly basic medium is the common method of hydrolyzing an alkyl halide. If any water molecules react to form HX, this is quickly removed from the scene of action by the hydroxyl ion, which in this way also aids the forward reaction.

The formation of RX (the reverse of Exp. 3-17) can likewise

be enhanced by the use of a large excess of one reagent. An excess of concentrated HX will be diluted very little by the water formed in this reaction, and hence the reaction

$$ROH + HX \rightleftarrows RX + HOH \qquad\qquad [3\text{-}18]$$

will continue to proceed toward the right. A second method of improving the yield of an alkyl halide is to remove the water as fast as it is formed. This can be accomplished by substances called dehydrating agents, which form very stable compounds with water. Sulfuric acid and anhydrous zinc chloride have a strong tendency to remove water from other substances. When added to the above reaction, either of these compounds effectively removes the water from the reaction mixture and thus shifts the equilibrium to the right. These two substances probably also increase the acid strength of the HX and in that way aid the formation of RX.

B. Formation of Alkyl Cyanides

Heating a primary alkyl halide with an alkali cyanide produces an alkyl cyanide. Secondary and tertiary alkyl halides react to some extent, but side reactions (such as the formation of alkenes)

$$RX + Na^+CN^- \longrightarrow R\text{-}C\equiv N + Na^+X^- \qquad\qquad [3\text{-}19]$$

$$CH_3\text{-}CH_2\text{-}Cl + Na^+CN^- \longrightarrow CH_3\text{-}CH_2\text{-}C\equiv N + Na^+Cl^- \qquad\qquad [3\text{-}20]$$

prevent good yields. This is a polar displacement reaction (see p. 22) in which CN^- displaces $: \overset{..}{\underset{..}{X}} :$ from RX.

Exercise 3-3. Write an equation for the reaction of isobutyl bromide and sodium cyanide.

C. Reactions with Metals

1. Sodium (Wurtz Reaction). The alkali metals attack alkyl halides to remove the halogen and give the ionic metal halide. This reaction undoubtedly involves a free alkyl radical at this point. The alkyl radical, which is at least as reactive as a sodium

$$R:X + Na\cdot \longrightarrow Na^+X^- + R\cdot \qquad\qquad [3\text{-}21]$$
$$\downarrow$$
$$\hookrightarrow R:R$$

atom, may then join with a second alkyl to yield $R:R$, a saturated hydrocarbon having double the number of carbons of the original alkyl halide. The other course for the reaction to take is for a second sodium to react with $R\cdot$ to yield $Na:R$. This sodium alkyl might then react (Exp. 3-22) to give the hydrocarbon. Expression 3-21 is clearly a free-radical reaction, whereas Expression 3-22

$$R:X + Na:R \longrightarrow R\text{-}R + Na^+X^- \qquad\qquad [3\text{-}22]$$

involves exchanges of pairs of electrons (polar mechanism). There is evidence to support both paths for the Wurtz reaction, but a free radical is certainly an intermediate during the formation of the reagent $Na:R$ in Expression 3-22. It is quite possible that both paths may be followed simultaneously.

$$2CH_3\text{-}CH_2\text{-}CH_2\text{-}CH_2\text{-}Br + 2Na\cdot \xrightarrow[53\%]{}$$
$$CH_3\text{-}CH_2\text{-}CH_2\text{-}CH_2\text{-}CH_2\text{-}CH_2\text{-}CH_2\text{-}CH_3 + 2Na^+Br^- \qquad [3\text{-}23]$$

Exercise 3-4.   Show a synthesis for 2,5-dimethylhexane by the Wurtz reaction.

2. Lead.   Lead and other heavy metals, such as mercury, form bonds with carbon that are much more stable than those of the alkali metals. Lead tetraethyl is an important item of commerce because of its use in Ethyl gasoline. It is made by heating a sodium-lead alloy having the approximate ratio NaPb with ethyl chloride under pressure. The lead tetraethyl can be removed by passing steam through the mixture. It is a liquid with boiling point 203°C.

$$2NaPb + 4CH_3\text{-}CH_2\text{-}Cl \longrightarrow Pb(CH_2\text{-}CH_3)_4 + 2NaCl + PbCl_2 \qquad [3\text{-}24]$$

3. Zinc (Frankland Reaction) and Magnesium (Formation of Grignard Reagent).   Metals such as zinc and magnesium, of intermediate character between the alkali metals and the heavy metals, also react with alkyl halides. Zinc dust combines with alkyl iodides to give zinc alkyls, which, being spontaneously inflammable in air, are difficult to handle in the laboratory and

$$2CH_3-CH_2-I + 2Zn(dust) \longrightarrow (CH_3-CH_2)_2Zn + ZnI_2 \qquad [3-25]$$
$$\text{diethyl zinc}$$

hence not very useful. Frankland discovered the zinc alkyls in 1849 and was probably the first to recognize the existence of a carbon-to-metal bond.

Exercise 3-5. Write an equation for the combustion of diethyl zinc in air.

Because of their ease of formation and their many uses, the organomagnesium compounds overshadow all the other organometallic compounds at the present time. Barbier in 1899 reported some work with carbon-magnesium compounds. Grignard put the reagents formed by magnesium and an alkyl halide to wide use, and the reagents and reactions are now associated with his name. Alkyl halides dissolved in dry ether react with magnesium turnings, evolving enough heat to keep the ether boiling. The products are called Grignard reagents. These reagents react as illustrated

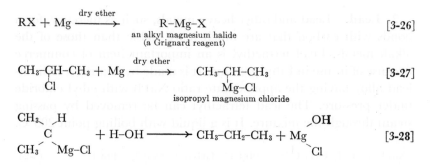

with water (Exp. 3-28) or with acids. They also react with a large number of compounds to be mentioned in subsequent chapters.

They readily hydrolyze to alkanes, a good laboratory method of preparing pure hydrocarbons in small amounts.

3-4. Uses of Alkyl Halides

The most important use of alkyl halides is in making other compounds. Methyl and ethyl chlorides are also used as refrigerants. Methyl bromide is a fumigant for packaged foods. Ethyl chloride

TABLE | *Uses of Polyhalides*
3-2

FORMULA	NAME	PRODUCTION IN POUNDS, 1953	USES
$ClCH_2$–CH_2Cl	ethylene chloride	528,646,000	solvent, fumigant
$ClCH_2$–$CHCl_2$	trichloroethane		cleaning fluid
$ClCH=CCl_2$	trichloroethylene	323,313,000	cleaning fluid
$CHCl_3$	chloroform	25,537,000	solvent, general anesthetic
CCl_4	carbon tetrachloride	259,705,000	Pyrene fire-extinguisher, fat-solvent, cleaning fluid
CH_3–CH_2Cl	ethyl chloride	520,078,000	refrigerant, local anesthetic
CH_3Cl	methyl chloride	40,520,000	refrigerant
CH_2Cl_2	methylene chloride	63,629,000	varnish-remover
$CH_2=CHCl$	vinyl chloride	401,701,000	polymers
$CH_3CH_2CHCl_2$	dichloropropane		soil-fumigant
(Variable)	chlorinated paraffin wax (70% Cl)	33,382,000	fire-retardant
CHI_3	iodoform		antiseptic
CCl_3–CCl_3	hexachloroethane		HC smoke screen (zinc dust reacts to give $ZnCl_2$ + C)
CCl_2F_2	dichlorodifluoromethane		Freon-12 refrigerant (see p. 564)

is a local anesthetic, especially handy for use in "freezing" tissues in minor surgery; liquefied and sprayed in a fine stream upon some tissue (the gums in dental surgery, for example), it evaporates so rapidly that it chills the tissue.

3-5. Polyhalides

Hydrocarbons with more than one hydrogen replaced by a halogen are referred to as polyhalides. They are prepared by special methods to be discussed later. Table 3-2 contains information about some alkyl chlorides and polyhalides.

SUMMARY

Alkyl halides, RX, are prepared by direct halogenation of alkanes and from alcohols. They can be used to place alkyl groups on carbon (Wurtz, NaCN), oxygen, nitrogen, and metals (Zn, Pb, Mg, etc.). Most of these reactions lie ahead in this text.

REFERENCE

Nickerson, "Tetraethyl Lead," J. Chem. Education, 31, 560 (1954).

EXERCISES

(Exercises 1–5 will be found within the text.)

6. Write structural formulas for

1. isobutyl iodide
2. tert-butyl bromide
3. sec-butyl chloride
4. 3-bromo-2-methylpentane

5. 2-iodo-4-methylpentane
6. 1-chloro-3,3-dimethylbutane
7. 2-bromo-2-methylpropane

7. Write structural formulas for all compounds of molecular formula $C_5H_{11}Cl$. State the class (degree) and the IUC name of each.

8. Write equations for the reaction, if any, between

   1. 1-bromopropane and sodium cyanide
   2. 2-bromo-3-methylbutane and magnesium (in dry ether)
   3. iodoethane and aqueous sodium hydroxide
   4. isopropyl alcohol and phosphorus tribromide
   5. sec-butyl alcohol and phosphorus pentachloride

9. Indicate a method of converting 2-bromopropane to each of the following. Show all reagents and conditions.

   1. propane
   2. 2,3-dimethylbutane
   3. isopropyl alcohol

10. Show by equations how 1-bromopropane could be converted into

    1. di-n-propyl zinc   4. n-propyl cyanide
    2. propane   5. n-propyl alcohol
    3. hexane

11. A compound, C_5H_{12}, reacts with chlorine to form one and only one product having the formula $C_5H_{11}Cl$. Write a structure for the original hydrocarbon and an equation for the indicated reaction. The *rate* at which the reaction occurs is *not* to be considered in this exercise.

12. Propane is treated with bromine, and a mixture of all isomeric products of formula C_3H_7Br is isolated. This mixture is then treated with excess sodium. Write the formulas of all the resulting products.

13. An alkyl iodide was used to make a Grignard reagent, which was treated with water to give 2-methylbutane. Write all the possible structures for the original alkyl iodide.

14. Trade names or common names are given for the polyhalides in Table 3-2. Supply the IUC names for these compounds.

4 Alkenes, or Olefins

In contrast to the inert saturated hydrocarbons, many hydrocarbons add reagents of various kinds and are therefore called *unsaturated*. There are several families of such compounds, three of which will be discussed in the next two chapters. These homologous series are the alkenes, or olefins, the alkynes, or acetylenes, and the alkadienes, or diolefins.

The alkenes are hydrocarbons in which two of the carbon atoms are joined by a double bond. The existence of two shared pairs of electrons between two atoms reduces the number of hydrogens of the corresponding alkane by two, so that the general formula for an alkene is C_nH_{2n}. The simplest alkene, containing two carbons, is called ethylene, and this homologous series is sometimes referred to as the ethylene series.

TABLE 4-1 | *First Members of the Alkene Series*

FORMULA	COMMON NAME	IUC NAME
$CH_2=CH_2$	ethylene	ethene
$CH_3-CH=CH_2$	propylene	propene
$CH_3-CH_2-CH=CH_2$	α-butylene	1-butene
$CH_3-CH=CH-CH_3$	β-butylene	2-butene
$\begin{array}{c}CH_3\\ \diagdown\\ \quad C=CH_2\\ \diagup\\ CH_3\end{array}$	isobutylene	2-methyl-1-propene

4-1. Nomenclature of Alkenes

We derive the common names of the alkenes from the common names of the corresponding alkanes by dropping the ending *ane* and adding *ylene*.

The following additions to the rules already given for IUC names will enable the student to name any alkene. The longest chain selected for the basic stem must contain the double bond. The number used to locate the double bond is the lower number of the two carbons to which it is attached. The ending of the corresponding alkane is changed to *ene* for the alkene.

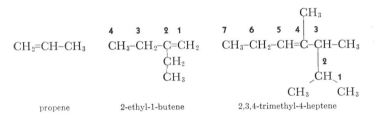

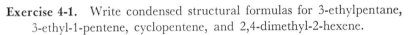

propene   2-ethyl-1-butene   2,3,4-trimethyl-4-heptene

Exercise 4-1.   Write condensed structural formulas for 3-ethylpentane, 3-ethyl-1-pentene, cyclopentene, and 2,4-dimethyl-2-hexene.

4-2. Preparation of Alkenes

1. Cracking Process.   The cracking process for increasing the gasoline fraction of crude petroleum is accompanied by the formation of short-chain alkenes in great abundance. Large quantities of ethylene and propylene and mixtures of the butylenes are available for commercial synthesis. These fractions are not obtained pure, however, and are too difficult to separate completely for many laboratory purposes.

2. Removal of HX from an Alkyl Halide.   A solution of KOH in alcohol will react with an alkyl halide to remove a molecule of HX and leave an alkene. This reaction (Exp. 4-1) is in sharp contrast to the effect of aqueous KOH on n-propyl bromide (p. 71).

$$CH_3-CH_2-CH_2-Br + OH^- \xrightarrow[36\%]{\text{alcohol}} CH_3-CH=CH_2 + Br^- + H_2O \qquad [4\text{-}1]$$

$$CH_3-CH_2-CH_2-Br + OH^- \xrightarrow{\text{water}} CH_3-CH_2-CH_2-OH + Br^- \qquad [4\text{-}2]$$

Expression 4-2 represents an attack by hydroxyl ions on the relatively positive carbon carrying the bromine atom. The second reaction is at the surface since n-propyl bromide is insoluble in the medium; so there is a low concentration of hydroxyl ions at the point of attack. In alcohol, in which all the reactants are soluble, there is a higher concentration of base at the point of attack, and the removal of the bromine is accompanied by elimination of a proton from the adjacent carbon atom. The role a solvent may play in a reaction is here clearly shown by the different paths which this reaction takes in water and in alcohol.

Exercise 4-2.   Indicate by an equation the effect of sodium hydroxide on isobutyl bromide in water as a solvent; in alcohol as a solvent.

3. Removal of X_2 from a 1,2-Dihalide.

When halogen atoms are located on adjacent atoms in a carbon chain, they may be removed by reagents such as zinc dust; if they are separated by an intervening carbon atom, a three-membered ring is formed.

$$\begin{array}{cc} CH_3-CH-CH-CH_3 + Zn(\text{dust}) \longrightarrow CH_3-CH=CH-CH_3 + ZnBr_2 \qquad [4\text{-}3] \\ \quad\ \ Br \quad\ Br \end{array}$$

Cyclopropane (p. 58) is prepared commercially by this type of reaction. If sodium metal is used for it, it is called an "internal" Wurtz reaction.

$$Br-CH_2-CH_2-CH_2-Br + Zn(\text{dust}) \xrightarrow[57\%]{210°} \begin{array}{c} CH_2-CH_2 \\ \diagdown CH_2 \diagup \end{array} + ZnBr_2 \qquad [4\text{-}4]$$

Exercise 4-3.   Show by an equation the effect of treating 2,3-dibromo-2-methylbutane with zinc dust.

4. Dehydration of Alcohols. The catalytic action of such a substance as aluminum oxide on an alcohol at temperatures of 360°C or less is sufficient to remove a molecule of water. Such chemical reagents as P_2O_5, H_3PO_4, and H_2SO_4 also act as strong

$$CH_3\text{--}CH_2\text{--}CH_2OH \xrightarrow[100\%]{Al_2O_3,\ 360°} CH_3\text{--}CH=CH_2 + H_2O \qquad [4\text{-}5]$$

dehydrating agents on alcohols. The mechanism of the reaction and the conditions for the use of these reagents will be discussed in Chapter 8.

4-3. Formal Charges

Before we discuss the reactions of alkenes, it is convenient to introduce the concept of "formal charges."

An arrow is frequently used to indicate a coordinate covalent bond. The arrow points from the donor atom to the acceptor atom. The formula for sulfur dioxide is then written as $O=S\rightarrow O$

$\left(\text{from } :\overset{..}{O}::\overset{..}{S}:\overset{..}{\underset{..}{O}}:\ \right)$. Sulfuric acid is written as $H\text{--}O\text{--}S\text{--}O\text{--}H$ and

nitric acid as $H\text{--}O\text{--}N=O$.

$$\begin{array}{c} O \\ \uparrow \\ H\text{--}O\text{--}S\text{--}O\text{--}H \\ \downarrow \\ O \end{array}$$

$$\begin{array}{c} H\text{--}O\text{--}N=O \\ \downarrow \\ O \end{array}$$

Another way of indicating a coordinate covalent bond is by counting formal charges on the atoms. If we grant ourselves the power to make the atoms and electrons stand still for a moment, what we call the "formal charge" on an atom may be counted, on the following assumption: all bonds between atoms are shared equally. (This has nothing to do with reality but is a convenience for balancing equations later and for talking about some mechanisms of reaction.) In SO_2, then, sulfur may be said to have full possession of five electrons (marked x below)—that is, half of the electrons in the double bond and coordinate covalent bond and both unshared electrons. The normal sulfur atom (see the Periodic Table) has six electrons in its outside shell; so its formal charge in this molecule is $+1$. The oxygen written at the right in the formula

$$:\overset{..}{O} :: \overset{..}{S} : \overset{..}{\underset{..}{O}} : $$

for SO_2 has full possession of seven electrons, one more than its normal six; so oxygen has a formal charge of -1. These are formal charges, not real charges. There is actually no complete separation of electrical charge in the SO_2 molecule; it is neutral. There is probably a partial separation since sulfur is more electropositive than oxygen. The real charge, then, is probably a ($+$ or $-$) fraction of 1. We have defined the formal charge as $+1$ for S and -1 for one O. If we used formal charges, sulfur dioxide would be written as $\overset{+}{O}=\overset{-}{S}-O$ instead of $O=S\rightarrow O$, and the formulas with formal charges for sulfuric acid, nitric acid, and sulfur trioxide would be written as follows:

$$
\begin{array}{ccc}
\overset{\displaystyle O^-}{\underset{\displaystyle O_-}{H-O-\overset{++}{S}-OH}} &
\overset{}{H-O-\overset{+}{N}\!\!\diagdown_{\!\!\searrow\!O^-}^{\!\!\nearrow O^{\bullet}}} &
\overset{}{\underset{\displaystyle \underset{O}{\overset{\|}{S}}}{\overset{-\ ++\ -}{O-S-O}}}
\end{array}
$$

4-4. Theory of Alkene Reactions

The reactivity of the alkene bond can be rationalized on the basis of the mechanical model of the ethylene molecule (see the ac-

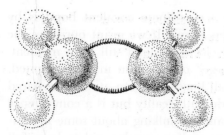

companying sketch), in which the double bond is made of metal springs. An alkene reacts by *adding* a reagent, breaking the double bond. The bonds "spring" to their normal angles of 109°28′, and the compound thereupon becomes saturated by forming the two new single bonds to the reagent.

Though the mechanical model may give a satisfactory picture of what happens in an alkene addition, it is fair to seek a submicroscopic, electronic picture. It seems probable that the first stage in an addition to an alkene is the momentary partial displacement of one pair of electrons from the double bond toward one of the two carbons carrying it:

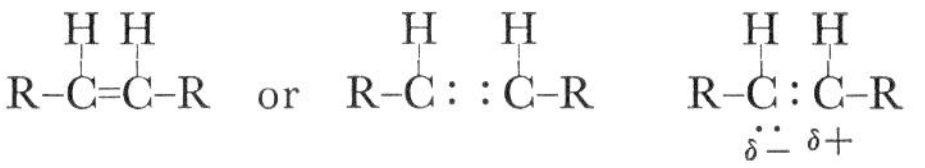

$$
\begin{array}{ccccc}
\text{H H} & & \text{H} \quad \text{H} & & \text{H H} \\
\text{R-C=C-R} & \text{or} & \text{R-C} : : \text{C-R} & & \text{R-C} : \text{C-R} \\
& & & & \delta- \quad \delta+
\end{array}
\qquad [4\text{-}6]
$$

This results in a formal positive charge on one carbon and a negative charge on the other. The real charges are no doubt zero. But *at the moment of reaction* the carbons may have δ+ or δ− (partial) charges as shown at the right in Expression 4-6. This "excitement" of the alkene molecule is shown in Expression 4-7. At the same

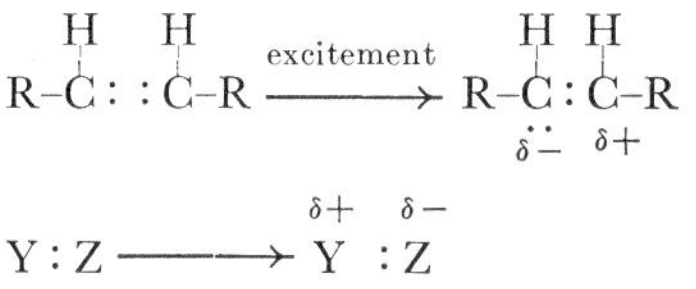

$$
\begin{array}{ccc}
\text{H} \quad \text{H} & \xrightarrow{\text{excitement}} & \text{H H} \\
\text{R-C} : : \text{C-R} & & \text{R-C} : \text{C-R} \\
& & \delta- \quad \delta+
\end{array}
\qquad [4\text{-}7]
$$

$$
\begin{array}{c}
\delta+ \quad \delta- \\
\text{Y} : \text{Z} \longrightarrow \text{Y} : \text{Z}
\end{array}
$$

time that the alkene molecule is excited, a partial separation of charges, in some cases due to a catalyst or to the conditions of the reaction mixture, may occur in the reagent.

From inspection only it would be assumed that the excited alkene would now add either a positively or a negatively charged group or perhaps both simultaneously. It has been shown in a few cases, however, and it is considered as a probability in others, that an alkene acts only as a donor of electrons. Hence a positively charged group is added first (Exp. 4-8). Then the carbonium ion (a carbon having three groups attached to it but only six electrons in the outside shell) adds the second group (Exp. 4-9). The positive charge

$$\underset{\substack{\delta- \quad \delta+}}{\overset{\text{H}\quad\text{H}}{\text{R-C:C-R}}} + \underset{}{\overset{\delta+ \quad \delta-}{\text{Y}\quad:\text{Z}:}} \longrightarrow \underset{\oplus}{\overset{\text{H}\quad\text{H}}{\text{R-C:C-R}}}_{\text{Y}} \qquad\qquad [4\text{-}8]$$

$$\underset{\text{Y}\quad\oplus}{\overset{\text{H}\quad\text{H}}{\text{R-C:C-R}}} + :\overset{..}{\text{Z}}:^{\ominus} \longrightarrow \underset{\text{Y}\quad\text{Z}}{\overset{\text{H}\quad\text{H}}{\text{R-C-C-R}}} \qquad\qquad [4\text{-}9]$$

on the carbonium ion is circled to indicate that this particle can have no independent existence, in contrast to Na^+, Cl^-, H^+, OSO_3H^-, etc. (See the footnote on p. 54.)

Besides addition reactions, alkenes also undergo oxidations. The first stage of these may be addition, but the final result is more than a simple addition.

4-5. Addition Reactions of Alkenes

1. Halogens. The ease of addition of the halogens to an alkene is in the order $F_2 > Cl_2 > Br_2 > I_2$. The addition of bromine is convenient to carry out in the laboratory since liquid bromine can be dissolved in CCl_4 and added to the alkene. If the alkene is a gas, it may be bubbled through such a solution. The process, according to the theory presented in section 4-4, involves three

$$CH_3\text{-}CH{=}CH_2 \xrightarrow{\text{excitement}} \overset{\delta+}{C}H_3\text{-}CH\text{-}\overset{\delta-}{C}H_2 \qquad\qquad [4\text{-}10]$$

$$Br_2 \xrightarrow{} \overset{\delta+}{Br} \, :\overset{\delta-}{Br}: $$

$$\overset{\delta+ \quad \delta-}{CH_3\text{-}CH\text{-}CH_2} + :\overset{\delta+}{Br} \longrightarrow \overset{\oplus}{CH_3\text{-}CH\text{-}CH_2} \\ \phantom{CH_3\text{-}CH\text{-}CH_2aaaaaaaaa}:Br: \qquad [4\text{-}11]$$

$$\underset{\oplusBr}{CH_3\text{-}CH\text{-}CH_2} + :\overset{..\ominus}{Br}: \longrightarrow \underset{BrBr}{CH_3\text{-}CH\text{-}CH_2} \qquad\qquad [4\text{-}12]$$

steps. The first is the excitement of the propylene molecule and the bromine molecule to partially charged states (Exp. 4-10). The second is the formation of the carbonium ion (Exp. 4-11). The third is the addition of the negatively charged particle to the carbonium ion (Exp. 4-12).

Exercise 4-4. Show the three distinct steps postulated for halogen additions in the reaction of 2-methyl-1-butene and bromine.

Additions of halogens to alkenes take place at ordinary temperatures. At very high temperatures chlorine will substitute rather than add. Propylene, for example, in the presence of a small

$$CH_3-CH=CH_2 + Cl_2 \xrightarrow{600°} ClCH_2-CH=CH_2 + HCl \qquad [4\text{-}13]$$
$$\text{allyl chloride}$$

amount of oxygen and at a temperature of approximately 600°, yields allyl chloride and HCl.

2. Hydrogen. Alkenes can be made to add hydrogen, in the presence of nickel or platinum, to form an alkane. The double bond breaks, and each carbon forms a new covalent bond with a hydro-

$$CH_3-CH_2-CH=CH_2 + H_2 \xrightarrow{Pt} CH_3-CH_2-CH_2-CH_3 \qquad [4\text{-}14]$$

gen atom. The reaction with hydrogen is frequently called a catalytic reduction or a catalytic hydrogenation.

3. Hydrogen Halides. Alkenes add the hydrogen halides and form alkyl halides in the steps shown in Expressions 4-15a and 4-15b.

$$CH_2=CH_2 + \overset{\delta+}{H} \overset{\delta-}{:}\overset{}{I}: \longrightarrow CH_3-\overset{\oplus}{C}H_2 \qquad [4\text{-}15a]$$

$$CH_3-\overset{\oplus}{C}H_2 + I^- \longrightarrow CH_3-CH_2-I \qquad [4\text{-}15b]$$

Exercise 4-5. Write an expression for the addition of HBr to 3-hexene. Show three steps.

4. Sulfuric Acid. Concentrated sulfuric acid dissolves unsaturated hydrocarbons readily. Cyclohexene, for example, adds sulfuric acid to yield cyclohexyl hydrogen sulfate.

$$
\begin{array}{c}
\text{CH} \\
\text{CH}_2 \qquad \text{CH} \\
\text{CH}_2 \qquad \text{CH}_2 \\
\text{CH}_2
\end{array}
+ \text{H OSO}_3\text{H} \longrightarrow
\begin{array}{c}
\text{CH}_2\text{-CH}_2 \\
\text{CH}_2 \qquad\qquad \text{CH-OSO}_3\text{H} \\
\text{CH}_2\text{-CH}_2
\end{array}
\qquad [4\text{-}16]
$$

cyclohexyl hydrogen sulfate

5. Hypohalous Acids. If an alkene is bubbled through a solution of hypochlorous acid, a new type of compound is formed, called

$$
\text{CH}_2{=}\text{CH}_2 + \overset{\delta-}{\text{HO}}\ \overset{\delta+}{\text{Cl}} \xrightarrow[85\%]{} \underset{\overset{|}{\text{OH}}\ \overset{|}{\text{Cl}}}{\text{CH}_2\text{-CH}_2} \qquad [4\text{-}17]
$$

a chlorohydrin. Though the formula of hypochlorous acid is often written as HClO, its structure is $\text{H} \!:\! \overset{..}{\underset{..}{\text{O}}} \!:\! \overset{..}{\underset{..}{\text{Cl}}} \!:$. When HOCl adds to an alkene, it splits into the two fragments $\overset{..}{\text{H}}\overset{..}{\text{O}} \!:^{\ominus}$ and $\overset{\oplus}{\text{Cl}}$, of which the chlorine is positive with respect to the hydroxyl group.*

Exercise 4-6. Write an equation for the addition of hypobromous acid to 2-butene.

6. Alkenes. One method of increasing the gasoline fraction in petroleum (Chap. 2) is to add one molecule of an alkene to another. Sulfuric acid is one of a number of acid catalysts that will accom-

$$
\begin{array}{c}
\text{CH}_3 \\
\qquad \text{C=CH}_2 + \text{H}^+\ \text{OSO}_3\text{H} \longrightarrow \\
\text{CH}_3
\end{array}
\begin{array}{c}
\text{CH}_3 \\
\qquad \text{C}^{\oplus}\text{-CH}_3 + (\text{OSO}_3\text{H})^{\ominus} \\
\text{CH}_3
\end{array}
$$

$$
\begin{array}{c}
\text{CH}_3\ \overset{\oplus}{} \\
\qquad \text{C-CH}_3 + \\
\text{CH}_3
\end{array}
\begin{array}{c}
\text{CH}_{3\ \delta+} \ \overset{\delta-}{} \\
\qquad \text{C=CH}_2 \\
\text{CH}_3
\end{array}
\longmapsto
\begin{array}{c}
\text{CH}_3\ \overset{\oplus}{} \qquad \text{CH}_3 \\
\qquad \text{C-CH}_2\text{-C-CH}_3 \\
\text{CH}_3 \qquad\quad \text{CH}_3
\end{array}
\qquad [4\text{-}19]
$$

* In strong bases like sodium hydroxide, hypochlorous acid can be made to give up a proton, the capacity one ordinarily expects of a substance called an acid.

$$
\text{HOCl} + \text{OH}^- \longrightarrow \text{OCl}^- + \text{HOH} \qquad\qquad\qquad\qquad\qquad [4\text{-}18]
$$

plish the addition on such an alkene as isobutylene. The first reaction, according to the carbonium ion theory, is the addition of a proton from the acid to produce a carbonium ion, which then adds to a second molecule of isobutylene. This eight-carbon carbonium ion is unstable and will stabilize itself by loss of a proton from carbon 1 or 3 to give a mixture of the two alkenes in Expression 4-20. Either of these "isooctenes" adds a molecule of hydrogen

$$
\begin{array}{c}
\underset{\displaystyle\underset{1}{\overset{\displaystyle CH_3}{}}}{CH_3}\!\!\overset{\oplus}{}
\end{array}
$$

in the presence of Ni to give the same product, "isooctane," the standard antiknock fuel (100 octane).

4-6. Markownikoff's Rule

In the examples of addition to alkenes given in the preceding paragraphs (except that in Exp. 4-19) either the reagents (Exp. 4-10, 4-14) or the alkenes (Exp. 4-15, 4-16, 4-17) were symmetrical in structure. When both the reagent and the alkene are unsymmetrical, two possible products may be written. Both products may actually be formed in the reaction, but one or the other predominates. In 1870, Markownikoff proposed the following rule for predicting the dominant product of additions to alkenes: if an unsymmetrical reagent adds to an unsymmetrical alkene, the more positive part of the reagent adds to the olefinic carbon that already has more hydrogens. For examples see Expressions 4-21–4-24.

$$CH_3-CH=CH_2 + \overset{+}{H}\overset{-}{I} \longrightarrow CH_3-\underset{\underset{I}{|}}{CH}-CH_3 \qquad [4\text{-}21]$$

$$\underset{CH_3}{\overset{CH_3}{\diagdown}}C=CH_2 + \overset{+}{H}\overset{-}{OSO_3H} \longrightarrow \underset{CH_3}{\overset{CH_3}{\diagdown}}\underset{\underset{OSO_3H}{|}}{C}\overset{CH_3}{\diagup} \qquad [4\text{-}22]$$

$$CH_3CH_2CH=CH_2 + \overset{\ominus}{HO}\overset{\oplus}{Cl} \longrightarrow CH_3-CH_2-\underset{\underset{OH}{|}}{CH}-\underset{\underset{Cl}{|}}{CH_2} \qquad [4\text{-}23]$$

$$\underset{CH_3}{\overset{CH_3}{\diagdown}}C=CH_2 + \underset{CH_3}{\overset{CH_3}{\diagdown}}\overset{\oplus}{C}-CH_3 \longrightarrow \underset{CH_3}{\overset{CH_3}{\diagdown}}\overset{\oplus}{C}-CH_2-\underset{\underset{CH_3}{|}}{\overset{\overset{CH_3}{|}}{C}}-CH_3 \qquad [4\text{-}24]$$

According to the electronic interpretation presented in section 4-4, the rule is reasonable if the electronic displacement is in the direction of the curved arrow. Later (pp. 458–463) we shall offer

$$\underset{\mathbf{3}}{CH_3} \longrightarrow \underset{\mathbf{2}}{\overset{\oplus}{CH}}\overset{\overset{\frown}{\cdot\cdot}}{-}\underset{\mathbf{1}}{\overset{\ominus}{CH_2}}$$

evidence to show that a methyl group has a greater tendency to release electrons than hydrogen. (Carbon 1 carries two H's, whereas carbon 2 carries one H and one CH_3- group.) If carbon 1, then, at the moment of reaction carries a formal negative charge, it would be expected to acquire the more positive part of the adding agent.

Exercise 4-7. What would you expect to happen if 2-pentene were allowed to react with HOCl? Write an equation.

4-7. Oxidation Reactions of Alkenes

1. Combustion in Air or Oxygen. Since all substances containing only carbon and hydrogen burn in air or oxygen, this property of alkenes is due, not to the double bond, but to the elements present. All carbon-carbon and carbon-hydrogen bonds are broken, and new carbon-oxygen and hydrogen-oxygen bonds are formed. The exact mechanism is not known, but the amount of heat evolved is known fairly accurately.

$$CH_3CH_2CH=CH_2 + 6O_2 \longrightarrow 4CO_2 + 4H_2O + \text{heat} \qquad [4\text{-}25]$$

2. Baeyer's Test for Unsaturation.   Mild oxidizing agents do not completely destroy an alkene. A solution of $KMnO_4$ (purple) is decolorized in the presence of an alkene in either acidic or basic solution. Generally, dilute Na_2CO_3 solution is used as the medium. Under these conditions the permanganate is reduced to a brown precipitate of manganese dioxide:

$$2MnO_4^- + 4H_2O + 3CH_2=CH_2 \longrightarrow 3\overset{\displaystyle OH}{\underset{\displaystyle |}{C}}H_2-\overset{\displaystyle OH}{\underset{\displaystyle |}{C}}H_2 + 2MnO_2 + 2OH^- \qquad [4\text{-}26]$$

The solution becomes more basic as the reaction continues. The first product of the reaction with ethylene is a substance called ethylene glycol, which contains two hydroxy groups. This reaction may be carried out in a test tube in the laboratory. The observation of the rapid disappearance of the purple color of the permanganate ion is a test for unsaturation in an organic compound. In this reaction the permanganate ion acts as the oxidizing agent.

The subject of oxidation-reduction will be discussed at length in Chapter 14.

Exercise 4-8.   Write equations for the complete combustion of isobutylene and for the mild oxidation of isobutylene with MnO_4^- in basic solution.

4-8. Uses of Alkenes

The alkenes obtained from the cracking process are used in synthesizing gasoline and as a source for many chemicals, including a number of alcohols.

Ethylene is a gas having a sweetish taste and capable of inducing sleep. It also causes a deepening in the color of citrus fruits. A carload of fruit can be picked green, sent to its destination, and then ripened. The cost is only a few cents per carload. The ripening is said to be due to an acceleration of the hydrolysis of the starches in the fruit to sugars. Ethylene is also used in the manufacture of the blistering agent called "mustard gas." This compound is actually a liquid, with boiling point 217°, but its vapors are extremely dangerous to the skin, even in small quantities, and it does have an appreciable vapor pressure at atmospheric temperatures.

$$2CH_2=CH_2 + S_2Cl_2 \xrightarrow{} Cl-CH_2-CH_2-S-CH_2-CH_2-Cl + S \qquad [4\text{-}27]$$
$$\underset{93\%}{} \qquad \underset{\text{mustard gas}}{}$$

SUMMARY

Alkenes are formed as by-products of the cracking process and may be alkylated back to branch-chain alkanes, useful in gasoline. In the laboratory alkenes are prepared from alcohols, alkyl halides, or dihalogen compounds.

The characteristic reaction of an alkene is addition. The role of the alkene is that of an electron donor for an electron-deficient agent. Substances that add to alkenes are halogens, H_2 with a catalyst, HX, HOX, $HOSO_3H$, and OH groups from oxidizing agents such as MnO_4^- in water (Baeyer).

Other uses and reactions of alkenes will become evident in later chapters.

REFERENCES

1. Egloff, "Modern Motor Fuels," J. Chem. Education. 18, 582 (1941).
2. O'Kelly and Sachanen, "Synthesis of Neohexane and Triptane," Ind. Eng. Chem., 38, 463 (1946).
3. Carson, "Industrial Catalysis—Gasoline," J. Chem. Education, 24, 151 (1947).

EXERCISES

(Exercises 1–8 will be found within the text.)

9. Write structural formulas for all alkenes of formula C_6H_{12}, and name them by the IUC system.

10. Write the equations for any reaction between
    1. 1-pentene and hydrogen chloride
    2. ethylene and concentrated sulfuric acid
    3. 2-bromo-2-methylpentane and alcoholic potassium hydroxide

11. Are 1-hexene and cyclohexane isomers? Compare their chemical properties.

12. Write equations showing two different methods of chemically distinguishing between 2-butene and butane with reactions that can be carried out in a test tube.

13. Indicate the means of accomplishing the following transformations. Show all reagents and conditions.
    1. 1,2-dibromopropane to 2-bromopropane
    2. n-propyl alcohol to isopropyl hydrogen sulfate
    3. 2-chloropropane to propane
    4. 1-butene to 3,4-dimethylhexane
    5. 1-butene to 2-butene
    6. chlorocyclopentane to

14. Two compounds, A and B, have the same molecular formula, C_5H_{10}. Compound A reacts readily with hydrogen in the presence of platinum to form the compound C_5H_{12}; it also reacts readily with bromine to form only the compound $C_5H_{10}Br_2$. B fails to react with hydrogen and a catalyst below 300°, but slowly reacts with bromine (with evolution of HBr) to form the compound C_5H_9Br. Write the formula and name of B and of all compounds that might be A.

15. What volume of oxygen is needed to burn three liters of 2-butene?

16. Five grams of gasoline decolorize 0.80 g of bromine. If the gasoline contains, on the average, compounds containing just six carbon atoms (alkanes and alkenes), what percentage is alkene?

5 Alkynes, or Acetylenes, and Alkadienes, or Diolefins

A second series of unsaturated hydrocarbons, called the alkynes, or acetylenes, contain a triple bond between two of the carbons. The degree of unsaturation in an alkyne is greater than in an alkene. Since more energy is released when the triple bond is broken than when a double bond breaks in an alkene, alkynes may be said to be more reactive than alkenes. The general formula for the alkynes is C_nH_{2n-2}. The simplest is called acetylene and has the formula $CH{\equiv}CH$. The first members of the series are listed in Table 5-1.

TABLE 5-1 | *Alkynes*

FORMULA	COMMON NAME	IUC NAME
$CH{\equiv}CH$	acetylene	ethyne
$CH_3-C{\equiv}CH$	methyl acetylene	propyne
$CH_3-CH_2-C{\equiv}CH$	ethyl acetylene	1-butyne
$CH_3-C{\equiv}C-CH_3$	dimethyl acetylene	2-butyne

5-1. Nomenclature of Alkynes

The alkynes are given common names as derivatives of acetylene. The names of the alkyl groups attached to the carbon-carbon triple bond are followed by the word *acetylene*.

The rules for the IUC names of the alkynes are the same as for the alkenes except that the ending is changed from *ene* to *yne*. Again the number locating the triple bond is the lower number of the two carbons containing it:

$$CH_3-CH_2-C\equiv C-\underset{\underset{CH_3}{|}}{C}H-CH_3 \qquad CH_2=CH-C\equiv CH$$

2-methyl-3-hexyne　　　　　　　1-butene-3-yne

Exercise 5-1. Write condensed structural formulas for cyclopropane, diethylacetylene, 3-methyl-1-butyne, and 2,3-dimethylcyclopentene.

5-2. Preparation of Alkynes

1. Special Method for Acetylene Only. Acetylene is prepared by a special method that does not apply to other members of the series. The action of water on calcium carbide results in a smooth evolution of acetylene gas and the formation of calcium hydroxide (slaked lime):

$$Ca^{++} : C\equiv C : ^= + 2H_2O \longrightarrow Ca(OH)_2 + HC\equiv CH \qquad [5\text{-}1]$$

The calcium carbide is prepared in an electric furnace, at high temperatures, from calcium oxide and carbon:

$$CaO + 3C \xrightarrow{3,000°C} CaC_2 + CO \qquad [5\text{-}2]$$

Calcium oxide is formed by heating limestone:

$$CaCO_3 \xrightarrow{550°C} CaO + CO_2 \qquad [5\text{-}3]$$

"Carbide" is used for forming acetylene in miners' lamps and also in many home lighting systems where electricity is not available.

2. Removal of 2HX from a Dihalide. There are three general methods of preparing alkynes. One involves the removal of two molecules of HX from adjacent carbons with alcoholic KOH.

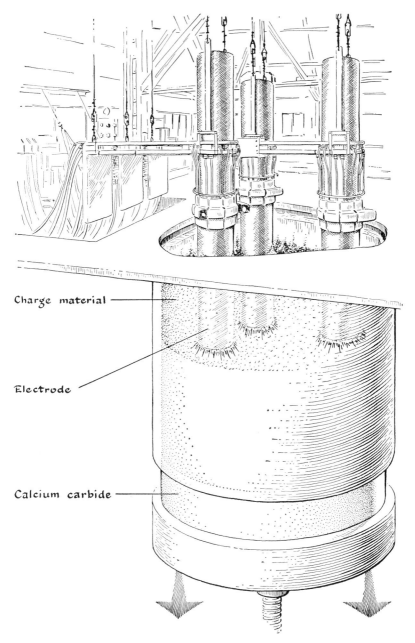

Charge material

Electrode

Calcium carbide

Drawn from a photograph supplied by the Pittsburgh Lectromelt Furnace Corpora-
tion.

The removal may be carried out stepwise as in Expression 5-4.

$$\begin{array}{ccc} \underset{\underset{Br}{|}}{\overset{\overset{H}{|}}{CH_3-C}}-\underset{\underset{Br}{|}}{\overset{\overset{H}{|}}{C}}-CH_3 & \xrightarrow[\text{KOH}]{\text{alc.}} & CH_3-CH=\underset{\underset{Br}{|}}{C}-CH_3 & \xrightarrow[\underset{65\%}{\text{KOH}}]{\text{alc.}} & CH_3-C{\equiv}C-CH_3 \end{array} \qquad [\textbf{5-4}]$$

An excess of alcoholic KOH gives an alkyne, two molecules of water, and two equivalents of KBr.

Exercise 5-2. What other product might be expected in the first reaction of Expression 5-4? Equation?

3. Removal of $2X_2$ from a Tetrahalide. If two pairs of halogen atoms lie on adjacent carbons in a molecule, they may be removed by treatment with zinc dust:

$$\underset{\underset{Cl}{|}}{\overset{\overset{Cl}{|}}{CH_3CH_2C}}-\underset{\underset{Cl}{|}}{\overset{\overset{Cl}{|}}{CH}} \xrightarrow{\text{Zn dust}} CH_3CH_2\underset{\underset{Cl}{|}}{C}{=}\underset{\underset{Cl}{|}}{CH} \xrightarrow{\text{Zn dust}} CH_3CH_2C{\equiv}CH + ZnCl_2 \quad [\textbf{5-5}]$$

4. Alkyl Halides on Sodium Acetylides. Some higher alkynes may be obtained from the sodium salt of an alkyne by treatment with various alkyl halides. Methyl iodide is frequently used to advantage with the sodium salt in liquid ammonia. Disodium acetylide (see p. 101) reacts with alkyl halides in liquid ammonia to yield symmetrical alkynes.

$$CH_3-CH_2-C{\equiv}C-Na + CH_3I \xrightarrow[40\%]{} CH_3-CH_2-C{\equiv}C-CH_3 + Na^+I^- \qquad [\textbf{5-6}]$$

$$2CH_3-CH_2-CH_2-Br + Na-C{\equiv}C-Na \xrightarrow[65\%]{}$$
$$CH_3-CH_2-CH_2-C{\equiv}C-CH_2-CH_2-CH_3 + 2NaBr \qquad [\textbf{5-7}]$$

Exercise 5-3. How would you synthesize 1-pentyne by the method just described? (Use Expression 5-20.)

5-3. Reactions of Alkynes

Those reactions of alkynes which are due to breaking of the triple bond are not different in character from the analogous reac-

tions of double-bonded compounds; they merely go one step further
since there is a triple bond to break and since they take place
because the bond is easy to break. The triple bond also imparts
another property to the carbons to which it is attached if there

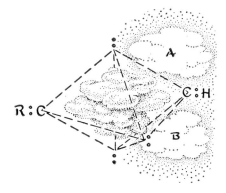

is a hydrogen on one or both of them. The high electron density
between the two carbons leaves the areas A and B somewhat defi-
cient in electrons. Consequently, the carbon-hydrogen electron
pair is drawn more closely to the carbon, allowing a proton to be
lost more readily than from a corresponding alkene. Alkynes do
act as weak acids of about the same strength as $H-C\equiv N$. Note the
analogous structures of $H-C\equiv C-R$ and $H-C\equiv N$.

Reactions involving displacement of the hydrogen on an alkyne
carbon are classed as replacement reactions.

5-4. Addition to Alkynes

1. Catalytic Reduction. Hydrogenation of an alkyne in the
presence of Ni or Pt yields an alkane through the intermediate
formation of an alkene. The addition can be stopped at the first
stage, but it is not a very useful means of preparing an alkene since
the alkyne is generally more difficult to synthesize than the expected
alkene.

$$R-C\equiv CH \xrightarrow[Pt]{H_2} R-CH=CH_2 \xrightarrow[Pt]{H_2} R-CH_2-CH_3 \qquad [5-8]$$

2. Halogens. The halogens add readily to alkynes to give di-halogen compounds first. The reaction with iodine stops at this stage, in part, at least, because there is not room for four large iodine atoms around adjacent carbons. Chlorine will react with acetylene explosively, but the addition can be controlled in the presence of $SbCl_5$. The disappearance of the red color of bromine

$$HC{\equiv}CH + \overset{\oplus}{Cl} : \overset{\ominus}{Cl} \xrightarrow{SbCl_5} CHCl{=}CHCl \xrightarrow[SbCl_5]{Cl_2} CHCl_2{-}CHCl_2 \qquad [5\text{-}9]$$

$$RC{\equiv}CH + \overset{\oplus}{Br} : \overset{\ominus}{Br} \longrightarrow RCBr{=}CHBr \xrightarrow{Br_2} RCBr_2{-}CHBr_2 \qquad [5\text{-}10]$$

$$HC{\equiv}CH + I_2 \xrightarrow{140\text{-}160°} CHI{=}CHI \qquad [5\text{-}11]$$

is used as a qualitative test for unsaturation, but this naturally does not distinguish an alkyne from an alkene. Iodine adds to acetylene only at 150° in a closed system. The product is an alkene that will not add more iodine. The iodine atom is too large or too unreactive (maybe both) to add to 1,2-di-iodoethene.

Exercise 5-4. Trichloroethylene, $CHCl{=}CCl_2$ (see p. 75), is made commercially from acetylene. Write a series of reactions that would accomplish this synthesis.

3. Hydrohalogen Acids. Vinyl halides and ethylidene halides are the products in the addition of HX to acetylene. The vinyl group is $CH_2{=}CH{-}$, and the ethylidene group is $CH_3{-}CH{<}$. Other alkynes also add HX according to Markownikoff's rule.

$$CH{\equiv}CH + \overset{\oplus}{H} \overset{\ominus}{Br} \longrightarrow \underset{\substack{\text{vinyl}\\\text{bromide}}}{CH_2{=}\underset{Br}{CH}} \xrightarrow{HBr} \underset{\substack{\text{ethylidene}\\\text{bromide}}}{CH_3{-}CHBr_2} \qquad [5\text{-}12]$$

4. Water. With mercury II sulfate in acid solution as a catalyst, acetylene will add a molecule of water to give acetaldehyde, a compound of a new homologous series. A reasonable mechanism for the reaction assumes that the water molecule adds in a manner similar to the addition of HX. However, the logical intermediate has never been isolated. It can be given the name "vinyl alcohol."

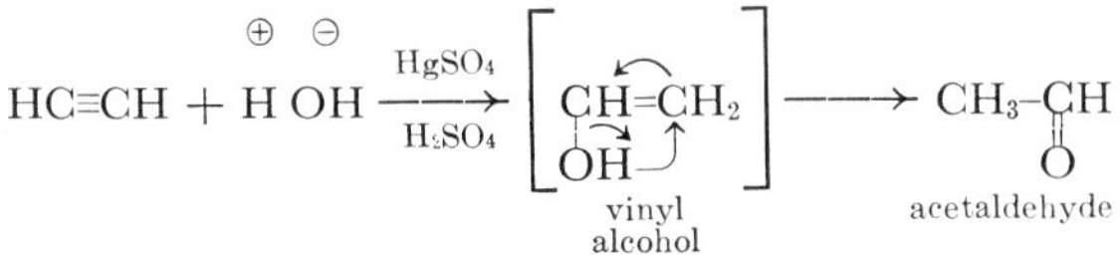

$$HC \equiv CH + \overset{\oplus}{H} \overset{\ominus}{O}H \xrightarrow[H_2SO_4]{HgSO_4} \left[\begin{array}{c} CH = CH_2 \\ OH \end{array} \right] \longrightarrow CH_3 - \underset{O}{CH} \qquad [5\text{-}13]$$

vinyl alcohol   acetaldehyde

The proton from the oxygen of the OH group migrates to the second carbon as the C=C bond breaks (curved arrow), and a pair of electrons on the oxygen shifts to give stable octets to both carbon and oxygen.

5. Self-addition (Acetylene Only). In the presence of copper I chloride, acetylene will add to itself. The hydrogen from one molecule of acetylene adds to a carbon in another acetylene molecule. The carbon that loses the hydrogen now adds to the second carbon in the other acetylene molecule. The new bonds indicated

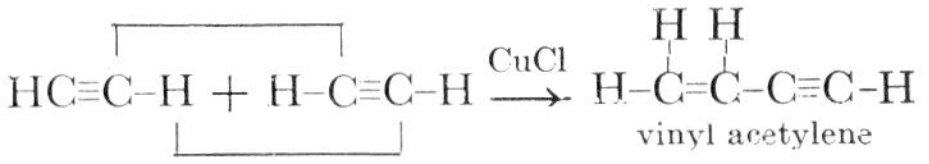

$$HC \equiv C - H + H - C \equiv C - H \xrightarrow{CuCl} \overset{H \; H}{H - C = C - C \equiv C - H} \qquad [5\text{-}14]$$

vinyl acetylene

at the left form the product shown on the right. Vinyl acetylene is used in making a type of synthetic rubber called neoprene. (See Exp. 5-33.)

6. Hydrogen Cyanide. Hydrogen cyanide, HCN (see p. 281), will add to acetylene in the presence of barium cyanide on carbon

$$HC \equiv CH + \overset{\oplus}{H} \overset{\ominus}{CN} \xrightarrow[C]{Ba(CN)_2} CH_2 = CH - C \equiv N \qquad [5\text{-}15]$$

acrylonitrile

as a catalyst. This addition reaction has been carried out only for acetylene. The product is acrylonitrile. (Note the resemblance between Exp. 5-14 and Exp. 5-15.)

The additions of HOX and HOSO$_2$–OH take place on alkynes.

but in general the products are mixtures of substances, and the reactions are not especially useful.

5-5. Oxidation of Alkynes

1. Combustion. The combustion of acetylene is a commercially important reaction because the burning is accompanied by a brilliant white light and a large amount of heat. The first property is utilized in miners' lamps and is due to the presence of unburned carbon particles heated to incandescence. The same reaction,

$$HC\equiv CH + \tfrac{5}{2}O_2 \xrightarrow[\text{air}]{\text{excess}} 2CO_2 + H_2O \qquad\qquad [5\text{-}16]$$

using oxygen, takes place in the oxyacetylene torch. If a diminished amount of oxygen is used, carbon monoxide is one of the products. Carbon monoxide burns in air (or oxygen) to form carbon dioxide.

$$HC\equiv CH + \tfrac{3}{2}O_2 \longrightarrow 2CO + H_2O \qquad\qquad [5\text{-}17]$$

$$2CO + O_2 \longrightarrow 2CO_2 \qquad\qquad [5\text{-}18]$$

In an oxyacetylene welding torch, the reaction shown in Expression 5-17 is desirable because carbon monoxide is a strong reducing agent at elevated temperature and the metal at the point at which it is being welded needs a reducing atmosphere to prevent its own oxidation. The carbon monoxide burns at the edge of the flame according to Expression 5-18 so that no danger of carbon monoxide poisoning exists for the welder.

2. Baeyer's test. Chemical oxidizing agents, such as the permanganate ion in alkaline or acid solution, have a milder oxidizing effect on alkynes than combustion, just as they do on alkenes in the Baeyer reaction. Until this reaction is discussed in Chapter 14, it may be pictured as analogous to that with alkenes (p. 89). It will be remembered, however, that an OH group on a carbon carrying a double bond is an unstable configuration (p. 99), and a hydrogen migrates to the second carbon to give an aldehyde.

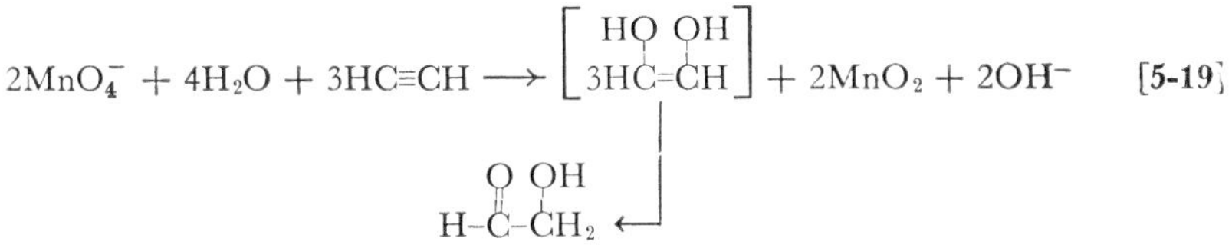

$$2MnO_4^- + 4H_2O + 3HC{\equiv}CH \longrightarrow \left[\begin{matrix} HO & OH \\ 3HC{=}CH \end{matrix} \right] + 2MnO_2 + 2OH^- \qquad [5\text{-}19]$$

$$\begin{matrix} O & OH \\ \| & | \\ H-C-CH_2 \end{matrix} \longleftarrow$$

The alcohol-aldehyde shown here is stable but in the presence of MnO_4^- is oxidized further. That reaction will be considered on page 247.

The disappearance of the purple color of the permanganate ion and the appearance of brown MnO_2 are the observable results of the unsaturation test. The Baeyer test does not distinguish an alkene from an alkyne.

5-6. Replacement in Alkynes: Acetylides

The metallic salts of alkynes are formed by replacement of a hydrogen on the alkynic carbon with a metal (see p. 96). If no hydrogen is present on the alkynic carbon, no salt can be formed. The triple bond is not affected during such a reaction.

Sodium will react with acetylene, and the reaction may be carried out conveniently by bubbling acetylene through a solution of sodium in liquid ammonia:

$$HC{\equiv}CH + Na \xrightarrow{\text{liq. }NH_3} \underset{\substack{\text{monosodium}\\\text{acetylide}}}{H-C{\equiv}C-Na} + \tfrac{1}{2}H_2 \qquad [5\text{-}20]$$

The second hydrogen also is displaced if acetylene is passed over heated sodium:

$$HC{\equiv}CH + 2Na \longrightarrow \underset{\substack{\text{disodium}\\\text{acetylide}}}{Na-C{\equiv}C-Na} + H_2 \qquad [5\text{-}21]$$

The products are white salts, stable in air, which readily hydrolyze back to acetylene (compare CaC_2, p. 94):

$$Na-C{\equiv}C-Na + 2H_2O \longrightarrow HC{\equiv}CH + 2NaOH \qquad [5\text{-}22]$$

The acetylides of the heavier metals copper, silver, and mercury are insoluble in water and explosive when dry. They hydrolyze only in acid solution and consequently are formed by precipitation from alkaline solution. The silver salt is made by bubbling acetylene through an ammoniacal solution of silver nitrate in which the silver is present as a silver-ammonia complex ion:

$$Ag(NH_3)_2^+ + HC\equiv CH \longrightarrow AgC\equiv CH + NH_4^+ + NH_3 \qquad [5\text{-}23]$$

or

$$2Ag(NH_3)_2^+ + HC\equiv CH \longrightarrow \underset{\substack{\text{white precipitate,} \\ \text{silver acetylide}}}{Ag\text{-}C\equiv C\text{-}Ag} + 2NH_4^+ + 2NH_3 \qquad [5\text{-}24]$$

(A brown precipitate of silver oxide is first formed when ammonium hydroxide is added to the silver ion. This Ag_2O then dissolves in excess ammonia.)

$$2Ag^+NO_3^- + 2NH_3 + H_2O \longrightarrow \underline{Ag_2O} + 2NH_4^+ + 2NO_3^- \qquad [5\text{-}25]$$

$$Ag_2O + 2NH_4^+ + 2NH_3 \longrightarrow 2Ag(NH_3)_2^+ + H_2O \qquad [5\text{-}26]$$

Red copper I acetylide is made in the same way from the copper I ammonia complex ion:

$$HC\equiv CH + 2Cu(NH_3)_2^+ \longrightarrow \underset{\text{red precipitate}}{\underline{Cu\text{-}C\equiv C\text{-}Cu}} + 2NH_4^+ + 2NH_3 \qquad [5\text{-}27]$$

These are three general reactions for all alkynes that have a hydrogen on at least one carbon. Other alkynes, $R\text{-}C\equiv C\text{-}R'$, cannot, of course, give replacement products.

Exercise 5-5. Write an equation for the reaction of propyne and $Ag(NH_3)_2^+$; of propyne and sodium; of the sodium salt of propyne and ethyl iodide.

5-7. Uses of Acetylene

Acetylene is the only alkyne of any commercial importance. Its uses have been mentioned in the discussion of its chemical properties: for light in miners' lamps, for heat in oxyacetylene cutting and welding torches, and as a starting compound for making numerous chemicals—acetaldehyde, acetic acid, neoprene, and others to be mentioned later. (See also pp. 495–497.)

Whereas, in the United States, petroleum has been the final source of a large fraction of the organic chemicals used in commerce, in Germany, where oil is scarce, acetylene has been the basis for virtually the entire aliphatic chemical industry.

5-8. Alkadienes

It is possible to write structures for four chain compounds having the formula C_4H_6. This empirical formula fits the general formula C_nH_{2n-2}, which will be recognized as that of an alkyne, of which there are two isomers. A third and a fourth structure may still be drawn, without violating the octet rule for any atom, if we join

$$
\begin{array}{cccc}
H & H\;H & H \\
\overset{..}{H:C}::\overset{..}{C}:\overset{..}{C}::\overset{..}{C}:H & & \text{or} & CH_2=CH-CH=CH_2 \\
\end{array}
$$

$$
\begin{array}{ccc}
H & H\;H \\
\overset{..}{H:C}::\overset{..}{C}::\overset{..}{C}:\overset{..}{C}:H & \text{or} & CH_2=C=CH-CH_3 \\
\quad\;\; \overset{..}{H} \\
\end{array}
$$

carbons with two double bonds. These are called alkadienes, or diolefins. The simplest alkadiene is $CH_2=C=CH_2$, called allene.

TABLE

5-2 | *Alkadienes*

STRUCTURE	COMMON NAME	IUC NAME
$CH_2=C=CH_2$	allene	propadiene
$CH_2=CH-CH=CH_2$	butadiene	1,3-butadiene
$CH_2=C=CH-CH_3$		1,2-butadiene
$CH_2=C-CH=CH_2$ $\quad\;\;\; CH_3$	isoprene	2-methyl-1,3-butadiene
$CH_2=C\!-\!-\!-\!C=CH_2$ $\quad\; CH_3 \;\; CH_3$		2,3-dimethyl-1,3-butadiene

The common names of the alkadienes follow no rules; most of them have no common names.

The rules in the IUC system are the same as for alkenes but with the suffix *diene* in place of *ene* on the stem derived from the corresponding saturated hydrocarbon. In CH_2=CH–CH=CH_2, *diene* is added to the stem *but-* from *butane*, but an *a* is interposed to make the name pronounceable. The compound is called 1,3-butadiene.

5-9. Conjugated Systems of Double Bonds

If two double bonds in a molecule are separated by several carbons, as in CH_2=CH–CH_2–CH_2–CH_2–CH=CH–CH_3, the properties of each double bond are the same as they would be if the other were not present. Both double bonds will add Br_2 or HX and undergo the other reactions that have been described for alkenes. If the double bonds are on the same carbon, as they are in allene, one has a profound effect on the other. Allene is an extremely reactive and unstable substance; it is so unstable, in fact, that it is difficult to prepare. When the double bonds are removed from each other by only one carbon, the result is still different from allene or

$$-C=C-C=C-$$
$$1 \quad 2 \quad 3 \quad 4$$

the other case discussed. This 1,3 system of double bonds is called a conjugated system of double bonds and has a greater stability than would be expected for an ordinary alkene. This stabler configuration is the one that commonly occurs in nature for substances containing two double bonds, although compounds containing isolated double bonds are known.

5-10. 1,4 Addition to Conjugated Systems of Double Bonds

Reactions that take place on a conjugated system of double bonds involve the system as a whole and not the individual multiple bonds. The addition of bromine to 1,3-butadiene will illustrate this. The result is a product in which the bromine appears on carbons 1 and 4 and one double bond has shifted to position 2 (F).

$$
\underset{\underset{4\ \ \ \ 3\ \ \ \ 2\ \ \ \ 1}{}}{CH_2=CH\overset{\delta+}{-}CH\overset{\delta-}{=}CH_2} \xrightarrow{\text{excitement}} \underset{(A)}{\overset{4\ \ \ \ 3\ \ \ \ 2\ \ \ \ 1}{CH_2=CH-CH-CH_2}}\overset{\oplus\ \ \ \ \ominus}{}
$$

$$
\underset{\oplus\ \ \ \ \ \ (B)\ \ \ \ \ominus}{\overset{4\ \ \ \ 3\ \ \uparrow 2\ \ \ \ 1}{CH_2-CH=CH-CH_2}}
$$

[5-28a]

$$
\begin{array}{c}
\overset{\oplus\quad\ominus}{CH_2=CH-CH-CH_2} \\
\updownarrow \\
\underset{\oplus\qquad\ominus}{CH_2-CH=CH-CH_2}
\end{array}
\ +\ \overset{\delta+\ \ \ \delta-}{:Br\ :Br:}\ \longrightarrow\
\begin{array}{c}
\overset{(C)\ \ \oplus\quad Br}{CH_2=CH-CH-CH_2} \\
\updownarrow \\
\underset{\oplus\ \ \ (D)\ \ Br}{CH_2-CH=CH-CH_2}
\end{array}
$$

[5-28b]

$$
\begin{array}{c}
\overset{\oplus\quad Br}{CH_2=CH-CH-CH_2} \\
\updownarrow \\
\underset{\oplus\qquad Br}{CH_2-CH=CH-CH_2}
\end{array}
\ +\ :\overset{\ominus}{Br}:\ \longrightarrow\
\begin{array}{c}
\overset{(E)\ \ Br\ \ Br}{CH_2=CH-CH-CH_2} \\
\\
\underset{Br\quad(F)\ \ Br}{CH_2-CH=CH-CH_2}
\end{array}
$$

[5-28c]

Expression 5-28 indicates how this may come about. If the excitation in 1,3-butadiene gives a little additional energy to the molecule, one might represent the excited state by the partial charge separation shown at A. Of almost equal energy content is the state represented at B. (B is obtained from A by the shift of a pair of electrons.) The excited 1,3-butadiene molecule, then, is probably best represented by some summation of the two pictures A and B that we cannot achieve on paper. The two positive fragments C and D contribute to the total picture of the intermediate in the reaction, but neither alone is a true representation of this state. The two fragments do not necessarily contribute equally to the picture. The fact that the product of the reaction is mostly F rather than E suggests that D makes a greater contribution than C.

The addition of bromine to 1,3-butadiene, resulting in 1,4-

dibromo-2-butene (F), is called a 1,4 addition to the conjugated system. The competing addition, giving a small amount of E, is called a 1,2 addition.

If an excess of bromine is added to the 1,3-butadiene, two molecules will add and give, as final product, 1,2,3,4-tetrabromobutane:

$$CH_2=CH-CH=CH_2 \xrightarrow{Br_2} \underset{Br}{CH_2-CH=CH-CH_2} \xrightarrow[100\%]{Br_2} \underset{Br \quad Br \quad Br \quad Br}{CH_2-CH-CH-CH_2} \qquad [5\text{-}29]$$

Exercise 5-6.   Write an equation showing intermediates to represent the addition of one molecule of HBr to 1,3-butadiene.

Many years ago, Thiele suggested an explanation for the observation that 1,3-dienes add 1,4. He suggested that each of the four carbons in the diene has a *partial valence* sticking out, ready to form a bond when an appropriate reagent is encountered. This

C=C—C=C

theory is not far removed from the explanation given above, though the language is slightly different. What we would now say is that the double bonds at carbons 1 and 4 have some single-bond character and that the single bond between carbons 2 and 3 has some double-bond character. This is implied in the two pictures A and B of Expression 5-28a.

There is strong evidence for this view from electron diffraction measurements of bond lengths. In a large number of compounds containing single bonds or only one multiple bond, the distance C–C is found to be 1.54 A (1.54 × 10⁻⁸ cm) and C=C to be 1.33 A. But in 1,3-butadiene those distances are 1.46 A and 1.35 A, respectively. This seems ample justification for saying that a double bond between carbons 2 and 3 is incipient in the compound.

This phenomenon, called resonance, will be encountered again and again in this text. (See, for example, p. 119 and Chap. 18.)

5-11. Rubber

Portuguese explorers in Brazil found Indians bouncing balls made of rubber, and for as long as 200 years after this discovery rubber was still used only for toys and playthings. In 1839 Goodyear made the accidental but momentous discovery that heating the rubber with sulfur produces a more durable, a more elastic, and a more useful substance than the sticky, unstable crude rubber. But the discovery of vulcanization still did not make the rubber industry grow rapidly. The growth of the automobile industry after 1900 gave the incentive for the expansion of the rubber industry. The price of rubber suddenly became exorbitant because of the large demand and the small supply. Transfer of the rubber tree from Brazil to Burma and the Straits Settlements introduced competition in the production of latex, and the price soon dropped to a very low point again. The low price of natural rubber discouraged any concerted attempts to produce a synthetic rubber until the Germans were driven to it during the First World War as a result of the Allied blockade.

Natural rubber is a milky white emulsion of latex in water. The latex is coagulated by acetic acid and then washed and rolled into sheets. This sheet rubber is vulcanized by heating with 4–5% powdered sulfur and with "accelerators" (organic sulfur-containing compounds that also generally contain nitrogen) that speed up the process. Rubber that is to be made into tires contains antioxidants (generally nitrogen-containing compounds) to reduce the rate of deterioration in sunlight and air and carbon black to decrease friction with concrete surfaces. Other substances are present as fillers and coloring matter.

Some idea of the structure of natural rubber may be obtained by examining the results of the experiments carried out on it. An analysis of rubber shows that only carbon and hydrogen are present in the molecule, and a quantitative analysis results in the empirical formula C_5H_8. Since rubber has a high molecular weight, the molecular formula could be represented by $(C_5H_8)_n$. Bromine will add to rubber, giving a substance in which the ratio of the various atoms is approximately $C_5H_8Br_2$. Heating natural rubber in the absence of air yields the diene isoprene. In the presence of sodium, isoprene will react with itself to give a substance of high molecular weight and of rubber-like properties. It is not exactly like natural rubber, but the two substances have similar structural formulas. This synthetic rubber will add one molecule of bromine per C_5H_8 unit and will yield isoprene again on heating. Since the addition of one molecule of bromine per C_5H_8 unit is implied in the reactions just described for natural rubber, there must be one double bond per C_5H_8 unit in the original molecule. But the isoprene obtained by heating natural rubber contains two double bonds.

$$(C_5H_8)_{n+} + nBr_2 \longrightarrow (C_5H_8Br_2)_n \qquad [5\text{-}30]$$

The disappearance of one double bond always accompanies a 1,4 addition to a conjugated system. This leads to Expression 5-31 as a picture of the reaction of isoprene in the presence of sodium.

The process of combining a large number of small molecules into one large molecule in a way that is capable of functional and

$$\underset{\underset{CH_3}{|}}{CH_2{=}CH{-}C}{=}CH_2 + n\underset{\underset{CH_3}{|}}{CH_2{=}CH{-}C}{=}CH_2 \overset{Na}{\longrightarrow}$$

$$-\underset{\underset{CH_3}{|}}{CH_2{-}CH{=}C}{-}CH_2 - \left[\underset{\underset{CH_3}{|}}{CH_2{-}CH{=}C}{-}CH_2 \right]_n - \qquad [\textbf{5-31}]$$

indefinite continuation is called polymerization, and the resulting compounds are called polymers.

5-12. Types of Synthetic Rubber

1. Buna. The first synthetic rubber made by Germany during the First World War involved a 1,4 addition of another diene. The starting compound was 2,3-dimethyl-1,3-butadiene, and the catalyst used for the polymerization was sodium. The product was called Buna rubber, the name coming from *bu*tadiene and the symbol for sodium, *Na*.

$$CH_2=C-C=CH_2 \xrightarrow{Na} -CH_2-C=C-CH_2-\left[CH_2-C=C-CH_2-\right]_n \qquad [5\text{-}32]$$
$$\overset{|}{CH_3}\overset{|}{CH_3} \qquad\qquad \overset{|}{CH_3}\overset{|}{CH_3} \qquad\qquad \overset{|}{CH_3}\overset{|}{CH_3}$$

2. Neoprene.   Acetylene is the starting compound for the synthetic rubber called neoprene. Vinyl acetylene, made by self-addition of acetylene in the presence of copper I chloride, adds HCl in the 1,4 manner. The final isolated product, chloroprene, results from the rearrangement of the 1,4 addition intermediate. Chloroprene may then be polymerized in the presence of a catalyst. Neoprene is insoluble in gasoline and other hydrocarbons,

$$HC{\equiv}CH + HC{\equiv}CH \xrightarrow{CuCl} HC{\equiv}C\text{-}CH{=}CH_2 \xrightarrow[CuCl]{HCl}$$

$$\text{vinyl acetylene}$$

$$CH_2{=}C{=}CH\text{-}CH_2Cl \xrightarrow{\text{rearranges}} CH_2{=}CH\text{-}\underset{\underset{Cl}{|}}{C}{=}CH_2 \xrightarrow{\text{catalyst}}$$

$$\text{chloroprene}$$

$$-CH_2\text{-}CH{=}\underset{\underset{Cl}{|}}{C}\text{-}CH_2\text{-}\left[CH_2\text{-}CH{=}\underset{\underset{Cl}{|}}{C}\text{-}CH_2\text{-}\right]_x \qquad [5\text{-}33]$$

$$\text{neoprene}$$

whereas natural rubber swells and disintegrates in contact with these substances. This property has made neoprene usable in many places where a flexible substance is necessary in contact with these hydrocarbons, as, for example, in the dispensing hose used for gasoline at filling stations. Neoprene heels are said to last many times longer than ordinary rubber heels on shoes of filling station attendants. The inner lining of self-sealing gasoline tanks, used in wartime in aircraft and tanks, was made of neoprene, while the outer layer was of ordinary rubber. When a bullet pierced the tank, gasoline leaking through the hole caused the natural rubber to swell. The hole closed or at least became smaller, enabling the pilot to limp back to safety.

3. Buna S.   The synthetic rubber developed for use in tires during the Second World War is a copolymer of butadiene and an aromatic hydrocarbon, styrene. (For the preparation of styrene, see p. 226.) A copolymer is a polymer formed by the polymerization of a mixture of two different monomers. It is not just a mixture of two polymers. The repeating unit is not exactly the one enclosed

$$CH_2=CH-CH=CH_2 + \underset{\underset{\text{styrene}}{\overset{|}{\underset{C_6H_5}{}}}}{CH=CH_2} \xrightarrow{\text{catalyst}}$$

$$-CH_2-CH=CH-CH_2-\left[\underset{\underset{\text{Buna S}}{\overset{|}{\underset{C_6H_5}{}}}}{-CH-CH_2-CH_2-CH=CH-CH_2-}\right]_x \qquad [5\text{-}34]$$

in brackets, for the starting mixture is 75 parts 1,3-butadiene and 25 parts styrene. Buna S was also given the name Government Rubber S, GR-S.

4. Buna N. A copolymer of butadiene and acrylonitrile, a nitrogen-containing compound, was also developed during the last war as a synthetic rubber for tank treads, where resistance to abrasion under heavy loads was very important. The ratio of starting compounds is 75 parts of butadiene to 25 parts of acrylonitrile.

$$CH_2=CH-CH=CH_2 + \underset{\underset{\text{acrylonitrile}}{\overset{|}{\underset{C\equiv N}{}}}}{CH=CH_2} \xrightarrow{\text{catalyst}}$$

$$-\left[\underset{\underset{\text{Buna N}}{\overset{}{\underset{C\equiv N}{}}}}{-CH_2-CH=CH-CH_2-CH-CH_2-}\right]_x \qquad [5\text{-}35]$$

5. Butyl Rubber. Butyl rubber is a copolymer of butadiene and isobutylene. It is finding wide use at present in inner tubes for tires because of one outstanding property: it is nearly impervious to gases. Tires containing butyl rubber inner tubes may be used for months without any substantial loss of air.

$$CH_2=CH-CH=CH_2 + \underset{\underset{\text{isobutylene}}{\overset{CH_3}{\underset{CH_3}{|}}}}{C=CH_2} \xrightarrow{\text{catalyst}}$$

$$-\left[\underset{\underset{\text{butyl rubber}}{\overset{CH_3}{\underset{CH_3}{|}}}}{-CH_2-CH=CH-CH_2-C-CH_2-}\right]_x \qquad [5\text{-}36]$$

6. Thiokol. All types of synthetic rubber mentioned so far have been polymers of dienes or copolymers of a diene and an alkene. Each of these polymers, as represented by our writing of its structure, still has one double bond in each unit of the molecule. It is natural to ask whether the double bond accounts in some way for the elasticity of the rubber. There is certainly some relation between them, for if HCl is added to the rubber, it finally becomes hard and brittle. This behavior was exhibited by tires in the early days of motoring, when it was a common experience to find that a spare tire carried on the back of a car, exposed to the air and sun, deteriorated about as rapidly as one in use. Oxidation took out the double bonds in the rubber molecule. This is now largely prevented by inhibitors added during the manufacture.

Although the double bond is present in all the polymers pictured, that does not appear to be the only way in which elasticity may be attained. Thiokol rubber is a product of the reaction of a dichloro compound (chlorines on either adjacent or separated carbons) and sodium polysulfide. The sodium polysulfide is easily made by dissolving sulfur in hot sodium hydroxide. It may contain 2–5 sulfur atoms, and Na_2S_4 is taken as representative.

$$6NaOH + 3S \longrightarrow 2Na_2S + Na_2SO_3 + 3H_2O \qquad [5\text{-}37]$$

$$Na_2S + S \longrightarrow Na_2S_{2-5} \qquad [5\text{-}38]$$

$$\underset{\text{1,2-dichloroethane}}{Cl\text{-}CH_2\text{-}CH_2\text{-}Cl} + \underset{\substack{\text{sodium}\\\text{tetrasulfide}}}{Na_2S_4} \longrightarrow$$

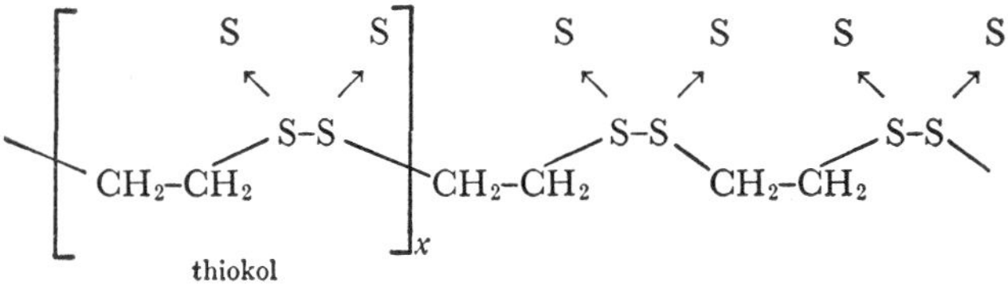

The fact that two sulfurs may be easily removed from the polymer by NaOH and restored by treatment with sulfur at 130° suggests the structure shown in Expression 5-39, in which two of the four sulfurs are held by coordinate covalent linkages. However, there is no evidence to exclude the following structure, which would show the same X-ray pattern:

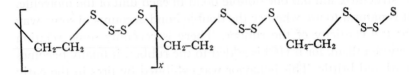

Thiokol rubber was used for a short time to recap tires, but its odor in heated garages soon discouraged the practice. Since thiokol is not soluble in hydrocarbons, it can be used in many places where neoprene also is useful.

7. Polythene.   A new polymer called polythene appeared during the Second World War as an insulator for radar equipment and in other places where a substance with excellent dielectric properties is needed. The structure is not known to be simply a long chain of methylene groups; but, since the substance is very unreactive, nothing is now known of its real structure. Ethylene can be polymerized at 200° at a pressure of about 2,000 atmospheres when a trace of oxygen is present.

$$CH_2=CH_2 \xrightarrow[\substack{200° \\ \text{trace } O_2}]{2{,}000 \text{ atm}} -[CH_2-CH_2]- \underset{\text{polythene}}{}_n \qquad [5\text{-}40]$$

A polyethylene of higher molecular weight and superior properties is now being made at room temperature and low pressure as a result of Ziegler's discovery of the proper catalyst [Al $(C_2H_5)_3$ + TiCl$_4$, for example].

SUMMARY

Besides the propensity for addition, which other unsaturated compounds also possess, alkynes have a looser hold on a hydrogen attached to carbon at the triple bond than other hydrocarbons do. Such a hydrogen may be displaced by Na, $Ag(NH_3)_2^+$, or $Cu(NH_3)_2^+$. Alkynes add all the reagents that alkenes add. Acetylene also adds HCN and can be dimerized.

Acetylene is prepared by the action of water on calcium carbide. Other alkynes are prepared by modification of methods applied to the synthesis of alkenes or by alkylating sodium acetylide. The difficulty of obtaining starting materials keeps the higher alkynes from being readily available. Double bonds in alkadienes may be cumulated, conjugated, or isolated. If two double bonds are on the same carbon (cumulated), the alkadiene is unstable and very reactive; if the double bonds are remote from each other, one influences the other very little; if the double bonds are conjugated, the alkadiene has a special stability, greater than that of either of the other two systems. In general, a 1,3-conjugated system adds 1,4.

Alkadienes are useful in the synthesis of various kinds of rubber.

REFERENCES

1. Hauser, "Synthetic Rubber and Plastics," J. Chem. Education, 21, 15 (1944).
2. Thomas, Lightbown, Sparks, Frolich, and Murphree, "Butyl Rubber," Ind. Eng. Chem., 32, 1283 (1940).
3. Williams, "Vulcanization of Rubber with Sulfur," Ind. Eng. Chem., 39, 901 (1947).

EXERCISES

(Exercises 1–6 will be found within the text.)

7. Write structural formulas for
   1. 3-ethyl-1-hexyne
   2. dimethylacetylene
   3. 2-chloro-1,3-butadiene
   4. 3-methyl-2-hexene-4-yne
   5. 1,3-cyclopentadiene

8. Write structural formulas and IUC names for all isomeric forms of C_5H_8
   1. that contain a triple bond
   2. that contain two double bonds

9. Write an equation for the reaction of ammoniacal silver nitrate with any substance in Exercise 7 that will react.

10. Write equations to show the reaction of 1-butyne with each of the following reagents:

1. $Cu(NH_3)_2^+$ 4. Br_2
2. HBr   5. H_2 in presence of Ni
3. $Ag(NH_3)_2^+$

11. Suggest chemical means of distinguishing between
1. 1-butyne and 2-butyne
2. 2-butyne and butane
3. 1-butyne and 1,3-butadiene

12. How do alkynes differ in chemical properties from alkenes? from alkadienes?

13. Write structural formulas for all the unsaturated compounds that can be reduced to 2-methylbutane by hydrogenation with a catalyst.

14. Combustion of 0.2048 g of an organic compound (not a hydrocarbon) gives 0.3890 g of CO_2 and 0.1271 g of H_2O. What are the percentages of C and H present]

15. Analysis of a gas reveals that it is 92.3% C and 7.7% H. What is the empirical formula of the compound? If the molecular weight is 26, what is the compound?

16. For what alkyne does the volume of gases formed in its combustion equal the volume of gases burned?

17. What is the longest alkyne molecule in which the carbon atoms are linear? (Make a model.) Do you think it would be possible to synthesize a cyclic alkyne?

6 Aromatic Hydrocarbons and Aryl Halides

Three different unsaturated cyclic compounds containing six carbons may exist, as indicated by the structures below. The cyclic compound called cyclohexatriene (and its homologues) has properties that make it unique among polyalkenes—properties distinctive enough to warrant separate study of this homologous series, called the aromatic hydrocarbons.

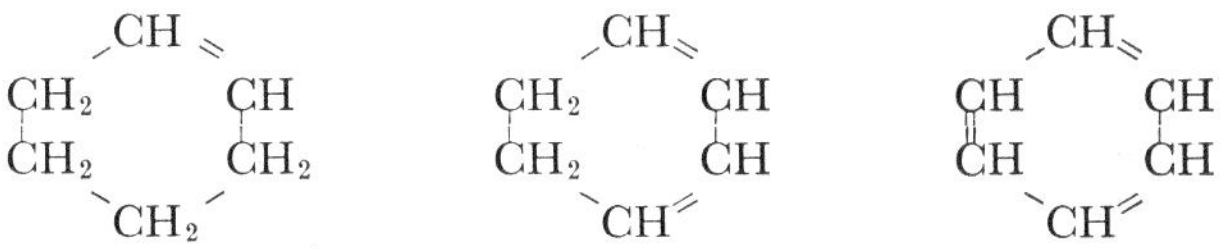

$$
\begin{array}{ccc}
\underset{\text{cyclohexene}}{\begin{array}{c} \text{CH} \\ \text{CH}_2 \quad\quad \text{CH} \\ \text{CH}_2 \quad\quad \text{CH}_2 \\ \text{CH}_2 \end{array}} &
\underset{\text{cyclohexadiene}}{\begin{array}{c} \text{CH} \\ \text{CH}_2 \quad\quad \text{CH} \\ \text{CH}_2 \quad\quad \text{CH} \\ \text{CH} \end{array}} &
\underset{\text{cyclohexatriene}}{\begin{array}{c} \text{CH} \\ \text{CH} \quad\quad \text{CH} \\ \text{CH} \quad\quad \text{CH} \\ \text{CH} \end{array}}
\end{array}
$$

6-1. Aromatic Hydrocarbons

Cyclohexene undergoes the reactions associated with the straight-chain alkenes. The six-membered ring is not strained to any extent, and the compound is neither more nor less reactive than 2-hexene. Cyclohexene adds bromine to yield cyclohexenedibromide:

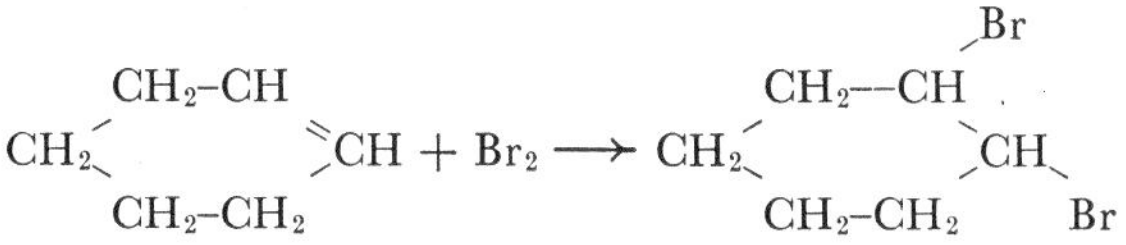

$$
\begin{array}{c} \text{CH}_2\text{–CH} \\ \text{CH}_2 \quad\quad \text{CH} \\ \text{CH}_2\text{–CH}_2 \end{array} + \text{Br}_2 \longrightarrow
\begin{array}{c} \quad\quad\quad \text{Br} \\ \text{CH}_2\text{—CH} \\ \text{CH}_2 \quad\quad \text{CH} \\ \text{CH}_2\text{–CH}_2 \quad \text{Br} \end{array}
\quad\quad [6\text{-}1]
$$

The product, like other 1,2-dibromo compounds, can be treated with alcoholic KOH solution to remove two molecules of HBr.

$$\begin{array}{c} \text{CH}_2\text{--CHBr} \\ \text{CH}_2 \diagdown \\ \quad\quad\quad \diagup \text{CHBr} + 2\text{KOH} \xrightarrow[\substack{\text{low} \\ \text{yield}}]{\text{alcohol}} \\ \text{CH}_2\text{---CH}_2 \end{array} \quad \begin{array}{c} \text{CH=CH} \\ \text{CH}_2 \diagdown \\ \quad\quad\quad \diagup \text{CH} + 2\text{KBr} + 2\text{H}_2\text{O} \\ \text{CH}_2\text{-CH} \end{array}$$

[6-2]

The product is cyclohexadiene. Since it has a conjugated system of double bonds, cyclohexadiene should add Br_2 and other reagents in the 1,4 positions. Since an alkynic bond gives linear character to the adjacent atoms, a six-membered ring cannot contain such a bond. Hence, in the reaction with alcoholic KOH, only a diene would be expected, and a diene is, in fact, obtained.

If cyclohexadiene is allowed to pass over a palladium catalyst at 100°, it spontaneously loses a molecule of hydrogen to become cyclohexatriene. A small amount of heat is evolved in the reaction:

$$\begin{array}{c} \text{CH =CH} \\ \text{CH}_2 \diagdown \\ \quad\quad\quad \diagup \text{CH} \xrightarrow[100\%]{\text{Pd}} \\ \text{CH}_2\text{-CH} \end{array} \quad \begin{array}{c} \text{CH=CH} \\ \text{CH} \diagdown \\ \quad\quad\quad \diagup \text{CH} + \text{H}_2 + \text{heat} \\ \text{CH-CH} \end{array}$$

[6-3]

The transition from cyclohexene through cyclohexadiene to cyclohexatriene increases the number of unsaturated linkages from one to three. Since the disappearance of the red color of bromine and the disappearance of the purple color of the permanganate ion are indications of unsaturation, it might reasonably be expected that these compounds would give both these indications with increasing rapidity from the first to the third compound.

If we try the tests, however, we make the startling discovery that cyclohexatriene does not add bromine at all and does not decolorize permanganate solutions. It has a peculiar saturation in spite of its three double bonds. Compounds exhibiting this inertness toward chemical reagents that ordinarily add to double bonds between carbon atoms are said to have aromatic character. Cyclohexatriene, more commonly called benzene, is the first member of the homologous series called the aromatic hydrocarbons.

At this point an abbreviated symbol for benzene will be introduced and used hereafter:

$$CH-CH$$
⬡ will represent $CH \underset{CH=CH}{\overset{}{\big<}} \quad \overset{}{\big>} CH$

At each corner of the hexagon is a carbon with a hydrogen attached to it. The bonds in the ring are alternate single and double bonds. In toluene, one hydrogen is replaced by a methyl group, and the condensed structural formula is written thus: ⬡-CH₃

TABLE
6-1 | *Homologues of Benzene*

FORMULA	COMMON NAME	BOILING POINT
⬡	benzene	78°
⬡-CH₃	toluene	111°
⬡-CH₂CH₃	ethylbenzene	136°
⬡-CH₃ ⟋CH₃	*ortho*-xylene	142°
⬡-CH₃ CH₃⟋	*meta*-xylene	139°
CH₃-⬡-CH₃	*para*-xylene	138°

6-2. Nomenclature of Aromatic Compounds

The common names of five homologues of benzene are given in Table 6-1. When two groups are adjacent on the ring, one is said to be in the *ortho* position with respect to the other. The isomer of dimethylbenzene having such groups is called *ortho*-dimethylbenzene, *o*-dimethylbenzene, or *o*-xylene. When written by hand, *ortho* or *o*- is underlined; in print it is italicized.

If two groups on the ring have between them one position on which no substitution has been made, the groups are *meta* (*m-*) to each other; if two groups are opposite each other on the ring, they are *para* (*p*-) to each other:

m-bromonitrobenzene

p-toluenesulfonic acid

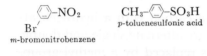

A numbering system may also be used and is generally used if more than two groups appear on the aromatic ring. By the numbering system, the compounds above would be known as 3-nitrobromobenzene or 3-bromonitrobenzene and 4-toluenesulfonic acid. In these two cases, the second substituent is understood to be on carbon 1 in the ring.

Attention is called to some points of common usage in nomenclature:

⌬–, or C_6H_5, is the phenyl group, not to be confused with ⌬–OH, or C_6H_5OH, which is phenol, a compound. The word *phenyl* is used in the following names:

$C_6H_5–Cl$	phenyl chloride, more commonly called chlorobenzene
$C_6H_5–CH_2–NO_2$	phenylnitromethane
$C_6H_5–Mg–Br$	phenylmagnesium bromide
$C_6H_5–C_6H_5$	biphenyl
$C_6H_5–CH_2–C_6H_5$	diphenylmethane

but not in the following:

$C_6H_5–NO_2$	nitrobenzene
$C_6H_5–SO_3H$	benzenesulfonic acid
$C_6H_5–CH_2–Cl$	benzyl chloride or, less commonly, phenylchloromethane

Exercise 6-1. Write structures for *m*-dinitrobenzene, 1,4-diphenylbutane, 1,3,5-trichlorobenzene, and 1,2-dimethylcyclohexane.

6-3. Stability of the Aromatic Ring

Aromatic compounds are remarkably unreactive in terms of the reactions that have already been discussed for double bonds. It is probably a good thing that the name "cyclohexatriene" was never used for what we call benzene. The name would have implied properties that the compound does not have. Would it not be better also to drop the present symbolism of double bonds in benzene if it is not an alkene? The answer is yes, but no one has come up with a better hieroglyphic; or, at least, none has been accepted. What the chemist has done, instead, is to keep the symbolism of double and single bonds in benzene but to modify considerably, in his mind, the meaning of the pictures he draws.

Before abandoning the symbolism of cyclohexatriene, one might inquire whether anything inherent in the cyclic system itself compels modification of the properties of the double bonds. A conjugated system is considerably more stable than a system of isolated double bonds. The alternate single and double bond system in the six-membered ring attains a symmetry that is impossible in a chain. This special property of the ring could perhaps be blamed for some of the increased saturation observed in aromatic hydrocarbons. The question we are raising is: "In this completely conjugated system, is there any vulnerable spot in the molecule where a reagent that attacks alkenes can catch hold?" Whatever your answer to this question is, it remains true that chemists have not been quite satisfied that the extrapolation of properties of a 1,3 system to complete conjugation is enough to account for the saturation of the aromatic ring.

Close examination of the structural formula of *ortho*-xylene reveals that it may be written in two different ways. Structure 1 con-

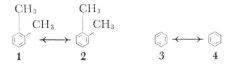

CH_3 CH_3
  CH_3 CH_3

1   2   3   4

tains a double bond between the carbons carrying the methyl side chains; structure 2 does not. In benzene itself the two structures are indistinguishable.

At present the hieroglyphic used for the aromatic nucleus carries the following interpretation: Neither picture 1 nor picture 2 is exactly correct, but the true picture of the molecule is some summation of the two that cannot be put on paper. The German word *Zwischenstufe* (intermediate stage) may also help to convey the correct idea, which is now called resonance. "Resonance" seems an unfortunate term because it implies a shift back and forth between two structures. There is no such shift. Benzene (or *o*-xylene) is not structure 3 (1) part of the time and 4 (2) the rest of the time. Instead, benzene (or *o*-xylene) has six equivalent bonds all the time. Each bond might be called a bond and a half or, better still, just a "benzene bond." Actually, while chemists continue to use the classical representation, they bear in mind the special properties that now annotate the old symbols. The double-headed arrow is frequently used to indicate resonance in two structures, as shown above for *o*-xylene and benzene.

Two rules for considering the phenomenon of resonance are useful here and in future cases:

1. Canonical structures* can differ only in the distribution of electrons and not in the position of atoms.
2. Resonance will be important only in canonical structures of nearly equal stability. If the contributing structures are exactly equivalent, resonance stabilization will be very important.

The structure of benzene presented here is a modern one and does not take historical developments into consideration. Part of the story of the development of ideas on the covalent bond in aromatic compounds is given in Chapter 18.

---

* Canonical, or lawful, structures are electronic structures that hold the Rule of Eight (p. 15) sacred most of the time and always conserve the total number of electrons available.

6-4. Substitution Reactions of Benzene

The reactions that best characterize benzene and its homologues are substitutions by electron-seeking agents. The four reactions described in this section are of the type in which a reagent XY may yield an electron-seeking fragment $X^{\oplus}$ at the moment of reaction (or before the reaction). A complex is formed with the

$$\bigcirc + X^{\oplus} \longrightarrow \overset{H}{\underset{\oplus}{\bigcirc}}\overset{X}{\underset{H}{\diagdown}} \overset{-H^{+}}{\longrightarrow} \bigcirc\text{-X} \qquad\qquad \text{[6-4]}$$

ring, which then kicks off a proton to stabilize itself in the restored aromatic ring. The initial attack is probably the same as that of an electron-seeking agent on an alkene, but the final step is not an addition, as it is in the alkene. The fragment $Y^{\ominus}$, the catalyst, or the substrate as a whole may aid in the removal of the proton.

1. Halogenation. In the presence of aluminum chloride or metallic iron, chlorine or bromine substitutes for hydrogen on the aromatic ring, forming chloro- or bromobenzene, both called aryl halides. The charge separation in chlorine indicated in Expression 6-5a is undoubtedly induced by the aluminum chloride. The second step is the formation of a complex between benzene and the positive fragment $Cl^{\oplus}$. The restoration of the very stable aro-

$$:Cl:Cl: + \overset{:Cl:}{\underset{:Cl:}{Al}} :Cl: \longrightarrow \left[\overset{\delta+}{:Cl} \quad \overset{:Cl:}{\underset{:Cl:}{:Cl:Al:Cl:}}\overset{\delta-}{}\right] \qquad \text{[6-5a]}$$

$$\left[\overset{H}{\underset{H'\delta-}{\bigcirc}}\overset{}{}_{\delta+} + :Cl^{\oplus} \longrightarrow \overset{H}{\underset{H}{\bigcirc}}\overset{Cl}{\diagdown}\right] \overset{-H^{+}}{\underset{62\%}{\longrightarrow}} \bigcirc\text{-Cl} \qquad \text{[6-5b]}$$

chlorobenzene

matic ring in the last step is the reason why the reaction takes the direction it does—that is, substitution instead of addition. The proton is removed by the $AlCl_4^{\ominus}$ ion, with the following result:

$$H^{\oplus} + AlCl_4^{\ominus} \longrightarrow HCl + AlCl_3 \qquad \text{[6-6]}$$

When iron is the catalyst, the first reaction of the halogen is with the iron to form the "carrier." If the halogen is bromine, it then reacts with the iron III bromide to give the complex $[Br^{\delta+}FeBr_4^{\delta-}]$.

$$3Br_2 + 2Fe \longrightarrow 2FeBr_3 \qquad \text{[6-7]}$$

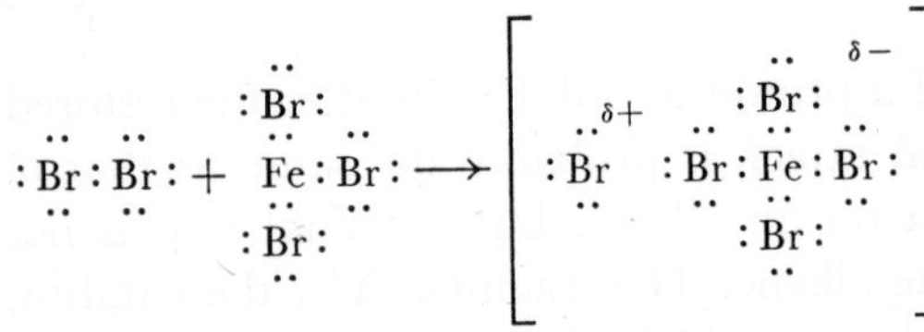

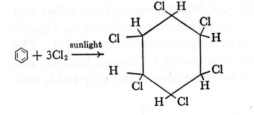

[6-8]

Under the influence of sunlight, chlorine will add to benzene although bromine will not. The product, hexachlorocyclohexane,

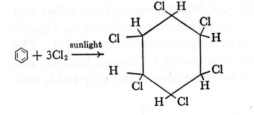

[6-9]

is an important insecticide, sold under the name BHC (benzenehexachloride). It can be made to lose three molecules of HCl fairly easily and thus give a trichlorobenzene, restoring the aromatic ring.

2. Nitration. Nitric acid attacks aromatic hydrocarbons at low temperature (50–100°) in comparison with alkanes (400°). The reaction with alkanes is at a high enough temperature to take place between gases; aromatic nitration takes place between liq-

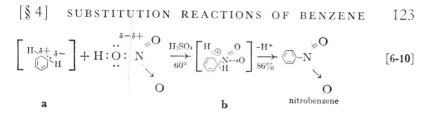

a   b   nitrobenzene

[6-10]

uids. The initial reaction is postulated as an attack by the relatively electropositive nitrogen atom on the activated ring (a). The electropositive nitrogen is aided in playing this role by the action of the sulfuric acid on nitric acid:

$$HONO_2 + HOSO_3H \longrightarrow NO_2^{\oplus} + OSO_3H^{\ominus} + H_2O \qquad [6\text{-}11]$$

The resulting intermediate (b) loses a proton to stabilize itself rather than add the $OSO_3H^{\ominus}$ group. (An OH^- group, of course, is not available in the strongly acid solution.) In the end, then, an aromatic ring is regenerated.

3. Sulfonation. Fuming sulfuric acid dissolves benzene at room temperature to form benzenesulfonic acid. Under the same conditions the alkanes are unaffected. The sulfur trioxide may react to

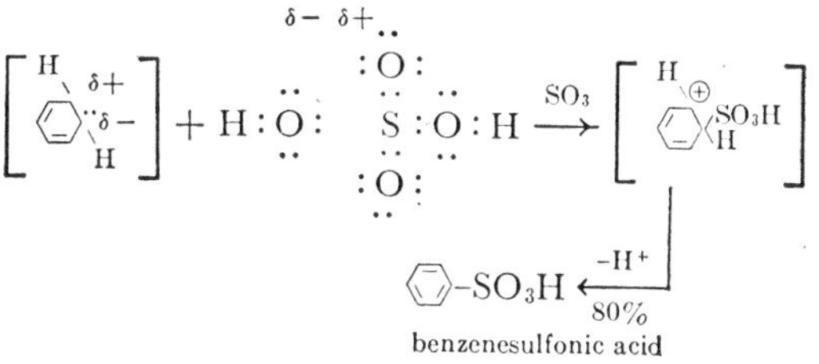

[6-12]

benzenesulfonic acid

a slight extent with the sulfuric acid, in which it dissolves, to give small amounts of the two ions shown in Expression 6-13. The evi-

$$\overset{\delta- \;\; \delta+}{HO\ SO_3H} + SO_3 \longrightarrow \overset{\oplus}{SO_3H} + HOSO_3^{\ominus} \qquad [6\text{-}13]$$

dence for the positive fragment is not as clear-cut as in the nitration; still, this is a useful way to view the reaction, for nitration and sulfonation solutions are similar in character.

4. Friedel-Crafts Reaction. Another reaction that distinguishes aromatic from aliphatic hydrocarbons is that of an alkyl halide with an aromatic hydrocarbon in the presence of $AlCl_3$.

$$RX + AlCl_3 \rightleftharpoons R^{\oplus}AlCl_3X^{\ominus} \qquad [6\text{-}14]$$

$$\left[\begin{array}{c} H \\ \delta - \cdots \bigcirc \\ \delta + H \end{array}\right] + R^{\oplus} \longrightarrow \left[\begin{array}{c} H \\ R \bigcirc \\ \oplus H \end{array}\right] \xrightarrow{-H^+} R-\bigcirc \qquad [6\text{-}15]$$

$$H^{\oplus} + AlCl_3X^{\ominus} \longrightarrow AlCl_3 + HX \qquad [6\text{-}16]$$

This Friedel-Crafts reaction is also catalyzed by other reagents, such as anhydrous HF, BF_3, and $AlBr_3$, and polyhalides such as CH_2Cl_2, $CHCl_3$, and CCl_4 will take the place of the alkyl halides. For example, diphenylmethane can be synthesized from benzene and methylene chloride (Exp. 6-17).

$$2\bigcirc + \begin{array}{c} Cl \\ CH_2 \\ Cl \end{array} \xrightarrow{AlCl_3} \bigcirc-CH_2 + 2HCl \qquad [6\text{-}17]$$
$$\qquad\qquad\qquad\qquad \bigcirc$$

6-5. Sources of Aromatic Hydrocarbons

1. Coal Tar. If coal is subjected to the process called destructive distillation (heating in the absence of air), it undergoes some decomposition. The products are coal gas (a fuel), a liquid called coal tar, and the residue, coke. From one ton of coal, 100 pounds of coal tar is obtained; 1–2 pounds of this is benzene, 0.3–0.5 pound is toluene, and 0.1–0.2 pound is a mixture of the three xylenes. After water and ammonia are removed from coal tar, it is fractionally distilled in much the same way as petroleum. The lowest-boiling fraction, boiling from 80° to 150°, includes the compounds given in Table 6-1. The higher-boiling fractions are made up of phenols, higher hydrocarbons, and pitch. These will be discussed in more detail later.

2. Petroleum. Before the Second World War the demand for toluene and other aromatic hydrocarbons was so great that a method of preparing them from petroleum was sought with great vigor and rather quickly found. The reaction known as "hydroforming" was for a time (1940 and later) run on a very large scale. The aromatic compounds are formed by cyclic dehydrogenation of saturated compounds. The compounds shown in brackets in Expressions 6-18 and 6-19 are not isolated in the process. Molybdenum and chromium oxides on a carrier are catalysts for the formation of toluene and the xylenes, but are not as good for synthesizing benzene. A platinum catalyst, more recently found, can be used at 100–500 pounds pressure and 500°C for the production of benzene, toluene, and the xylenes.

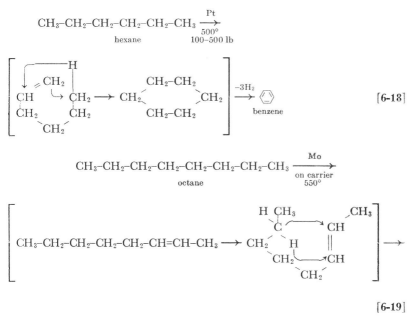

[6-18]

[6-19]

$+ 3H_2$

o-xylene

6-6. Aryl Halides

Aryl halides are compounds in which one hydrogen of the aromatic ring has been replaced by a halogen. The principal method of preparing them, direct substitution, has been discussed as a reaction of the aromatic hydrocarbons (p. 121). A new situation arises if there is a side chain on the ring. In toluene, for example, does the halogen attack the side chain or the ring? The answer depends on the conditions. It will be recalled that the catalyst for halogenation of an alkane is sunlight. If toluene is treated with chlorine at 110° in the sunlight, the aliphatic side chain is attacked, and an alkyl halide called benzyl chloride is obtained:

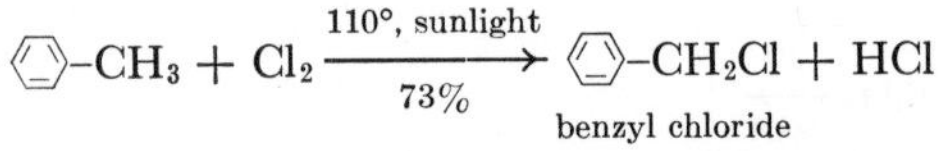

$$\langle \bigcirc \rangle\text{-CH}_3 + \text{Cl}_2 \xrightarrow[73\%]{110°, \text{ sunlight}} \langle \bigcirc \rangle\text{-CH}_2\text{Cl} + \text{HCl} \qquad [6\text{-}20]$$
$$\text{benzyl chloride}$$

In the dark, and with an iron carrier as catalyst, Cl_2 substitutes on the ring. There are, however, still three positions to attack on the

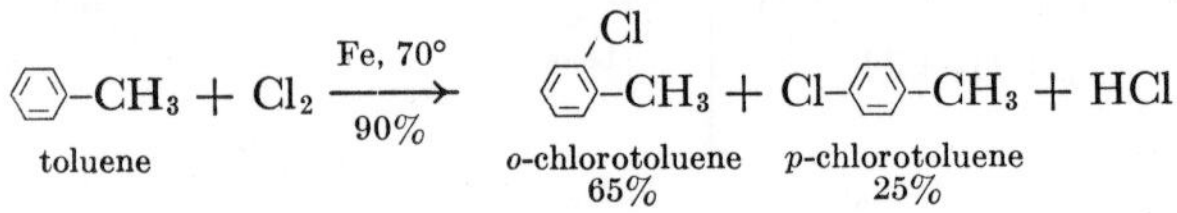

$$\langle \bigcirc \rangle\text{-CH}_3 + \text{Cl}_2 \xrightarrow[90\%]{\text{Fe, } 70°} \langle \bigcirc \rangle\overset{\text{Cl}}{\text{-CH}_3} + \text{Cl-}\langle \bigcirc \rangle\text{-CH}_3 + \text{HCl} \qquad [6\text{-}21]$$
$$\text{toluene} \qquad\qquad \underset{65\%}{o\text{-chlorotoluene}} \quad \underset{25\%}{p\text{-chlorotoluene}}$$

ring. They appear to be, but are not, equally vulnerable. It has been found that the three positions (*o*, *m*, *p*) are not equally available to substituents. As shown here for toluene, the entering atom, Cl in this case, goes to *o* and *p*, and only a negligibly small amount goes to the *m* position. This rule may be used for all alkyl groups on the benzene ring. No matter what the entering group, the alkyl group already on the ring always directs it to positions that are *o* and *p* to the alkyl group. (For an explanation, see p. 386.) These two isomers are not necessarily produced in equal amounts.

Exercise 6-2. By referring to the electronic pictures drawn for the mechanism of chlorination in benzene (p. 121), indicate the electronic structures that contribute to the structure of toluene on the assumption that the methyl group orients *o*, *p*.

6-7. Physical Properties of the Aryl Halides

The boiling points of an aryl halide and an alkyl halide with the same number of carbon atoms are not separated by many degrees. There are greater differences in boiling points among the benzyl halides (structure, p. 118) and the halogenated toluenes than among the o, m, and p isomers of $CH_3-C_6H_4-X$.

TABLE

6-2   *Halides*

		Cl		Br		I	
		B.P.	M.P.	B.P.	M.P.	B.P.	M.P.
$C_6H_{13}-$	n-hexyl	134		157		180	
$C_6H_{11}-$	cyclohexyl	142		165			
C_6H_5-	phenyl	132	−45	156	−31	189	−31
$C_6H_5-CH_2-$	benzyl	179	−43	198	−4		24
$CH_3-C_6H_4-$	o-tolyl	159	−34	181	−26	211	
	m-tolyl	162	−48	183	−40	204	
	p-tolyl	162	7	185	28	211	55

6-8. Reactions of Aryl Halides

A halogen atom on an aromatic ring is much more tightly held than a halogen in an aliphatic chain. This inertness of aryl halides is depicted in the following comparison of the reactions of the two series.

A. Hydrolysis

The alkyl halides are hydrolyzed readily in water or basic solution; the aryl halides take strong heating with sodium hydroxide. If the six-carbon compounds are chosen for comparison, the differences are those shown in Expressions 6-22–6-25. The reaction between chlorobenzene and sodium hydroxide is actually a fusion

$$CH_3CH_2CH_2CH_2CH_2CH_2Cl + HOH \xrightarrow{100°}$$
$$CH_3CH_2CH_2CH_2CH_2CH_2OH + HCl \qquad [6\text{-}22]$$

$$\text{⬡}-CH_2-Cl + HOH \xrightarrow[76\%]{100°} \text{⬡}-CH_2OH + HCl \qquad [6\text{-}23]$$
benzyl chloride   benzyl alcohol

$$\begin{array}{c} CH_2-CH_2 \\ CH_2 \qquad CHCl \\ CH_2-CH_2 \end{array} + HOH \xrightarrow{140°} \begin{array}{c} CH_2-CH_2 \\ CH_2 \qquad CHOH \\ CH_2-CH_2 \end{array} + HCl \quad [6\text{-}24]$$
cyclohexyl chloride

$$\text{⬡}-Cl + NaOH \xrightarrow[\substack{4,000\ lb \\ 90\%}]{375°} \text{⬡}-OH + NaCl \qquad [6\text{-}25]$$
chlorobenzene

under pressure. (It cannot be carried out in glass, of course.) Benzyl chloride, on the other hand, is quite readily hydrolyzed in boiling water and is more reactive than n-hexyl chloride. The ease of hydrolysis also varies with the halogen: iodide > bromide > chloride.

B. With Sodium Cyanide

In contrast to an alkyl halide, an aryl halide will not react at all with an alkali cyanide. (See p. 72.)

C. With Metals

1. Wurtz-Fittig Reaction. Sodium will remove a halogen from an aryl halide, but much more slowly than from an alkyl halide.

$$2\text{⬡}-Cl + 2Na \xrightarrow{15\%} \text{⬡}-\text{⬡} + 2NaCl \qquad [6\text{-}26]$$

The yield is low in this reaction, and it has not proved to be very useful. However, a mixture of an alkyl and an aryl halide gives a fairly good yield of an alkyl benzene. This has been useful since some of the alkyl benzenes are not easy to obtain in any other way.

$$CH_3CH_2CH_2CH_2-Br + \langle\!\bigcirc\!\rangle-Br + 2Na \xrightarrow[45\%]{}$$

$$CH_3-CH_2-CH_2-CH_2-\langle\!\bigcirc\!\rangle + 2NaBr \qquad [6\text{-}27]$$
$$\text{n-butyl benzene}$$

The mechanism is similar to that given earlier for the Wurtz reaction for alkyl halides.

Exercise 6-3. By referring to the mechanism given for the Wurtz reaction (p. 72), show what happens to the appropriate electrons in the bond-breaking and bond-making to form n-butylbenzene.

2. Grignard Reagent. Aryl bromides in ether solution react with Mg more slowly than alkyl bromides and give aryl magnesium halides:

$$\langle\!\bigcirc\!\rangle-Br + Mg \xrightarrow{\text{ether}} \langle\!\bigcirc\!\rangle-Mg-Br \qquad [6\text{-}28]$$

An aryl chloride forms a Grignard reagent only under special conditions.

D. Halogenation

The aryl halides undergo all the reactions described as characteristic of aromatic hydrocarbons in contrast to those of aliphatic hydrocarbons. The halogen atom is *o*- and *p*-directing for further substitution, and it makes substitution more difficult than in benzene itself.

$$Cl-\langle\!\bigcirc\!\rangle + Cl_2 \xrightarrow{Fe} \underset{\substack{o\text{-dichlorobenzene}\\(30\ \text{parts})}}{Cl-\langle\!\bigcirc\!\rangle^{Cl}} + \underset{\substack{p\text{-dichlorobenzene}\\(66\ \text{parts})}}{Cl-\langle\!\bigcirc\!\rangle-Cl} \qquad [6\text{-}29]$$

E. Nitration

$$Cl-\langle\!\bigcirc\!\rangle + \overset{\ominus}{HO}\ \overset{\oplus}{NO_2} \xrightarrow[97\%]{H_2SO_4} \underset{\substack{o\text{-chloronitrobenzene}\\(30\ \text{parts})}}{Cl-\langle\!\bigcirc\!\rangle^{NO_2}} + \underset{\substack{p\text{-chloronitrobenzene}\\(70\ \text{parts})}}{Cl-\langle\!\bigcirc\!\rangle-NO_2} \qquad [6\text{-}30]$$

F. Sulfonation

Aryl halides dissolve in fuming sulfuric acid to give substituted sulfonic acids. In Expression 6-31 the amount of *o* substitution is quite small, probably because $-Br$ and $-SO_3H$ are both large

$$Br\text{-}\langle\bigcirc\rangle + HO\overset{\ominus}{}\overset{\oplus}{}SO_2OH \xrightarrow{SO_3} \underset{85\%}{Br\text{-}\langle\bigcirc\rangle\text{-}SO_3H} + \underset{<5\%}{\langle\bigcirc\rangle\text{-}Br}^{SO_3H} \qquad [6\text{-}31]$$

groups. There is scarcely enough room for $-Br$ and $-SO_3H$ groups to lie adjacent to each other in the molecule. The space considerations are very important in determining the ratio of *o* to *p* isomers in these ring substitutions. The carbon-carbon distance in benzene

$$\overset{Br}{\underset{}{\langle\bigcirc\rangle}}$$

is 1.39 A; the carbon-bromine distance in bromobenzene is 1.84 A. The radius of the bromine atom is approximately 1.14 A. If the angle between carbons in the ring is 120°, the boundary of the bromine atom will be close enough to the *ortho* carbon atoms to make it difficult for large groups to enter these positions. Measurements of the sulfonic acid group are not available.

G. Friedel-Crafts Reaction

$$2X\text{-}\langle\bigcirc\rangle + RCl \xrightarrow[]{\overset{\delta+\delta-}{}AlCl_3} X\text{-}\langle\bigcirc\rangle\text{-}R + X\text{-}\langle\bigcirc\rangle + 2HCl \qquad [6\text{-}32]$$

R may be CH_3-, CH_3-CH_2-, $CH_3-\underset{|}{CH}-CH_3$, or $CH_3-\underset{|}{CH}-CH_2-CH_3$.

The n-propyl and n-butyl halides both give branch-chain products.

$$2Br\text{-}\langle\bigcirc\rangle + 2 \overset{CH_3}{\underset{CH_3}{\diagdown}}CHCl \xrightarrow{AlCl_3} Br\text{-}\langle\bigcirc\rangle\text{-}CH\overset{CH_3}{\underset{CH_3}{\diagup}} + \langle\bigcirc\rangle\text{-}CH\overset{CH_3}{\underset{CH_3}{\diagup}} + 2HCl \qquad [6\text{-}33]$$

6-9. Reactions of Benzene Homologues

After learning that a halogen as well as an alkyl group is *o*- and *p*-directing, we shall do well to review the last four reactions of the previous section with respect to benzene homologues. The reactions have already been discussed for benzene itself (pp. 121ff.).

1. Halogenation.   In Expression 6-35 the attack on the side

$$2CH_3\text{-}CH_2\text{-}\bigcirc + Cl_2 \xrightarrow{\text{Fe}} CH_3CH_2\text{-}\bigcirc\text{-}Cl + CH_3CH_2\text{-}\bigcirc\overset{Cl}{} + 2HCl \quad [6\text{-}34]$$

$$CH_3\text{-}CH_2\text{-}\bigcirc + Cl_2 \xrightarrow[70\%]{\text{sunlight}} \underset{Cl}{CH_3CH}\text{-}\bigcirc + ClCH_2CH_2\text{-}\bigcirc + HCl \quad [6\text{-}35]$$
$$\phantom{CH_3\text{-}CH_2\text{-}\bigcirc + Cl_2 \xrightarrow{sunlight} CH_3CH} 35\% \qquad\quad 35\%$$

chain is about equally centered on each carbon.

2. Nitration.   Nitration goes more readily with toluene than with benzene. It is a general rule for most reagents that a side chain on an aromatic ring modifies the ring in such a way that the homologue is more easily attacked than benzene itself. The nitration of toluene proceeds readily until three nitro groups are introduced. The final product of the nitration is a well-known high explosive, 2,4,6-trinitrotoluene, abbreviated as TNT.

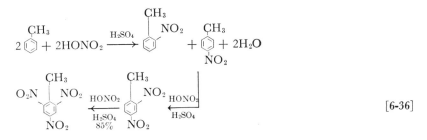

3. Sulfonation

$$2\,\underset{}{\overset{CH_3}{\bigcirc}} + 2SO_3 \xrightarrow{H_2SO_4} \underset{\underset{95\%}{SO_3H}}{\overset{CH_3}{\bigcirc}} + \underset{SO_3H}{\overset{CH_3}{\bigcirc}} \qquad [6\text{-}37]$$

4. Friedel-Crafts Reaction

$$2 \; C_6H_5CH_2\text{-}CH_3 + 2CH_3Br \xrightarrow{AlBr_3} \underset{CH_2\text{-}CH_3}{\overset{CH_3}{C_6H_4}} + \underset{}{\overset{CH_3}{\underset{}{C_6H_4}}} CH_2\text{-}CH_3 + 2HBr \quad [6\text{-}38]$$

At the present time half of all the detergents (defined on p. 277) sold are alkyl aryl sulfonates. Propylene is the starting compound in one process for making the alkyl side chain on the aromatic ring. It is polymerized to the tetramer stage (on the average), and then a Friedel-Crafts reaction on the tetramer with benzene yields an alkyl benzene. A specific compound, 2,4,6-trimethyl-1-nonene,

$$CH_3\text{-}CH=CH_2 \xrightarrow{\text{catalyst}} CH_3\text{-}CH_2\text{-}CH_2\text{-}\underset{CH_3}{\overset{CH_3}{CH}}\text{-}CH_2\text{-}\underset{CH_3}{\overset{CH_3}{CH}}\text{-}CH_2\text{-}\underset{CH_3}{\overset{CH_3}{C}}=CH_2 \xrightarrow{HF+C_6H_6}$$

a tetramer of propylene

$$CH_3\text{-}CH_2\text{-}CH_2\text{-}\underset{CH_3}{\overset{CH_3}{CH}}\text{-}CH_2\text{-}\underset{CH_3}{\overset{CH_3}{CH}}\text{-}CH_2\text{-}\underset{C_6H_5}{\overset{CH_3}{C}}\text{-}CH_3 \xrightarrow{H_2SO_4}$$

a dodecylbenzene

$$C_{12}H_{25}\text{-}C_6H_4\text{-}SO_3H \xrightarrow{NaOH} C_{12}H_{25}\text{-}C_6H_4\text{-}SO_3Na \qquad [6\text{-}39]$$

a dodecylbenzenesulfonic acid　　　a sodium dodecylbenzenesulfonate

is shown in Expression 6-39 as a representation of the isomers that are present in the tetramer. Sulfonation of the Friedel-Crafts product, dodecylbenzene, then yields a dodecylbenzenesulfonic acid, probably mostly a *para* isomer, as shown. Neutralization of this sulfonic acid will yield the sodium salt, a detergent.

Earlier the same type of compound was made with kerosene as the source of the alkyl side chain. Chlorination of kerosene gave a "keryl" chloride, undoubtedly a mixture of isomers, which may again be represented by a C_{12} chain. Toluene was generally the aromatic hydrocarbon used in the Friedel-Crafts reaction. Sulfonation is shown as taking place *ortho* to the methyl group on a

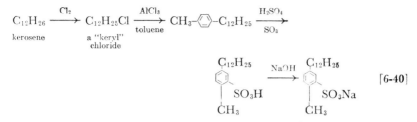

representative *para* isomer, though there is little reason to suppose that this would be more probable than orientation *ortho* to the C_{12} group.

SUMMARY

The peculiar saturation of the aromatic ring is accounted for by the phenomenon of resonance. The common reagents attacking the aromatic ring are electron-seekers, as $Br^{\oplus}$ from Br_2, $\overset{\oplus}{N}O_2$ from $HONO_2 + H_2SO_4$, $\overset{\oplus}{S}O_3H$ from fuming sulfuric acid, and $R^{\oplus}$ from $RCl + AlCl_3$. Homologues of benzene give electron-seeking substitutions in which the new group enters o and p to the R group.

The supply of aromatic hydrocarbons from coal tar is now supplemented by synthesis (high-temperature catalytic dehydrogenation and cyclization) from aliphatic hydrocarbons.

The halogen is held much more tightly in aryl halides than in alkyl halides. A halogen atom orients o and p for entering groups and retards substitution.

EXERCISES

(Exercises 1–3 will be found within the text.)

4. Name the following compounds:

1. Cl–⬡–CH_3
2. Cl–⬡–NO_2
3. Cl–⬡–SO_3H
   NO_2
4. O_2N–⬡–CH_2–Cl
5. CH_3–⬡–SO_3H
6. CH_3–⬡–$CH(CH_3)_2$

5. Write structures for the following compounds:

   1. *p*-bromoethylbenzene
   2. 1-bromo-2-phenylethane
   3. *o*-bromobenzenesulfonic acid
   4. sodium *p*-toluenesulfonate

6. Write the formulas for the organic products of the reaction between each of the four hydrocarbons

   1. hexane
   2. methylcyclohexane
   3. 1-hexene
   4. toluene

   and each of the following reagents:

   a. nitrating acid (at room temperature)
   b. dilute aqueous $KMnO_4$
   c. bromine at 60° (iron nail)
   d. concentrated or fuming H_2SO_4

7. Suggest a test tube reaction that will enable you to distinguish cyclohexene from benzene and 1-hexyne from ethylbenzene. Write an equation for each reaction suggested.

8. Contrast and compare the behavior of alkyl and aryl halides by writing equations for all reactions that occur when

   1. 1-bromo-2-ethylbenzene
   2. 1-bromo-2-phenylethane

   is each treated separately with

   a. warm aqueous NaOH
   b. magnesium (dry ether)
   c. sodium
   d. alcoholic KOH
   e. potassium cyanide

9. Starting with benzene and using any necessary aliphatic as well as inorganic substances, indicate how to prepare each of the following substances. Show all reagents, conditions, and intermediate compounds, including all isomers expected in appreciable amount.

   1. *p*-chlorobenzenesulfonic acid
   2. *o*-nitrotoluene
   3. *p*-bromobenzyl bromide

10. Showing all reagents, conditions, and intermediate organic products, indicate a means of accomplishing the following indicated transformations (none can be run in one step):

    1. bromobenzene $\longrightarrow$ benzene (a Grignard reagent is involved)
    2. 2-butene $\longrightarrow$ 2-butyne
    3. ethyl alcohol $\longrightarrow$ 1,1,2,2-tetrabromoethane
    4. n-propyl alcohol $\longrightarrow$ 2,3-dimethylbutane

11. What is meant by the term "resonance"? What are the principal structures contributing to *m*-xylene?

12. The addition of chlorine to benzene in the presence of sunlight to give hexachlorocyclohexane is often called an "abnormal" reaction. Explain.

13. How do aromatic hydrocarbons differ from each of the aliphatic hydrocarbon families studied so far?

14. Write the reaction you would expect to take place between cyclohexadiene and one molecule of bromine.

15. a. Write structural formulas for the nine isomers of formula $C_8H_9NO_2$ in which the nitro group is attached to an aromatic ring.
    b. Name each isomer.

7 Alcohols and Phenols

The student is now acquainted with some of the properties of molecules containing C-H and C-X bonds and also with some of the effects that multiple bonds have on these two groups in various situations. It is time to turn to an investigation of the carbon-oxygen bond. Carbon-oxygen bonds exhibit a large number of functions depending on the additional environment of each of the two atoms. The following bonds, for example, all have different properties:

$$-\overset{|}{\underset{|}{C}}-O-H \qquad -\overset{|}{\underset{|}{C}}-O-C \qquad \underset{\diagdown}{\overset{\diagup}{C}}{=}O \qquad -C\overset{\diagup O}{\underset{\diagdown O-H}{}} \qquad -C\overset{\diagup O}{\underset{\diagdown O-C}{}}$$

All of these and some other carbon-oxygen functions will be discussed in Chapters 7–14.

The $:\overset{.}{\underset{..}{O}}:H$ group, with one less electron than the hydroxyl ion $\left(:\overset{..}{\underset{..}{O}}:H^- \right)$, is called the hydroxy group. When a hydroxy group replaces a hydrogen on a carbon in an aliphatic compound, the resulting compound is an alcohol, $R:\overset{..}{\underset{..}{O}}:H$ or R–OH. When a hydroxy group replaces a hydrogen on an aromatic ring, the new compound is called a phenol, Ar–OH.

136

TABLE
7-1 | *Alcohols and Phenols*

FORMULA	COMMON NAME	IUC NAME	BOILING POINT
CH_3-OH	methyl alcohol	methanol	66
CH_3-CH_2-OH	ethyl alcohol	ethanol	78
$CH_3-CH_2-CH_2-OH$	n-propyl alcohol	1-propanol	87
$CH_3-CH-CH_3$ $\quad\vert$ $\quad OH$	isopropyl alcohol	2-propanol	83
$CH_3-CH_2-CH_2-CH_2-OH$	n-butyl alcohol	1-butanol	116
$CH_3-CH_2-CH-CH_3$ $\qquad\vert$ $\qquad OH$	sec-butyl alcohol	2-butanol	99
CH_3 $\quad\diagdown$ $\qquad CH-CH_2OH$ $\quad\diagup$ CH_3	isobutyl alcohol	2-methyl- 1-propanol	108
CH_3 $\quad\diagdown$ $\qquad C-OH$ $\quad\diagup\ \vert$ $CH_3\ \ CH_3$	tert-butyl alcohol	2-methyl- 2-propanol	83
CH_2-CH_2 $CH_2\diagup\qquad\diagdown CHOH$ $\quad CH_2-CH_2$	cyclohexyl alcohol	cyclohexanol	160
⬡-OH	phenol	phenol	180
⬡-CH_2OH	benzyl alcohol	phenylmethanol	205
$\diagup CH_3$ ⬡-OH	o-cresol	2-methylphenol	190
⬡-OH $CH_3\diagup$	m-cresol	3-methylphenol	202
CH_3-⬡-OH	p-cresol	4-methylphenol	202

The common name of an alcohol is derived from the common name of the corresponding saturated hydrocarbon plus "alcohol." Secondary butyl alcohol and tertiary butyl alcohol are so designated from the condition of the carbon carrying the alcohol group. (Compare, on p. 66, the way the alkyl halides are named.)

The IUC ending for an alcohol is *-ol*. A number designating the position of the hydroxy group precedes the name of the hydrocarbon stem. The cresols are named as substituted phenols in the IUC system, with a number showing the position of the methyl group.

Exercise 7-1. Write structures for all secondary alcohols of formula $C_5H_{11}OH$.

7-1. Physical Properties of the Alcohols

The alcohols are the first organic compounds to be considered that are appreciably soluble in water. All alcohols containing one, two, or three carbons and tert-butyl alcohol are completely miscible with water—that is, are soluble in all proportions; but alcohols with four carbons (with one exception) are only partially miscible—that is, are soluble to a limited extent. Beyond four carbons, the alcohols are still less soluble in water. At room temperature all of the alcohols mentioned are colorless liquids with a specific gravity of about 0.8, whereas the four-carbon alkanes, alkenes, alkynes, and cycloalkanes boil at temperatures below room temperature (except 2-butyne, whose boiling point is 29°C). The presence of the hydroxy group in both water and alcohols is reflected in the similarity of the physical properties of these substances. For an explanation of the high boiling point of alcohols, see p. 617ff.

7-2. Preparation of Alcohols and Phenols: General Methods

1. Hydration of Alkenes (for Alcohols Only). Alkenes of low molecular weight dissolve readily in sulfuric acid to form alkyl hydrogen sulfates. These addition compounds are readily hydrolyzed to alcohols by dilution with water and boiling. If the alkene is unsymmetrical, the product predicted by Markownikoff's rule predominates.

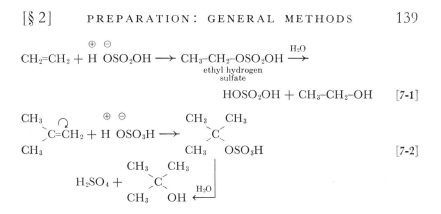

$$CH_2=CH_2 + \overset{\oplus}{H} \ \overset{\ominus}{OSO_2OH} \longrightarrow CH_3-CH_2-OSO_2OH \overset{H_2O}{\longrightarrow}$$

ethyl hydrogen
sulfate

$$HOSO_2OH + CH_3-CH_2-OH \qquad [7\text{-}1]$$

$$\begin{array}{c} CH_3 \\ \diagdown \\ \diagup \\ CH_3 \end{array} C=CH_2 + \overset{\oplus}{H} \ \overset{\ominus}{OSO_3H} \longrightarrow \begin{array}{cc} CH_3 & CH_3 \\ \diagdown \ \diagup \\ C \\ \diagup \ \diagdown \\ CH_3 & OSO_3H \end{array} \qquad [7\text{-}2]$$

$$H_2SO_4 + \begin{array}{cc} CH_3 & CH_3 \\ \diagdown \ \diagup \\ C \\ \diagup \ \diagdown \\ CH_3 & OH \end{array} \overset{H_2O}{\longleftarrow}$$

2. Hydrolysis of Alkyl and Aryl Halides. Alkyl halides are converted to alcohols by alkaline hydrolysis. The alkali may react directly with the alkyl halide, or it may reduce the concentration

$$R\text{-}X + HOH \overset{OH^-}{\longrightarrow} ROH + HX \qquad [7\text{-}3]$$

of H^+ (Exp. 7-3), either process driving the reaction to the right. A mixture of five-carbon alcohols called Pentasol is made commercially by this process from a mixture of pentanes. The reaction is carried out by one company on such a scale that 100,000 gallons of hydrocarbons pass through the reaction zone in a day. The pentanes are first chlorinated, and the resulting mixture of chloro-alkanes is hydrolyzed to the isomeric alcohols.

$$C_5H_{12} + Cl_2 \longrightarrow C_5H_{11}Cl \overset{NaOH}{\longrightarrow} C_5H_{11}OH \text{ (isomeric mixture)} \qquad [7\text{-}4]$$

Though the hydrolysis of an alkene is not applicable to the preparation of a phenol, aryl halides can be hydrolyzed to phenols at elevated temperature. Phenol itself is prepared commercially by this reaction (Exp. 6-25).

3. Alkali Fusion of Sulfonic Acids (for Phenols Only). A second general method of preparing phenols (which cannot be applied to alcohols) is the reaction of fused sodium hydroxide with a sodium aryl sulfonate. The contrast between this dry reaction at elevated temperatures and the wet reactions of hydrolysis used in the making of alcohols is noteworthy. We prepare the salts of sulfonic acids by neutralizing the sulfonation products of the aromatic hydrocarbons with a base:

$$\underset{\substack{\text{a sodium} \\ \text{aryl sulfonate}}}{\text{Ar-SO}_3\text{Na}} + \text{NaOH} \xrightarrow[\text{200–275°}]{\text{fuse}} \text{ArOH} + \text{Na}_2\text{SO}_3 \qquad [7\text{-}5]$$

Exercise 7-2.   Write a reaction for the sulfonation of toluene and reactions for the preparation of the sodium salts of the two main products of this sulfonation.

7-3. Preparation of Alcohols and Phenols: Industrial Methods

1. Methyl Alcohol.   Since 1924, methyl alcohol has been made synthetically from carbon monoxide and hydrogen. The starting substances are obtained from the water-gas process, in which steam

$$\underset{\substack{\text{coke} \quad \text{steam}}}{\text{C} \ + \text{HOH}} \xrightarrow{1,000°} \underset{\substack{\text{water} \\ \text{gas}}}{\text{CO} + \text{H}_2} \qquad [7\text{-}6]$$

$$\text{CO} + 2\text{H}_2 \xrightarrow[\substack{\text{200 atm} \\ \text{ZnO–ZnCrO}_4}]{300°} \text{CH}_3\text{OH} \qquad [7\text{-}7]$$

is passed over coke heated to about 1,000°C. The water gas resulting from this reaction is then subjected to a pressure of 200–250 atmospheres at a temperature of 200–400° in the presence of a catalyzing mixture of zinc oxide and zinc chromate. The product consists mainly of methyl alcohol. When the ratio of hydrogen to carbon monoxide is 6, the conversion to methanol is 98%; when the ratio is 2, the conversion is 83%. With a different catalyst and a higher temperature a mixture of alcohols containing 1–8 carbons per molecule results. Note that extra hydrogen must be added to the water-gas mixture to convert all the carbon monoxide to the alcohol.

Methanol is used as a denaturant for ethyl alcohol and as a solvent. One of its most important uses is as an antifreeze in liquid-cooled engines.

2. Ethyl Alcohol. Ethyl alcohol is made by the hydrolysis of ethylene, which is obtained in large quantities from the cracking process. (The reaction was mentioned as one of the general methods of synthesizing alcohols.) The fermentation of sugars obtained from molasses and starch is another important method of preparing ethanol. (Rice, corn, and potatoes are sources of starch.)

Fermentations involve the use of enzymes obtained from living matter, either animal or plant. If starch from any of the sources just mentioned is warmed to 63° for one hour in a dilute solution containing malt (germinated barley), the enzyme called diastase, from the malt, is able to bring about the hydrolysis of starch to maltose. The maltose is further hydrolyzed enzymatically, by maltase, to glucose. Sucrose from molasses may be hydrolyzed

$$(C_6H_{10}O_5)_n + \frac{n}{2} H_2O \xrightarrow{\text{diastase}} \frac{n}{2} C_{12}H_{22}O_{11} \qquad [7\text{-}8]$$
$$\text{starch} \qquad\qquad\qquad\qquad\qquad \text{maltose}$$

$$C_{12}H_{22}O_{11} + H_2O \xrightarrow{\text{maltase}} 2C_6H_{12}O_6 \qquad [7\text{-}9]$$
$$\text{maltose} \qquad\qquad\qquad \text{glucose}$$

without enzymes in acid solution to two sugars, glucose and fructose, or it may be treated directly with a yeast. In the second case the first reaction is still the hydrolysis to two six-carbon sugars.

$$C_{12}H_{22}O_{11} + H_2O \xrightarrow[\text{or yeast}]{\text{acid}} C_6H_{12}O_6 + C_6H_{12}O_6 \qquad [7\text{-}10]$$
$$\text{sucrose} \qquad\qquad\qquad \text{glucose} \qquad \text{fructose}$$

By enzymatic degradation glucose or the mixture of glucose and fructose is converted at room temperature to ethyl alcohol, with the evolution of carbon dioxide. The product is a dilute solution

$$C_6H_{12}O_6 \xrightarrow[20\text{-}25°]{\text{zymase}} 2CH_3\text{--}CH_2\text{--}OH + 2CO_2 \qquad [7\text{-}11]$$

of ethyl alcohol (about 10%) which can be concentrated to about 95% alcohol by fractional distillation.

The last 5% of water can be removed and "absolute" alcohol (100%) obtained by treatment first with CaO and then with calcium metal to remove the last traces of water:

$$CaO + H_2O \longrightarrow Ca(OH)_2 \tag{7-12}$$

$$Ca + 2H_2O \longrightarrow Ca(OH)_2 + H_2 \tag{7-13}$$

The water is removed commercially by a different method.

Among pure organic compounds only cane sugar and ethylene are produced in larger quantities than ethyl alcohol. Although the largest use of ethyl alcohol is in beverages, it is second only to water in importance as a solvent. Alcoholic solutions of various substances such as green soap and iodine are called tinctures. The alcohol contributes to the antiseptic property of these solutions. Ethyl alcohol is used as an antifreeze and has been used as a motor fuel where gasoline is not plentiful. It is also a raw material for certain synthetic rubber processes.

3. Propyl Alcohols. Variation of the conditions used in synthesizing methyl alcohol from water gas will yield a larger percentage of the higher alcohols, of which one is n-propyl alcohol:

$$3CO + 6H_2 \xrightarrow[\substack{200\,\text{atm} \\ \text{catalyst}}]{400°} CH_3-CH_2-CH_2OH + 2H_2O \tag{7-14}$$

Hydrolysis of the addition product of propylene and sulfuric acid yields isopropyl alcohol:

$$CH_3-CH\overset{\delta+}{=}\overset{\delta-}{CH_2} + \overset{\delta+}{H}\ \overset{\delta-}{OSO_3H} \longrightarrow \underset{OSO_3H}{CH_3-CH-CH_3} \xrightarrow{HOH} \underset{OH}{CH_3-CH-CH_3} + H_2SO_4 \tag{7-15}$$

4. Butyl Alcohols. Another fermentation process (Weizmann process), using a different type of enzyme from that used in making ethyl alcohol, was developed before the First World War for the production of acetone. Corn starch was the initial starting material, and the bacterium was named *Clostridium acetobutylicum.* n-Butyl alcohol was the main product of the fermentation; but, since no uses involving its consumption in large quantities were known at the time, it was considered a by-product with respect to acetone, which is a solvent for smokeless powder. After the war, n-butyl alcohol became the main product when the discovery was made that it could be used in preparing an excellent lacquer solvent, which was a boon to the automobile industry of the 1920's.

$$\text{corn starch} \xrightarrow[\textit{acetobutylicum}]{\textit{Clostridium}} \begin{cases} CH_3-CH_2-CH_2-CH_2OH \\ CH_3-CH_2-OH \\ \text{acetone} \\ CO_2 \text{ and } H_2 \end{cases} \qquad [7\text{-}16]$$

This is one of a number of chemical processes in which economic change has altered the relative importance of various products of a chemical reaction. The principal liquid products, n-butyl alcohol, ethyl alcohol, and acetone (Exp. 7-16), are obtained in the approximate ratio 6:1:3. Two gases, carbon dioxide and hydrogen, are evolved in appreciable quantities.

The pressure process previously described for the synthesis of methanol is also the commercial means of preparing isobutyl alcohol, and the hydrolysis of alkenes available from the cracking process is used for the commercial preparation of both sec- and tert-butyl alcohol.

Exercise 7-3.   What alkenes could be used to make secondary butyl alcohol by this method? Write the equations. Write an equation for the preparation of tertiary butyl alcohol by this method.

5. Phenol.   Phenol and the cresols are available from the refining of coal tar. Phenol is also made by the hydrolysis of chlorobenzene (p. 128). Phenol is used in large quantities for making plastics such as Bakelite.

6. Cyclohexanol. Catalytic hydrogenation of phenol in the presence of nickel at 160° under pressure gives a good yield of cyclohexanol, the only cyclic alcohol synthesized in appreciable quantities industrially.

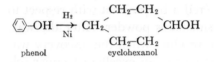

$$\text{phenol} \qquad \text{cyclohexanol} \qquad\qquad [7\text{-}17]$$

7-4. Reactions of Alcohols and Phenols

Many of the reactions of alcohols—for example, with Na, PCl_3, and $CaCl_2$—are, as might be expected, analogous to those of water.

1. With Sodium. Sodium displaces hydrogen from water, an alcohol, or a phenol, with the formation of a base. The bases formed in the three reactions are sodium hydroxide, a sodium

$$\text{H--OH} + \text{Na} \longrightarrow \text{NaOH} + \tfrac{1}{2}\text{H}_2 \qquad\qquad [7\text{-}18]$$

$$\text{R--OH} + \text{Na} \longrightarrow \text{NaOR} + \tfrac{1}{2}\text{H}_2 \qquad\qquad [7\text{-}19]$$

$$\text{Ar--OH} + \text{Na} \longrightarrow \text{NaOAr} + \tfrac{1}{2}\text{H}_2 \qquad\qquad [7\text{-}20]$$

$$\underset{\text{OH}}{\text{CH}_3\text{--CH--CH}_3} + \text{Na} \longrightarrow \underset{\substack{\text{ONa}\\\text{sodium}\\\text{isopropoxide}}}{\text{CH}_3\text{--CH--CH}_3} + \tfrac{1}{2}\text{H}_2 \qquad [7\text{-}21]$$

alkoxide, and a sodium phenoxide, respectively. The alkoxides are designated by the name of the metal and the stem of the appropriate alkyl group, followed by *-oxide*.

The aromatic ring has a strong influence on the –OH group. Alcohols are such weak acids that they will not react with sodium hydroxide, but phenols are strong enough acids to dissolve readily in bases. (Phenol itself is often called carbolic acid.) This difference

$$\text{CH}_3\text{--}\bigcirc\text{--OH} + \text{NaOH} \longrightarrow \text{CH}_3\text{--}\bigcirc\text{--ONa} + \text{H}_2\text{O} \qquad [7\text{-}22]$$

in behavior will be considered in more detail in Chapter 9.

2. With Phosphorus Halides. The phosphorus halides react analogously with water and the alcohols to form the hydrogen or alkyl halides, but phenols react to such a small extent that the reactions are not worth while as a method of preparing the aryl halides. The reactions and the mechanism were described on pages 68–69 as a method of preparing alkyl halides. The alkyl iodides and bromides are formed in good yields by this method. With PCl_3

$$\begin{array}{c} CH_3 \\ \diagdown \\ CH-CH_2OH + PCl_5 \xrightarrow[75\%]{} \\ \diagup \\ CH_3 \end{array} \begin{array}{c} CH_3 \\ \diagdown \\ CH-CH_2-Cl + POCl_3 + HCl \\ \diagup \\ CH_3 \end{array} \qquad [7\text{-}23]$$

$$3ROH + PCl_3 \longrightarrow P(OR)_3 + 3HCl \qquad [7\text{-}24]$$

the reaction frequently stops at the stage where HCl splits out instead of RCl. An alkyl phosphite, $P(OR)_3$, is then formed, diminishing the yield of RCl (Exp. 7-24). (See also p. 571.)

3. With $CaCl_2$ and Other Salts. Many inorganic salts crystallize from water in the solid state combined with water molecules. Some salts hold this water of crystallization so tightly that heating decomposes them rather than merely driving off the water. The dihydrate of magnesium chloride is such a compound. Others, such as $CuSO_4 \cdot 5H_2O$, lose water readily on heating and give an anhydrous salt. If the anhydrous copper II sulfate is left in moist

$$MgCl_2 \cdot 6H_2O \longrightarrow MgCl_2 \cdot 2H_2O + 4H_2O \qquad [7\text{-}25]$$

$$MgCl_2 \cdot 2H_2O \longrightarrow Mg(OH)_2 + 2HCl \qquad [7\text{-}26]$$

$$\underset{\text{blue}}{CuSO_4 \cdot 5H_2O} \longrightarrow \underset{\text{white}}{CuSO_4} + 5H_2O \qquad [7\text{-}27]$$

air, it slowly takes on water, changing to a blue color again. Still other salts, such as $CaCl_2$, take up water from moist air to such an extent and so rapidly that they can be used as drying agents:

$$CaCl_2 + 6H_2O \longrightarrow CaCl_2 \cdot 6H_2O \qquad [7\text{-}28]$$

Calcium chloride, as well as some other compounds, behaves in the same way toward alcohols:

$$CaCl_2 + 4CH_3OH \longrightarrow CaCl_2 \cdot 4CH_3OH \qquad\qquad [7\text{-}29]$$

The size of the alcohol (in part) determines the number of molecules that will coordinate per molecule of salt. Such compounds with water are called hydrates, and those with alcohols are called alcoholates. As the length of the carbon chain in the alcohol increases, its resemblance to water decreases, and the property of forming alcoholates is lost.

Some other alcoholates are $MgCl_2 \cdot 6CH_3OH$, $CuSO_4 \cdot 2CH_3OH$, and $BaO \cdot 2CH_3OH$.

4. With HX. Reaction of the hydrohalogen acids with alcohols has been mentioned (p. 67) as a means of preparing alkyl halides. A comparison of the rates of reaction of various alcohols with a given HX shows that the order of activity is $3° R > 2° R > 1° R$. The yields, however, are generally in reverse order because the tertiary alcohols (especially) are subject to dehydration as a side reaction.

$$ROH + HX \longrightarrow RX + H_2O \qquad\qquad [7\text{-}30]$$

Water is one of the products of this reaction. The addition of some reagent that will hasten the release of water speeds up the reaction and is generally needed with HCl and HBr, but not with HI.

$$R\text{-}CH_2OH + HCl \xrightarrow{\text{ZnCl}_2} RCH_2Cl + H_2O \qquad\qquad [7\text{-}31]$$

$$R\text{-}CH_2OH + HBr \xrightarrow{\text{H}_2\text{SO}_4} RCH_2Br + H_2O \qquad\qquad [7\text{-}32]$$

$$R\text{-}CH_2OH + HI \longrightarrow RCH_2I + H_2O \qquad\qquad [7\text{-}33]$$

These reagents are very strong acids, which release protons to coordinate with oxygen in the alcohol. As the water is lost, the

$$R\text{-}CH_2OH \xrightarrow{H^+} R\text{-}CH_2\overset{+}{\underset{H}{O}}H \xrightarrow{Cl^-} RCH_2Cl + H_2O \qquad\qquad [7\text{-}34]$$

halide becomes attached to the alkyl group. Sulfuric acid is commonly used with HBr. The reactivity of HCl is generally enhanced by dissolving anhydrous zinc chloride in the concentrated acid. These strong acids, which are also dehydrating agents, are generally necessary with primary and secondary alcohols; tertiary alcohols give better yields without them. (Tertiary alcohols are too easily dehydrated.)

5. With HOSO$_2$OH. Alcohols dissolve in sulfuric acid, and most of them react on warming (and some without warming) to give alkyl hydrogen sulfates. Further heating may then split off a molecule of sulfuric acid to give an alkene. With long-chain alcohols (12–18 carbons) the alkyl sulfates are useful in making detergents ("soapless soaps"). Sodium lauryl sulfate, for example, is sold as Dreft or Drene.

$$ROH + HOSO_2OH \longrightarrow \left[R:\overset{..}{\underset{H}{\overset{..}{O}}}:H:\overset{..}{\underset{..}{O}}:SO_3H \right] \longrightarrow R\text{-}OSO_2OH + H_2O \qquad [7\text{-}35]$$

$$\underset{\substack{\text{lauryl} \\ \text{alcohol}}}{C_{12}H_{25}OH} + HOSO_3H \longrightarrow \underset{\substack{\text{lauryl hydro-} \\ \text{gen sulfate}}}{C_{12}H_{25}OSO_3H} \xrightarrow[\text{with NaOH}]{\text{neutralize}}$$

$$H_2O + \underset{\substack{\text{sodium} \\ \text{lauryl sulfate}}}{C_{12}H_{25}OSO_3Na} \qquad [7\text{-}36]$$

Dehydration of alcohols by catalysts and by chemical reagents such as H$_2$SO$_4$ and P$_2$O$_5$ was mentioned on page 81. Under certain

$$CH_2 \underset{CH_2\text{-}CH_2}{\overset{CH_2\text{-}CH_2}{\diagdown}} CHOH \xrightarrow[80\%]{H_2SO_4} CH_2 \underset{CH_2\text{-}CH_2}{\overset{CH_2\text{-}CH}{\diagdown}} CH + H_2O \qquad [7\text{-}37]$$

conditions two molecules of alcohol lose a molecule of water to form an ether, a reaction that will be discussed in Chapter 8.

6. Oxidation. The oxidation products of alcohols depend on the degree of the alcohol. (Oxidation will be discussed in Chap. 12.) Primary alcohols yield aldehydes and then, with more complete oxidation, organic acids. For present purposes the oxidation may be thought of as an interposition of oxygen between a carbon atom and a hydrogen atom. Secondary alcohols yield ketones upon

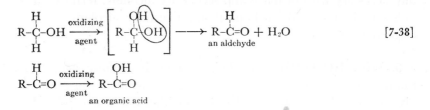

$$\text{R--C=O} \xrightarrow[\text{agent}]{\text{oxidizing}} \text{R--C=O}$$
an organic acid

oxidation. Tertiary alcohols do not react unless the oxidation is vigorous enough to break a carbon-carbon bond.

$$\text{R--C-OH} \xrightarrow[\text{agent}]{\text{oxidizing}} \left[\text{R--C-OH} \right] \longrightarrow \text{R--C-R} \qquad [7\text{-}39]$$
a ketone

$$\text{R--C-OH} \xrightarrow[\text{agent}]{\text{oxidizing}} \text{no reaction} \qquad [7\text{-}40]$$

7. FeCl₃ Test for Phenols. Phenols and alcohols that are insoluble in water may be distinguished by their solubility in sodium hydroxide. Alcohols are neutral, but phenols are weak acids and will dissolve in the base.

A test used to distinguish phenols that are at least slightly soluble in water from alcohols and other kinds of compounds is based on the use of an iron III chloride solution. A deep red color is obtained with phenols through the formation of a complex ion, the final stage of which is shown in Expression 7-41. Alcohols give no color with iron III chloride solution.

$$\text{Fe}^{+3} + 6 \bigcirc\text{-OH} \longrightarrow [\text{Fe}(-\text{O-}\bigcirc)_6]^{-3} + 6\text{H}^+ \qquad [7\text{-}41]$$
deep red solution

8. With Br₂ (Phenols Only). The hydroxy group on an aromatic ring is a very strong *ortho-* and *para*-directing group. According to the resonance theory (pages 120 and 375ff.) the reason for this orientation will be clear if the canonical structures in Expression 7-42-c,d,e make an appreciable contribution to the total picture of the phenol molecule. If they do, the *ortho* and *para* positions have the highest electron densities of the six carbons in the

$$[7\text{-}42]$$

| a | b | c | d | e |

ring. Consequently, any reagent seeking electrons will find them at the *o* and *p* positions. The common reagents that react with phenols are electron-seeking agents—for example, $Br^{\oplus}$ from bromine and $NO_2^{\oplus}$ from nitric-sulfuric acid mixture.

The hydroxy group is such a strong activating group that halogens substitute readily in all three available positions.

Exercise 7-4. Write out electronic pictures to show how the monobromo-, the dibromo-, and the tribromophenol are formed.

If an *ortho* or *para* position is occupied in the original phenol, of course only the other *ortho* or *para* position is free for substitution. When the hydroxy and alkyl groups compete for orienta-

$$[7\text{-}43]$$

tion of entering groups, the hydroxy group determines the entering position.

9. Nitration of Phenols. Some phenols can be nitrated at controlled temperatures, although oxidation prevents clean-cut substitutions and tars are frequently obtained. At 15°C, phenol itself can be nitrated to give a mixture of *o*- and *p*-nitrophenol in good yield. Only the *ortho* isomer is shown in Expression 7-44.

$$\overset{\delta+}{O}H \quad\quad\quad\quad \overset{OH}{\underset{}{\big|}} NO_2$$

$$\underset{}{\bigcirc} + \overset{\delta-}{HO} \overset{\delta-}{} \overset{\delta+}{NO_2} \xrightarrow[15°]{H_2SO_4} p\text{-} + \bigcirc + H_2O \quad\quad [7\text{-}44]$$

$$\underset{25\%}{} \quad \underset{36\%}{}$$

Exercise 7-5. Write electronic pictures to show the mechanism of the formation of *p*-nitrophenol in the nitration above.

Exercise 7-6. Write reactions showing the products and mechanism of the bromination of *m*-nitrophenol.

7-5. Glycols

A carbon atom carrying two hydroxy groups results, in general, in an unstable configuration; but, if the hydroxy groups are on different carbons, a stable substance obtains. Such compounds are called glycols, and the first is called simply glycol or, more specifically, ethylene glycol. The other important industrial glycols are named in Table 7-2. Aromatic rings with two or more OH groups are discussed in Chapter 28.

TABLE

7-2 | *Glycols*

FORMULA	COMMON NAME	IUC NAME	BOILING POINT
CH_2-CH_2 $\vert \quad \vert$ $OH \ OH$	ethylene glycol	1,2-ethanediol	197
$CH_2-CH_2-CH_2$ $\vert \quad\quad\quad \vert$ $OH \quad\quad OH$	trimethylene glycol	1,3-propanediol	216
$CH_3-CH-CH_2$ $\quad\ \vert \quad \vert$ $\quad OH \ OH$	propylene glycol	1,2-propanediol	188

1. Nomenclature. The common names for 1,2-glycols consist of the names of the parent alkenes and, separately, the word *glycol*.

Unfortunately, the names have not been dropped from general usage in spite of the contradiction inherent in the method. The name "ethylene glycol," for example, adds *glycol* (two OH groups) to *ethylene* ($CH_2=CH_2$) in spite of the fact that the substance is not unsaturated. The name "trimethylene" is used because three (*tri*) methylene ($-CH_2-$) groups are linked together in a chain.

The IUC method of naming two groups is to insert the Latin *di* in front of the ending for the group. The IUC system is consistent in treating the glycols as saturated substances.

2. Properties. Having a second hydroxyl group in the molecule raises the boiling point of a glycol considerably above that of the corresponding alcohol and also increases the viscosity. Glycols absorb water, to a limited extent, from the atmosphere and are therefore used as humectants for tobacco. High boiling points and miscibility with water make the glycols useful as antifreeze solutions. Ethylene glycol is used for this purpose in both automobile and aircraft cooling systems.

The chemical properties of the glycols are what would be expected from two alcohol groups, each of which is without effect on the properties of the other.

$$\begin{matrix} CH_2OH \\ | \\ CH_2OH \end{matrix} + 2Na \longrightarrow \begin{matrix} CH_2-ONa \\ | \\ CH_2-ONa \end{matrix} + H_2 \qquad\qquad [7\text{-}45]$$

$$\begin{matrix} CH_2OH \\ | \\ CH_2OH \end{matrix} + 2HBr \xrightarrow[Na_2CO_3]{H_2SO_4} \begin{matrix} CH_2-Br \\ | \\ CH_2-Br \end{matrix} + 2H_2O \qquad [7\text{-}46]$$
$$\xleftarrow{}$$
$$53\%$$

3. Preparation. Ethylene glycol is made from ethylene in two ways: (a) by the addition of hypochlorous acid and hydrolysis (Exp. 7-47); (b) by catalytic oxidation to ethylene oxide and hydrolysis (Exp. 7-48).

Soda lime, which is a mixture of sodium hydroxide and calcium oxide, removes a molecule of HCl from the chlorohydrin to give a ring compound containing oxygen. The ring is easily broken to produce the glycol.

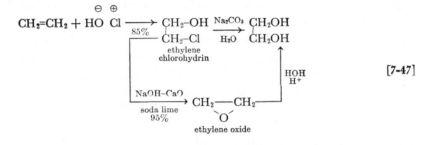

A silver catalyst may be used for the direct addition of oxygen to ethylene. The ethylene oxide resulting is easily hydrolyzed to

$$CH_2{=}CH_2 + O_2 \xrightarrow[\substack{1 \text{ atm, } 235° \\ 60\%}]{\text{Ag catalyst}} CH_2{-}CH_2 \xrightarrow[H^+]{H_2O} CH_2{-}CH_2 \qquad [7\text{-}48]$$
$$\phantom{CH_2{=}CH_2 + O_2 xxxxxxxxxxxx} O OH \; OH$$

the glycol in acid solution. The second method is now replacing the first because it is much cheaper.

7-6. Glycerol

One important trihydroxy alcohol is glycerol, $CH_2OH{-}CHOH{-}CH_2OH$. Glycerol is a viscous oil with boiling point 290°C; it has a sweet taste and is very hygroscopic.

1. Preparation.   Glycerol is a by-product in the manufacture of soap from fats:

fat + NaOH $\longrightarrow$ a soap + glycerol   [7-49]

It is now manufactured from the propylene obtained from the petroleum-cracking process. High-temperature chlorination of propylene produces a good yield of allyl chloride, which readily hydrolyzes to allyl alcohol. Addition of chlorine and water (hypochlorous acid) to the alkenic alcohol gives a mixture of mono- and dichlorohydrins, all of which can be hydrolyzed to glycerol.

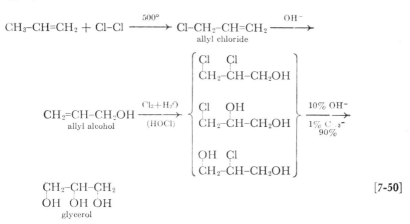

$$CH_3-CH=CH_2 + Cl-Cl \xrightarrow{500°} \underset{\text{allyl chloride}}{Cl-CH_2-CH=CH_2} \xrightarrow{OH^-}$$

[7-50]

$$\underset{\text{glycerol}}{\underset{OH\ \ OH\ \ OH}{CH_2-CH-CH_2}}$$

Exercise 7-7. Which of the intermediate products in Expression 7-50 is a result of an addition contrary to Markownikoff's rule?

2. Uses. Glycerol is used as an antifreeze and as a humectant for tobacco, but its principal use is in manufacturing dynamite.

If glycerol is added to a mixture of nitric and sulfuric acids below room temperature, it forms a highly explosive substance called

$$\begin{array}{l} CH_2OH \\ CHOH \\ CH_2OH \end{array} + 3H \begin{array}{c} + \ - \\ O-N \end{array} \begin{array}{c} \nearrow O \\ \searrow O \end{array} \xrightarrow[\substack{15° \\ 93\%}]{H_2SO_4} \underset{\substack{\text{glyceryl trinitrate} \\ \text{(nitroglycerine)}}}{\begin{array}{l} CH_2-O-NO_2 \\ CH-O-NO_2 \\ CH_2-O-NO_2 \end{array}} + 3H_2O$$

[7-51]

nitroglycerine. Nitroglycerine is very unstable to shock and detonates with violence. It is a soupy liquid at room temperature and melts at 13°C. Nobel (1867) found that the hazards of handling nitroglycerine could be greatly reduced by allowing it to soak into a solid such as sawdust or kieselguhr. In this form the product is called dynamite and is a safe and useful explosive.

Methanol reacts with nitric acid under controlled conditions to give methyl nitrate, an explosive of greater energy than nitroglycerine. When prepared with care, it is described as safe to handle. It is a liquid, with boiling point 63°, and has been manufactured in Germany since 1943 under the name Myrol.

Exercise 7-8.   Write an equation for the preparation of methyl nitrate.

Exercise 7-9.   Compare the electronic structures of methyl nitrate and nitromethane.

SUMMARY

Alcohols of five or fewer carbons and cyclohexanol are of industrial importance. The primary alcohols come from catalytic combination of carbon monoxide and hydrogen at elevated temperature and pressure; secondary and tertiary alcohols are formed in the acidic hydrolysis of alkenes. The exceptions are the pentanols, obtained from hydrolysis of chlorinated hydrocarbons; 1-butanol, obtained by the Weizmann fermentation; and ethanol, obtained by fermentation, hydrolysis of an alkene, and the high-pressure method. Phenols occur in coal tar, but the supply is enhanced by synthetic methods.

Phenols are easily distinguished from alcohols since they are weak acids, give deep colors with Fe^{+3}, and react readily with bromine water. Alcohols are neutral, but hydrogen can be displaced by Na, acyl halides, and acid anhydrides (Chap. 14); in other reactions (with PCl_3, HX, etc.) the OH group is displaced. The OH group in phenols is very difficult to displace, and the characteristic reactions of phenols are due either to weak acid properties or to ease of substitution on the aromatic ring.

Glycols and glycerols exhibit the chemical properties of ordinary alcohols but the extra OH groups make important changes in physical properties—for example, increased water solubility and much higher boiling points.

The uses of alcohols in synthesis will be apparent in nearly every chapter in this text. Methanol is a solvent for shellac and varnishes and a denaturant for ethanol. Ethanol is an antifreeze for cooling systems, a solvent in tinctures, a source of heat (canned heat) and power (internal-combustion engines), an antiseptic and germicide. Ethanol is perhaps the least toxic of the alcohols but in large doses has a narcotic effect followed by paralysis. Methanol may cause blindness in small quantities. Long-chain alcohols are used to make sulfate detergents (Drene, Dreft).

REFERENCES

1. Corson, "Industrial Catalysis—Methanol," J. Chem. Education, 24, 153 (1947).
2. Graw, Lamprey, and Sommer, "Properties vs. Performance of Present Day Antifreeze Solutions," J. Chem. Education, 18, 488 (1941).
3. Haynes, "Wood Chemicals Meet Synthetic Competition," J. Chem. Education, 24, 109 (1947).
4. Kastens, "Synthetic Methanol Production," Ind. Eng. Chem., 40, 2230 (1948).

EXERCISES

(Exercises 1–9 will be found within the text.)

10. Write structural formulas and IUC names for all eight alcohols of formula $C_5H_{11}OH$, and indicate whether each is primary, secondary, or tertiary.

11. Using R to represent any alkyl group, write a general formula that can be only a secondary alcohol; one that can be only a primary alcohol.

12. What reagents react with alcohols in a manner similar to their reactions with water? What reagents that do not react with alcohols will react with phenols?

13. Compare and contrast alcohols and phenols by writing equations for any reactions that occur when

1. 1-hexanol
2. cyclohexanol
3. phenol

are separately treated with

a. aqueous NaOH

   b. concentrated HBr + H_2SO_4
   c. hydrogen + Ni
   d. hot concentrated H_2SO_4

14. Distinguish between hexane and 1-hexanol by a test-tube reaction.

15. Which alcohols of four carbons can be dehydrated to a single alkene, and which should give mixtures of alkenes?

16. Write a structural formula for a five-carbon alcohol that could not be dehydrated.

17. Of the alcohols of fewer than five carbons, which are made commercially by a fermentation process? by hydrolysis of an alkene? by a hydrogenation process under high pressure?

18. Showing all reagents, conditions, and intermediate organic compounds, indicate how to accomplish the following conversions:

   1. n-propyl bromide to isopropyl alcohol
   2. benzenesulfonic acid to cyclohexanol
   3. 1-butanol to n-butyl cyanide
   4. 1-propanol to 2-propanol
   5. toluene to benzyl alcohol

19. Two different compounds, A and B, react with HBr to give the same product, C, of formula C_4H_9Br. Write possible structures for compounds A, B, and C and equations for the two reactions.

20. Hydrogen bromide was prepared in solution by treating sodium bromide with sulfuric acid. How much sodium bromide must be used to prepare 47 g of 1,2-dibromoethane from the appropriate glycol?

21. A phenol of formula C_7H_8O reacts with excess bromine to yield a compound of formula $C_7H_6OBr_2$. Write a possible structure for the original phenol and an equation for its reaction with bromine.

22. A certain glycol gave a dibromide when it was allowed to react with an excess of $HBr-H_2SO_4$ mixture, but the dibromide did not give an alkene when treated with zinc dust. Write a possible structure for the glycol.

23. Why is "nitroglycerine" a misnomer for glyceryl trinitrate?

24. If methanol and 1-propanol were the only alcohols produced in a particular application of the pressure method of synthesis from CO and H_2, would the formation of one rather than the other be favored by increasing the pressure (Le Châtelier's principle)?

8 Ethers

Replacing one of the hydrogens of water by an alkyl group produces an alcohol, which has many of the properties of water. If both hydrogens are replaced by alkyl groups, the resulting substance is called an ether, and the resemblance to water in both physical and chemical properties diminishes sharply. Some of the physical properties of dimethyl ether are compared with those of water and ethyl alcohol in Table 8-1.

TABLE 8-1 | *Some Properties of Water, Ethyl Alcohol, and Dimethyl Ether Compared*

SUBSTANCE	M.P., °C	B.P., °C	SOL. IN H_2O	SOL. IN C_2H_5OH	SOL. IN BENZENE
H–OH	0	100	——	miscible	.06% at 25°
CH_3-CH_2-OH	−112	78	miscible	——	miscible
CH_3-O-CH_3	−138	−24	3% at 18°	miscible	miscible

The properties of the ethers closely resemble those of the alkanes of corresponding molecular weight. The molecular weights of

TABLE 8-2 | *Some Properties of n-Pentane and Diethyl Ether Compared*

SUBSTANCE	MOL. WT.	B.P., °C	SOL. IN H_2O	INDEX OF REFRACTION n_D	DENSITY
$CH_3-CH_2-CH_2-CH_2-CH_3$	74	36	insol.	1.3570	0.63
$CH_3-CH_2-O-CH_2-CH_3$	76	35	6%	1.3497	0.72

n-pentane and diethyl ether, for example, are very close, and the boiling points, solubility in water, density, and index of refraction are not far apart. Both substances are light, colorless liquids, and they have similar odors.

8-1. Nomenclature of Ethers

The first members of the homologous series of aliphatic ethers, with their common names and boiling points, are listed in Table 8-3. We derive the common names by naming each group attached to the oxygen, the shorter or continuous-chain group generally first. If the two groups are the same, as in CH_3–O–CH_3, the compound may be called either dimethyl ether or methyl ether; the second name implies that the groups are the same.

The IUC names have not attained popular usage, and the rules for that system will not be given here.

TABLE

8-3 | *Ethers*

FORMULA	B.P., °C	COMMON NAME
CH_3–O–CH_3	−24	methyl ether
CH_3–O–CH_2CH_3	10	methyl ethyl ether
CH_3CH_2–O–CH_2CH_3	35	ethyl ether
CH_3–O–CH $\diagup^{CH_3}_{\diagdown CH_3}$	32	methyl isopropyl ether
CH_3–O–$CH_2CH_2CH_3$	41	methyl n-propyl ether
CH_3–O–$CH_2CH_2CH_2CH_3$	70	methyl n-butyl ether
CH_3CH_2–O–$CH_2CH_2CH_3$	64	ethyl n-propyl ether
CH_3CH_2–O–$\overset{\displaystyle CHCH_3}{\underset{\displaystyle CH_3}{\vert}}$	54	ethyl isopropyl ether
C_6H_5–O–CH_3	155	anisole
C_6H_5–O–C_6H_5	259	diphenyl ether

8-2. Functional Isomerism

The general formula of ethers is the same as that of alcohols, $C_nH_{2n+2}O$. In terms of alkyl groups the two formulas are R–OH and R–OR′, respectively. Ethyl alcohol and methyl ether have the same molecular formula and are therefore isomers, but the isomerism exhibited by these compounds is different from that exhibited by the hydrocarbons and alkyl halides. The two compounds are functional isomers; that is, they differ in functional group. The functional group of a molecule is the group that determines the chemical properties of the compound (the reactions it undergoes). All members of a homologous series have the same functional group. In alcohols, the –OH group determines the chemical properties; in ethers, the functional group is –O–.

Exercise 8-1. What is the functional group in an alkyne? Write a functional isomer of $CH_3–CH_2–C\equiv CH$; of cyclohexane.

8-3. Preparation of Ethers

1. Direct Preparation from Alcohols. Ethyl alcohol may be dehydrated with sulfuric acid at 160° to give ethylene and water (Chap. 4). The reaction also goes in another direction at the same time, however, splitting out one molecule of water from two molecules of ethyl alcohol to give diethyl ether. The first reaction is favored at 160° or above, the second at 145°. An excess of alcohol

$$2CH_3CH_2OH \xrightarrow{H_2SO_4} CH_3CH_2OCH_2CH_3 + H_2O \qquad [8\text{-}1]$$

over H_2SO_4 (about 12 to 1) also favors ether formation. This has led to the assumption that the reaction may take place in two steps:

$$CH_3CH_2–OH + \overset{\oplus}{H} \overset{\ominus}{O}–SO_2OH \rightarrow CH_3–CH_2–O–SO_2OH + H_2O \qquad [8\text{-}2]$$

$$CH_3CH_2–OSO_2OH + \overset{\oplus}{H} \overset{\ominus}{O}–CH_2CH_3 \xrightarrow{145°} CH_3–CH_2–O–CH_2–CH_3 + H_2SO_4 \qquad [8\text{-}3]$$

The reaction may be carried out with other catalysts, such as Al_2O_3 and HCl, and the fact that HCl catalyzes it casts considerable doubt on the assumption, for the reactants corresponding to those of Expression 8-3—CH_3–CH_2–Cl and ethanol—do not react to yield diethyl ether.

Exercise 8-2. Show how di-isopropyl ether may be made directly from an alcohol.

2. Williamson Synthesis. The dehydration reactions are run commercially to make ethyl and isopropyl ethers but will not give pure mixed ethers. The Williamson synthesis, in which an alkyl halide is heated with the sodium salt of an alcohol or phenol, is a

$$RONa + R'X \longrightarrow R\text{–}O\text{–}R' + NaX \qquad [8\text{-}4]$$

general method of preparing ethers and may be applied to any ether, symmetrical or mixed. (Aryl halides are too inert for this process.)

Anisole is prepared by this method from sodium phenoxide and methyl iodide. Since the alcohols and phenols are readily avail-

$$\langle\!\!\!\bigcirc\!\!\!\rangle\text{–ONa} + CH_3I \xrightarrow[74\%]{} \langle\!\!\!\bigcirc\!\!\!\rangle\text{–OCH}_3 + NaI \qquad [8\text{-}5]$$
$$\text{anisole}$$

able, the synthesis of an ether may begin with an alcohol or phenol.

$$CH_3\text{–}\underset{\underset{OH}{|}}{C}H\text{–}CH_3 + HBr \xrightarrow{H_2SO_4} CH_3\text{–}\underset{\underset{Br}{|}}{C}H\text{–}CH_3 + H_2O \qquad [8\text{-}6]$$

$$CH_3\text{–}CH_2\text{–OH} + Na \longrightarrow CH_3\text{–}CH_2\text{–ONa} + \tfrac{1}{2}H_2 \qquad [8\text{-}7]$$

$$CH_3\text{–}\underset{\underset{Br}{|}}{C}H\text{–}CH_3 + CH_3\text{–}CH_2\text{–ONa} \longrightarrow \underset{CH_3}{\overset{CH_3}{\diagup}}CH\text{–O–}CH_2\text{–}CH_3 \qquad [8\text{-}8]$$
$$\text{ethyl isopropyl ether}$$

Exercise 8-3. Write a synthesis, starting with an alcohol and a phenol, for methyl p-tolyl ether.

8-4. Reactions of Ethers

1. Combustion. The ethers are nearly as inert chemically as the alkanes. They are also quite as inflammable as the alkanes:

$$CH_3-CH_2-O-CH_2-CH_3 + 6O_2 \longrightarrow 4CO_2 + 5H_2O \qquad [8-9]$$

2. Splitting. The presence of the oxygen atom in the chain does make some difference in character between an ether and a saturated hydrocarbon. Ethers can be split upon refluxing with

$$R-O-R' + 2HI \xrightarrow{\Delta} RI + R'I + H_2O \qquad [8-10]$$

$$Ar-O-R + HI \xrightarrow{\Delta} ArOH + RI \qquad [8-11]$$

concentrated HI into two molecules of an alkyl iodide (or two different alkyl iodides in the case of an unsymmetrical ether) and a molecule of water. A mixed aromatic-aliphatic ether will react with HI under like conditions to yield an alkyl iodide and a phenol. To split an ether with HCl or HBr requires the application of both

$$C_2H_5-O-CH\begin{smallmatrix}CH_3\\\\CH_3\end{smallmatrix} + 2HI \xrightarrow[100\%]{} C_2H_5I + \begin{smallmatrix}CH_3 & H\\\\ \diagdown C \diagup \\\\ CH_3 & I\end{smallmatrix} + H_2O \qquad [8-12]$$

pressure and heat, and so these reagents are not often used. Some phenolic ethers can be split with $AlCl_3$ as a catalyst.

3. Nitration. Mixed aliphatic-aromatic ethers may be nitrated readily with a mixture of nitric and sulfuric acids. The reaction goes more easily than the nitration of an aryl halide.

Exercise 8-4. Referring to the discussion on bromination of phenols (p. 149), write the most probable canonical structures for $CH_3-O-C_6H_5$.

Exercise 8-5. Using the results of Expression 8-4, predict the position that a nitro group would take on entering the anisole molecule.

Write a mechanism for the nitration of $CH_3-O-C_6H_5$ in terms of the electronic theory given on p. 123.

8-5. Uses of Ethers

Diethyl and di-isopropyl ethers are the only ones used commercially in large quantities. They are both used as solvents for fats and oils.

Diethyl ether was first used as an anesthetic in 1842. Specially purified ether to which small amounts of ethyl alcohol are added to prevent oxidation is used for this purpose. Other ethers are now often used when different anesthetic effects are desirable; for example, vinyl ether is used for quick but short anesthesia in minor operations.

8-6. Cyclic Ethers

If the oxygen atom is joined to two carbons in a ring, the resulting compound is a cyclic ether. The simplest cyclic ether contains two carbons and is called ethylene oxide, CH_2-CH_2. Another useful

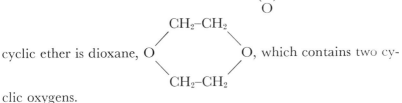

cyclic ether is dioxane, O   O, which contains two cyclic oxygens.

A. Preparation

Ethylene oxide may be prepared from ethylene chlorohydrin by treatment with soda lime or by catalytic oxidation of ethylene (p. 152). With an alkaline catalyst the three-membered ring can be opened up and closed again with a second molecule to give a six-membered ring containing two oxygens.

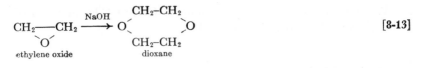

$$[8-13]$$

B. *Reactions of Ethylene Oxide*

1. With H_2O. The ease with which the three-membered ring opens has just been suggested in the preparation of dioxane. The student who recalls the chemistry of cyclopropane will be able to guess the behavior of ethylene oxide toward many reagents. Water will add to the broken ring in the presence of acid at 90–95°. The carbon-oxygen bond is the weak one; and, since carbon is

$$\overset{\delta+}{CH_2}\text{———}\overset{\delta+}{CH_2} + \overset{\delta-}{H}\ \overset{H^+}{OH} \longrightarrow \underset{\underset{\text{ethylene glycol}}{OH\ \ OH}}{CH_2\text{-}CH_2} \qquad [8\text{-}14]$$

more electropositive than oxygen (notice their positions in the Periodic Table), hydrogen joins the oxygen and the OH group joins the carbon. A glycol is the resulting product.

2. With ROH. An alcohol, in the presence of sulfuric acid as a catalyst, would be expected to behave in the same manner as water. Such a reaction results in a new class of compounds, the

$$\overset{\delta+}{CH_2}\text{———}CH_2 + \overset{\delta-}{RO}\ \overset{\delta+}{H} \overset{H_2SO_4}{\longrightarrow} \underset{\underset{\text{a cellosolve}}{OR\ \ OH}}{CH_2\text{-}CH_2} \qquad [8\text{-}15]$$

cellosolves, which may be considered both ethers and alcohols, for they show the properties expected of the functional groups of both ethers and alcohols. Since a cellosolve is an alcohol, it is soluble in water; since it is an ether, it acts as a solvent for hydrocarbons. Its chemical behavior also follows that of the respective functional groups. The OH group in the cellosolve will react with sodium and the other reagents described in Chapter 7, but the ether linkage is almost as inert as the aliphatic chain itself. Can you predict the behavior of butyl cellosolve (Exp. 8-16) toward concentrated HI?

When n-butyl alcohol reacts with ethylene oxide, the product is called butyl cellosolve.

$$CH_3\text{-}CH_2\text{-}CH_2\text{-}CH_2\text{-}OH + CH_2\underset{\diagdown O \diagup}{\text{——}}CH_2 \xrightarrow{HOSO_3H}$$

$$CH_3\text{-}CH_2\text{-}CH_2\text{-}CH_2\text{-}O\text{-}CH_2\text{-}CH_2\text{-}OH \qquad [8\text{-}16]$$
$$\text{butyl cellosolve}$$

Exercise 8-6.　Write an equation for the reaction of butyl cellosolve with each of the following reagents: Na, PCl_5, HI.

Ethylene glycol in large excess will add to ethylene oxide to yield a compound called diethylene glycol, a high-boiling viscous

$$HOCH_2CH_2OH + CH_2\underset{\diagdown O \diagup}{\text{——}}CH_2 \xrightarrow[90\text{-}95°]{H_2SO_4} HOCH_2CH_2OCH_2CH_2OH \qquad [8\text{-}17]$$
$$\text{diethylene glycol}$$

liquid resembling glycerol.

Exercise 8-7.　What result would you expect if ethylene glycol were heated with a twenty-fold excess of ethylene oxide at 95° and with a little H_2SO_4?

C.　Uses of Cyclic Ethers

Ethylene oxide is used as a fumigant and insecticide for tobacco, grain, and dried fruits. It leaves no odor and is readily removed by currents of air since it boils (13.5°C) below room temperature. Unlike other ethers, ethylene oxide is very reactive and is used in making medicinals and at least one perfume, synthetic attar of roses.

Dioxane is soluble in water. Since it dissolves many organic substances, it can be used to increase the compatibility of water and organic substances not soluble in water. Dioxane is a good solvent for resins, gums, fats and oils (p. 275), and cellulose acetate (p. 418).

SUMMARY

Ethers bear a closer resemblance to alkanes than they do to the alcohols or water of which they may be considered derivatives. Ethers are generally inert, but they may be split by HI. An ether linkage on an aromatic ring is an *o,p*-directing group.

The Williamson synthesis can be used to prepare mixed or symmetrical ethers.

The three-membered ring in ethylene oxide can be opened, by acid-catalyzed reactions with water, alcohols, ammonia, and amines, to give useful emulsifying agents.

REFERENCE

Bowden, "The Etherification Process," **J.** Chem. Education, 24, 432 (1947).

EXERCISES

(Exercises 1–7 will be found within the text.)

8. Write structures for all the isomers of $C_4H_{10}O$, and name them.

9. Complete and balance the following:

1. $CH_3OH + Na \longrightarrow$
2. $Br\text{-}\hexagon\text{-}OH + NaOH \longrightarrow$
3. $(CH_3)_2CHONa + CH_3\text{-}CH_2\text{-}Br \longrightarrow$
4. $\underset{\diagdown O \diagup}{CH_2\text{——}CH_2} + CH_3\text{-}CH_2\text{-}CH_2OH + H^+ \longrightarrow$
5. $\underset{\diagdown O \diagup}{CH_3\text{-}CH\text{——}CH_2} + HOH + H^+ \longrightarrow$
6. $\underset{\diagdown O \diagup}{CH_2\text{——}CH_2} + \underset{OH}{CH_3\text{-}CH\text{-}CH_3} + H^+ \longrightarrow$

10. Write reactions for two methods of preparing isopropyl ether.

11. Write a synthesis for each of the following:

    1. C_2H_5–O–⬡

    2. ethyl isobutyl ether

    3. CH_3–O–CH_2–CH_2–OH

    4. C_2H_5–O–CH_2–CH_2–O–C_2H_5

12. Write equations for the following reactions:

    1. combustion of di-isopropyl ether

    2. HI on C_2H_5–O–⬡

    3. HI on C_2H_5–O–CH_2–CH_2–O–C_2H_5

13. Why should ethyl cellosolve dissolve both water and fats?

9 Acids and Bases

The terms "acid" and "base" have been used so far in this text without definition since some familiarity with the terms was assumed. Among carbon compounds there are several functional groups classed as acidic and basic, and before these are examined closely it will be well to look at the phenomena of acids and bases in completely general terms. The various functional groups in carbon compounds called acidic have many properties not associated with the general phenomenon, and these will be deferred until the particular functional group is studied.

One acidic functional group, the carboxyl group, $-C\underset{\diagdown OH}{\overset{\diagup\!\!\diagup O}{}}$, is present in acetic acid, $CH_3-C\underset{\diagdown OH}{\overset{\diagup\!\!\diagup O}{}}$, which occurs in vinegar. Its acidic properties will be discussed here, but its other properties will be deferred to Chapter 10.

9-1. Definition of Acid and Base

For many years the definition of acid and base has been changing, not because chemists are uncertain what an acid or a base is, but because they have been finding out more and more about acid-base reactions. In trying, during the last twenty-five years, to include all acid-base phenomena within one concept, they have

168

broadened the definitions to such an extent that some of them seem to include practically every known chemical substance as either an acid or a base.

It is probably best to present a "phenomenological" definition of acids and bases, one based on empirical observations of their particular properties, and from this to go on to present the various concepts of acids and bases that attempt to explain their behavior. An acid is a substance that has a sour taste, will react, usually rather readily, with a base, and will cause certain colored compounds, called indicators, to change color. A base is a substance that has a bitter taste (or will at least overcome the sour taste of an acid), will react, usually readily, with an acid, and will cause indicators to assume different colors than they have in the presence of acids. Acids and bases frequently catalyze chemical reactions. One of the products of the reaction between an acid and a base is always a salt.

9-2. Theories to Account for the Behavior of Acids and Bases

1. Arrhenius.   From the behavior of acids, bases, and salts in aqueous solution, deductions as to the cause of their behavior have been made, and from these came the Arrhenius definition of an acid as a substance that contains hydrogen, part or all of which will dissociate in aqueous solution to yield hydrogen ions, H^+ (or hydronium ions, H_3O^+). A base is a substance that contains the hydroxyl group, which will dissociate in aqueous solution to yield hydroxide ions, OH^-. Some typical acids and bases conforming to this concept are listed here. When one of these acids is neutralized

Acids	*Bases*
Hydrochloric acid, HCl	Sodium hydroxide, NaOH
Hydrosulfuric acid, H_2S	Potassium hydroxide, KOH
Sulfuric acid, $HOSO_2OH$	Magnesium hydroxide, $Mg(OH)_2$
Phosphoric acid, $HOPO(OH)_2$	Calcium hydroxide, $Ca(OH)_2$
Acetic acid, CH_3COOH	Ammonium hydroxide, NH_4OH

by one of these bases, the products of the neutralization are a salt and water. Examples of such neutralization are shown in Expressions 9-1–9-3. A weak acid or base is one that dissociates only to a

$$NaOH + HCl \rightleftharpoons NaCl + H_2O \qquad [9\text{-}1]$$

$$2KOH + H_2SO_4 \rightleftharpoons K_2SO_4 + 2H_2O \qquad [9\text{-}2]$$

$$NH_4OH + CH_3COOH \rightleftharpoons CH_3COONH_4 + H_2O \qquad [9\text{-}3]$$

small extent when in solution; a strong acid or base dissociates to a large extent.

This concept of acid and base is adequate for the elementary study of general chemistry.

2. Brönsted. To avoid limiting the concept to aqueous solutions, Brönsted broadened it further by defining an acid as a proton donor, a base as a proton acceptor. When such an acid dissociates a proton, the remaining radical is a base. Such an acid and the base formed are called a conjugate acid-base pair, and one speaks of the conjugate base of a given acid, the conjugate acid of a given base. A number of conjugate acid-base pairs are listed in Table 9-1 in the approximate order of decreasing strength of acid

TABLE 9-1 | *Some Conjugate Acid-Base Pairs, According to the Brönsted System*

ACID =	PROTON +	BASE
HCl hydrochloric acid	H^+	Cl^- chloride ion
H_2SO_4 sulfuric acid	H^+	HSO_4^- hydrogen sulfate ion
HNO_3 nitric acid	H^+	NO_3^- nitrate ion
HSO_4^- hydrogen sulfate ion	H^+	$SO_4^=$ sulfate ion

Table 9-1 continued

ACID =	PROTON +	BASE
CH₃COOH ⇌ acetic acid	H⁺ +	CH₃COO⁻ acetate ion
Al(H₂O)₆⁺³ ⇌ hexa-aquoaluminum ion	H⁺ +	Al(H₂O)₅(OH)⁺⁺ hydroxopenta-aquoaluminum ion
H₂CO₃ ⇌ carbonic acid	H⁺ +	HCO₃⁻ hydrogen carbonate ion
H₂S ⇌ hydrosulfuric acid	H⁺ +	HS⁻ hydrosulfide ion
C₆H₅NH₃⁺ ⇌ anilinium ion	H⁺ +	C₆H₅NH₂ aniline
NH₄⁺ ⇌ ammonium ion	H⁺ +	NH₃ ammonia
CH₃NH₃⁺ ⇌ methyl ammonium ion	H⁺ +	CH₃NH₂ methyl amine
C₆H₅OH ⇌ phenol	H⁺ +	C₆H₅O⁻ phenoxide ion
HCO₃⁻ ⇌ hydrogen carbonate ion	H⁺ +	CO₃⁼ carbonate ion
H₂O ⇌ water	H⁺ +	OH⁻ hydroxide ion
CH₃OH ⇌ methanol	H⁺ +	CH₃O⁻ methoxide ion
NH₃ ⇌ ammonia	H⁺ +	NH₂⁻ amide ion

and increasing strength of base. It will be noticed that the stronger the acid, the weaker its conjugate base, and the stronger the base, the weaker its conjugate acid. Any reaction between an acid and a base may be considered merely as the transfer of a proton from one base to a second base, the reaction being greater in the direction in which the proton becomes associated with the stronger of the two bases. In the first example of Table 9-2 the two bases,

TABLE

9-2 | *Neutralization Reactions*

ACID$_1$ + BASE$_2$		BASE$_1$ + ACID$_2$
HCl + OH$^-$	$\longrightarrow$ $\leftarrow$	Cl$^-$ + HOH
NH$_4{}^+$ + OH$^-$	$\longrightarrow$ $\leftarrow$	NH$_3$ + HOH
H$_3$O$^+$ + CO$_3{}^-$	$\longrightarrow$ $\leftarrow$	H$_2$O + HCO$_3{}^-$
CH$_3$COOH + HS$^-$	$\rightleftarrows$	CH$_3$COO$^-$ + H$_2$S
HOSO$_3$H + HONO$_2$	$\rightleftarrows$	*HOSO$_3{}^-$ + H$_2$O + NO$_2{}^+$

* See pp. 54 and 123.

OH$^-$ and Cl$^-$, compete for the proton. Since the hydroxyl ion is a much stronger base than the chloride ion, the direction of the reaction is predominantly to the right, as indicated. This concept has the advantage of being applicable to organic bases and to solvents other than water, and at the same time it includes all acids and bases defined by the Arrhenius concept. However, while it admits that many substances other than the hydroxyl ion may act as bases, the hydrogen atom remains a necessary constituent of all acids.

The Brönsted theory has been modified and extended by the solvent system of acids and bases. This generalizes the concept of acids in a way that roughly parallels the Brönsted generalization of the concept of bases. An acid is defined as a substance that will dissociate the same cation as the solvent cation, and a base is a substance that will dissociate the same anion as the solvent anion. Liquid sulfur dioxide, water, and liquid ammonia are examples of

$2SO_2$	$\longleftarrow$ $\longrightarrow$	$SO^{++} + SO_3^{=}$	[9-4]
$2H_2O$	$\longleftarrow$ $\longrightarrow$	$H_3O^+ + OH^-$	[9-5]
$2NH_3$	$\longleftarrow$ $\longrightarrow$	$NH_4^+ + NH_2^-$	[9-6]

the solvent system of acids and bases. Typical acids and bases, as defined by this system, are shown in Expressions 9-7–9-9. The

SOLVENT	ACID + BASE	SALT + SOLVENT	
SO_2	$SOCl_2 + K_2SO_3 \longrightarrow$	$2KCl + 2SO_2$	[9-7]
H_2O	$H_3OCl + NaOH \longrightarrow$	$NaCl + 2H_2O$	[9-8]
NH_3	$NH_4Cl + NaNH_2 \longrightarrow$	$NaCl + 2NH_3$	[9-9]

system assigns too important a role to the solvent and is therefore limited in its application.

3. Lewis. The most general of all acid-base concepts is that of Lewis, according to which an acid is any substance that is able to accept a pair of electrons to form a coordinate covalent bond, and a base is any substance that has an unshared pair of electrons with which it can form a coordinate covalent bond. This concept includes all of the less generalized concepts previously discussed. Reactions between acids and bases according to this concept are given in Expressions 9-10–9-16. Expression 9-13 is of interest

BASE   ACID

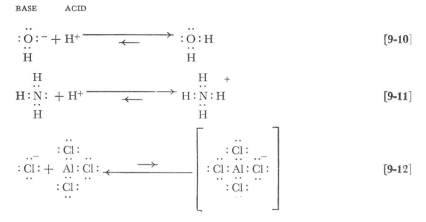

BASE   ACID

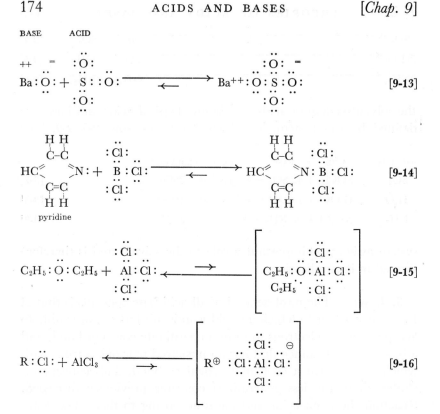

$$[9\text{-}13]$$

$$[9\text{-}14]$$

pyridine

$$[9\text{-}15]$$

$$[9\text{-}16]$$

because it is a typical "dry" acid-base reaction, which takes place between an acidic oxide and a basic oxide. No protons or solvent is involved. Expression 9-10 is an acid-base reaction by either the Arrhenius or the Brönsted concept, and Expression 9-11 by the Brönsted concept. Expressions 9-14, 9-15, and 9-16 are of interest in organic chemistry.

None of the acid-base concepts discussed above is right or wrong. The chemist thinks in terms of the one that fits his problem best. He may be compared to a carpenter choosing a tool for a particular job. Any one of several tools may do the job, but the carpenter will select the one tool that is best fitted for it. For most work in general chemistry, the Arrhenius concept is sufficient. For work in nonaqueous solvents, the Brönsted concept is more often applicable. The Lewis concept is valuable quite frequently in carbon chemistry; we have already applied the idea in the chemistry of the alkenes without calling it by that name; alkenes are Lewis bases in all the addition reactions discussed.

9-3. Strength of Acids

The Brönsted concept suggests that any compound containing a hydrogen atom might conceivably lose it to the proper second molecule and hence be called an acid. Learning something of the relative ease with which a proton is lost from a given molecule gives a means of comparing the relative acidic property of various compounds. A review of reactions already known will reveal that the student is acquainted with many of these relationships. The mineral acids are all strong acids; that is, they are highly ionized or completely ionized (lose protons readily to water) in dilute solution.

$$HCl + HOH \rightleftharpoons H_3O^+ + Cl^- \quad\quad [9\text{-}17]$$
chloride ion

$$HOSO_3H + HOH \rightleftharpoons H_3O^+ + OSO_3H^- \quad\quad [9\text{-}18]$$
hydrogen sulfate ion

$$OSO_3H^- + HOH \rightleftharpoons H_3O^+ + OSO_3^= \quad\quad [9\text{-}19]$$
sulfate ion

$$HONO_2 + HOH \rightleftharpoons H_3O^+ + ONO_2^- \quad\quad [9\text{-}20]$$
nitrate ion

$$HOPO(OH)_2 + HOH \rightleftharpoons H_3O^+ + OPO(OH)_2^- \quad\quad [9\text{-}21]$$
dihydrogen phosphate ion

$$HOSO_2NH_2 + HOH \rightleftharpoons H_3O^+ + OSO_2NH_2^- \quad\quad [9\text{-}22]$$
sulfamate ion

$$\bigcirc\text{-}SO_3H + HOH \rightleftharpoons H_3O^+ + \bigcirc\text{-}SO_3^- \quad\quad [9\text{-}23]$$
benzenesulfonate ion

The sour taste in vinegar is due to acetic acid, which has the formula $CH_3\text{-}C{\overset{\displaystyle O}{\underset{\displaystyle OH}{}}}$. It is ionized only slightly in water solution.

$$CH_3-C\overset{O}{\underset{OH}{}} + HOH \rightleftharpoons H_3O^+ + CH_3-C\overset{O}{\underset{O^-}{}} \qquad [9\text{-}24]$$

acetate ion

The strongest carboxylic acid, formic acid, is ionized only about 5% in dilute solution. Other organic acids ionize 1–5%.

$$H-C\overset{O}{\underset{OH}{}} + HOH \rightleftharpoons H_3O^+ + H-C\overset{O}{\underset{O^-}{}} \qquad [9\text{-}25]$$

formate ion

An acid that is still weaker than any of these is carbonic acid, made by dissolving carbon dioxide in water. This is the substance that makes most soft drinks attractive by adding a certain zest to them. It is ionized only 0.17% or less. The second step to car-

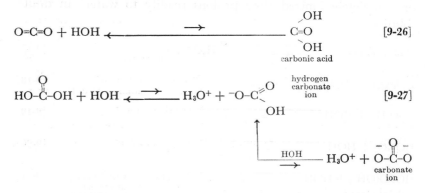

$$O{=}C{=}O + HOH \rightleftharpoons \overset{OH}{\underset{OH}{C{=}O}} \qquad [9\text{-}26]$$

carbonic acid

bonate ion takes place to an even smaller degree than the first step.

Still weaker than any of these acids is phenol. It is so slightly ionized that it will react only with strong bases such as NaOH. The weaker base, $NaHCO_3$, sodium hydrogen carbonate, will not react with it.

$$\bigcirc{-}OH + :\overset{..}{\underset{..}{O}}:H^- \rightleftharpoons \bigcirc{-}\overset{..}{\underset{..}{O}}:^- + H_2O \qquad [9\text{-}28]$$

The alcohols are comparable in acid strength to water and are called neutral compounds. The homologous series, alkanes, alkenes, alkynes, aromatic hydrocarbons, and ethers all contain hydrogen atoms, but these are so strongly held by covalent bonds to carbon that there is no tendency at all for them to ionize.

Even though alcohols and alkynes are neutral substances, however, the hydrogen on the oxygen atom of an alcohol or on a carbon atom of an alkyne is replaceable, and in a sense those compounds may be considered as very weak acids. Sodium will displace hydrogen from an alcohol and some alkynes. There is an important distinction, however, between Expressions 9-29 and 9-30 and the

$$CH_3\text{--}CH_2 : \overset{..}{\underset{..}{O}} : H + Na \cdot \longrightarrow CH_3\text{--}CH_2 : \overset{..}{O} : Na + H \cdot \qquad [9\text{-}29]$$

$$CH_3\text{--}C{\equiv}C : H + Na \cdot \longrightarrow CH_3\text{--}C{\equiv}C : Na + H \cdot \qquad [9\text{-}30]$$

ionization of acids described previously. The former are not acid-base reactions. No *pairs* of electrons are donated or accepted. Instead, a sodium atom with its single electron displaces a hydrogen atom (not a proton) with its single electron.

Alcohols and alkynes are neutral to most bases, but the hydrogen atoms are less tightly bound than the hydrogens in the C-H bonds that exist in alkanes, alkenes, aromatic hydrocarbons, and ethers. These last homologous series may be considered at the bottom of the acid strength scale among the compounds mentioned so far. Hydrogen cannot even be replaced by sodium in these compounds.

Strong acids:

Mineral acids such as HCl, $HONO_2$, and $HOSO_2NH_2$

$\langle\text{—}\rangle$-SO_3H, benzenesulfonic acid

Weak acids:

$H\text{--}C\overset{O}{\underset{OH}{\diagdown}}$, formic acid

$CH_3\text{--}C\overset{O}{\underset{OH}{\diagdown}}$, acetic acid

R--COOH and other carboxylic acids

Very weak acids:

$\underset{\text{OH}}{\overset{\text{OH}}{\text{C=O}}}$, carbonic acid

⬡–OH, phenol

Neutral, but H displaced by Na:

R–OH, an alcohol
R–C≡CH, an alkyne

Neutral, H not displaced by Na:

R–O–R, an ether
RH, an alkane
R–CH=CH$_2$, an alkene
ArH, an aromatic hydrocarbon

9-4. Bases

The strong bases called alkaline (from Group Ia) and alkaline earth (Group IIa) hydroxides ionize completely or nearly completely in water to yield hydroxyl ions. These bases taste bitter, have a slick or soapy feel, turn litmus blue, and neutralize acids.

$$\text{NaOH} \longrightarrow \text{Na}^+ + \text{OH}^- \qquad\qquad [\textbf{9-31}]$$

$$\text{Ca(OH)}_2 \longrightarrow \text{Ca}^{++} + 2\text{OH}^- \qquad\qquad [\textbf{9-32}]$$

$$\text{Ba(OH)}_2 \longrightarrow \text{Ba}^{++} + 2\text{OH}^- \qquad\qquad [\textbf{9-33}]$$

Almost all the other hydroxides are only very slightly soluble in water and are weak bases as well. The following may be cited as examples: Al(OH)_3, Fe(OH)_2, Ti(OH)_4.

The commonest weak base is a solution of ammonia in water, which reacts to a slight extent according to Expression 9-34. The

$$\text{NH}_3 + \text{H}_2\text{O} \rightleftarrows \text{NH}_4^+ + \text{OH}^- \qquad\qquad [\textbf{9-34}]$$

hydrogens in the NH_3 may be replaced by R groups, and the resulting organic bases are called amines. They have many of the chemical properties of ammonia. Amines dissolve in water and react to yield substituted ammonium ions and hydroxyl ions. If the R group is short (1–4 carbons), the amine has about the same

$$RNH_2 + H_2O \; \overset{\longrightarrow}{\longleftarrow} \; RNH_3^+ + OH^- \qquad [9\text{-}35]$$

basic strength as ammonia. The other two hydrogens in RNH_2 may also be replaced by the same or different alkyl groups.

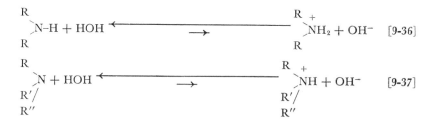

$$\begin{aligned} &\text{R} \\ &\quad \searrow \text{N–H} + \text{HOH} \; \overset{\longrightarrow}{\longleftarrow} \; \text{R} \\ &\text{R} \end{aligned} \qquad \begin{aligned} &\text{R} \quad + \\ &\quad \searrow \text{NH}_2 + \text{OH}^- \quad [9\text{-}36] \\ &\text{R} \end{aligned}$$

$$\begin{aligned} &\text{R} \\ &\quad \searrow \text{N} + \text{HOH} \; \overset{\longrightarrow}{\longleftarrow} \\ &\text{R}' / \\ &\text{R}'' \end{aligned} \qquad \begin{aligned} &\text{R} \quad + \\ &\quad \searrow \text{NH} + \text{OH}^- \quad [9\text{-}37] \\ &\text{R}' / \\ &\text{R}'' \end{aligned}$$

9-5. Neutralization and Titration

The process of letting a base react with an acid or an acid with a base is called neutralization. Heat is evolved during the reaction.

$$Na^+OH^- + H^+Cl^- \longrightarrow Na^+Cl^- + H_2O \qquad [9\text{-}38]$$

To examine more closely what is taking place during a neutralization, let us review the meaning of an equation (Exp. 9-33).

1. The equation is a shorthand method of saying that sodium hydroxide will react with hydrochloric acid to yield sodium chloride and water.

2. But the equation also has a quantitative meaning as well as this qualitative one. One "molecule" of sodium hydroxide will react with one "molecule" of hydrochloric acid to yield one "molecule" of sodium chloride and one molecule of water. The quotation marks indicate that molecules of the designated substances do not exist as such in these solutions.

3. A third implication, also of a quantitative nature, is that one formula weight of sodium hydroxide in grams (or other unit of weight) will react with one formula weight of hydrochloric acid in grams to yield one formula weight of sodium chloride in grams and one formula weight of water in grams.

If the solution described above is examined at the end of the neutralization, both the sour taste of the acid and the soapy feel of the original sodium hydroxide solution will have disappeared. Furthermore, the solution will have exactly the properties of a solution of an equivalent amount of sodium chloride. If the solution is evaporated, only sodium chloride will remain.

It has also been observed that a certain amount of heat—13,700 calories—is liberated during the neutralization of any strong acid by a strong base for each formula weight of water (in grams) formed during the reaction. This is true regardless of the strong acid or base used.

This should be convincing evidence that the same essential reaction is taking place in all Arrhenius acid-base neutralizations. Since the one thing that Arrhenius acids have in common is a proton and all bases have a hydroxyl group in common, neutralization must be a reaction between these two ions.

$$H^+ + OH^- \longrightarrow H_2O \qquad\qquad [9\text{-}39]$$

The quantitative nature of the neutralization process enables the chemist to use a known quantity of an acid for determining how much base is present in an unknown solution. This laboratory procedure is called titration. The question arises: How is it possible to tell when a titration is complete—that is, at the neutral point?

A number of organic substances have different colors in acidic and basic solutions. Litmus, for example, is red in acid solution and blue in basic solution. Phenolphthalein is colorless in acid solution and red in basic solution. Methyl orange is yellow in basic solution and red in acid solution. (See Chap. 30.) These changes in color enable the experimenter to see when he has reached the neutral point in titrating an acid with a base.

9-6. Salt Formation

Neutralization of an acid by a base is a well-known method of producing a salt and water. If the reaction is between a strong

$$KOH + CH_3\text{-}C\overset{O}{\underset{OH}{\diagdown}} \longrightarrow CH_3\text{-}C\overset{O}{\underset{OK}{\diagdown}} + H_2O \qquad\qquad [9\text{-}40]$$

potassium
acetate

$$Ca(OH)_2 + H_2SO_4 \longrightarrow CaSO_4 + 2H_2O \qquad\qquad [9\text{-}41]$$

acid and a strong base in dilute solution, only hydroxyl and hydrogen ions actually take part, and the equation for the reaction is $H^+ + OH^- \longrightarrow H_2O$. If the complete picture is desired, the other ions may be shown as spectators. In the equation below, for example, Na^+ and Cl^- appear on both sides but take no active part in the reaction.

$$Na^+ + OH^- + H^+ + Cl^- \longrightarrow Na^+ + Cl^- + H_2O \qquad\qquad [9\text{-}42]$$

Acid-base reactions may be carried out between two solids, two gases, two liquids, or combinations of these, and it is convenient to show the reactions as taking place between molecules. Water may or may not be formed, depending on the reactants.

solid-solid: $NaOH + HOSO_2NH_2 \longrightarrow NaOSO_2NH_2 + H_2O$ $\qquad$ [9-43]

gas-gas: $\;:NH_3 + H:Cl \longrightarrow NH_4Cl$ $\qquad$ [9-44]

solid-gas: $Na_2O + SO_3 \longrightarrow Na_2SO_4$ $\qquad$ [9-45]

liquid-liquid: $\;:NH_3 + HO\text{-}\overset{O}{\overset{\|}{C}}\text{-}CH_3 \longrightarrow NH_4\text{-}O\text{-}\overset{O}{\overset{\|}{C}}\text{-}CH_3$ $\qquad$ [9-46]

ammonium acetate

gas-liquid: $\;:NH_3 + \langle\bigcirc\rangle\text{-}SO_3H \longrightarrow \langle\bigcirc\rangle\text{-}SO_3NH_4$ $\qquad$ [9-47]

ammonium
benzenesulfonate

liquid-liquid: $CaO + SiO_2 \xrightarrow[\substack{blast \\ furnace}]{fuse\ in} CaSiO_3$ $\qquad$ [9-48]

calcium
silicate

In the accompanying equations, basic oxides, sodium oxide and calcium oxide, are shown reacting with acidic oxides, sulfur trioxide and silicon dioxide. These are acid-base reactions in the Lewis sense. In Expression 9-45, electrons from the oxygen in the

$$
+\ \overset{..}{\underset{..}{.}}\ +\qquad \overset{O}{\underset{O}{\overset{\nwarrow}{\underset{\swarrow}{S=O}}}}\ \longrightarrow\ \overset{+\ -\ \uparrow\ -\ +}{Na\ \ O-S-O\ \ Na}
$$

Na : O : Na +   S=O → Na O-S-O Na

base form a new O-S bond, while one sodium atom shifts to another oxygen. In this case the sulfur atom is the acceptor, the oxygen from the base is the donor, and one sodium atom merely changes its allegiance from one oxygen atom to another.

9-7. Ionization of Water

By accurate measurement of the electrical conductivity of pure water at $25°C$, it has been found that the total concentration of conducting ions is 2×10^{-7} mole per liter (that is, 0.0000002). Hence the concentration of H^+ ions and of OH^- ions must each be 1×10^{-7} mole per liter. A mole (M) of hydrogen ions is one replaceable hydrogen atom in a formula weight of an acid (see the next section). In Expression 9-49 one may change the equilibrium position by altering the concentration of H^+ and OH^- (Le Châte-

$$H_2O \underset{r_2}{\overset{r_1}{\rightleftarrows}} H^+ + OH^- \tag{9-49}$$

lier's principle), but the concentration of water will remain nearly unchanged if only dilute solutions are considered. Now let us develop an expression for the quantitative change in these concentrations. The rate r_1 of the left-to-right reaction depends only on the concentration of water, and, since this is practically constant, the rate must be practically constant:

$$\therefore r_1 = k_1 \tag{9-50}$$

The rate r_2 of the reverse reaction is proportional to the concentrations of H^+ and OH^- ions, which are subject to variation:

$$r_2 \propto [H^+][OH^-] \qquad\qquad [9\text{-}51]$$
$$r_2 = k_2[H^+][OH^-] \qquad\qquad [9\text{-}52]$$

(Brackets indicate "concentration expressed as moles per liter.") At equilibrium, the two rates must be equal; that is,

$$r_1 = r_2$$
$$\therefore k_2[H^+][OH^-] = k_1$$

Dividing by k_2, we get

$$[H^+][OH^-] = \frac{k_1}{k_2} \qquad\qquad [9\text{-}53]$$

Since the quotient of two constants is a constant,

$$[H^+][OH^-] = K \qquad\qquad [9\text{-}54]$$

This expression indicates that in all dilute water solutions the product of $[H^+]$ and $[OH^-]$ is constant. Substituting the value of each determined for pure water into this equation, we can find the value of the constant K at $25°C$:

$$(1 \times 10^{-7})(1 \times 10^{-7}) = K$$
$$K = 1 \times 10^{-14}$$
$$[H^+][OH^-] = 1 \times 10^{-14} \qquad\qquad [9\text{-}55]$$

Notice that the concentration of one of the ions would become zero only if the concentration of the other were infinite; in other words, both ions are present *in all water solutions*, but not necessarily in equal concentration. For 0.01M HCl, in which $[H^+]$ is 0.01M, the $[OH^-]$ is obtained as follows:

$$[OH^-] = \frac{1 \times 10^{-14}}{[H^+]} = \frac{1 \times 10^{-14}}{.01} = \frac{10^{-14}}{10^{-2}} = 1 \times 10^{-12} \text{ mole/l}$$

This is a very small but real value. Water solutions are said to be *neutral*, *acidic*, or *basic* according to the following definitions:

neutral solution: $[H^+] = [OH^-]$

acid solution: $[H^+] > [OH^-]$

basic solution: $[H^+] < [OH^-]$

9-8. Formal Solutions

$$NaOH + HCl \longrightarrow NaCl + H_2O \qquad \qquad [9\text{-}56]$$

$$2NaOH + H_2SO_4 \longrightarrow Na_2SO_4 + 2H_2O \qquad \qquad [9\text{-}57]$$

One formula weight (in grams) of NaOH in any amount of water will neutralize one formula weight (in grams) of HCl in water solution, but the same weight of NaOH will neutralize only half of a formula weight (in grams) of H_2SO_4. Sulfuric acid contains two hydrogen atoms that can be replaced by sodium atoms; HCl possesses only one such hydrogen.

When one formula weight of a substance in grams (also called one gram-formula weight) is dissolved in a liter of solution, the solution is said to have a concentration of one formal, and the symbol **F** is used to designate it. A 0.1F solution of NaOH then contains 0.1 gram formula weight (0.1×40 g) of NaOH per liter of solution.

9-9. The pH

The concentration of hydrogen ions in a solution may be expressed as moles of hydrogen ions per liter of solution. A mole (M) of hydrogen ions is one replaceable hydrogen atom in a formula weight of an acid. A 1F HCl solution is also 1M in H^+ concentration, assuming that it is completely ionized:

$$HCl \rightleftarrows H^+ + Cl^-$$

A solution of H_2SO_4 that is 0.1F is 0.2M in $[H^+]$:

$H_2SO_4 \rightleftarrows 2H^+ + SO_4^-$

1F H_2SO_4 yields 2M [H^+]

The pH of a solution is defined as the negative logarithm to the base 10 of the hydrogen ion concentration in moles per liter. Hence, if [H^+] is 0.01M, the pH is 2.

What is the pH of a 0.02F solution of $HONO_2$, assuming that it is completely ionized?

$pH = -\log [H^+]$

$.02F = .02M = 2 \times 10^{-2}M$

$-\log [H^+] = -(\log 2 + \log 10^{-2}) = -(0.30 - 2) = 1.7$

$pH = 1.7$

Water is ionized (see p. 182) to the extent of about one molecule in 10 million. The hydrogen ion concentration in water, then, is 10^{-7} mole per liter. The pH of water is 7, and solutions that have pH 7 may be said to be neutral.

In examining the following table of corresponding values of [H^+], [OH^-], and pH, notice (1) that a change of one unit in pH represents a ten-fold change in concentration of H^+ ions, (2) that pH 7 represents a neutral solution, and (3) that progressively smaller values of pH correspond to increasing acidity, whereas progressively larger values of pH correspond to increasing basicity.

[H^+]	0.1	0.01	10^{-3}	...	10^{-6}	10^{-7}	10^{-8}	...	10^{-11}	10^{-12}	10^{-13}
[OH^-]	10^{-13}	10^{-12}	10^{-11}	...	10^{-8}	10^{-7}	10^{-6}	...	10^{-3}	0.01	0.1
pH	1	2	3	...	6	7	8	...	11	12	13

$\longleftarrow$ acidic __ neutral __ basic $\longrightarrow$

Careful control of the pH is essential to certain industries and in living organisms. In the brewing industry and in the preparation of penicillin, for example, the pH of the medium is adjusted to encourage growth of the microorganisms (yeasts or molds). The pH of the blood and other body fluids is normally held closely to values slightly above 7 (the stomach fluids are an exception, having a pH around 2). Sudden changes in the pH arising either from disease or from ingestion of large amounts of strong acids or hydroxides may cause death.

9-10. Hydrolysis of Salts

At the beginning of this chapter we presented a number of acids in which the acid character is due to the ionized proton. A number of substances that act as bases because they ionize to give hydroxyl ions were also shown. Many other ions also, dissolved in water, will form acidic or basic solutions.

In water solution (as well as in the crystalline state) sodium acetate is ionized as in Expression 9-58. A solution of sodium ace-

$$CH_3-COONa \longrightarrow CH_3COO^- + Na^+ \qquad [9-58]$$

tate turns litmus blue, indicating a basic solution. The acetate ion has reacted with water (a very weak acid) to give acetic acid (a

$$\underset{\text{base}_1}{CH_3COO^-} + \underset{\text{acid}_2}{HOH} \xrightarrow{\hspace{3cm}} \underset{\text{acid}_1}{CH_3COOH} + \underset{\text{base}_2}{OH^-} \quad [9\text{-}59]$$

weak acid) and hydroxide ions (a strong base). The hydroxide ion is a stronger base than the acetate ion, and the equilibrium is less than 2% in favor of the right-hand side of the equation. Still the resulting solution is basic toward litmus.

Many other ionized salts behave in a similar fashion. Solutions of the salts Na_2CO_3, $NaCN$, CH_3-CH_2-ONa, and C_6H_5ONa are

$$[9\text{-}60]$$

$$NaCN \longrightarrow Na^+ + CN^-$$

$$CN^- + H_2O \xrightleftharpoons{\quad\longrightarrow\quad} HCN + OH^- \qquad [9-61]$$

$$CH_3CH_2ONa \longrightarrow CH_3CH_2O^- + Na^+$$

$$\xrightarrow[\longleftarrow]{HOH} CH_3CH_2\text{-}OH + OH^- \qquad [9-62]$$

all alkaline to litmus paper because carbonate, cyanide, alkoxide, and phenoxide ions are strong bases. The alkoxide ion is such a strong base that it accepts a proton extremely rapidly from water molecules. The phenoxide ion is a somewhat weaker base than the alkoxide ion. Salts like $NaCl$, $BaCl_2$, and K_2SO_4 in solution are

$$\langle\bigcirc\rangle\text{-ONa} \longrightarrow \langle\bigcirc\rangle\text{-O}^- + Na^+$$

$$\xrightarrow[\longleftarrow]{H_2O} \langle\bigcirc\rangle\text{-OH} + OH^- \qquad [9-63]$$

neutral to litmus because Cl^- and $SO_4^=$ are very weak bases and have little tendency to accept a proton from a weak acid like water.

Some salts, on the other hand, give an acid reaction to litmus when they are dissolved in water. The ammonium ion is a strong enough acid to transfer a proton to a molecule of water. Though

$$NH_4Cl \longrightarrow NH_4^+ + Cl^- \qquad [9-64]$$

$$NH_4^+ + H_2O \xrightleftharpoons{\quad\longrightarrow\quad} NH_3 + H_3O^+ \qquad [9-65]$$

this reaction goes less than 5% to the right, the solution is still made acid to litmus by the small amount of the very strong acid H_3O^+. Other salts, such as $AlCl_3 \cdot 6H_2O$, $CuSO_4 \cdot 5H_2O$, $Al(NO_3)_3$, and $SiCl_4$, exhibit the same behavior. The hexahydrated aluminum

$$[Al(H_2O)_6]Cl_3 \longrightarrow [Al(H_2O)_6]^{+3} + 3Cl^- \qquad [9-66]$$

$$\underset{acid_1}{[Al(H_2O)_6]^{+3}} + \underset{base_2}{HOH} \xrightarrow{\quad\longrightarrow\quad} \underset{base_1}{[Al(H_2O)_5(OH)]^{++}} + \underset{acid_2}{H_3O^+} \quad [9-67]$$

ion is a weak acid and loses a proton to water (a weak base in this case). The new acid, a hydrated proton, turns litmus paper red.

In salts like sodium hydrogen carbonate, $NaHCO_3$, and sodium hydrogen sulfite, $NaHSO_3$, the effect on the pH of water is complicated by more competition. The first reaction tends to lower the

$$HCO_3^- + H_2O \rightleftarrows H_3O^+ + CO_3^= \qquad [9\text{-}68]$$

$$HCO_3^- + H_2O \rightleftarrows H_2CO_3 + OH^- \qquad [9\text{-}69]$$

pH by increasing $[H^+]$ (or H_3O^+ as actually shown here), while the second tends to raise the pH because $[OH^-]$ increases. One must go to the laboratory to find which wins. As measured in the laboratory, the pH of 0.1F $NaHCO_3$ is 8.4.

Exercise 9-1. Which reaction predominates in the hydrolysis of $NaHCO_3$? The pH of 0.1F $NaHSO_3$ being 4.5, which reaction with water predominates? Write the equations.

9-11. Hydrolysis of Covalent Compounds

The hydrolysis of a salt has been pictured as a reaction between ions or between ions and molecules. The hydrolysis of covalent compounds, according to one theory, can be described by means of a geometric picture in which the formation of a coordinate covalent bond is followed (1) by ionization or (2) by electron shifts to more stable configurations or (3) by loss of a small molecule within the complex.

This geometric picture has already been given as a possible mechanism for the hydrolysis of PCl_3 (p. 68). It is repeated here for SO_3 and $SOCl_2$. The reaction of SO_3 with H_2O is shown as the coordination of a molecule of water with the sulfur atom, momentarily expanding its electron shell to ten electrons.

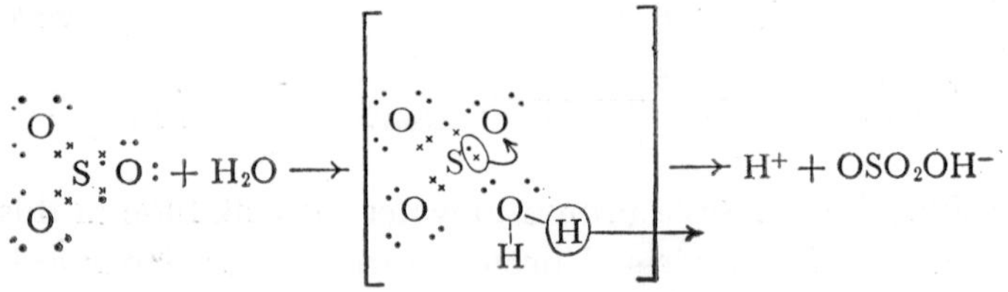

$$\rightarrow H^+ + OSO_2OH^- \qquad [9\text{-}70]$$

This unstable configuration is relieved by the loss of a proton (ionization) and a simultaneous shift of a shared pair between S and O to oxygen. The result is the formation of a proton and a hydrogen sulfate ion.

Exercise 9-2. Indicate the electronic shifts that take place in the hydrolysis of SO_2.

In thionyl chloride and phosphorus oxychloride the chlorine atoms have considerable electrovalent character, and hydrolysis of these compounds is therefore simply a competition between the two bases, H_2O and Cl^-. Water is a stronger base than chloride ion and so displaces the halogen. The acid in Expression 9-71, $ClSO_2H_2^+$, can lose a proton to the solvent, H_2O, and then react

$$
\begin{array}{c}
\ddot{C}l \\
\ddot{S} \\
\ddot{C}l\ddot{O}
\end{array}
+ H_2O \longrightarrow
\left[
\begin{array}{c}
\ddot{C}l \\
S\!:\!\ddot{O}\!: \\
\ddot{O} \\
H\quad H
\end{array}
\right]^{+}
+ Cl^-
\qquad [9\text{-}71]
$$

again with water to displace the second chlorine in the same way as the first. The over-all result is given in Expression 9-72. The

$$
SOCl_2 + 2H_2O \longrightarrow 2HCl + HO\text{-}S\text{-}OH \qquad [9\text{-}72]
$$
$$
\downarrow
$$
$$
O
$$

reaction is a displacement of one base, Cl^-, by a stronger base, H_2O.

Exercise 9-3. Write acid-base reactions to show the complete hydrolysis of $POCl_3$.

9-12. Hydrolysis of Alkyl Halides and Aryl Halides

The hydrolysis of an alkyl halide may also be pictured in the same way as the hydrolysis of the inorganic salts. The reaction is much slower, for the alkyl halides are not soluble in aqueous media and hence the reaction occurs only where the two surfaces meet. Since fewer collisions between molecules and ions can occur, the reaction is generally slower than one between two ions or between an ion and a strongly polar molecule.

$$CH_3-CH_2-Cl + OH^- \longrightarrow CH_3-CH_2-OH + Cl^-$$
$$\text{acid}_1 \qquad \text{base}_2 \qquad\qquad \text{acid}_2 \qquad \text{base}_1$$
$$[9\text{-}73]$$

The halogen of an aryl halide is even more tightly held by carbon than that of an alkyl halide, and the reaction must be carried out at fusion temperatures. Expression 9-75 shows a neutralization in

$$\langle \rangle\text{-Cl} + OH^- \xrightarrow{\text{fuse}} \langle \rangle\text{-OH} + Cl^-$$
$$\text{acid}_1 \qquad \text{base}_2 \qquad \text{acid}_2 \qquad \text{base}_1$$
$$[9\text{-}74]$$

$$\langle \rangle\text{-OH} + OH^- \longrightarrow \langle \rangle\text{-O}^- + H_2O$$
$$\text{acid}_1 \qquad \text{base}_2 \qquad \text{base}_1 \qquad \text{acid}_2$$
$$[9\text{-}75]$$

which only the OH^- and H^+ (from phenol) actually take part.

9-13. Hydrolysis of Alkenes and Alkynes

The hydrolysis of an alkene is likewise an acid-base reaction in the Lewis sense; the alkene acts as an electron donor.

$$CH_3-CH_2-CH{=}CH_2 + H^+Br^- \longrightarrow CH_3-CH_2-\overset{\oplus}{CH}-CH_3$$
$$[9\text{-}76a]$$

There is evidence that the proton or positive part of a reagent always adds first to an ordinary alkene. The reagent then always behaves like an acid, and the alkene is an electron donor, or base. The resulting intermediate, an ion having a positive charge, has been named the carbonium ion. It immediately or simultaneously adds the negative part of the reagent as the second step. It must be emphasized that the intermediate carbonium ion has a very

$$CH_3-CH_2-\underset{\oplus}{CH}-CH_3 + Br^- \longrightarrow CH_3-CH_2-\underset{Br}{CH}-CH_3$$
$$[9\text{-}76b]$$

short life, indeed, in comparison with the ionic reactions of inorganic compounds. It has been estimated to have a life of 10^{-5} second. (The example given, of HBr adding to an alkene, is, of course, not a hydrolysis.)

1. In the hydrolysis of an alkene in the presence of a catalytic amount of sulfuric acid, there is competition for the carbonium

$$[9\text{-}77]$$

ion between the two bases, H_2O and OSO_3H^-. Both are very weak bases, and the result is decided largely on a statistical basis; that is, the product will be mainly tert-butyl alcohol when the water is present in an overwhelming excess.

2. The hydrolysis of acetylene to produce acetaldehyde may be pictured in an analogous manner. The intermediate alkene alcohol is not stable and rearranges to acetaldehyde, a compound bearing a new functional group. (See Chap. 11.)

$$[9\text{-}78]$$

Exercise 9-4. Select the conjugate acid-base pairs in Expressions 9-77 and 9-78.

SUMMARY

Acid and base are defined in terms of the phenomena that they exhibit. Their behavior is accounted for by various theories (those of Arrhenius, Brönsted, and Lewis), each of which has advantages and limitations in practical application.

Qualitative ideas of the acid and base strengths of commonly encountered substances are related to arbitrary standards. The following terms are defined: neutralization, titration, pH, formal solution. Quantitative aspects of hydrolysis problems are left for another course, but organic and inorganic hydrolyses are discussed in general terms.

REFERENCE

Hall, "Systems of Acids and Bases," J. Chem. Education, 17, 124 (1940).

EXERCISES

(Exercises 1–4 will be found within the text.)

5. Define Arrhenius base, Brönsted acid, Lewis acid.

6. Write an equation to indicate that HSO_4^- is an acid; that HSO_4^- is a base.

7. Write equations for reactions that occur between the following pairs of substances, and indicate the conjugate acid-base pairs. Also indicate which reactions go nearly to completion.

1. HCO_3^- and OH^-
2. HCO_3^- and H_3O^+
3. NH_4^+ and OH^-
4. CH_3COOH and OH^-
5. CN^- and H_2O
6. CH_3CH_2OH and OH^-

8. a. By proper choice of reagents, write reactions to show that the following compounds can be divided into acids, bases, and neutral substances:

1. $CH_3-CH_2-CH_2OH$ 3. phenol   5. $NaHCO_3$
2. H_2SO_4 4. NaOH   6. CH_3CH_2COOH

b. By further choice of reagents, arrange the six compounds in *decreasing* order of acid strength.

c. If your previous experience does not allow you to separate any two (or more) of the six compounds, can you suggest a way of determining the proper order in the laboratory?

d. In answering part *b* of this question, a student placed $NaHCO_3$ in two positions in his scale. Can this be defended?

9. Omitting ions that do not react, write equations to account for the following observations and explain each. The pH of water is

1. raised when CH_3CH_2COOK is added
2. raised when CH_3CH_2ONa is added
3. lowered when NH_4Br is added
4. lowered when $FeCl_3 \cdot 6H_2O$ is added
5. unchanged when NaCl is added
6. unchanged when CH_3COONH_4 is added

10. Apply the Lewis theory to the following reactions by labeling the acid and the base in each and showing the electron configuration:

1. $BCl_3 + NH_3$
2. $AlCl_3 + CH_3CH_2Cl$
3. nitration of benzene (both steps of proposed mechanism)

11. Does anhydrous $AlCl_3$ act as an acid or as a base in the Friedel-Crafts reaction of ethyl chloride with benzene? Is it an acid (or a base) in the Arrhenius sense? in the Brönsted sense? in the Lewis sense?

12. Write two reactions that will show that acetylene is more acidic than ethylene.

13. Calculate the pH of 0.1F hydrochloric acid, assuming it is 100% ionized. Calculate the pH of 0.1F acetic acid, assuming it is 1.4% ionized.

14. The pH of an unknown solution was found to be 3.2. What is the hydrogen ion concentration? In another solution the pH is 10.8. What is the hydrogen ion concentration? May a solution have a pH of zero?

15. The pH of a sample of tap water was found to be 7.8. What is the hydrogen ion concentration? What is the hydroxyl ion concentration?

16. Write the equation for the reaction between ethanol and methyl magnesium bromide. Is this an acid-base reaction in the Arrhenius sense? in the Brönsted sense? in the Lewis sense? If so, which reactant is acid and which base?

10 Carboxylic Acids

The functional group in a carboxylic acid is $-\overset{\displaystyle O}{\overset{\|}{C}}-OH$,

and the general formula is $R-\overset{\displaystyle O}{\overset{\|}{C}}\diagdown_{OH}$ or $R-COOH$. Acids contain-

ing one or more carboxyl groups may be further designated as

TABLE
10-1 | *Carboxylic Acids*

FORMULA	COMMON NAME	IUC NAME	OCCURRENCE
$H-COOH$	formic acid	methanoic acid	red ants
CH_3-COOH	acetic acid	ethanoic acid	vinegar
CH_3-CH_2-COOH	propionic acid	propanoic acid	
$CH_3(CH_2)_2COOH$	n-butyric acid	butanoic acid	rancid butter
$CH_3(CH_2)_3COOH$	n-valeric acid	pentanoic acid	
$CH_3(CH_2)_4COOH$		hexanoic acid	goat-milk fat
$CH_3(CH_2)_6COOH$		octanoic acid	goat-milk fat
$CH_3(CH_2)_8COOH$		decanoic acid	goat-milk fat
$CH_3(CH_2)_{10}COOH$	lauric acid	dodecanoic acid	coconut oil
$CH_3(CH_2)_{12}COOH$	myristic acid	tetradecanoic acid	coconut oil, nutmeg oil
$CH_3(CH_2)_{14}COOH$	palmitic acid	hexadecanoic acid	palm oil
$CH_3(CH_2)_{16}COOH$	stearic acid	octadecanoic acid	lard, tallow
$CH_3(CH_2)_7CH=CH(CH_2)_7COOH$	oleic acid	9-octadecenoic acid	olive oil
⬡-COOH	benzoic acid	benzoic acid	

monocarboxylic, dicarboxylic, tricarboxylic, etc. Acids in the aliphatic series are also referred to as fatty acids since many of them occur in fats. It is a notable fact that the naturally occurring fatty acids are chain compounds containing an even number of carbon atoms.

10-1. Nomenclature of Carboxylic Acids

The common names are frequently taken from Latin names that indicate the source from which the acid was first obtained. The symbol *n-* is seldom used with acids of six carbons or more since the normal compound is the only one that is common in nature or the laboratory.

The IUC names are derived from the original hydrocarbon stem name by addition of the suffix *-oic* and the word *acid*. No number is needed to designate the position of the carboxyl group since it must of necessity be on the end of the chain.

One unsaturated acid, oleic, is given at the end of Table 10-1 because it is a common one, occurring in many fats and oils. (General discussion of unsaturated acids is reserved for Chap. 23.)

10-2. Formic Acid and Its Reactions

Formic acid, the first in the homologous series, is a colorless liquid with a biting taste and a sharp odor. It was first prepared by grinding red ants in a mortar and extracting with water.

Commercially, it is prepared by passing carbon monoxide into molten sodium hydroxide at a temperature of 210°C and under 8 atmospheres of pressure. The resulting salt, called sodium formate, can be made to liberate the acid by treatment with a strong inorganic acid.

$$CO + NaOH \xrightarrow[\text{8 atm}]{210°} H-C\overset{\displaystyle O}{\underset{\displaystyle ONa}{\diagdown}} \qquad\qquad [10\text{-}1]$$
$$\text{sodium formate}$$

$$HCOONa + H_3PO_4 \longrightarrow H-COOH + NaH_2PO_4 \qquad\qquad [10\text{-}2]$$

Formic acid is readily dehydrated by concentrated sulfuric acid. This reaction must be avoided in the preparation of the acid. (Since phosphoric acid is a noncharring acid, its use was designated for the reaction in Expression 10-2. Dilute sulfuric can be used in place of phosphoric acid, however.) The reaction with

$$H\text{-}COOH \xrightarrow[100\%]{H_2SO_4} CO + H_2O \qquad\qquad [10\text{-}3]$$

concentrated H_2SO_4 is used as a laboratory method for preparing carbon monoxide.

Formic acid is readily oxidized even by mild oxidizing agents, since it is itself a strong reducing agent.

$$H\text{-}C{\overset{O}{\underset{OH}{\diagup}}} \xrightarrow[\substack{agent}]{oxidizing} \underset{\text{carbonic acid}}{HO\text{-}\overset{O}{\overset{\|}{C}}\text{-}OH} \longrightarrow CO_2 + H_2O \qquad [10\text{-}4]$$

Formic acid is used in large quantities for dehairing hides in the tanning industry. It is also used for coagulating rubber latex and as an acid for dye baths in the textile industry.

10-3. Acetic Acid

Vinegar is a dilute solution $(4\text{--}5\%)$ of acetic acid in water and is commonly prepared by a fermentation process. Ethyl alcohol (obtained by the enzymatic degradation of the sugars in cider or grape juice by zymase) may be oxidized to acetic acid by the bacterium *Mycoderma aceti* ("mother of vinegar") in the presence of air.

Glacial acetic acid $(100\%$ acetic acid) is so called because it freezes at $16.5°$ and in that form looks like ice. It is made commercially from acetylene as a starting material. The hydrolysis of acetylene yields acetaldehyde (Exp. 5-13), which is easily oxidized to the acid at $75°$ in the presence of air and a catalyst of cobalt II acetate.

$$CH_3-\overset{\overset{\displaystyle H}{|}}{C}=O + air \xrightarrow[95\%]{\underset{\text{catalyst, 75°}}{Co(OOCCH_3)_2}} CH_3-C\overset{\displaystyle O}{\underset{\displaystyle OH}{\diagdown}} \qquad [10\text{-}5]$$

10-4. General Methods of Preparation of Carboxylic Acids

1. Oxidation of a Primary Alcohol or an Aldehyde.   The

$$R-\overset{\overset{\displaystyle H}{|}}{\underset{\underset{\displaystyle H}{|}}{C}}-OH \xrightarrow[\text{agent}]{\text{oxidizing}} \left[R-\overset{\overset{\displaystyle OH}{|}}{\underset{\underset{\displaystyle H}{|}}{C}}-OH \right] \longrightarrow \underset{\text{an aldehyde}}{R-\overset{\overset{\displaystyle H}{|}}{C}=O} \xrightarrow[]{\text{oxidizing}\,\text{agent}} \underset{\text{an acid}}{R-C\overset{\displaystyle O}{\underset{\displaystyle OH}{\diagdown}}} \qquad [10\text{-}6]$$

oxidation of any primary alcohol by most oxidizing agents will yield an acid. It is often possible, if desired, to isolate an aldehyde as an intermediate in the oxidation. The stability of the aromatic ring toward oxidizing agents makes it possible to oxidize an alcohol or aldehyde group on a side chain without affecting the ring.

$$\underset{\text{benzyl alcohol}}{\langle\bigcirc\rangle-CH_2OH} \xrightarrow[\text{agent}]{\text{oxidizing}} \underset{\text{benzaldehyde}}{\langle\bigcirc\rangle-\overset{\overset{\displaystyle H}{|}}{C}=O} \xrightarrow[93\%]{\underset{\text{agent}}{\text{oxidizing}}} \underset{\text{benzoic acid}}{\langle\bigcirc\rangle-C\overset{\displaystyle O}{\underset{\displaystyle OH}{\diagdown}}} \qquad [10\text{-}7]$$

Even an R group can be oxidized by vigorous oxidizing agents without rupturing the ring. The side chain is always degraded to one carbon, regardless of the original length of the R group.

$$\langle\bigcirc\rangle-CH_2-CH_3 \xrightarrow[\text{agent}]{\text{oxidizing}} \langle\bigcirc\rangle-C\overset{\displaystyle O}{\underset{\displaystyle OH}{\diagdown}} + CO_2 + 2H_2O \qquad [10\text{-}8]$$

2. Hydrolysis of a Cyanide.   (Review Chap. 9.) Cyanides are hydrolyzed in acidic or even more readily in basic solution. Partial

$$\underset{}{R-\overset{\delta+}{C}\equiv\overset{\delta-}{N}} + :\overset{..}{\underset{..}{O}}:H^- \longrightarrow \left[R-C=\overset{..}{N}:\overset{\diagdown}{OH} \right]^- \longrightarrow \left[R-\overset{..}{\underset{\diagdown O}{C}}-NH \right]^- \xrightarrow{H_2O}$$

$$R-\overset{}{\underset{\diagdown O}{C}}-NH_2 + OH^- \qquad [10\text{-}9]$$

hydrolysis of a cyanide may end with the formation of an amide (Exp. 10-9). When the hydrolysis is complete in acid solution, the products are ammonium ion and the carboxylic acid (Exp. 10-10); in basic solution ammonia is liberated and a carboxylate ion remains in solution (Exp. 10-11). The carboxylic acid may be ob-

$$R-C\equiv N + H_2O + H^+ \longrightarrow RCOOH + NH_4^+ \qquad [10\text{-}10]$$

$$R-C\equiv N + H_2O + OH^- \longrightarrow RCOO^- + NH_3 \qquad [10\text{-}11]$$

tained from the carboxylate ion by treatment with a strong acid.

In acid the path of the cyanide to the carboxylic acid undoubtedly includes $RCONH_2$ as an intermediate and probably some of the other structures in Expression 10-12. If the cyanide structure includes any contribution from $R-\overset{\delta+}{C}=\overset{\delta-}{\overset{..}{N}}:$, the first fragments are reasonable.

[10-12]

In the aliphatic series, one method of lengthening a carbon chain by one carbon makes use of the cyanide as an intermediate. Starting with a primary alcohol, the corresponding acid containing one more carbon atom may be made by the transformations of Expression 10-13 (only main products are shown).

$$R\text{-}OH \xrightarrow[\text{H}_2\text{SO}_4]{\text{HBr}} R\text{-}Br \xrightarrow{\text{NaCN}} R\text{-}C\equiv N \xrightarrow[\text{H}_2\text{O}]{\text{NaOH}} R\text{-}C\overset{O}{\underset{\text{ONa}}{\diagdown}} \xrightarrow{\text{H}^+} RCOOH \qquad [\textbf{10-13}]$$

3. Grignard Reaction. A Grignard reagent will add to CO_2 to give a product that readily hydrolyzes to an acid. The reaction is conveniently carried out in the laboratory by adding Dry Ice to an ether solution of the Grignard.

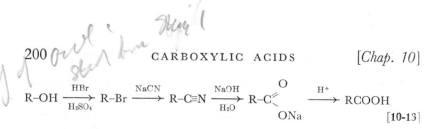

$$[\textbf{10-14}]$$

10-5. Reactions of the Acidic Hydrogen

1. Salt Formation

$$R\text{-}C\overset{O}{\underset{\text{OH}}{\diagdown}} + NaOH \longrightarrow R\text{-}C\overset{O}{\underset{\text{ONa}}{\diagdown}} + H_2O \qquad [\textbf{10-15}]$$

With a base like $Ca(OH)_2$, two molecules of the carboxylic acid are neutralized per calcium ion. When acetic acid is used, calcium acetate is the product. A saturated solution of calcium acetate is

$$2CH_3\text{-}C\overset{O}{\underset{\text{OH}}{\diagdown}} + Ca(OH)_2 \longrightarrow \begin{matrix} CH_3\text{-}C\overset{O}{\diagdown} \\ \overset{O}{\diagdown} \\ {}_{O}^{} Ca \\ \overset{O}{\diagup} \\ CH_3\text{-}C\overset{}{\underset{O}{\diagup}} \end{matrix} + 2H_2O \qquad [\textbf{10-16}]$$

calcium acetate

$$CH_3CH_2CH_2COOH + NH_4OH \longrightarrow \underset{\text{ammonium n-butyrate}}{CH_3CH_2CH_2COONH_4} + H_2O \qquad [\textbf{10-17}]$$

capable of forming a jelly-like solid with ethyl alcohol, used for "canned heat."

2. Metals. Zinc will slowly dissolve in the lower acids of the series. Hydrogen is liberated. The more active metals Li, K, and

$$
\text{Zn} + 2\text{HCOOH} \longrightarrow \quad
\begin{array}{c}
\text{O} \\
\text{H-C}{\displaystyle \Big\langle} \\
\text{O} \\
\phantom{\text{H-C}}\Big\rangle\text{Zn} + \text{H}_2 \\
\text{O} \\
\text{H-C}{\displaystyle \Big\langle} \\
\text{O}
\end{array}
\qquad [10\text{-}18]
$$

Na will liberate hydrogen from longer-chain acids; they must be heated with those of high molecular weight.

$$
\text{CH}_3(\text{CH}_2)_4\text{COOH} + \text{Na} \longrightarrow \text{CH}_3\text{-}(\text{CH}_2)_4\text{-COONa} + \tfrac{1}{2}\text{H}_2 \qquad [10\text{-}19]
$$

SUMMARY

Carboxylic acids of an even number of carbon atoms occur widely in nature as constituents of fats, oils, and waxes. The higher members of this homologous series are more important than in any other series so far studied in relation to the lower members. The carboxylic acids are weak in comparison with mineral acids; the carboxyl group is m-directing in aromatic substitutions.

Carboxylic acids are prepared by oxidation of an alcohol, of an aldehyde, or of the side chain on an aromatic ring, by hydrolysis of a cyanide, and by the action of carbon dioxide on a Grignard reagent. The reactions of the acidic hydrogen in a carboxyl group are discussed here, but the properties of the OH group in the carboxyl group are reserved for Chapter 14.

EXERCISES

1. Indicate two methods for converting ethyl chloride to propanoic acid.

2. Write equations for the reactions, if any, between the following substances:

1. *p*-chlorobenzoic acid and $NaHCO_3$
2. *o*-hydroxybenzoic acid and NaOH
3. sodium *o*-nitrobenzoate and aqueous HCl

3. Indicate how to accomplish the following conversions:

1. 1-propanol $\longrightarrow$ propanoic acid
2. 2-propanol $\longrightarrow$ 2-methylpropanoic acid
3. toluene $\longrightarrow$ *p*-chlorobenzoic acid (the COOH group orients *meta*)

4. Synthesize $CH_3-CH_2-O-CH_2-COOH$. Hint: start with ethylene oxide.

5. How could you use sodium hydroxide and sodium hydrogen carbonate to distinguish the three compounds benzoic acid, benzyl alcohol, and *p*-cresol?

6. One gram of a substance from an unlabeled bottle on the shelf containing organic acids is found to neutralize 35.4 ml of a 0.1F solution of KOH. If the acid is a pure substance, what is it likely to be? How many grams of KOH were used in the titration?

7. How would you distinguish formic from acetic acid?

8. When a compound, $C_{13}H_{12}O$, is treated with concentrated HI, two products, A and B, are produced. Compound A gives a color with $FeCl_3$, and B contains iodine. Upon oxidation, B is converted to C, an acid of formula $C_7H_6O_2$. What is the structure of the original compound? Write equations for each reaction described.

11 Carbonyl Compounds: Aldehydes

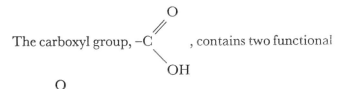

The carboxyl group, $-C\overset{\displaystyle O}{\underset{\displaystyle OH}{}}$, contains two functional

groups, $-OH$ and $-\overset{\displaystyle O}{\underset{\displaystyle}{C}}-$. The disguise is nearly perfect in that the properties of a carboxyl group are not the sum of the properties of the two functions. The whole carboxyl group is different from, if not greater than, the sum of its parts. This justifies studying the

carboxyl and *carbonyl* $\left(-\overset{\displaystyle O}{\underset{\displaystyle}{C}}- \right)$ groups separately. If the atoms attached immediately to the carbonyl are hydrogen and carbon, the function is called an aldehyde, $R-\overset{\displaystyle H}{\underset{\displaystyle}{C}}{=}O$; if both atoms are carbon, the compound is called a ketone, $R-\overset{\displaystyle O}{\underset{\displaystyle}{C}}-R'$ (Chap. 12). Even these two groups behave differently toward some reagents, for in general the hydrogen makes the aldehydes considerably more reactive than the ketones.

The chemistry of the carbonyl group is most readily accounted for if it is assumed that one of the pairs of electrons in the double bond is displaced toward the oxygen atom. The formula for an aldehyde, then, might better be represented by some combination of the following structures:

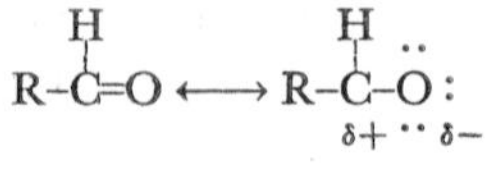

$$\underset{\begin{array}{c}H\\ |\end{array}}{R-C=O} \longleftrightarrow \underset{\delta+ \quad \delta-}{\overset{\begin{array}{c}H\\ |\end{array}}{R-C-\ddot{O}:}}$$

In contrast to an alkene, which acts as an electron-pair donor (a Lewis base) in all its addition reactions, the carbonyl double bond acts as an electron-pair acceptor (a Lewis acid).

TABLE

11-1 | *Aldehydes*

STRUCTURAL FORMULA	COMMON NAME	IUC NAME
H–CHO	formaldehyde	methanal
CH₃–CHO	acetaldehyde	ethanal
CH₃–CH₂–CHO	propionaldehyde	propanal
CH₃–CH₂–CH₂–CHO	n-butyraldehyde	butanal
CH₃⟍CH–CHO⟋CH₃	isobutyraldehyde	2-methylpropanal
CH₃–CH₂|CH₃–CH₂–CH–CHO	———	2-ethylbutanal
Cl₃C–CHO	chloral	2,2,2-trichloroethanal
CH₃ CH₃| |CH₂=C(CH₂)₃CH–CH₂–CHO	citronellal	3,7-dimethyl-7-octenal
⬡-CHO	benzaldehyde	benzaldehyde

11-1. Nomenclature of Aldehydes

We derive the common names of the lower members of the series of aldehydes from the common names of the acids having the same number of carbons by dropping the *-ic* ending and adding the suffix *-aldehyde*. The IUC ending for an aldehyde is *-al* and is suffixed to the stem of the corresponding hydrocarbon. No number is needed to designate the position of the aldehyde group since the group is always on the No. 1 carbon in the chain.

Exercise 11-1.   Name the following compounds:

1. CH$_3$–CH–CH$_2$–CHO
      CH$_2$–CH$_3$

2. CH$_3$–CH=CH–CHO

3. CH$_3$–CH$_2$–⟨_⟩–CHO

4. CH$_3$–CH–CH$_2$–CHO
      OH

Exercise 11-2.   Write structural formulas for the following compounds: 3-ethylhexanal, *p*-hydroxybenzaldehyde, 2-chloropropanoic acid, *o*-bromobenzaldehyde, 3-butenal.

11-2. Preparation of Aldehydes

Four general methods of preparing aliphatic aldehydes are in common use. (None of these can be used for aromatic aldehydes. The reasons are practical rather than theoretical. In general, the temperatures required for the production of aromatic aldehydes by these methods are so high that decomposition would accompany, if not supersede, formation of the compound. Methods of making aromatic aldehydes are given on pp. 625–626.)

1. Oxidation of a Primary Alcohol.   Aldehydes may be prepared from primary alcohols either by oxidation or by dehydrogenation. The common chemical oxidizing agents may be used for the first method, which is feasible only for low-boiling aldehydes

$$\underset{\overset{|}{H}}{\overset{\overset{H}{|}}{R-C-OH}} \xrightarrow[\text{agent}]{\text{oxidizing}} \left[\underset{\overset{|}{H}}{\overset{\overset{O-H}{|}}{R-C-OH}}\right] \longrightarrow \overset{\overset{H}{|}}{R-C=O} + H_2O \qquad [\textbf{11-1}]$$

that can be removed before they are oxidized to acids. A catalyst of copper gauze kept at 250° is used in the manufacture of "formalin" (a 40% solution of formaldehyde in water). A mixture of air and methanol vapor is passed over the heated copper. A com-

bination of dehydrogenation and oxidation of the hydrogen results, the latter keeping the temperature at the desired point;
that is, both free hydrogen and oxidized hydrogen (water) are
formed during the reaction.

2. Dehydrogenation of a Primary Alcohol. Catalytic dehydrogenation by a copper catalyst in the absence of air is the industrial process for making aldehydes from alcohols, except for $HCHO$
and CH_3CHO.

$$\underset{\underset{H}{|}}{\overset{\overset{H}{|}}{R-C}}-OH \xrightarrow[200-300°]{Cu} \overset{\overset{H}{|}}{R-C}=O + H_2 \qquad [11\text{-}2]$$

3. From Acids in the Vapor State. Between temperatures of
200° and 300°C a mixture of acid vapors from formic acid and any
other volatile acid will decompose, when passed over MnO, to
an aldehyde, CO_2, and H_2O.

$$R-COOH + H-COOH \xrightarrow[200-300°]{MnO} \overset{\overset{H}{|}}{R-C}=O + H_2O + CO_2 \qquad [11\text{-}3]$$

4. From Salts of Acids. If an alkali metal salt of an acid is
heated with sodium formate, an aldehyde will distill away from
the reaction mixture and leave sodium carbonate.

$$R-C{\overset{O}{\underset{ONa}{}}} + \underset{\underset{\text{sodium formate}}{ONa}}{H-C{\overset{O}{}}} \longrightarrow \overset{\overset{H}{|}}{R-C}=O + Na_2CO_3 \qquad [11\text{-}4]$$

5. Special Method for Acetaldehyde. Acetaldehyde is not
made commercially by any of the above processes, but rather by
hydrolysis of acetylene in the presence of sulfuric acid and mercury
II sulfate (Exp. 5-13).

11-3. Reactions of Aldehydes

The reactions of aldehydes will be divided into the following types for the purpose of showing similarities and differences among them: oxidation, addition, polymerization, condensation, and substitution. The first member of the series sometimes behaves somewhat differently from other aldehydes; when it does, the difference will be specifically mentioned.

A. *Oxidation of Aldehydes*

It was shown in Chapter 10 that aldehydes are easily oxidized to organic acids. Formaldehyde is the most easily oxidized of all

$$R-CHO \xrightarrow[\text{agent}]{\text{oxidizing}} \underset{\text{an acid}}{RCOOH} \qquad [11\text{-}5]$$

the aldehydes, and the first product of oxidation still contains an aldehyde function, which may be oxidized further to carbonic acid ($CO_2 + H_2O$).

$$\underset{H}{\overset{H}{H-C=O}} \xrightarrow[\text{agent}]{\text{oxidizing}} \overset{O}{\underset{}{H-C-OH}} \xrightarrow[\text{agent}]{\text{oxidizing}} \overset{O}{\underset{}{HO-C-OH}} \qquad [11\text{-}6]$$

Various oxidizing agents may be used, and the selection of a particular one depends in part on one's purpose. Mild oxidizing agents may be used as tests for aldehydes; more vigorous ones are used for preparative purposes. Oxygen or air in the presence of many oxide catalysts, such as MnO, V_2O_5, and CuO, bring about the oxidation. Chemical reagents such as permanganate ion in alkaline solution or dichromate ion in acid solution are effective.

Tollens' reagent, an ammoniacal solution of silver oxide, gives a very sensitive test for the aldehyde group by depositing a silver mirror on the walls of a test tube. This reagent is made by adding ammonium hydroxide to a suspension of Ag_2O until it just dissolves upon shaking. The Ag_2O is the precipitate initially obtained at the first addition of ammonium hydroxide to a silver nitrate

$$2Ag^+ + 2OH^- \longrightarrow Ag_2O + H_2O$$
$$4NH_3 + H_2O + Ag_2O \longrightarrow 2Ag(NH_3)_2OH$$
$$\Updownarrow$$
$$2Ag(NH_3)_2^+ + 2OH^- \qquad\qquad [11\text{-}7]$$

$$\underset{\displaystyle R\text{-}\overset{\displaystyle H}{\underset{|}{C}}\text{=}O}{} + 2Ag(NH_3)_2^+ + 2OH^- \longrightarrow R\text{-}C\!\!\begin{array}{c} \nearrow O \\ \searrow ONH_4 \end{array} + 3NH_3 + \underline{2Ag} + H_2O \qquad [11\text{-}8]$$

solution. The complex silver ion oxidizes the aldehyde readily in the presence of base and is itself reduced to metallic silver. The ammonium salt of the acid is the organic product obtained.

Fehling's solution is another oxidizing agent frequently used to test for the aldehyde group. Since the reagent consists of an unstable complex ion, it is prepared just before use by mixing two solutions kept in separate bottles. Solution No. 1 is a copper II sulfate solution, and No. 2 is an alkaline solution of sodium potassium tartrate. The complex copper II tartrate ion is soluble in basic solution. Without some complexing agent, solution is impossible, for otherwise $Cu(OH)_2$ would be precipitated:

$$CuSO_4 + 2NaOH \longrightarrow \underline{Cu(OH)_2} + Na_2SO_4 \qquad [11\text{-}9]$$

The reaction between the soluble copper II complex ion and an aldehyde in basic solution results in a red-brown precipitate of Cu_2O, copper I oxide.

$$\underset{\displaystyle R\text{-}\overset{\displaystyle H}{\underset{|}{C}}\text{=}O}{} + 2Cu^{++} + 5OH^- \longrightarrow R\text{-}C\!\!\begin{array}{c} \nearrow O^- \\ \searrow O \end{array} + \underline{Cu_2O} + 3H_2O \qquad [11\text{-}10]$$

B. *Addition to Aldehydes*

The addition reactions of a carbonyl group are readily understood if the displacement of electrons in the group is assumed to be represented by the following picture:

$$\begin{array}{ccc} H & & H \\ | & & | \\ R-C=O & \longleftrightarrow & R-C-\overset{..}{\underset{..}{O}}: \\ & & \underset{\delta+\ \delta-}{} \end{array}$$

It has been shown for some reactions—for example, the addition of HCN—that the negative part of the reagent adds first to the carbon of the carbonyl group. The carbonyl group, therefore, is an electron-seeking group, at least in this particular reaction.

1. Hydrogen Cyanide

$$\begin{array}{ccccccc} H & & H & & & H & & H \\ | & & | & & & | & & | \\ R-C=O & \longleftrightarrow & R-C-\overset{..}{O}: + H & :CN & \longrightarrow & R-C-\overset{..}{O}:{}^{\ominus} & \xrightarrow{H^+} & R-C-OH \\ & & \underset{\delta+\ \delta-}{} & \overset{\delta+\ \delta-}{} & & & CN & & CN \end{array}$$

$$\text{a cyanohydrin}$$

$$[11\text{-}11]$$

2. Hydrogen.   Hydrogen will add to the aldehyde group in the presence of metals such as platinum or nickel under low or high pressure. The products are primary alcohols.

$$\begin{array}{cc} H & H \\ | & | \\ R-C=O + H_2 & \xrightarrow[\text{or Ni}]{Pt} R-C-OH \\ & | \\ & H \end{array} \qquad [11\text{-}12]$$

3. Sodium Hydrogen Sulfite.   The addition of sodium hydrogen sulfite to an aldehyde yields a white crystalline substance, which can be recrystallized from water. It has been proved that the hydrogen sulfite addition products (as they are called) bear a C-to-S linkage rather than a C-to-O   Pure aldehyde may be

$$\begin{array}{cc} H & H \\ | & | \\ R-C=O + NaHSO_3 & \underset{H^+ or OH^-}{\rightleftharpoons} R-C-OH \\ & | \\ & SO_2ONa \end{array} \qquad [11\text{-}13]$$

obtained from the recrystallized product by a reversal of the reaction in Expression 11-13, either an acid or a base being added. This is the principal use of the reaction.

Exercise 11-3. Write an equation for the reaction of $C_6H_5-CHOH-SO_3^-$ with H^+.

4. Ammonia. Aldehydes also add ammonia to give unstable aldehyde ammonias. Formaldehyde reacts in the same way, but the reaction continues further (by splitting out water) to yield a

$$\underset{\text{H}}{\overset{\text{H}}{\text{R-C=O}}} + \overset{\delta+\ \delta-}{\text{H : NH}_2} \longrightarrow \underset{\text{NH}_2}{\overset{\text{H}}{\text{R-C-OH}}} \tag{11-14}$$

$$\underset{\text{H}}{\overset{\text{H}}{\text{H-C=O}}} + \overset{\delta+\ \delta-}{\text{H : NH}_2} \longrightarrow \underset{\text{NH}_2}{\overset{\text{H}}{\text{H-C-OH}}} \xrightarrow[\text{3 NH}_3]{\text{5HCHO}} (CH_2)_6N_4 + 6H_2O \tag{11-15}$$

complex ring compound called hexamethylenetetramine. Acids will reverse this reaction. Hexamethylenetetramine is used as an accelerator in the vulcanizing of rubber, in incendiary mixtures, and in the manufacture of Cyclonite (RDX, Hexogen), a high explosive more powerful than TNT. It is sold as a drug under the name Urotropin and before the advent of sulfa drugs was widely used in the treatment of cystitis.

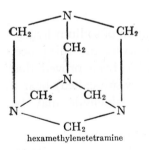

hexamethylenetetramine

Benzaldehyde, like formaldehyde, also deviates from the normal behavior of an aldehyde with ammonia. The first reaction may be an addition of a molecule of ammonia, but the reaction continues until three molecules of benzaldehyde have condensed with two molecules of ammonia. The final product is called hydrobenzamide.

$$C_6H_5\text{-CHO} + NH_3 \longrightarrow \underset{NH_2}{\overset{H}{C_6H_5\text{-}\underset{|}{\overset{|}{C}}\text{-OH}}} \xrightarrow[\underset{71\%}{NH_3}]{2C_6H_5CHO}$$

$$\underset{\underset{\text{hydrobenzamide}}{C_6H_5}}{C_6H_5\text{-CH=N-}\underset{|}{CH}\text{-N=CHC}_6H_5} + 3H_2O \qquad [11\text{-}16]$$

5. Alcohols. Alcohols will add reversibly to aldehydes to form hemiacetals. In the presence of dry HCl (1%) and an excess of the alcohol there is a further reaction, in which a molecule of water is lost. (NH_4Cl and $CaCl_2$ also catalyze acetal formation.) The final product is an acetal, which is stable in basic solution but decomposes on standing in acid solution or upon warming. In the pres-

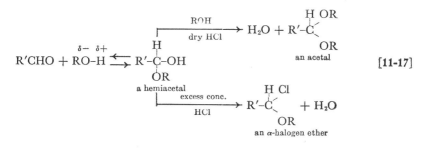

ence of an excess of concentrated HCl, a hemiacetal may be thought of as reacting like an alcohol; that is, the OH group is replaced by a Cl, and the resulting compound is an α-halogen ether. (For the notation, see p. 215.)

Exercise 11-4. Write an equation with structural formulas for the reaction of n-butyraldehyde in excess ethanol containing dissolved HCl.

6. Grignard Reagents. Grignard reagents can be used to prepare primary or secondary alcohols from the proper aldehydes. The intermediate addition compounds are easily hydrolyzed to a basic magnesium salt and the desired alcohol. Formaldehyde yields a primary alcohol, and all other aldehydes yield secondary alcohols.

$$\underset{H}{\overset{H}{H-C}}=O + \overset{\delta-}{R} \overset{\delta+}{-MgX} \longrightarrow \underset{R}{\overset{H}{H-C}}\overset{\ominus}{-O} \overset{\oplus}{MgX} \overset{\overset{\oplus}{H}\overset{\ominus}{OH}}{\longrightarrow} Mg\overset{OH}{\underset{X}{\diagdown}} + \underset{\text{a primary alcohol}}{R-CH_2OH}$$

[11-18]

$$\underset{H}{\overset{H}{R'-C}}=O + \overset{\delta-}{R} \overset{\delta+}{-MgX} \longrightarrow \underset{R}{\overset{H}{R'-C}}\overset{\ominus}{-O} \overset{\oplus}{MgX} \overset{\overset{\oplus}{H}\overset{\ominus}{OH}}{\longrightarrow} Mg\overset{OH}{\underset{X}{\diagdown}} + \underset{\underset{\text{a secondary alcohol}}{H}}{\overset{R'}{R-C}-OH}$$

[11-19]

The use of formaldehyde with a Grignard makes a good laboratory procedure for lengthening the carbon chain in an alcohol by one carbon. The transformations of Expression 11-20 are involved.

$$R-OH \xrightarrow[H_2SO_4]{HPr} R-Br \xrightarrow[\text{ether}]{Mg} R-MgX \xrightarrow{HCHO}$$
$$R-CH_2O-Mg-X \xrightarrow{HOH} RCH_2OH \quad [11-20]$$

Exercise 11-5. Show how to transform 2-propanol into 2-methyl-1-propanol.

C. Polymerization of Aldehydes

The polymerization of aldehydes is actually a reaction involving self-addition and might have been considered in section B. The carbonyl group adds to itself, and whether it forms a ring or a linear polymer depends on the particular aldehyde and the conditions.

Acetaldehyde forms a cyclic trimer called paraldehyde in the presence of strong acids. Since acetaldehyde boils (21°C) below room temperature, it is ordinarily sold as paraldehyde because the trimer is easily depolymerized to the monomer. Both reactions are acid-catalyzed. How is it possible, then, to go in one desired

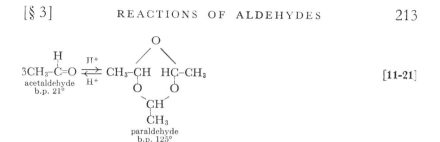

$$3CH_3-\overset{\overset{H}{|}}{C}=O \underset{H^+}{\overset{H^+}{\rightleftharpoons}} \quad [11\text{-}21]$$

acetaldehyde
b.p. 21°

paraldehyde
b.p. 125°

direction? To prepare paraldehyde, one adds a little acid to cold acetaldehyde and allows it to stand. After equilibrium is established, the acid is neutralized, and the paraldehyde may be dried and distilled. To obtain acetaldehyde from paraldehyde, one may simply add a trace of acid and warm the solution to allow slow volatilization of the acetaldehyde.

Formaldehyde also forms a cyclic trimer in the presence of strong acids, and it forms a linear polymer if a water solution is evaporated to dryness on the steam bath:

$$HOH + (x + 2)(H-\overset{\overset{H}{|}}{C}=O) \longrightarrow HOCH_2-O-[CH_2-O]_x-CH_2OH \quad [11\text{-}22]$$

paraformaldehyde

It has been estimated that x (Exp. 11-22) ranges from 4 to 100. Either of these polymers will evolve gaseous formaldehyde with strong heating. Trioxymethylene (the cyclic trimer) or paraformaldehyde candles are conveniently burned to yield formaldehyde as a fumigant.

D. Condensation Reactions of Aldehydes

Condensation reactions are of two types: one in which the reaction between the aldehyde and the reagent involves a splitting out of some simple unit (for example, H_2O, HCl, or NH_3) and one in which a new carbon-to-carbon linkage results.

The first type of condensation is made with certain nitrogen-containing compounds that are related to water and ammonia. One of the reagents, hydroxylamine (NH_2OH), may be considered as a hydroxy derivative of ammonia and has properties of both water and ammonia. Hydroxylamine melts at 33°C, has a density of 1.2 at this temperature, and decomposes above 50°C, sometimes with explosive violence. It is miscible with water and methyl and ethyl alcohols, and its aqueous solutions are stable up to a concentration of 60%. It resembles ammonia in that it forms salts with HCl, $HOSO_3H$, and other inorganic acids. These salts are stable, and this fact provides a convenient way of handling hydroxylamine in its reactions. Free hydroxylamine is readily released in a reaction when the solution is made alkaline.

$$NH_2OH \cdot HCl + NaOH \longrightarrow NH_2OH + NaCl + H_2O \qquad [11\text{-}23]$$

Hydrazine, H_2N-NH_2, is a derivative of ammonia and closely resembles it in its reactions. The hydrogens may be replaced by other groups, and one of the important substitution products is phenylhydrazine. This is a colorless liquid that freezes at 19°C and boils, with some decomposition, at 243°C. Phenylhydrazine is a strong enough base to form a hydrochloride. Another important property of phenylhydrazine, hydrazine, and hydroxylamine (see also p. 297) is that they are good reducing agents.

$$\text{⟨⟩}-NH-NH_2 + HCl \longrightarrow \text{⟨⟩}-NH-NH_2 \cdot HCl \qquad [11\text{-}24]$$

1. Hydroxylamine, NH_2OH. The condensation reactions of aldehydes with these nitrogen-containing compounds may be considered additions that are followed by the splitting out of water.

$$\begin{array}{c} H \qquad {}^{\delta+\delta-} H \\ R-C{=}O + H{-\!\!-\!\!-}N-OH \longrightarrow \left[\begin{array}{c} H\,\widehat{OH} \\ R-C{\diagdown}\,H \\ N \\ {}^{\diagdown}OH \end{array} \right] \longrightarrow \begin{array}{c} H \\ R-C{=}NOH \\ \text{an aldoxime} \end{array} \qquad [11\text{-}25] \end{array}$$

The products are called aldoximes. Most of them are readily purified solids, and their melting points may be used as one means of identifying the aldehyde used in the reaction.

2. Phenylhydrazine, $C_6H_5NHNH_2$. Phenylhydrazine also re-acts with many aldehydes to give yellow crystalline compounds having characteristic melting points, and these products also are frequently used for the identification of different aldehydes. The products are called phenylhydrazones.

$$\overset{H}{\underset{R-C=O}{|}} + \overset{\delta+\,\delta-}{H}-\overset{H}{\underset{N}{|}}\overset{H}{\underset{N}{|}}-\bigcirc \longrightarrow \left[R-\overset{H(OH)}{\underset{\underset{N-N-\bigcirc}{N-N-\bigcirc}}{C(H/H)}} \right] \longrightarrow \overset{H}{\underset{R-C=N-N}{|}}\overset{H}{\underset{}{|}}-\bigcirc + H_2O$$

a phenylhydrazone   [11-26]

3. Aldol Condensation. It is frequently necessary, in talking about the behavior of a compound, to mention a certain atom in the chain attached to the functional group. For a particular com-pound we may do this by numbering the atoms. If we wish to refer to the general class of compounds, another system is conventional.

$$\overset{\epsilon\ \ \delta\ \ \gamma\ \ \beta\ \ \alpha}{C-C-C-C=C-}\overset{H}{\underset{C=O}{|}}$$

The first carbon away from the functional group is called the α-carbon, the second the β-carbon, etc. The above chain is the skeleton of an α,β-unsaturated aldehyde.

The aldol condensation takes place between the carbonyl group of one aldehyde molecule and the α-carbon of a second aldehyde molecule. The reaction takes place only if there is at least one hydro-gen atom on this α-carbon. A mild alkaline catalyst is necessary for an appreciable rate of reaction, and this gives a key to the probable mechanism.

$$\overset{\delta+\ H}{\underset{\delta-\ |\ \ H}{H-C-C=O}} + OH^- \rightleftarrows \overset{..\ominus}{H-C-CHO} + H_2O \qquad [11\text{-}27a]$$

$$\overset{H\ ..\ \ \ ..\ominus}{CH_3-C:O:} + CH_2-CHO \rightleftarrows \overset{H\ ..\ \ominus}{CH_3-C-O:} \qquad [11\text{-}27b]$$

$$CH_3-CH-CH_2-CHO + HOH \rightleftarrows CH_3-CH-CH_2-CHO + OH^- \qquad [11\text{-}27c]$$

The first step is pictured as the formation of an anion with a negative charge on the α-carbon by reaction of the aldehyde with the base. In the second step, this anion adds to the carbonyl group of a free molecule of the aldehyde. This is a reasonable step inasmuch as the carbonyl group is an electron-seeking group. The product of this second step, an anion, is a strong base and might be expected to pick up a proton readily from the solvent, water. This is shown in the third step, which results in a β-hydroxyaldehyde and a regeneration of the alkaline catalyst. When acetaldehyde is the starting compound, as in Expression 11-27, the product is known as aldol, and the general reaction has come to be known as the aldol condensation. With KOH at 0°, the yield is 85%. When propionaldehyde is the starting compound (Exp. 11-28), α-methyl-β-hydroxyvaleraldehyde is formed.

$$2CH_3-CH_2-CHO \xrightarrow[60\%]{\text{dil. KOH, 10°}} CH_3-CH_2-\underset{\underset{\displaystyle OH}{|}}{C}\diagdown\underset{\underset{\displaystyle CH_3}{|}}{\overset{\overset{\displaystyle H}{|}}{C}}-CHO \xrightarrow{70\%}$$

$$\text{\footnotesize α-methyl-β-hydroxyvaleraldehyde}$$

$$CH_3CH_2-CH=\underset{\underset{\displaystyle CH_3}{|}}{C}-CHO \qquad [11\text{-}28]$$

Exercise 11-6. Write a three-step mechanism to show how α-methyl-β-hydroxyvaleraldehyde is formed from propionaldehyde by the aldol condensation.

The class of compounds formed in the aldol condensation, β-hydroxyaldehydes, readily lose a molecule of water to give an α,β-unsaturated aldehyde (Exp. 11-28). Loss of water will seem reasonable if it is noted that the α,β-unsaturated aldehyde contains a conjugated system of double bonds, already noted to be a fairly stable system (see p. 104). Of course, if the α-carbon in the hydroxyaldehyde bears no hydrogen, loss of water cannot occur. Grignard in 1910 obtained 70% 2-methyl-2-pentenal from propionaldehyde when the aldol product was not isolated first.

If an aldehyde has no α-hydrogen atoms, it cannot undergo the aldol condensation. If the concentration of the alkali is increased to about 50%, an entirely different reaction proceeds on such aldehydes. The Cannizzaro reaction is an oxidation-reduction reaction in which one molecule of the aldehyde is oxidized while another is reduced. Oxidation of an aldehyde, of course, results in an acid, and reduction yields a primary alcohol.

$$
\underset{\underset{CH_3}{|}}{\overset{\overset{CH_3}{|}}{CH_3-C-CHO}} + \underset{\underset{CH_3}{|}}{\overset{\overset{CH_3}{|}}{CH_3-C-CHO}} + OH^- \xrightarrow[\text{alcohol}]{H_2O \text{ and}}
$$

$$
\underset{\underset{\underset{55\%}{CH_3\ \ O^-}}{|}}{\overset{\overset{CH_3\ \ O}{|}}{CH_3-C-C\overset{\diagup}{\diagdown}}} + \underset{\underset{\underset{59\%}{CH_3}}{|}}{\overset{\overset{CH_3}{|}}{CH_3-C-CH_2OH}} \qquad [11\text{-}29]
$$

All aldehydes in which the carbonyl group is attached to an aromatic ring necessarily carry no hydrogen on the α-carbon and hence are subject to the Cannizzaro reaction.

Exercise 11-7.   Write a Cannizzaro reaction for benzaldehyde, for form-aldehyde, and for 4-ethylbenzaldehyde.

E. Substitution in Aldehydes

Substitution of a halogen for hydrogen may take place on the alkyl group in an aliphatic aldehyde. The substitution always takes place on the α-carbon atom with a loss of HX. Acetaldehyde will react with excess chlorine to give chloral (2,2,2-trichloro-ethanal). Chloral is a liquid, soluble in water, which reacts with water to form a solid crystalline compound called chloral hydrate.

$$
CH_3-CHO + 3Cl_2 \longrightarrow \underset{\underset{\underset{\text{chloral}}{Cl}}{|}}{\overset{\overset{Cl\ \ H}{|\ \ \ |}}{Cl-C-C}}=O + 3HCl \qquad [11\text{-}30]
$$

$$
Cl_3C-CHO + HOH \xrightarrow[21\%]{-10^\circ} \underset{\underset{\underset{\text{chloral hydrate}}{OH}}{|}}{\overset{\overset{H}{|}}{Cl_3C-C-OH}} \qquad [11\text{-}31]
$$

The structure for it shown in Expression 11-31 violates the rule that two OH groups on the same carbon give such an unstable configuration that water will split out. The presence of the strongly electronegative chlorines on the α-carbon apparently modifies the properties of the carbonyl group enough to stabilize the hydrate.

Chloral hydrate is a hypnotic, though not a very safe one to use. (A hypnotic is any agent that tends to produce sleep.)

11-4. Uses of Aldehydes

Formaldehyde is a gas in the pure state and is handled commercially as a 40% water solution or as trioxymethylene. It is used as a disinfectant, as a preservative for animal tissue (for example, in embalming fluid), to make resins with urea, casein, or phenol, and to make the internal antiseptic Urotropin. It is also the essential ingredient in liquid deodorants that function by desensitizing the olfactory nerve endings (for example, Airwick).

Acetaldehyde boils at 21°C and is sold as the trimer, paraldehyde, which boils at 125°C. Its main use in organic chemistry is in preparing other compounds, such as acetic acid. Acetaldehyde is present in green apples and accounts in part for the sharp taste. It is lost when the apple ripens.

In contrast to the sharp odors of the aldehydes of lower molecular weight, the aldehydes with 8–12 carbons have pleasant odors and are used in synthetic perfumes. Mixtures of these with other substances imitate the odors of oils of rose, lemon grass, coriander, cassia, violet, and orange flower.

Chloral takes part in a condensation reaction with chlorobenzene to form a very important insecticide, DDT. Sulfuric acid can be used as the condensing agent to remove water. DDT was first prepared in 1874 but was not recognized as an insecticide until 1939.

$$Cl_3C-\overset{\underset{\displaystyle H}{|}}{C}=O + 2\langle\bigcirc\rangle-Cl \xrightarrow[70\%]{H_2SO_4} Cl_3C-\underset{\underset{\displaystyle \underset{\displaystyle DDT}{Cl}}{\overset{\displaystyle \bigcirc}{|}}}{\overset{\underset{\displaystyle H}{|}}{C}}-\langle\bigcirc\rangle-Cl + H_2O \qquad [11\text{-}32]$$

In 1951, the production of DDT in the United States was 106,-000,000 pounds, used chiefly to kill flies, moths, and mosquitoes. It promises to wipe out malaria. The name DDT is an abbreviation for dichlorodiphenyltrichloroethane.

SUMMARY

See the end of Chapter 12.

EXERCISES

(Exercises 1-7 will be found within the text.)

8. Write structural formulas for propionaldehyde, 2,2-dimethyl heptanal, aldol, the dimethyl acetal of formaldehyde, the hydrogen sulfite addition compound of isobutyraldehyde.

9. Name the following compounds:

1. $CH_3-CH=CH-CH_2OH$

2. ⬡$-CH_2-\underset{\underset{CH_3}{|}}{C}H-CHO$

3. $CH_3-CH_2-\underset{\underset{OH}{|}}{C}H-CN$

4. $(CH_3)_3C-CHO$

10. How many different functional groups can you write for a compound of formula $C_9H_{10}O$?

11. Write equations for the indicated reactions:

1. $(CH_3)_2CH-CHO + OH^- (10\%)$
2. $(CH_3)_3C-CHO + OH^- (50\%)$
3. ⬡$-CHO + NaHSO_3$
4. $(CH_3-CH_2)_2CH-CHO + $⬡$-NHNH_2$
5. $CH_3-CH_2-CH_2-CHO + Ag(NH_3)_2^+$
6. $H-CHO + 1\text{-propanol} + \text{concentrated HCl}$

12. Write equations showing the action of ethyl magnesium bromide on H_2O, formaldehyde, 1-propanol, propanal.

13. Indicate how to accomplish the following conversions:

1. ethanol to propanoic acid
2. acetaldehyde to 2-propanol
3. benzaldehyde to 1-phenyl-1-ethanol
4. acetaldehyde to 3-methyl-2-butanol
5. propanoic acid to propanal
6. acetylene to 2-butenal

14. Write reactions of the following reagents with propanal:

1. ethanol and concentrated HCl
2. ethanol and dry HCl
3. hydrogen and Raney nickel (see p. 283)
4. hydrogen cyanide
5. hydroxylamine
6. sodium and ethanol
7. sodium hydrogen sulfite
8. dilute sodium hydroxide
9. phenylhydrazine
10. Fehling's solution
11. ammonia
12. chlorine

15. Write specific reactions to show how formaldehyde behaves differently from other aldehydes.

12 Carbonyl Compounds: Ketones

The general formula for a ketone is $R-\overset{\displaystyle O}{\overset{\|}{C}}-R'$, and the functional group is the carbonyl group, $-\overset{\displaystyle O}{\overset{\|}{C}}-$. Ketones are very closely related to aldehydes, the only difference being that the former have an R group instead of an H atom on the carbonyl group. This makes the ketones, in general, somewhat less reactive than the aldehydes. Ketones are functional isomers of the aldehydes containing the same number of carbon atoms.

TABLE
12-1 | *Ketones*

STRUCTURE	COMMON NAME	IUC NAME
$CH_3-CO-CH_3$	acetone	propanone
$CH_3-CO-CH_2-CH_3$	methyl ethyl ketone	butanone
$CH_3-CH_2-CO-CH_2-CH_3$	diethyl ketone	3-pentanone
$CH_3-CH_2-CH_2-CO-CH_3$	methyl n-propyl ketone	2-pentanone
$\begin{array}{c}CH_3\\ \diagdown\\CH-CO-CH_3\\ \diagup\\CH_3\end{array}$	methyl isopropyl ketone	3-methyl-2-butanone
$\begin{array}{c}CH_2-CH_2\\ \diagup \diagdown\\CH_2C=O\\ \diagdown \diagup\\CH_2-CH_2\end{array}$	cyclohexanone	cyclohexanone
$\langle\bigcirc\rangle-CO-CH_3$	methyl phenyl ketone or acetophenone	————

12-1. Nomenclature of Ketones

We form many of the common names of ketones by naming the two radicals attached to the carbonyl group in succession and adding the word *ketone*. The shorter radical is conventionally named first. We derive the IUC names by adding the suffix *-one* to the original hydrocarbon stem designating the longest chain containing the carbonyl group. A number must be given to show where the carbonyl group is located in cases that are not unique.

Exercise 12-1. Write structural formulas for the following compounds: isopropyl sec-butyl ketone, 1-buten-3-one, isopropyl phenyl ketone, butanal-2-one.

12-2. Preparation of Ketones

The general methods of preparation of aliphatic ketones are the same as those used for aldehydes. Consequently the discussion will be a review of the reactions already learned for aldehydes. Aromatic and mixed ketones can be synthesized by the oxidation of a secondary alcohol, but the other three general methods cannot be used. (Methods of preparation of these ketones are given on pp. 261, 265, and 625–626.)

1. Oxidation of a Secondary Alcohol. The only change necessary for preparing a ketone rather than an aldehyde by oxidation is the use of a secondary alcohol in place of a primary alcohol.

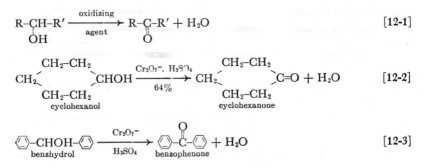

$$\text{R–CH–R'} \xrightarrow[\text{agent}]{\text{oxidizing}} \text{R–C–R'} + H_2O \qquad [12\text{-}1]$$
$$\underset{\text{OH}}{} \qquad \underset{O}{}$$

$$\begin{array}{ccc}
\text{CH}_2\text{–CH}_2 & & \text{CH}_2\text{–CH}_2 \\
\text{CH}_2 \diagdown & \text{CHOH} \xrightarrow[64\%]{\text{Cr}_2\text{O}_7{}^-,\ \text{H}_2\text{SO}_4} \text{CH}_2 \diagdown & \text{C=O} + \text{H}_2\text{O} \\
\text{CH}_2\text{–CH}_2 & & \text{CH}_2\text{–CH}_2
\end{array} \qquad [12\text{-}2]$$

cyclohexanol   cyclohexanone

$$\text{⬡–CHOH–⬡} \xrightarrow[\text{H}_2\text{SO}_4]{\text{Cr}_2\text{O}_7{}^-} \text{⬡–C–⬡} + \text{H}_2\text{O} \qquad [12\text{-}3]$$

benzhydrol   benzophenone

2. Dehydrogenation of a Secondary Alcohol. Acetone is prepared commercially by dehydrogenation of isopropyl alcohol,

$$R\text{-}\underset{\underset{OH}{|}}{C}H\text{-}R' \xrightarrow[200\text{-}300°]{Cu} R\text{-}\underset{\underset{O}{\|}}{C}\text{-}R' + H_2 \qquad [12\text{-}4]$$

which is made by the acid hydrolysis of propylene, obtained in abundance from the petroleum industry via the cracking process. Acetone is also a principal product of the Weizmann process (pp. 142–143).

$$CH_3\text{-}CH\text{=}CH_2 + HOH \xrightarrow{H_2SO_4} CH_3\text{-}\underset{\underset{OH}{|}}{C}H\text{-}CH_3 \xrightarrow[300°]{Cu} CH_3\text{-}\underset{\underset{O}{\|}}{C}\text{-}CH_3 + H_2 \qquad [12\text{-}5]$$

3. From Acid Vapors. Good yields of symmetrical ketones may be obtained by the passage of a volatile acid over MnO at elevated temperature. If a mixture of two acids is used, three ketones are possible, and hence the reaction is not a desirable one for preparing an unsymmetrical ketone.

$$2R\text{-}\underset{\underset{OH}{}}{\overset{O}{C}} \xrightarrow[200\text{-}300°]{MnO} R\text{-}\underset{\underset{}{\|}}{\overset{O}{C}}\text{-}R + CO_2 + H_2O \qquad [12\text{-}6]$$

$$R\text{-}COOH + R'\text{-}COOH \longrightarrow \begin{Bmatrix} R\text{-}CO\text{-}R \\ R\text{-}CO\text{-}R' \\ R'\text{-}CO\text{-}R' \end{Bmatrix} \text{mixture} + CO_2 + H_2O \quad [12\text{-}7]$$

4. From Salts of Acids. If the salt of a carboxylic acid is heated in the dry state, a ketone is obtained at some elevated tempera-

$$\begin{array}{c} CH_3\text{-}CH_2\text{-}\overset{O}{C}\diagdown_{O}\diagdown \\ Ca \xrightarrow[86\%]{\Delta} CH_3\text{-}CH_2\text{-}\overset{O}{\underset{\|}{C}}\text{-}CH_2\text{-}CH_3 + CaCO_3 \qquad [12\text{-}8] \\ CH_3\text{-}CH_2\text{-}\overset{}{C}\diagup_{O}\diagup \end{array}$$

ture. Besides the sodium salt of an organic acid, other salts, especially those of calcium, barium, and magnesium, are used. In this reaction, as in the one just described, a mixture of ketones will result if a mixture of salts of different acids is used. The carbonate of the metal is always the second product.

5. Hydrolysis of Substituted Alkynes.   The hydrolysis of any alkyl alkyne results in a methyl ketone. Dialkyl alkynes may be hydrolyzed in the same fashion. The reaction is rarely used because it is much more difficult to make the alkynes than to make the resulting ketones by other methods. The only reason for mentioning this reaction is to show the complete analogy of the methods of preparing aldehydes and ketones.

$$R-C\equiv CH + \overset{\delta+}{H} : \overset{\delta-}{OH} \xrightarrow[H^+]{HgSO_4} \left[\begin{array}{c} R-\underset{\underset{OH}{|}}{C}=CH_2 \end{array} \right] \xrightarrow{H\ shift} R-\underset{\underset{O}{\|}}{C}-CH_3 \qquad [\mathbf{12\text{-}9}]$$

a methyl ketone

$$R-C\equiv C-R + \overset{\delta+}{H} : \overset{\delta-}{OH} \xrightarrow[H^+]{HgSO_4} \left[\begin{array}{c} R-\underset{\underset{OH}{|}}{C}=CH-R \end{array} \right] \xrightarrow{H\ shift} R-\underset{\underset{O}{\|}}{C}-CH_2-R \qquad [\mathbf{12\text{-}10}]$$

12-3. Reactions of Ketones

A. *Oxidation*

The difference between an aldehyde and a ketone is easily detected in test-tube reactions by oxidizing agents. Aldehydes react with Tollen's reagent and Fehling's solution, but ketones do not. As a matter of fact, ketones such as acetone can be used as solvents in which to carry out the tests for aldehydes.

When vigorous oxidizing agents are used to oxidize ketones, carbon-carbon bonds are finally broken, and products containing fewer carbon atoms than the starting compound are obtained. The oxidation of a ketone is seldom a useful reaction.

B. Addition to Ketones

Since ketones are less reactive than aldehydes, not all of the reagents that add to aldehydes will add to ketones. Some other reagents attack only the ketones of low molecular weight. The properties of a ketone can best be accounted for if $R—\overset{\delta+}{C}—\overset{\delta-}{\overset{..}{O}}:$ makes

$$\begin{array}{c} \diagup \\ R \end{array}$$

an important contribution to the structure of the carbonyl group.

1. Hydrogen Cyanide. Cyanohydrins are formed nearly as readily with ketones as with aldehydes. Acetone, for example, gives a good yield of acetone cyanohydrin.

$$CH_3-\overset{O}{\overset{\|}{C}}-CH_3 + H \overset{\delta+}{} \overset{\delta-}{CN} \xrightarrow[77\%]{} CH_3-\underset{\underset{CH_3}{|}}{\overset{OH}{\overset{\diagup}{C}}}{\underset{CN}{}}$$

[12-11]

acetone cyanohydrin

2. Hydrogen. Reductions of ketones, either catalytic or with chemical agents, yield secondary alcohols, with two notable exceptions. These special reactions are the pinacol reaction and the Clemmensen reduction.

a. Secondary Alcohols. Addition of hydrogen to a ketone in the presence of a catalyst such as platinum or nickel yields a secondary

$$R-\overset{O}{\overset{\|}{C}}-R' + H_2 \xrightarrow{Ni} R-CHOH-R'$$

[12-12]

a secondary alcohol

alcohol. Reductions with alkaline reagents (exception, *b*, below) such as lithium aluminum hydride (p. 499) also end at the same stage. In neutral or acid solution also (exception. *c*, below), ketones are reduced to secondary alcohols.

A new synthesis of styrene (p. 110) includes the catalytic reduction of a ketone as one step. This commercial synthesis is also of interest because the controlled oxidation of a side chain on a benzene ring ends in a ketone (however, see p. 198). This method is not generally applicable to the preparation of ketones (p. 222) on a small scale in the laboratory because rigorous temperature control is essential. Ethyl benzene is oxidized, at low air pressure, in the

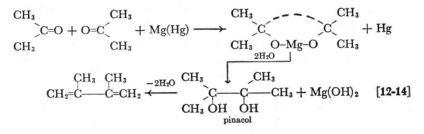

presence of manganous acetate, and at 130°C, to acetophenone. The ketone can be reduced, with a nickel, palladium, or copper chromite catalyst, at low hydrogen pressure, and at a temperature of 150°, to the corresponding secondary alcohol. Dehydration of this alcohol then yields styrene, an alkene.

b. Pinacol Reaction. When the reduction of a ketone is carried out in alkaline solution (with sodium in alcohol, for instance), a bimolecular reduction proceeds to some extent. Perhaps the best reagent for the pinacol reaction is magnesium amalgam and water.

Pinacol is easily dehydrated to 2,3-dimethyl-1,3-butadiene, which is easily polymerized with Na as a catalyst. This was the basis of the synthetic rubber industry in Germany during the First World War. (See p. 108.)

c. Clemmensen Reduction. A specific reducing agent, zinc (or zinc amalgam) and hydrochloric acid, has the property of reducing a ketone to the corresponding hydrocarbon. The method is known as the Clemmensen reduction and is particularly useful for ketones bearing an aromatic ring.

$$\underset{\text{propiophenone}}{\bigcirc\text{-}\overset{\overset{\text{O}}{\|}}{\text{C}}\text{-CH}_2\text{-CH}_3} + 2\text{Zn} + 4\text{HCl} \xrightarrow{90\%} \underset{\text{n-propylbenzene}}{\bigcirc\text{-CH}_2\text{-CH}_2\text{-CH}_3} + 2\text{ZnCl}_2 + \text{H}_2\text{O} \qquad [12\text{-}15]$$

3. Sodium Hydrogen Sulfite. Only methyl ketones of low molecular weight and cyclic ketones are reactive enough to add $NaHSO_3$. This is sometimes useful in distinguishing methyl ketones from others.

$$\underset{\overset{|}{\text{O}}}{\text{CH}_3\text{-}\overset{|}{\text{C}}\text{-R}} + \text{NaHSO}_3 \longrightarrow \text{CH}_3\text{-}\overset{\overset{\text{OH}}{|}}{\underset{\text{R}\quad\text{SO}_3\text{Na}}{\text{C}}} \qquad [12\text{-}16]$$

4. Ammonia. Ammonia adds to some ketones to give complex addition compounds. These reactions will not be considered here.

5. Alcohols. Ketones are so much less reactive than aldehydes that the direct addition of alcohols to form hemiacetals and acetals is not practical. The ketone acetals are made by indirect methods that will not be discussed here.

6. Grignard Reagents. Ketones will add the Grignard reagent to give an addition product that is easily hydrolyzed to a tertiary

$$\underset{\overset{|}{\text{O}}}{\text{R-}\overset{|}{\text{C}}\text{-R}'} + \text{R}''\text{-MgX} \longrightarrow \text{R-}\overset{\overset{\text{R}'}{|}}{\underset{\overset{|}{\text{OMgX}}}{\text{C}}}\overset{}{\underset{\text{R}''}{}} \xrightarrow{\text{HOH}} \underset{\overset{\text{a tertiary}}{\text{alcohol}}}{\text{R-}\overset{\overset{\text{R}'}{|}}{\underset{\overset{|}{\text{OH}}}{\text{C}}}\text{-R}''} + \text{Mg}\overset{\overset{\text{X}}{\diagup}}{\underset{\diagdown\text{OH}}{}} \qquad [12\text{-}17]$$

$$\underset{}{\bigcirc\text{-}\overset{\overset{\text{O}}{\|}}{\text{C}}\text{-CH}_3} + \text{CH}_3\text{-CH}_2\text{-Mg-I} \longrightarrow \bigcirc\text{-}\overset{\overset{\text{CH}_3}{|}}{\underset{\underset{\text{CH}_3}{\overset{|}{\text{CH}_2}}}{\underset{|}{\text{C}}}}\text{-OMgI} \qquad [12\text{-}18]$$

$$\underset{\overset{\text{a tertiary}}{\text{alcohol}}}{\bigcirc\text{-}\overset{\overset{\text{CH}_3}{|}}{\underset{\overset{|}{\text{CH}_2\text{-CH}_3}}{\text{C}}}\text{-OH}} \xleftarrow{\quad\text{HOH}\quad}$$

alcohol. This method gives excellent yields from a large number of ketones and a variety of Grignard reagents.

Exercise 12-2. Write a synthesis for 2-methyl-2-butanol.

C. Polymerization of Ketones

Ketones do not undergo self-addition in acid solution and therefore will not polymerize in the reversible way that aldehydes do.

D. Condensations of Ketones

1. Hydroxylamine. Ketones as well as aldehydes react readily with hydroxylamine, NH_2OH, to give a general class of compounds called oximes, which may be further classified into aldoximes from aldehydes and ketoximes from ketones.

$$\begin{array}{c} R \\ \diagdown \\ R' \end{array} C{=}O + \overset{\delta+}{H}\ \overset{\delta-}{NHOH} \longrightarrow \begin{array}{c} R \\ \diagdown \\ R' \end{array} \overset{\text{(OH)}}{\underset{N H OH}{C}} \overset{-H_2O}{\longrightarrow} \begin{array}{c} R \\ \diagdown \\ R' \end{array} C{=}NOH \ . \qquad [12\text{-}19]$$

Exercise 12-3. Write an equation for the preparation of methyl ethyl ketoxime.

2. Phenylhydrazine. Ketones react with phenylhydrazine, $C_6H_5{-}NH{-}NH_2$, to form phenylhydrazones, compounds that are generally yellow crystalline solids, easily purified by recrystallization. The melting points are ordinarily sharp, and hence the phenylhydrazones may be used as one means of helping to identify an unknown ketone.

$$R{-}\underset{\underset{O}{\|}}{C}{-}R' + H{-}\underset{\underset{}{}}{\overset{H}{N}}{-}\overset{H}{N}{-}\langle\rangle \longrightarrow \left[\begin{array}{c} R \\ \diagdown \\ R' \end{array} \overset{\text{(OH H)}H}{\underset{}{C}}{-}N{-}N{-}\langle\rangle\right] \longrightarrow \begin{array}{c} R \\ \diagdown \\ R' \end{array} C{=}N{-}\overset{H}{N}{-}\langle\rangle \qquad [12\text{-}20]$$

a phenylhydrazone

Exercise 12-4. Write an equation for the preparation of the phenylhydrazone of methyl phenyl ketone.

3. Aldol Condensation. Acetone and a few other lower members of the homologous series will undergo the aldol condensation.

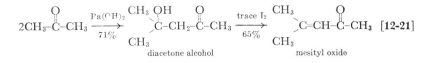

$$2CH_3-\overset{O}{\overset{\|}{C}}-CH_3 \xrightarrow[71\%]{Ba(OH)_2} \overset{CH_3}{\underset{CH_3}{\diagdown}}\overset{OH}{\underset{}{\diagup}}C-CH_2-\overset{O}{\overset{\|}{C}}-CH_3 \xrightarrow[65\%]{trace\ I_2} \overset{CH_3}{\underset{CH_3}{\diagdown}}C=CH-\overset{O}{\overset{\|}{C}}-CH_3 \quad [12\text{-}21]$$

diacetone alcohol   mesityl oxide

Diacetone alcohol readily undergoes dehydration upon distillation with only a crystal of iodine as a catalyst.

Exercise 12-5. Write the aldol condensation of acetone by a three-step mechanism (see p. 215).

E. Substitution in Ketones

1. Direct Substitution with Chlorine. Ketones readily undergo substitution with chlorine or bromine. It is a reaction to be avoided, in many cases, for the α-halogen ketones are lachrymators. Some of them, indeed, are used as tear gas, and at least one has been used as a chemical warfare agent. The reaction goes

$$CH_3-\overset{O}{\overset{\|}{C}}-\langle\bigcirc\rangle + Cl_2 \longrightarrow ClCH_2-\overset{O}{\overset{\|}{C}}-\langle\bigcirc\rangle + HCl \qquad\qquad [12\text{-}22]$$

$$CH_3-\overset{O}{\overset{\|}{C}}-CH_3 + Cl_2 \longrightarrow ClCH_2-\overset{O}{\overset{\|}{C}}-CH_3 + HCl$$

$$\underset{\text{trichloroacetone}}{Cl_3C-\overset{O}{\overset{\|}{C}}-CH_3} \overset{2Cl_2}{\longleftarrow} \qquad\qquad\qquad [12\text{-}23a]$$

so rapidly with acetone that it is difficult to stop with the introduction of one atom of chlorine.

Chloral (p. 217) and trichloroacetone are readily decomposed

$$\underset{\overset{\|}{O}}{Cl_3C-\overset{}{\underset{}{C}}-CH_3} + NaOH \xrightarrow[45\%]{} CH_3-C\overset{\diagup O}{\underset{\diagdown ONa}{}} + CHCl_3 \qquad [12\text{-}23b]$$

by bases in a reaction in which a carbon-carbon bond is broken.

$$\underset{\text{H}}{Cl_3C-\overset{\text{O}}{C}}=O + NaOH \longrightarrow H-C\overset{\diagup O}{\underset{\underset{\text{chloroform}}{ONa}}{\diagdown}} + CHCl_3 \qquad [12\text{-}24]$$

The same series of reactions can be carried out with solutions of alkali hypohalites, such as NaOBr, NaOCl, and NaOI. These are called haloform reactions because a haloform is one of the products.

2. Iodoform Reaction. The haloform reaction is used as a test for CH_3-CO-R, CH_3-CHO, CH_3-CH_2OH, and $CH_3-CHOH-R$ because these are the only compounds and classes of compounds that will give the yellow precipitate of iodoform (melting point $120°$) with NaOI reagent.

$$CH_3\overset{\overset{\text{O}}{\|}}{C}-R + 3NaOI \longrightarrow I_3C-\overset{\overset{\text{O}}{\|}}{C}-R + 3NaOH \qquad [12\text{-}25a]$$

$$I_3C-\overset{\overset{\text{O}}{\|}}{C}-R + NaOH \longrightarrow \underset{\text{iodoform}}{CHI_3} + R-C\overset{\diagup O}{\underset{ONa}{\diagdown}} \qquad [12\text{-}25b]$$

$$CH_3-CHO + 3NaOI \longrightarrow I_3C-CHO + 3NaOH \qquad [12\text{-}26a]$$

$$I_3C-CHO + NaOH \longrightarrow CHI_3 + H-C\overset{\diagup O}{\underset{ONa}{\diagdown}} \qquad [12\text{-}26b]$$

Ethyl alcohol does not give the haloform reaction as such but, since NaOI is a strong oxidizing agent, is first converted to acetaldehyde. Then the reaction goes as in Expression 12-26. Methyl sec-

$$CH_3-CH_2OH + NaOI \longrightarrow CH_3-CHO + NaI + H_2O \qquad [12\text{-}27]$$

ondary alcohols treated with NaOI first oxidize to methyl ketones, which then give the haloform reaction according to Expression 12-25.

$$CH_3-CHOH-R + NaOI \longrightarrow CH_3-CO-R + NaI + H_2O \qquad [12\text{-}28]$$

Exercise 12-6. Write an iodoform reaction on methyl phenyl ketone; on 2-pentanol.

12-4. Uses of Ketones

The only ketones available in large quantities are acetone and methyl ethyl ketone, both of which are used as solvents. Acetone is a solvent for many lacquers, nitrocellulose (smokeless powder), nail polish, acetylene, and many plastics. Cyclic ketones with very large rings, such as civetone and muscone, which have disgusting odors in the pure state, have very pleasant odors in great dilution

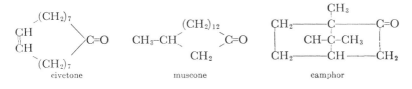

civetone   muscone   camphor

and are used in expensive perfumes. Camphor is another cyclic ketone, used as a mild analgesic, a mild antiseptic, and a temporary heart stimulant.

SUMMARY (Chaps. 11 and 12)

The reactions of aldehydes and ketones are summarized concisely in Table 12-2. Aldehydes are distinguished from ketones by their greater reactivity; in the laboratory one may take advantage of their oxidation reactions (Tollen's reagent, Fehling's solution, permanganate ion) for qualitative and even quantitative determinations of aldehydes. Aldehydes and some ketones may be purified by the use of sodium hydrogen sulfite addition products. Aldehydes and ketones are frequently identified by conversion to oxime or phenylhydrazone derivatives since many of these are solids whose melting points are recorded. These reactions give good yields, and the condensation products are easily recrystallized. The aldol condensation is one of the most important reactions studied, for the mechanism is applicable to numerous other base-catalyzed reactions (see p. 3 0). The addition reactions of the carbonyl group with hydrogen, hydrogen cyanide, and Grignard reagents are important in synthesis.

TABLE

12-2 | *Summary of Reactions of Aldehydes and Ketones*

	HCHO	RCHO	R–CO–R′
OXIDATION	first H–COOH, then $CO_2 + H_2O$	R–COOH	no reaction
ADDITION			
H_2	CH_3OH	$R–CH_2OH$	R–CHOH–R′ acid or neutral, pinacols in basic media
HCN	+ (positive reaction)	+	+
$NaHSO_3$	+	+	CH_3–CO–R only, adds complex derivatives
NH_3	$(CH_2)_6N_4$	+	
RMgX, then H_2O	primary alcohol	secondary alcohol	tertiary alcohol
ROH	+	+	no reaction
POLYMERIZA-TION	trimers and polymers	trimer	no reaction
CONDENSATION			
NH_2OH	+	+	+
⬡–NHNH₂	+	+	+
dil. NaOH	no reaction	an aldol	lower ones give an aldol
conc. NaOH	Cannizzaro	Cannizzaro, if no α–H	no reaction
SUBSTITUTION			
Cl_2		on α-carbon	on α-carbon
Haloform		CH_3–CHO only	all methyl ketones

EXERCISES

(Exercises 1–6 will be found within the text.)

7. Write structural formulas for di-n-propyl ketone, ethyl isopropyl ketone, 3-hexanone, 4-methyl-3-octanone, and butanone oxime.

8. Name the following compounds:

1. $CH_3-CH-C-CH-CH_3$
   $\;\;\;\;\;\;\;\; CH_3 \;\; O \;\; CH_2$
   $\; CH_3$

2. $CH_3-CH_2-C-CHO$
   $\;\;\;\;\;\;\;\;\;\;\;\;\;\;\;\; O$

3. $CH_3-CH-CH=CH-C-CH_3$
   $\;\;\;\;\;\;\;\; CH_3 \;\;\;\;\;\;\;\;\;\;\; O$

4. ⬡$-C-$⬡
   $\;\;\;\; O$

5. $CH_3-C-CH=CH_2$
   $\;\;\;\;\;\;\;\; O$

6. $CH_3-C-CH_2-C-CH_3$
   $\;\;\;\;\;\; OH \;\;\;\;\;\; O$
   (with CH_3 above the first C)

7. ⬡$-C-CH-CH_3$
   $\;\;\;\; O \;\; CH_3$

9. Write test-tube reactions for three ways of distinguishing aldehydes from ketones.

10. Indicate how to accomplish the following conversions:

    1. 1-bromopropane to acetone
    2. ethanol to butanone
    3. acetone to tert-butyl alcohol
    4. propanoic acid to 3-pentanol
    5. toluene to diphenyl ketone

11. Indicate how to distinguish between acetaldehyde, acetone, 2-propanol, and 3-pentanone.

12. Write equations showing the action of ethyl magnesium bromide on acetone, on 2,4-dimethyl-3-pentanone, and on isobutyraldehyde.

13. Two alcohols, A and B, both of formula $C_4H_{10}O$, react with acidic $KMnO_4$. From the reaction mixture using A, $\begin{matrix} CH_3 \\ CH_3 \end{matrix}\!\!\!\!> CH–COOH$ is isolated; from the reaction mixture using B, $CH_3–\overset{\overset{\displaystyle O}{\|}}{C}–CH_2CH_3$ is isolated. Identify alcohols A and B.

14. A liquid that gives an iodoform test but will not reduce Fehling's solution gives a phenylhydrazone that contains 10.0% nitrogen. Write a possible structure for the original compound.

15. How could one distinguish between 1-butanol, 2-butanol, and 2-methyl-2-propanol by oxidation?

16. A compound, $C_{14}H_{12}O$, reacts with phenylhydrazine but not with Tollens' reagent. Benzoic acid is the only organic product of oxidation of the compound. Deduce the structure of the compound.

17. How may acetone be transformed into acetic acid? acetic acid into acetone?

18. A substance, $C_5H_{10}O$, forms a phenylhydrazone but will not reduce Tollens' reagent or give an iodoform test. Write a possible structure for the compound.

19. Outline methods of preparing the following compounds, using any inorganic reagents and starting with organic reagents having only one or two carbon atoms.

    1. 2-propanol
    2. 3-pentanone
    3. pinacol
    4. 2-methyl-2-butanol
    5. iodoform

20. What product would you expect to get by treating propionaldehyde with ethylene glycol in the presence of dry HCl?

21. What is the theoretical yield of iodoform when 4.6 g of ethanol is treated with an excess of sodium hypoiodite?

22. What primary alcohol cannot be synthesized by the Grignard method?

23. Write equations for two methods of preparing n-propylbenzene.

13 Oxidation-Reduction

Reactions called oxidations and reductions have been introduced in nearly every chapter of this text, and it has been assumed that some concept of oxidation and reduction is familiar to the student. The generalized concept of oxidation-reduction will now be presented more fully.

13-1. Oxidation Involving Partial Transfer of Electrons

One of the characteristics of carbon compounds is that [except in carbon monoxide, for which three bonds are written to carbon ($C\equiv O$), and in the isocyanides ($R-N\equiv C$; see p. 305)] there are always four bonds to every carbon. Nevertheless, a difference in the character of the carbon atom in such compounds as $H-\overset{\overset{\displaystyle H}{|}}{\underset{\underset{\displaystyle H}{|}}{C}}-H$, $\overset{H}{\underset{H}{>}}C=O$, and $H-\overset{\nearrow O}{\underset{\searrow OH}{C}}$ is apparent in the properties of these compounds. Methane will burn to form CO_2 and H_2O, yet the number of bonds attached to the carbon in CH_4 and CO_2 is the same. The reaction is called a combustion, or rapid oxidation, and the carbon does change the character of its bonds during the reaction. In $H : \overset{\overset{\displaystyle H}{\cdot\cdot}}{\underset{\underset{\displaystyle H}{\cdot\cdot}}{C}} : H$ the hydrogen is an electropositive atom with respect to carbon and may be said to hold the pair of electrons less strongly than the carbon with which the pair is shared. The idea

235

that electron pairs are not necessarily equally shared has been used implicitly in our accounting for most of the organic reactions we have written so far. Now it is time to emphasize that unequal sharing is the general rule, rather than the exception, between unlike atoms. There is a way to estimate the relative control that either atom of a pair has over the pair of electrons. For present purposes it will be assumed that one can make a gross estimate merely by referring to the Periodic Table. Elements on the left-hand side of the table have a tendency to release electrons and become positive ions; elements on the right-hand side have a tendency to attract electrons and become negative ions. In the $C:H$ bond, the hydrogen atom is said arbitrarily to have a formal charge of $+1$ (called formal because the bond itself is actually neutral and hence the charge is not real) and the carbon one of -1. The assumption is that the carbon atom has the lion's share of a shared pair. (The lion's share is all of it.) If an atom from the right-hand side of the Periodic Table (such as O, S, N, or Cl) is attached to carbon, the carbon is positive and hence will have a formal charge of $+1$ with respect to that atom. The total formal charge of an atom in a molecule is called its oxidation number. In methane, then, the four $C:H$ bonds each having a formal charge of -1 for the carbon give carbon an oxidation number of -4. For hydrogen the oxidation number would be $+1$. In

$:\overset{..}{O}::C::\overset{..}{O}:$ the formal charge on carbon is $+4$ and on each oxygen -2; so the oxidation number of carbon in this compound

is $+4$, and that of oxygen is -2. In
$$\begin{array}{ccc} & H & H \\ & | & | \\ H- & C- & C-H \\ & | & | \\ & H & H \end{array}$$
the pair of electrons forming the carbon-carbon bond is assumed to be equally shared, and hence the formal charge on each carbon from this bond is zero. Each $C:H$ bond gives a formal charge of -1 to the carbons, and hence each carbon in ethane has a formal charge of -3 and an oxidation number of -3.

Examination of the following list indicates that carbon may have any oxidation number of any integral value from -4 to $+4$, including zero. When it is not obvious, the carbon to which the oxidation number applies is starred (*).

−4 CH_4

−3 $CH_3\text{–}CH_3$

−2 $CH_3\text{–}\overset{*}{C}H_2\text{–}CH_3$ $\underset{H}{\overset{H}{H\text{–}C\text{–}OH}}$ $\underset{H}{\overset{H}{H\text{–}C\text{–}Cl}}$

−1 $\underset{CH_3}{\overset{CH_3\;\;CH_3}{\diagdown\overset{*}{C}\text{–}H}}$ $\underset{H}{\overset{H}{CH_3\text{–}C_*\text{–}OH}}$

0 $\overset{H}{H\text{–}C\text{=}O}$ $\underset{CH_3\;\;OH}{\overset{CH_3\;\;H}{\diagdown\overset{*}{C}\diagdown}}$ $\underset{Cl}{\overset{H}{H\text{–}C\text{–}Cl}}$ $\overset{CH_3}{\underset{*}{\bigcirc}}$

+1 $\overset{H}{R\text{–}C\text{=}O}$ $\underset{R\;\;R}{\overset{R}{\diagdown C\text{–}OH}}$ $\overset{OH}{\underset{*}{\bigcirc}}$

+2 $\underset{O}{\overset{O}{R\text{–}C\text{–}R}}$ $\overset{O}{\underset{OH}{H\text{–}C\diagup}}$ $\underset{Cl}{\overset{Cl}{H\text{–}C\text{–}Cl}}$

+3 $\overset{O}{\underset{OH}{R\text{–}C\diagup}}$ $\overset{Cl}{\underset{Cl}{\bigcirc\text{–}\overset{*}{C}\text{–}Cl}}$

+4 $O\text{=}C\text{=}O$ $\underset{Cl}{\overset{Cl}{Cl\text{–}C\text{–}Cl}}$

The oxidation numbers of other atoms appearing in known structures may be obtained in the same way or from the following considerations. An oxidation is a loss of electrons from an atom or radical; a reduction is a gain of electrons by an atom or radical. The charge on an ion, then, is a measure of the number of electrons that the ion will gain or lose in being reduced or oxidized. The oxidation number of an atom in a monatomic ion is the electrical charge it bears.

The oxidation numbers of some common oxidizing and reducing agents and the changes they commonly undergo are shown in the accompanying table. The values are easily obtained if

REACTION	CHANGE IN OXIDATION NUMBER	MEDIUM
$MnO_4^- \longrightarrow MnO_2$	From +7 to +4	basic solution
$MnO_4^- \longrightarrow Mn^{++}$	From +7 to +2	acidic solution
$Cr_2O_7^- \longrightarrow Cr^{+3}$	From +6 to +3	acidic solution
$SO_3^- \longrightarrow SO_4^-$	From +4 to +6	

we assume that oxygen in a group always has an oxidation number of -2. There are only a few exceptions to this statement, the most prominent being the peroxides (such as hydrogen peroxide, $H : \overset{\cdot\cdot}{\underset{\cdot\cdot}{O}} : \overset{\cdot\cdot}{\underset{\cdot\cdot}{O}} : H$), in which the value obtained by the method applied to carbon is -1. This method is always available when structures are known.

In the reactions indicated below, selected from those of the compounds listed on page 237, it is apparent that there is a change in formal charge. The reactions are written so that oxidation proceeds to the right and reduction to the left. The addition of H_2 to an aldehyde or any other organic compound is a reduction since it

$$\overset{-4}{CH_4} \longrightarrow \overset{+4}{CO_2}$$

$$\overset{-1}{R-CH_2OH} \rightleftarrows \overset{+1}{R-CHO} \longrightarrow \overset{+3}{R-COOH}$$

$$\overset{-4}{CH_4} \overset{Cl_2}{\longrightarrow} \overset{-2}{CH_3Cl} \overset{OH^-}{\longrightarrow} \overset{-2}{CH_3OH}$$

decreases the electron-seeking character or oxidation number of the carbon. Removal of H_2 (dehydrogenation) from a primary alcohol to give an aldehyde or from a secondary alcohol to give a ketone is an oxidation since the electron-seeking character of the carbon is enhanced.

The oxidation numbers of a starting compound and the end product of a required synthesis may often suggest a possible method of synthesis. If one has forgotten the necessary reagents for transforming an alcohol into an acid, the oxidation numbers of the two will at least suggest that an oxidation is a necessity in the over-all process. Conversely, if we go from an acid to an alcohol, reduction must occur.

13-2. Oxidations Involving Complete Transfer of Electrons

Electrons are completely transferred from one atom to another during the following reactions. The results only, but not the mechanism of the transfer, will be discussed here.

A. Displacement Reactions

1. Displacement of Iodine.   Gaseous chlorine bubbled through a solution containing iodide ion will displace the iodide ion as

$$Cl_2 + 2I^- \longrightarrow I_2 + 2Cl^-$$

$$\overset{0}{:\overset{..}{\underset{..}{Cl}}:\overset{..}{\underset{..}{Cl}}:} + 2\overset{-1}{:\overset{..}{\underset{..}{I}}:^-} \longrightarrow \overset{0}{:\overset{..}{\underset{..}{I}}:\overset{..}{\underset{..}{I}}:} + 2\overset{-1}{:\overset{..}{\underset{..}{Cl}}:^-} \qquad [13\text{-}1]$$

iodine and give chloride ion. (The oxidation numbers are shown in small figures above the reagents.)

2. Displacement of Hydrogen.   In the displacement of hydrogen from an acid by metals such as Zn, Fe, and Mg, the metal is

$$\overset{0}{Zn}\cdot + 2\overset{+1}{H^+} \longrightarrow \overset{+2}{Zn^{++}} + \overset{0}{H:H} \qquad [13\text{-}2]$$

oxidized, since it loses two electrons, while the hydrogen is reduced, each atom gaining one electron.

B. *Direct Combination of Elements*

Chlorine gas will react violently with metallic sodium to give sodium chloride, which is ionized as shown. During this reaction

$$2Na + Cl_2 \longrightarrow 2NaCl$$

$$\overset{0}{2Na} \cdot + \overset{0}{:\!Cl\!:\!Cl\!:} \longrightarrow \overset{+1}{2Na} \overset{-1}{:\!Cl\!:^{-}} \tag{13-3}$$

the sodium atom loses an electron to become an ion (is oxidized) while the chlorine atom (which possesses seven electrons in Cl_2) gains one to become Cl^- (is reduced).

When the direct combination of elements results in a compound in which the bonds are covalent, the oxidation-reduction involves only a partial transfer of electrons. Whereas the electron pairs in

$$H_2 + Cl_2 \xrightarrow{\text{sunlight}} 2HCl$$

$$\overset{0}{H\!:\!H} + \overset{0}{:\!Cl\!:\!Cl\!:} \xrightarrow{\text{sunlight}} \overset{+1}{2H} \overset{-1}{:\!Cl\!:} \tag{13-4}$$

molecular hydrogen and chlorine are probably equally shared, the same is not true in molecular hydrogen chloride, for the chlorine atom is strongly electronegative and hence holds the electron pair more strongly than the hydrogen. The bond has considerable ionic character in gaseous HCl and is completely ionic in aqueous solution.

The two reactions between sulfur and oxygen are interesting In Group VI of the Periodic Table the lower atoms are more electropositive than those above them. In SO_2, the sulfur atom may be regarded as having lost two electrons to the more electronegative oxygen atom in the double bond and also to have lost both electrons to oxygen in the coordinate covalent bond. Therefore, the sulfur atom in SO_2 has complete possession of only two of its orig-

$$\overset{0}{:\!\ddot{S}} + \overset{0}{:\!\ddot{O}}\!:\!\ddot{O}\!: \longrightarrow \overset{-2\ +4\ -2}{O\!\leftarrow\!S\!=\!O}$$

$$:\!\ddot{S} + :\!\ddot{O}\!:\!\ddot{O}\!: \longrightarrow \overset{\overset{-2}{\underset{\displaystyle O}{}}}{\underset{+6}{O\!\leftarrow\!\overset{-2}{S}\!\rightarrow\!O}}^{\ -2} \qquad\qquad [13\text{-}5]$$

inal six electrons, and hence its oxidation number is $+4$. In SO_3, the sulfur atom has partially transferred all six of its electrons to oxygen—two in covalent bonds and four in two coordinate covalent bonds. Its oxidation number is, therefore, $+6$. Each oxygen in the compound has eight electrons, two more than it holds in the free state, so that its oxidation state is again -2.

C. Decomposition

Heating $KClO_3$ to a temperature above its melting point causes its decomposition to KCl and gaseous oxygen. Although chlorine

$$2K^+ \left[\overset{-2\ +5\ -2}{:\!\ddot{O}\!:\!\ddot{Cl}\!:\!\ddot{O}\!:} \atop {:\!\underset{-2}{\ddot{O}}\!:} \right]^- \overset{\Delta}{\longrightarrow} 2K^+ \overset{+1\ -1}{:\!\ddot{Cl}\!:} + \overset{0}{3O_2}$$

$$2K^+ \left[\begin{matrix} & & O \\ & \nearrow & \\ -O\!-\!Cl & & \\ & \searrow & \\ & & O \end{matrix} \right]^- \longrightarrow 2KCl + 3O_2 \qquad\qquad [13\text{-}6]$$

has eight electrons in its outside shell in both $KClO_3$ and KCl, the number of bonds is different in the two compounds, namely 3 and 1, and the oxidation number is different in the two, namely $+5$ and -1. The chlorine atom is reduced in oxidation number by 6 while oxygen is oxidized from -2 to 0.

In the thermal decomposition of NH_4NO_3 into nitrous oxide and water, the following changes of oxidation number take place.

$$\begin{bmatrix} & -3 & \\ & H & \\ & | & \\ H- & N & -H \\ & | & \\ & H & \end{bmatrix}^+ \begin{bmatrix} +5 & O \\ & \parallel \\ O-N & \\ & \searrow \\ & O \end{bmatrix}^- \quad \xrightarrow{\Delta} \quad \overset{0\ +2}{N \equiv N \rightarrow O} + 2H_2O \qquad [13\text{-}7]$$

Each N-H bond in the NH_4^+ has an oxidation number of -1 for N and $+1$ for H; so the oxidation number of N is $-4 + 1 = -3$ ($+1$ is added because the ion has a $+1$ charge). The nitrate nitrogen has an oxidation number of $+5$. In nitrous oxide the two nitrogens have oxidation numbers of 0 and $+2$. The total change in the decomposition, then, is $-3 + 5$ on the left, which is equal to $0 + 2$ on the right. One nitrogen is oxidized while the other is reduced. One may also say that the average oxidation number of N in nitrous oxide is $+1$.

13-3. Balancing Oxidation-Reduction Equations

Example I

Many oxidation-reduction reactions are somewhat more complicated than the ones just described in that several reactants and products are involved.

1. Manganese dioxide in acid solution will oxidize chloride ion to gaseous chlorine and produce water and manganous ions in the

$$MnO_2 + Cl^- + H^+ \longrightarrow Mn^{++} + Cl_2 + H_2O$$

process. This is a method commonly used to prepare chlorine in the laboratory. This qualitative statement is given in symbols above, but the expression is not balanced chemically since the number of atoms in the reactants is not equal to the number of atoms in the products. It is also not balanced electrically. In the equations written previously the process of balancing has been self-evident on inspection. In more complicated cases, a systematic method of balancing equations is desirable. The first step is to write the correct formulas for reactants and products. This has already been accomplished in the expression above.

2.   Select the atoms or ions in which changes in oxidation numbers have taken place, and write partial equations for them:

$$\overset{+4}{MnO_2} + 2e^- \longrightarrow \overset{+2}{Mn^{++}}$$

$$\overset{-1}{2Cl^-} - 2e^- \longrightarrow \overset{0}{Cl_2}$$

The change in the oxidation number of manganese from $+4$ to $+2$ means that Mn gains two electrons in the process. We indicate this in the partial equation by adding two electrons to the left-hand side. The change in the oxidation number of chlorine from -1 to 0 means that Cl loses an electron in changing from an anion to free chlorine. Consequently an electron is subtracted from the left-hand side of the second partial equation. Since chlorine exists as Cl_2 in the free state, the left-hand side is written to accommodate this fact.

Next we balance the partial equations chemically by adding H^+, OH^-, and H_2O where necessary:

$$MnO_2 + 2e^- + H^+ \longrightarrow Mn^{++} + H_2O$$

$$2Cl^- - 2e^- \longrightarrow Cl_2$$

It will usually be known whether a reaction is carried out in acid or basic solution. If it is known, as it is here, that the reaction mixture must be acidic to go, chemical intuition will suggest that H^+ be added to the left side and, furthermore, that OH^- will appear on neither side in the final equation. This reasoning will generally prevent fumbling for reaction products or reagents in this part of the balancing process.

Then we balance electrically, ordinarily by changing the number in front of the H^+ or OH^-, whichever occurs. Balance the

$$MnO_2 + 2e^- + 4H^+ \longrightarrow Mn^{++} + 2H_2O$$

$$2Cl^- - 2e^- \longrightarrow Cl_2$$

number of hydrogens in the two partial equations. The oxygens should be correct at this point also.

We call all of this step 2 because we generally accomplish all of it without rewriting the partial equations. This step actually involves three parts, which should be carried out in the order shown here.

3. Add the resulting equations together, first making the electrons gained equal to the electrons lost:

$$MnO_2 + 4H^+ + 2Cl^- \longrightarrow Cl_2 + Mn^{++} + 2H_2O \qquad \text{[13-8]}$$

4. Check by counting the oxygens, which should be in balance at this point.

Example II. Baeyer's Unsaturation Test

The method just described may also be applied to the oxidation of organic compounds.

1. A test for unsaturation in an alkene is the disappearance of the purple color of the permanganate ion in alkaline solution when the two are shaken together:

$$CH_2{=}CH_2 + MnO_4^- \longrightarrow \underset{\underset{OH}{|}\underset{OH}{|}}{CH_2{-}CH_2} + MnO_2$$

$$MnO_4^- \longrightarrow MnO_2$$

$$CH_2{=}CH_2 \longrightarrow \underset{\underset{OH}{|}\underset{OH}{|}}{CH_2{-}CH_2}$$

We obtain the number of electrons transferred by noting the change in oxidation number of the atoms in the two partial equations. The manganese atom changes in oxidation number from $+7$ to $+4$, which represents a gain of $3e^-$. In ethylene the change in

$$\overset{+7}{MnO_4^-} + 3e^- \longrightarrow \overset{+4}{MnO_2}$$

$$\overset{-2}{CH_2}=\overset{-2}{CH_2} - 2e^- \longrightarrow \overset{-1}{CH_2}-\overset{-1}{CH_2} \\ \quad\qquad\qquad\qquad OH \;\; OH$$

oxidation number of each carbon is from -2 to -1, a loss of $1e^-$ for each carbon.

2. Now we proceed to balance the equation chemically. Since the reaction is run in alkaline solution, we might start by adding OH^- to the left side of the first partial equation:

$$MnO_4^- + OH^- + 3e^- \longrightarrow MnO_2$$

It will be noted at once, however, that this results in an impasse (only negative charges on one side of the equation), which we can overcome by transferring the OH^- to the right side (never by adding H^+ to the left). Water is now added to the left side, and the equation is balanced electrically and chemically:

$$MnO_4^- + 3e^- + 2H_2O \longrightarrow MnO_2 + 4OH^-$$

Can we in any way justify this transfer of OH^- to the right-hand side of the equation? The answer is yes; for, if one uses an alkaline solution of permanganate ion as an oxidizing agent, the solution becomes still more alkaline as the oxidation proceeds.

3. We add the resulting equations together, first making the electrons gained equal to the electrons lost. We accomplish this by

$$2(MnO_4^- + 3e^- + 2H_2O \longrightarrow MnO_2 + 4OH^-)$$

$$3\left(CH_2{=}CH_2 - 2e^- + 2OH^- \longrightarrow \begin{array}{c} CH_2{-}CH_2 \\ OH \;\; OH \end{array}\right)$$

multiplying the first partial equation by 2 and the second by 3 and then adding to get

$$2MnO_4^- + 3CH_2=CH_2 + 4H_2O \longrightarrow 2MnO_2 + 3CH_2-CH_2 + 2OH^- \quad \text{[13-9]}$$
$$\qquad\qquad\qquad\qquad\qquad\qquad\qquad\qquad\qquad\quad OH \;\; OH$$

4. We check by counting the oxygens. If it is desirable to write the equation with molecular formulas instead of ions, we may fill these in as a final step:

$$2KMnO_4 + 3CH_2=CH_2 + 4H_2O \longrightarrow 2MnO_2 + 3CH_2-CH_2 + 2KOH \quad \text{[13-10]}$$
$$\qquad\qquad\qquad\qquad\qquad\qquad\qquad\qquad\qquad\qquad OH \;\; OH$$

13-4. Exercises for Practice

In most organic oxidations there is only a partial transfer of electrons or a change in the electronegative character of a particular carbon atom. For such cases the electron transfer must be worked out in terms of the change in formal charge. The inorganic reagent, however, generally exhibits a complete transfer of electrons. Several oxidations with two very important oxidizing agents, permanganate ion in basic solution and dichromate ion in acid solution, are given below. It is wise to carry through the four steps previously outlined for each of these oxidations. The final result and the changes in oxidation numbers are given for each example.

Exercise 13-1. Oxidation of a primary alcohol to an acid:

$$3CH_3CH_2\overset{*}{C}H_2OH + 2MnO_4^- \longrightarrow$$
$$\;\; -1$$

$$3CH_3CH_2\overset{*}{C}HO + 2MnO_2 + 2OH^- + 2H_2O$$
$$\qquad\qquad\qquad\qquad +1$$

$\overset{*}{M}$n gains $3e^-$. C loses $2e^-$.

This stage of oxidation of an alcohol is an impractical one to run in alkaline solution. Why?

Exercise 13-2. The oxidation of a primary alcohol goes fairly readily to the carboxylate ion, as shown for 1-propanol.

$$3CH_3CH_2\overset{*}{\underset{-1}{CH_2}}OH + 4MnO_4^- \longrightarrow$$

$$3CH_3CH_2\overset{*}{\underset{+3}{COO}}^- + 4MnO_2 + OH^- + 4H_2O$$

Mn gains 3e⁻. $\overset{*}{C}$ loses 4e⁻.

Exercise 13-3. Baeyer's test on acetylene:

$$3H\overset{-1}{C}\equiv\overset{-1}{C}H + 2\overset{+7}{Mn}O_4^- + 4H_2O \longrightarrow 3H-\overset{+1}{\underset{}{C}}-\overset{-1}{\underset{}{C}}H_2 + 2\overset{+4}{Mn}O_2 + 2OH^-$$

with the product showing O and OH groups.

Mn gains 3e⁻. One C loses 2e⁻; the other undergoes no change in oxidation number.

Exercise 13-4. The oxidation-reduction just written does not actually stop at this stage. The aldehyde-alcohol written as a product here is finally oxidized to the oxalate ion by permanganate ion in the cold.

$$H-\overset{+1}{\underset{}{C}}-\overset{-1}{\underset{}{C}}H_2 + 2\overset{+7}{Mn}O_4^- \longrightarrow \overset{+3}{C}OO^- + \overset{}{C}OO^- + 2\overset{+4}{Mn}O_2 + 2H_2O$$

Mn gains 3e⁻. One C loses 2e⁻; the other loses 4e⁻.

Exercise 13-5. Preparation of an aldehyde or a ketone (aldehyde given here):

$$3CH_3CH_2\overset{*}{\underset{-1}{CH_2}}OH + Cr_2O_7^= + 8H^+ \longrightarrow$$

$$3CH_3CH_2\overset{*}{\underset{+1}{CHO}} + 2Cr^{+3} + 7H_2O$$

Each Cr gains 3e⁻. $\overset{*}{C}$ loses 2e⁻.

Exercise 13-6

$$3CH_2\begin{matrix} \diagup CH_2{-}CH_2 \diagdown \\ \diagdown CH_2{-}CH_2 \diagup \end{matrix}\overset{*}{\underset{0}{C}}HOH + Cr_2O_7^{=} + 8H^+ \longrightarrow$$

$$3CH_2\begin{matrix} \diagup CH_2{-}CH_2 \diagdown \\ \diagdown CH_2{-}CH_2 \diagup \end{matrix}\underset{+2}{C}{=}O + 2Cr^{+3} + 7H_2O$$

*

Cr gains $3e^-$. C loses $2e^-$.

13-5. Oxidation Tests for Aldehydes

The reactions called oxidations or reductions in previous chapters will now be considered in the light of the generalized concepts of the two terms used in this chapter.

The oxidation of an aldehyde to an acid was indicated by the following simplified expression:

$$R{-}CHO \xrightarrow[\text{agent}]{\text{oxidizing}} R{-}COOH$$

Such an oxidation may be accomplished by a number of reagents.

1. Oxygen Gas with a Catalyst

$$2R{-}\overset{\overset{+1}{H}}{\overset{|}{C}}{=}O + \overset{0}{O_2} \xrightarrow{V_2O_5} 2R{-}\overset{+3}{C}\begin{matrix} {=}\overset{-2}{O} \\ {-}OH \end{matrix} \qquad [13\text{-}11]$$

This equation is balanced by inspection. The changes in oxidation number are from +1 to +3 for carbon and from 0 to −2 for oxygen. It should not be inferred that the oxygen marked with oxidation number −2 necessarily came from the reagent, O_2.

2. KMnO₄ in Alkaline Solution

1. $R-\overset{H}{\underset{|}{C}}{=}O + MnO_4^- \xrightarrow{OH^-} R-C\overset{O}{\underset{O^-}{\diagdown}} + MnO_2$

2. $\underset{+7}{2(2H_2O + MnO_4^-} + 3e^- \longrightarrow \underset{+4}{MnO_2} + 4OH^-)$

$3(\overset{+1}{R-\overset{H}{\underset{|}{C}}{=}O} + 3OH^- - 2e^- \longrightarrow \underset{+3}{R-C}\overset{O}{\underset{O^-}{\diagdown}} + 2H_2O)$

Adding, we get

3. $3R-\overset{H}{\underset{|}{C}}{=}O + 2MnO_4^- + OH^- \longrightarrow 2MnO_2 + 3R-C\overset{O}{\underset{O^-}{\diagdown}} + 2H_2O$ [13-12]

Decolorizing an alkaline solution of permanganate ion is a test for unsaturation as well as a test for such reducing agents as aldehydes. It cannot, therefore, be used alone as a conclusive test for an aldehyde.

3. Tollens' Reagent.

A solution of silver nitrate to which ammonium hydroxide has been added yields a brown precipitate of silver oxide; if more ammonia is then added, the silver salt goes back into solution as a silver-ammonia complex. This gives a very sensitive oxidizing agent for aldehydes, known as Tollens' reagent (see p. 208).

1. $Ag(NH_3)_2^+ + R-\overset{H}{\underset{|}{C}}{=}O \xrightarrow{OH^-} R-C\overset{O}{\underset{O^{\ominus}\ NH_4^{\oplus}}{\diagdown}} + \underline{Ag}$

2. $\underset{+1}{2(Ag(NH_3)_2^+} + H_2O + 1e^- \longrightarrow \underset{0}{\underline{Ag}} + NH_4^+ + OH^- + NH_3)$

$\underset{-1}{R-\overset{H}{\underset{|}{C}}{=}O} + 3OH^- - 2e^- \longrightarrow \underset{+3}{R-C}\overset{O}{\underset{O^-}{\diagdown}} + 2H_2O$

Adding, we get

3. $2Ag(NH_3)_2^+ + OH^- + R-\overset{H}{\underset{|}{C}}=O \longrightarrow R-C\overset{O}{\underset{O^-}{\diagdown}} + 2NH_4^+ + 2Ag + 2NH_3$ [13-13]

4. Fehling's Solution (see p. 208)

1. $R-\overset{H}{\underset{|}{C}}=O + Cu^{++} \xrightarrow{OH^-} R-C\overset{O}{\underset{O^-}{\diagdown}} + Cu_2O$
   (in tartrate complex)

2. $\overset{+2}{2Cu^{++}} + 2OH^- + 2e^- \longrightarrow \underline{\overset{+1}{Cu_2O}} + H_2O$
   (in tartrate complex)

$\underset{+1}{R-\overset{H}{\underset{|}{C}}=O} + 3OH^- - 2e^- \longrightarrow R-C\overset{\overset{+3}{O}}{\underset{O^-}{\diagdown}} + 2H_2O$

Adding, we get

3. $\underset{\text{(in tartrate complex)}}{2Cu^{++}} + R-\overset{H}{\underset{|}{C}}=O + 5OH^- \longrightarrow \underline{Cu_2O} + R-C\overset{O}{\underset{O^-}{\diagdown}} + 3H_2O$ [13-14]

5. Strong Oxidizing Agent and Formaldehyde.

This reaction goes one step further than with any other aldehyde. This will be demonstrated with alkaline permanganate, in which the first oxidation,

$3H-\overset{H}{\underset{|}{C}}=O + 2MnO_4^- + OH^- \longrightarrow 2MnO_2 + 3H-C\overset{O}{\underset{O^-}{\diagdown}} + 2H_2O$

is the same as that given in Expression 13-12 for other aldehydes.

1. $H-C\overset{O}{\underset{O^-}{\diagdown}} + MnO_4^- \longrightarrow CO_3^= + MnO_2$

$$2.\ 2(2H_2O + \overset{+7}{MnO_4^-} + 3e^- \longrightarrow \overset{+4}{MnO_2} + 4OH^-)$$

$$3(3OH^- + H-\overset{\overset{+2}{}\ O}{\underset{O^-}{C{\Large\diagup}}} - 2e^- \longrightarrow \overset{+4}{CO_3^=} + 2H_2O)$$

Adding, we get

$$3.\ 3H-\overset{O}{\underset{O^-}{C{\Large\diagup}}} + 2MnO_4^- + OH^- \longrightarrow 2MnO_2 + 3CO_3^= + 2H_2O \qquad [13\text{-}15]$$

The reactions of formic acid that are different from those of other acids are best accounted for if the substance is looked upon as both an acid and an aldehyde. The structure gives some justification for this. One of the properties of aldehydes exhibited by

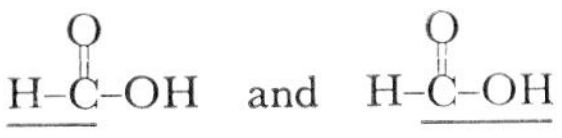

$$H-\overset{O}{\underset{}{\overset{\|}{C}}}-OH \quad \text{and} \quad H-\overset{O}{\underset{}{\overset{\|}{C}}}-OH$$

formic acid, their strong reducing property, is manifest in the following reactions: it is easily oxidized to CO_2 and H_2O (1) by Tollen's reagent, (2) by Fehling's solution, and (3) by permanganate ion, the last of which it decolorizes rapidly. It might be added that the carbon in formic acid has the oxidation number of a ketone ($+2$) rather than that of other carboxylic acids ($+3$).

13-6. Reactions from Previous Chapters

1. Preparation of an Organic Acid by Oxidation of a Side Chain

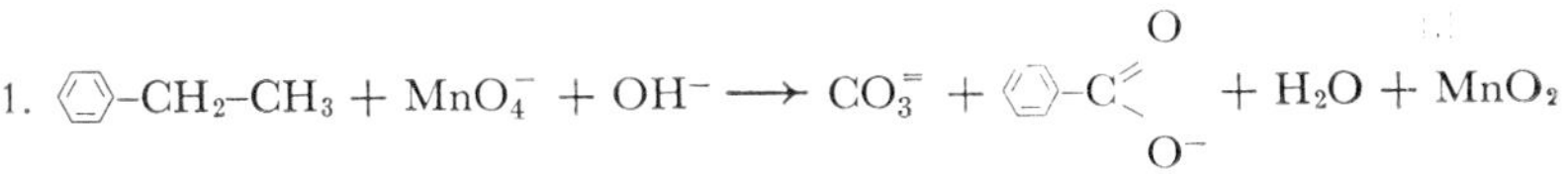

$$1.\ \bigcirc\!\!-CH_2-CH_3 + MnO_4^- + OH^- \longrightarrow CO_3^= + \bigcirc\!\!-\overset{O}{\underset{O^-}{C{\Large\diagup}}} + H_2O + MnO_2$$

$$2.\ 4(2H_2O + \overset{+7}{MnO_4^-} + 3e^- \longrightarrow \overset{+4}{MnO_2} + 4OH^-)$$

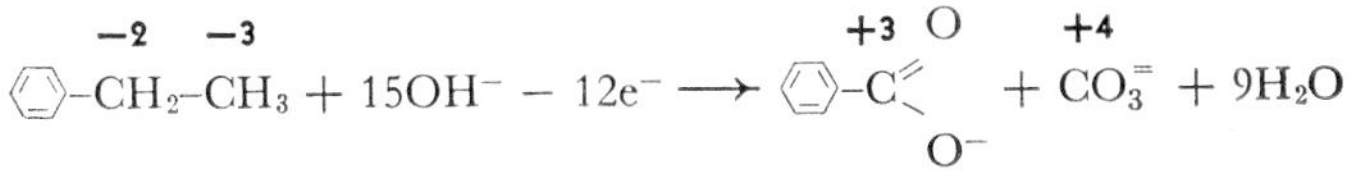

$$\bigcirc\!\!-\overset{-2}{C}H_2-\overset{-3}{C}H_3 + 15OH^- - 12e^- \longrightarrow \bigcirc\!\!-\overset{\overset{+3}{}\ O}{\underset{O^-}{C{\Large\diagup}}} + \overset{+4}{CO_3^=} + 9H_2O$$

Adding, we get

3. $\bigcirc\!\!-CH_2-CH_3 + 4MnO_4^- \longrightarrow$

$$4MnO_2 + \bigcirc\!\!-C\!\!\begin{array}{c}O\\ \diagdown\\ O^-\end{array} + CO_3^= + OH^- + 2H_2O \qquad [13\text{-}16]$$

2. The Pinacol Reduction (see p. 226)

1. $\begin{array}{c}CH_3\\ \diagdown\\ CH_3\end{array}\!\!C{=}O + Mg(Hg) + H_2O \longrightarrow \begin{array}{cc}CH_3 & CH_3\\ \diagup & \diagup\\ C\!\!-\!\!C\!\!-\!\!CH_3\\ CH_3 & OH\ OH\end{array} + Mg(OH)_2 + Hg$

2. $2\ \begin{array}{c}CH_3\ ^{+2}\\ \diagdown\\ CH_3\end{array}\!\!C{=}O + 2H_2O + 2e^- \longrightarrow \begin{array}{cc}CH_3\ ^{+1} & ^{+1}\ CH_3\\ \diagup\ |\quad |\ \diagdown\\ C\!\!-\!\!C\\ CH_3\ | \quad |\ CH_3\\ OH\quad OH\end{array} + 2OH^-$

$\overset{0}{Mg(Hg)} + 2OH^- - 2e^- \longrightarrow \overset{+2}{Mg(OH)_2} + Hg$

Adding, we get

3. $2\ \begin{array}{c}CH_3\\ \diagdown\\ CH_3\end{array}\!\!C{=}O + 2H_2O + Mg(Hg) \longrightarrow \begin{array}{cc}CH_3 & CH_3\\ \diagup\ |\quad |\ \diagdown\\ C\!\!-\!\!C\\ CH_3\ | \quad |\ CH_3\\ OH\quad OH\end{array} + Mg(OH)_2 + Hg$

$$\qquad\qquad [13\text{-}17]$$

3. Cannizzaro Reaction (see p. 217).

In the Cannizzaro reaction, one aldehydic carbon is oxidized at the expense of another. The equation is easily balanced by inspection since the changes in oxidation number are the same.

$$\underset{0}{H\!-\!\overset{H}{\underset{|}{C}}{=}O} + \underset{\overset{*}{0}}{H\!-\!\overset{H}{\underset{|}{C}}{=}O} + NaOH \longrightarrow \underset{+2}{H\!-\!C\!\!\begin{array}{c}O\\ \diagdown\\ O\!-\!Na\end{array}} + \underset{-2}{\overset{*}{C}H_3OH} \qquad [13\text{-}18]$$

$\overset{*}{C}$ loses $2e^-$. $\overset{}{C}$ gains $2e^-$.

4. Nitration of Benzene. In the nitration of benzene, the

$$\overset{-1}{\bigcirc} + \underset{\underset{O}{\downarrow}}{\overset{+5}{HO-N=O}} \xrightarrow{H_2SO_4} \overset{+3}{\underset{\bigcirc}{+1}N}\overset{O}{\underset{O}{\diagdown}} + H_2O \qquad\qquad [13\text{-}19]$$

carbon atom to which the nitrogen atom becomes attached changes from an electron seeker (oxidation number -1) to an electron donor atom ($+1$). The nitrogen changes its oxidation number from $+5$ in nitric acid (three covalent bonds to oxygen and a donated bond to oxygen) to $+3$ in nitrobenzene (two covalent bonds to oxygen [$+2$], one covalent bond to carbon [-1], and one donated bond to oxygen [$+2$]).

SUMMARY

Part of this chapter is concerned with bookkeeping—that is, with the mechanics of balancing equations. This is chemistry too. Some tricks for accomplishing this in a painless way are given in some detail, making use of the concept of oxidation number. A painful four-step process of balancing equations may quickly be reduced by practice to a smooth system in which you will have confidence.

The concept of oxidation-reduction is presented in a very generalized picture, which may be applied to all reactions—those involving complete transfer of electrons (mostly inorganic) and those involving partial transfer of electrons (largely organic).

REFERENCES

1. Vander Werf, "Balancing Oxidation-Reduction Equations in Organic Chemistry," J. Chem. Education, 22, 218 (1945).

2. Swinehart, "More on Oxidation Numbers," J. Chem. Education, 29, 284 (1952).

3. Gregg, "A Simplified Electronic Interpretation of Oxidation-Reduction," J. Chem. Education, 22, 548 (1945).

4. Ferguson, "Balancing Equations for Organic Oxidation-Reduction Reactions," J. Chem. Education, 23, 550 (1946).

5. Glasstone, "Oxidation Numbers and Valence," J. Chem. Education, 25, 278 (1948).

EXERCISES

(Exercises 1–6 will be found within the text.)

7. Determine the oxidation numbers of the atoms indicated:

1. P in PH_3
2. Cr in Cr_2O_3
3. Cl in ClO_4^-
4. S in $S_2O_3^=$
5. B in $Na_2B_4O_7$
6. N in $C_6H_5NO_2$
7. S in $C_6H_5-SO_3H$

8. any C in benzene
9. C No. 1 in CH_3COO^-
10. both chain C's in $C_6H_5-CH_2-CH_3$
11. C No. 1 in $CH_3CH_2NO_2$
12. C No. 1 in $CH_3-CH\overset{\diagup OH}{\diagdown NH_2}$

8. Obtain balanced equations for the following indicated reactions:

1. $H_2SO_4 + HI \longrightarrow H_2S + I_2$
2. $Cr_2O_7^= + Cl^- \longrightarrow Cl_2 + Cr^{+3}$ (acid solution)
3. $CH_2{=}CH_2 + Cr_2O_7^= \longrightarrow CO_2 + Cr^{+3}$ (acid solution)
4. $C_6H_5-CH_2OH + MnO_4^- \longrightarrow C_6H_5-COO^- + MnO_2$ (basic solution)
5. $\underset{COOH}{\overset{COOH}{|}} + MnO_4^- \longrightarrow CO_2 + Mn^{++}$ (acid solution)

9. Write balanced equations in ionic form for the following indicated reactions:

1. methyl phenyl ketone + zinc + H^+
2. propionaldehyde + Cu^{++} (complexed in basic solution)
3. propene + very dilute $KMnO_4$ (in basic solution)

10. In the preparation of 50 g of cyclohexanone by oxidation of cyclo-hexanol with $Na_2Cr_2O_7$ in H_2SO_4, how much sodium dichromate is actually consumed?

11. How much $KMnO_4$ is required to oxidize 100 g of oxalic acid ($HOOC-COOH$) to CO_2 in acid solution? to $CO_3^=$ in basic solution? Is it more economical to use permanganate solution in acid or in basic solution?

12. Write balanced half-reactions for the indicated changes (see section 13-3, Example I, step 2):

    A. In acid solution:

    1. $NO_3^- \longrightarrow NO$
    2. $SbO^+ \longrightarrow Sb_2O_5$
    3. $CH_3COCH_3 \longrightarrow CH_3CHOHCH_3$
    4. $C H_5-NO_2 \longrightarrow C_6H_5-NH_3^+$
    5. $CH_2=CH_2 \longrightarrow CO_2$

    B. In basic solution:

    1. $OCl^- \longrightarrow Cl^-$
    2. $N_2O_4 \longrightarrow NO_3^-$
    3. $CH_3CHO \longrightarrow CH_3COO^-$
    4. $CH_2OH-CH_2OH \longrightarrow \overset{\displaystyle COO^-}{\underset{\displaystyle COO^-}{|}}$
    5. $C_6H_5-CH_2CH_3 \longrightarrow C_6H_5-COO^- + CO_3^-$

13. In making the $HBr-H_2SO_4$ mixture for use in preparing n-butyl bromide from n-butyl alcohol, 8.5 ml of bromine (specific gravity 3.3) was poured into ice, and SO_2 was bubbled through the mixture until all the bromine reacted. If all the solution was used on 23 ml of n-butyl alcohol (specific gravity 0.80), was the alcohol or the HBr in excess?

14 Acid Derivatives

The carbonyl group has certain distinct properties when it is attached to carbon and hydrogen (the aldehyde function), other properties when it is attached to two carbons (the ketone function), and still others when it is attached to a hydroxyl group (the carboxyl function). This is only the beginning of the modifications that can be made in the carbonyl group, of which the following are examples:

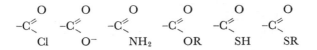

The classes of compounds containing these groups are referred to collectively as acid derivatives since they may be made directly from carboxylic acids. The first four of these derivatives, along with one more, $-C\equiv N$, which can also be obtained from an acid, will be discussed here.

The properties of acids involve three types of reaction: displacement of the ionized proton, replacement of the hydroxyl radical of the carboxyl group, and replacement of a hydrogen on an α-carbon. Reactions of the first type have already been mentioned in Chapter 10; reactions of the second type are discussed here; reactions of the third type are reserved for later consideration (Chap. 21).

Salts

14-1. Nomenclature of Salts

Organic acids dissolve in soluble inorganic bases to form salts. The salts are named in the following way: name of the metal and

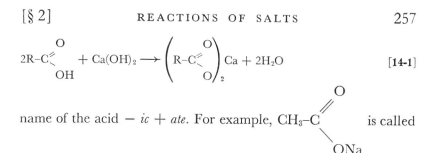

$$2R-C\underset{OH}{\overset{O}{\lessgtr}} + Ca(OH)_2 \longrightarrow \left(R-C\underset{O}{\overset{O}{\lessgtr}}\right)_2 Ca + 2H_2O \qquad [14\text{-}1]$$

name of the acid $-$ *ic* $+$ *ate*. For example, $CH_3-C\overset{O}{\underset{ONa}{\diagup}}$ is called

sodium acetate (*acetic* $-$ *ic* $+$ *ate*). This is similar to the naming of salts of inorganic oxygen acids.

Exercise 14-1. Write structural formulas for sodium isobutyrate, potassium *p*-nitrobenzoate, sodium phenoxide, sodium *p*-bromobenzenesulfonate, sodium methoxide, barium formate, and aluminum isopropoxide.

14-2. Reactions of Salts

1. With Acids. Acids that are stronger (more highly ionized) than an organic acid will liberate that organic acid from any of its salts. This reaction takes place for the same reason for which any neutralization reaction takes place: the formation of a compound that is only slightly ionized in comparison with the starting compounds.

$$R-C\underset{O^-}{\overset{O}{\lessgtr}} Na^+ + H^+OSO_3H^- \longrightarrow R-C\underset{OH}{\overset{O}{\lessgtr}} + Na^+OSO_3H^- \qquad [14\text{-}2]$$

2. Acid Anhydrides. Acid chlorides react with salts to yield acid anhydrides. This reaction is discussed on p. 262.

Acyl Halides, or Acid Halides

14-3. Nomenclature of Acyl Halides

The general formula for an acyl halide is $R-C\overset{O}{\underset{X}{\diagup}}$, in which the hydroxyl group of an acid has been replaced by a halogen atom. We name the acyl halides by dropping the *ic* ending of the name of the acid and adding *yl* along with the name of the particular halide.

For example, $CH_3-C\overset{O}{\underset{Br}{\diagdown}}$ is called (*acetic − ic + yl*) acetyl bro-

mide, and $CH_3-CH_2-CH_2-C\overset{O}{\underset{Cl}{\diagdown}}$ is called n-butyryl chloride.

Exercise 14-2. Write structural formulas for benzoyl chloride, isobutyryl bromide, and propionyl iodide.

14-4. Preparation of Acyl Halides

Compare the preparation of alkyl halides (pp. 66–70).

The hydroxyl group in a carboxyl group can be replaced by a halogen with such reagents as $SOCl_2$, PCl_3, PCl_5, PBr_3, and $P + I_2$. Thionyl chloride ($SOCl_2$) is the most frequently used with liquid acids. The acyl halide formed may be separated easily from the

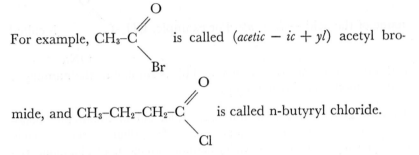

$$R-C\overset{O}{\underset{OH}{\diagdown}} + SOCl_2 \longrightarrow R-C\overset{O}{\underset{Cl}{\diagdown}} + SO_2 + HCl \qquad [14\text{-}3]$$

$$3R-C\overset{O}{\underset{OH}{\diagdown}} + PCl_3 \longrightarrow 3R-C\overset{O}{\underset{Cl}{\diagdown}} + H_3PO_3 \qquad [14\text{-}4]$$

$$R-C\overset{O}{\underset{OH}{\diagdown}} + PCl_5 \longrightarrow R-C\overset{O}{\underset{Cl}{\diagdown}} + POCl_3 + HCl \qquad [14\text{-}5]$$

$$6R-C\overset{O}{\underset{OH}{\diagdown}} + 2P + 3I_2 \longrightarrow 6R-C\overset{O}{\underset{I}{\diagdown}} + 2H_3PO_3 \qquad [14\text{-}6]$$

other products of the reaction since both of them are gases. Phosphorus trichloride may be used in the same way if the acyl chloride produced can be distilled without decomposition. Acids in the solid state are more frequently treated with PCl_5; the HCl and $POCl_3$ are removed by distillation at fairly low temperatures, leaving the desired product to be recrystallized.

Formic acid cannot be treated to form an acyl chloride, for the expected product is so unstable that it does not exist. The only

$$H-C\overset{O}{\underset{OH}{\diagdown}} + SOCl_2 \longrightarrow \left[H-C\overset{O}{\underset{Cl}{\diagdown}} \right] + SO_2 + HCl \qquad \text{[14-7]}$$
$$\phantom{H-C\overset{O}{\underset{Cl}{\diagdown}}} \longrightarrow CO + HCl$$

products isolated from this reaction are SO_2, CO, and HCl.

14-5. Reactions of Acyl Halides

1. Hydrolysis. [Compare the hydrolysis of $SOCl_2$ etc. (p. 189), and see also p. 288.] Since the acyl halides of low molecular weight react violently with water, the presence of water or moist air must be carefully avoided in their preparation. Acyl halides containing six or more carbons, however, are not readily hydrolyzed at room temperature, and heat must be applied to effect the reaction. Bases increase the speed of the reaction, as might be anticipated. As an example, benzoyl chloride reacts very slowly with water and must be shaken for some time with dilute NaOH before the reaction to form benzoate ion is complete. Acetyl chloride, on the

$$\text{C}_6\text{H}_5-C\overset{O}{\underset{Cl}{\diagdown}} + 2OH^- \xrightarrow[H_2O]{\Delta} \text{C}_6\text{H}_5-C\overset{O}{\underset{O^-}{\diagdown}} + Cl^- + H_2O \qquad \text{[14-8]}$$

other hand, reacts with water to form two acids. This chloride is of the same character as PCl_3, which also hydrolyzes to form two

$$CH_3-COCl + HOH \longrightarrow CH_3-COOH + HCl \qquad \text{[14-9]}$$

acids, H_3PO_3 and HCl. The reaction of acetyl chloride with water, like that of PCl_3 with water, is violent and proceeds with a hissing sound.

2. Ester Formation. (See pp. 271ff.) Though acyl halides in general are lachrymators, they react with alcohols to form sweet-smelling compounds called esters (Exp. 14-10).

$$\underset{X}{\overset{O}{R-C}} + R'OH \longrightarrow \underset{\underset{\text{an ester}}{OR'}}{\overset{O}{R-C}} + HX \qquad\qquad \text{[14-10]}$$

3. Preparation of an Acid Anhydride. A salt of an organic acid will react with an acyl halide to form an acid anhydride. The electrovalent product, the sodium halide, can be dissolved out from the mixture of products if hydrolysis of the acid anhydride is avoided by the use of cold solvent.

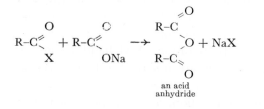

$$\text{[14-11]}$$

Exercise 14-3. Write equations for the preparation of propionic anhydride, starting with propionic acid.

4. Ammonolysis. An acyl halide will react with ammonia about as violently as with water. The products are an ammonium halide and the ammonia analogue corresponding to the reaction with water. The latter is called an amide.

$$\underset{X}{\overset{O}{R-C}} + 2NH_3 \longrightarrow \underset{\underset{\text{an amide}}{NH_2}}{\overset{O}{R-C}} + NH_4X \qquad\qquad \text{[14-12]}$$

5. Friedel-Crafts Reaction.   Acyl halides, as well as alkyl halides, undergo the Friedel-Crafts reaction with aromatic hydrocarbons in the presence of $AlCl_3$. The product is a mixed aliphatic-aromatic ketone (Exp. 14-13). Alkyl groups or halogen atoms on the aromatic ring yield mixtures of *o* and *p* isomers. For example,

$$R-\overset{O}{\underset{}{C}}-Cl + \langle \bigcirc \rangle \xrightarrow{AlCl_3} R-\overset{O}{\underset{}{C}}-\langle \bigcirc \rangle + HCl \qquad\qquad [14\text{-}13]$$

acetyl chloride reacting with bromobenzene gives a mixture *o*-bromoacetophenone and *p*-bromoacetophenone (Exp. 14-14; cf. Exp. 14-35). When benzoyl chloride reacts with toluene in a

$$CH_3-C\overset{O}{\underset{Cl}{\big\langle}} + Br-\langle \bigcirc \rangle \xrightarrow{AlCl_3} \overset{Br\ COCH_3}{\langle \bigcirc \rangle} + Br-\langle \bigcirc \rangle-\overset{O}{\underset{70\%}{C}}-CH_3 + HCl \qquad [14\text{-}14]$$

Friedel-Crafts reaction, a similar mixture of *o* and *p* products is obtained (Exp. 14-15).

$$\langle \bigcirc \rangle-C\overset{O}{\underset{Cl}{\big\langle}} + CH_3-\langle \bigcirc \rangle \xrightarrow{AlCl_3} CH_3-\langle \bigcirc \rangle-\underset{72\%}{\overset{O}{\underset{}{C}}}-\langle \bigcirc \rangle + \langle \bigcirc \rangle-\underset{\underset{CH_3}{\overset{|}{\underset{18\%}{}}}}{\overset{O}{\underset{}{C}}}-\langle \bigcirc \rangle + HCl \qquad [14\text{-}15]$$

Acid Anhydrides

Acid anhydrides react with water to form acids. The anhydrides may or may not be carbon compounds.

$$\underset{\substack{\text{sulfur}\\\text{trioxide}}}{O\leftarrow\overset{O}{\underset{}{S}}\rightarrow O} + H_2O \longrightarrow \underset{\substack{\downarrow\\O\\\text{sulfuric acid}}}{\overset{\underset{\uparrow}{O}}{H-O-S-O-H}} \qquad\qquad [14\text{-}16]$$

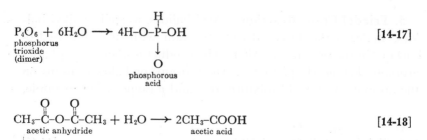

$$P_4O_6 + 6H_2O \longrightarrow 4H\text{-}O\text{-}\overset{\underset{\displaystyle O}{\displaystyle |}}{\overset{\displaystyle H}{P}}\text{-}OH \qquad [14\text{-}17]$$

phosphorus
trioxide
(dimer)

phosphorous
acid

$$\underset{\text{acetic anhydride}}{CH_3\text{-}\overset{O}{\overset{||}{C}}\text{-}O\text{-}\overset{O}{\overset{||}{C}}\text{-}CH_3} + H_2O \longrightarrow \underset{\text{acetic acid}}{2CH_3\text{-}COOH} \qquad [14\text{-}18]$$

Mixed acid anhydrides exist but are of no practical importance since a mixture of two acids would be obtained by hydrolysis of

$$R\text{-}\overset{O}{\overset{||}{C}}\text{-}O\text{-}\overset{O}{\overset{||}{C}}\text{-}R' + H_2O \longrightarrow R\text{-}COOH + R'COOH \qquad [14\text{-}19]$$

the compounds. Other reactions as well as hydrolysis would give mixtures not easy to separate.

Acyl halides also react with water to give two acids, one of which is a hydrohalogen acid. Since HCl has no anhydride, an acyl

$$R\text{-}COCl + H_2O \longrightarrow RCOOH + HCl \qquad [14\text{-}20]$$

$$PCl_3 + H_2O \longrightarrow H_3PO_3 + HCl \text{ (see p. 68)} \qquad [14\text{-}21]$$

halide cannot strictly be called a mixed anhydride. The analogy is useful, however, for the acyl halides and acid anhydrides have the same chemical properties in a qualitative way.

14-6. Preparation of Acid Anhydrides

Acid anhydrides of monocarboxylic acid cannot be prepared directly from acids by dehydration. Instead the acyl halide and the salt of the acid must be heated together. Formic anhydride has never yet been prepared and presumably does not exist.

$$R\text{-}\overset{O}{\underset{OH}{C}} + SOCl_2 \longrightarrow R\text{-}\overset{O}{\underset{Cl}{C}} + SO_2 + HCl \qquad [14\text{-}22]$$

$$R\text{-}\overset{O}{\overset{\|}{C}}\text{-OH} + NaOH \longrightarrow R\text{-}C\overset{O}{\underset{ONa}{\diagdown}} + H_2O \qquad\qquad [14\text{-}23]$$

$$R\text{-}C\overset{O}{\underset{Cl}{\diagdown}} + R\text{-}C\overset{O}{\underset{ONa}{\diagdown}} \longrightarrow R\text{-}\overset{O}{\overset{\|}{C}}\text{-O-}\overset{O}{\overset{\|}{C}}\text{-R} + NaCl \qquad [14\text{-}24]$$

Attempts to prepare it lead to CO as the only product containing carbon. Acid anhydrides of mineral acids also cannot be prepared by dehydration of the acid. In contrast to the carboxylic anhydride, however, many of these anhydrides may be prepared by direct combination. As examples, SO_2, SO_3, P_2O_3, and P_2O_5 will be discussed.

When sulfur burns in air, sulfur dioxide is formed, with only small amounts (about 3%) of SO_3. The student is probably already familiar with the choking odor of sulfur dioxide, which boils at $-10°C$. The gas can be oxidized to SO_3 in the presence of

$$S + O_2 \longrightarrow SO_2 \qquad\qquad\qquad\qquad\qquad [14\text{-}25]$$

spongy platinum, of V_2O_5, and of other oxides. This reaction is the basis of the "contact process" of manufacturing sulfuric acid. Sulfur trioxide is a liquid boiling at $44°C$; it also exists in white solid forms as dimer, trimer, and perhaps polymers.

$$2SO_2 + O_2 \underset{650°}{\overset{V_2O_5}{\rightleftarrows}} 2SO_3 + heat \qquad\qquad\qquad [14\text{-}26]$$

By controlled burning of white phosphorus at 50°, the trioxide, P_2O_3 (actual formula P_4O_6), a colorless crystalline compound, with melting point 24°C and boiling point 175°C, is obtained. An excess of oxygen results in a complete oxidation to P_2O_5 (actual formula P_4O_{10}), a comparatively nonvolatile white solid.

$$2P + 3O_2 \longrightarrow P_2O_3 \qquad\qquad\qquad\qquad [14\text{-}27]$$
$$\text{phosphorus trioxide}$$

$$2P + 5O_2 \text{ (excess)} \longrightarrow P_2O_5 \qquad\qquad\qquad [14\text{-}28]$$
$$\text{phosphorus pentoxide}$$

14-7. Reactions of Acid Anhydrides

The reactions of the organic acid anhydrides may be written entirely by analogy with the reactions of the acyl halides.

1. Hydrolysis. By definition, the hydrolysis of an acid anhydride leads to the acid. The mixed acid anhydrides give mixtures of two organic acids.

$$R-\overset{O}{\overset{\|}{C}}-O-\overset{O}{\overset{\|}{C}}-R + HOH \longrightarrow 2RCOOH \qquad\qquad [14\text{-}29]$$

The reaction shown in Expression 14-29 is rather slow, but the acid anhydrides of mineral acids react with considerable evolution of heat. Sulfur trioxide (Exp. 14-16) reacts so violently that a hissing sound is produced by the steam generated at the spot where it hits the water. Hydrolysis is the most important reaction of acid anhydrides.

$$SO_2 + HOH \rightleftharpoons \underset{\underset{O}{\downarrow}}{HO-S-OH} \qquad\qquad [14\text{-}30]$$

$$P_2O_5 + 3HOH \longrightarrow \underset{\underset{OH}{|}}{2HO-\overset{\overset{O}{\uparrow}}{P}-OH} \qquad\qquad [14\text{-}31]$$

2. Alcoholysis. (See p. 272.) Esters are formed from acid anhydrides along with a molecule of the organic acid. The corresponding reaction with sulfur trioxide or phosphorus pentoxide is impractical because of the large amount of heat evolved.

$$R-\overset{O}{\overset{\|}{C}}-O-\overset{O}{\overset{\|}{C}}-R + R'OH \longrightarrow R-C\overset{\diagup O}{\underset{\diagdown OR'}{}} + R-COOH \qquad\qquad [14\text{-}32]$$
$$\text{an ester}$$

3. Ammonolysis.   Again the analogy between acyl halides and acid anhydrides is close in the reaction with ammonia. When sulfur

$$R-\overset{O}{\underset{}{C}}-O-\overset{O}{\underset{}{C}}-R + 2NH_3 \longrightarrow R-C\overset{O}{\underset{NH_2}{}} + R-C\overset{O}{\underset{ONH_4}{}} \qquad [14\text{-}33]$$

an amide   ammonium salt

trioxide or phosphorus pentoxide reacts with ammonia, the great heat liberated is hard to dissipate rapidly, and the reaction becomes hard to control. Besides, the corresponding ammonium salts, products of these reactions, are easily obtained by direct reaction with the acids themselves. Ammonolysis of these inorganic anhydrides, then, is of no practical value.

4. Friedel-Crafts Reaction.   Ketones can be prepared by the Friedel-Crafts reaction between acid anhydrides and aromatic hydrocarbons or aryl halides.

$$\bigcirc + \begin{matrix} CH_3-C\nearrow^O \\ \searrow O \\ CH_3-C\searrow_O \end{matrix} \xrightarrow[85\%]{AlCl_3} \bigcirc-\overset{O}{\underset{}{C}}-CH_3 + CH_3-COOH \qquad [14\text{-}34]$$

$$Br-\bigcirc + \begin{matrix} CH_3-C\nearrow^O \\ \searrow O \\ CH_3-C\searrow_O \end{matrix} \xrightarrow{AlCl_3} Br-\bigcirc-\underset{\underset{69\text{-}79\%}{O}}{C}-CH_3 + \bigcirc-\overset{O}{\underset{Br}{C}}-CH_3 + CH_3COOH \qquad [14\text{-}35]$$

Amides

Amides are chemically neutral compounds having the general formula $R-C\overset{O}{\underset{NH_2}{}}$ (see structures *l*, *m*, and *n* on p. 286). The homologous series of amides have higher melting points than the other homologous series. The only liquid amide known, in fact, is the simplest one, $H-CONH_2$, called formamide.

14-8. Nomenclature of Amides

We name amides as derivatives of the acids from which they come by dropping the *-ic* ending and adding the suffix *-amide*;

$$CH_3-C\overset{\displaystyle O}{\underset{\displaystyle NH_2}{\big<}}$$, for example, is called acetamide (*acetic — ic +*

amide). With some acids, such as benzoic, we drop the ending *-oic*

and add *-amide* to get the name of the amide; $C_6H_5-C\overset{\displaystyle O}{\underset{\displaystyle NH_2}{\big<}}$ is

benzamide.

We obtain the IUC names from the original hydrocarbon of the same number of carbons by dropping *-e* and adding *-amide*; thus

$$H-C\overset{\displaystyle O}{\underset{\displaystyle NH_2}{\big<}}$$ is called methanamide (*methane — e + amide*).

FORMULA	COMMON NAME	IUC NAME
$H-CONH_2$	formamide	methanamide
CH_3-CONH_2	acetamide	ethanamide
$CH_3-CH_2-CONH_2$	propionamide	propanamide
$CH_3-CH-CONH_2$ $\quad\ \ \|$ $\quad\ CH_3$	isobutyramide	2-methylpropanamide
⬡-$CONH_2$	benzamide	benzamide

14-9. Preparation of Amides

1. Ammonolysis. Acyl halides, acid anhydrides, and esters react with ammonia to produce amides. The reaction, like hydrolysis, is much more violent with acyl halides than with acid anhydrides. The ammonolysis of an ester generally requires heating. An acid catalyst, such as NH_4Cl, will also speed up the reaction with esters of higher molecular weight.

$$R-C\underset{X}{\overset{O}{\lessgtr}} + 2NH_3 \longrightarrow R-C\underset{NH_2}{\overset{O}{\lessgtr}} + NH_4X \qquad [14\text{-}36]$$

$$R-\overset{O}{\overset{\|}{C}}-O-\overset{O}{\overset{\|}{C}}-R + 2NH_3 \longrightarrow R-C\underset{NH_2}{\overset{O}{\lessgtr}} + R-C\underset{ONH_4}{\overset{O}{\lessgtr}} \qquad [14\text{-}37]$$

$$R-C\underset{OR'}{\overset{O}{\lessgtr}} + NH_3 \longrightarrow R-C\underset{NH_2}{\overset{O}{\lessgtr}} + R'OH \qquad [14\text{-}38]$$

2. Heating an Ammonium Salt. If the ammonium salt of an organic acid is isolated in the dry state and then heated, a molecule of water may be driven off, leaving an amide.

$$R-C\underset{ONH_4}{\overset{O}{\lessgtr}} \overset{\Delta}{\longrightarrow} R-CONH_2 + H_2O \qquad [14\text{-}39]$$

3. Partial Hydrolysis of an Alkyl Cyanide. Complete hydrolysis of a cyanide gives an organic acid, but the hydrolysis of some cyanides in acid solution may be stopped before completion, resulting in an amide. It has been found that one good way to carry out such a reaction is to dissolve the alkyl cyanide in sulfuric acid containing one molecule of water for each molecule of cyanide. Refluxing this solution then produces the amide. The mechanism of the reaction has already been shown (p. 199).

$$R-C\equiv N + HOH \overset{H_2SO_4}{\longrightarrow} \left[R-\underset{OH}{\overset{}{C}}=NH \right] \longrightarrow R-C\underset{NH_2}{\overset{O}{\lessgtr}} \qquad [14\text{-}40]$$

The intermediate is shown as having an OH group **on a carbon** carrying a double bond; this arrangement would probably change to that of the more stable amide, in which the hydrogen is attached to nitrogen rather than to oxygen.

14-10. Reactions of Amides

1. Hydrolysis. (See also pp. 288–290.) As we have just said, an amide will hydrolyze more or less readily to an acid. This reaction may be carried out in either acidic or basic solution.

$$\underset{\underset{NH_2}{|}}{R-C}\overset{O}{\underset{}{\diagdown}} + OH^- \longrightarrow R-C\overset{O}{\underset{O^-}{\diagdown}} + NH_3 \qquad\qquad [14\text{-}41]$$

$$R-\overset{O}{\overset{\|}{C}}-NH_2 + H_3O^+ \longrightarrow R-COOH + NH_4^+ \qquad\qquad [14\text{-}42]$$

2. Reduction. High-pressure hydrogenation may be used advantageously for the preparation of long-chain amines from the corresponding amide. Octadecylamine is prepared commercially by this method, with copper chromite as a catalyst at 250° and 200 atmospheres pressure. When sodium and alcohol are used as

$$CH_3\text{-}(CH_2)_{16}\text{-}\underset{\underset{NH_2}{|}}{C}\overset{O}{\underset{}{\diagdown}} + H_2 \xrightarrow{\;CuCr_2O_4\;} \underset{\text{octadecylamine}}{CH_3(CH_2)_{16}CH_2NH_2} \qquad [14\text{-}43]$$

a source of hydrogen for the reduction, some amine is formed, but the main reaction is the reduction and hydrolysis to an alcohol. The yield of the alcohol, however, is only about 25% in the case given (Exp. 14-44).

$$CH_3(CH_2)_4CONH_2 + 4Na + 4C_2H_5OH \longrightarrow$$
$$\underset{25\%}{CH_3(CH_2)_4CH_2OH} + 4NaOC_2H_5 + NH_3 \qquad [14\text{-}44]$$

$$CH_3(CH_2)_4CONH_2 + 4Na + 4C_2H_5OH \longrightarrow$$
$$CH_3(CH_2)_4CH_2NH_2 + 4NaOC_2H_5 + H_2O \qquad [14\text{-}45]$$

3. Dehydration. An amide may be dehydrated to form a cyanide, which is the reverse of one of the reactions for preparing an amide. This removal of water requires a strong dehydrating agent, such as P_2O_5 or PCl_5.

$$R-C\overset{O}{\underset{NH_2}{\diagdown}} + P_2O_5 \longrightarrow 2H_3PO_4 + R-C\equiv N \qquad\qquad [14\text{-}46]$$

$$R-CONH_2 + PCl_5 \longrightarrow POCl_3 + 2HCl + R-C\equiv N \qquad\qquad [14\text{-}47]$$

Esters

The organic reaction product of an acid and an alcohol is an ester. The acid may be either inorganic or organic. The reaction products of alcohols with HX have already been studied as a separate homologous series. It is helpful, however, to consider the alkyl halides also as esters. With sulfuric acid and an alcohol, two

$$ROH + HX \longrightarrow RX + H_2O \qquad\qquad [14\text{-}48]$$

possible esters exist. The acid ester, or alkyl hydrogen sulfate, has already been encountered as an addition product of an alkene and sulfuric acid. The acid sulfates are also obtained by dissolving the

$$R-CH=CH_2 + \overset{\oplus}{H}\ \overset{\ominus}{OSO_3H} \longrightarrow \underset{\underset{\text{an alkyl hydrogen}}{\underset{\text{sulfate}}{OSO_3H}}}{RCH-CH_3} \qquad\qquad [14\text{-}49]$$

corresponding alcohol in sulfuric acid. Dimethyl sulfate is obtained by distillation from methyl hydrogen sulfate at reduced pressure.

$$CH_3OH + HOSO_3H \xrightarrow[90\%]{} CH_3OSO_3H + H_2O \qquad\qquad [14\text{-}50]$$

$$2CH_3OSO_3H \xrightarrow[83\%]{\text{red. press.}} CH_3O\overset{\uparrow}{\underset{\downarrow}{\overset{O}{\underset{O}{-S-}}}}OCH_3 + H_2SO_4 \qquad\qquad [14\text{-}51]$$

Other inorganic acids, such as phosphorous, phosphoric, nitrous, and nitric (for example see p. 153), also form esters by treatment with alcohols.

14-11. Nomenclature of Esters

The compounds commonly referred to as esters are represented by the general formula $R-C{\overset{\displaystyle O}{\underset{\displaystyle OR'}{}}}$ and are the products of the reaction between a carboxylic acid, RCOOH, and an alcohol, R'OH. The order of the name reverses the order in which the

TABLE

14-1 | *Esters*

FORMULA	NAME
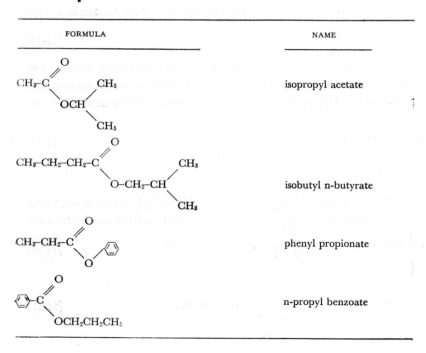	isopropyl acetate
	isobutyl n-butyrate
	phenyl propionate
	n-propyl benzoate

formula is written; that is, the name of the alkyl group, R', is followed by the name of the acid minus *ic* plus *ate*. The simplest ester, $H-C\overset{\displaystyle O}{\underset{\displaystyle OCH_3}{\big\langle}}$, is called methyl formate. The formulas and common names of a few esters are given in Table 14-1. The IUC names are seldom used.

Exercise 14-4. Write structural formulas for ethyl benzoate, benzyl acetate, phenyl α-bromoacetate, ethyl *o*-bromobenzoate, tert-butyl acetate.

14-12. Preparation of Esters

1. Direct Esterification. By direct treatment of an acid with an alcohol, refluxing over a long period of time, a small amount of an ester may be obtained. The equilibrium point in the reaction may be reached very much faster if a little H^+ is added to the mixture. If we use an excess of one substituent (generally the alcohol), the number of alcohol molecules with which each acid molecule can collide is increased; this decreases the final number of acid molecules and effectively shifts the equilibrium to the right. The yield of ester may also be increased by removal of one of the products of the reaction. It is ordinarily easier to remove the water by means of a dehydrating agent. Sulfuric acid is an excellent dehydrating agent and may therefore act in two capacities in an esterification (to catalyze the reaction and to shift the equilibrium).

The equation written as Expression 14-52 does not give a hint as to whether the OH group in the water comes from the OH group in the acid or from the alcohol. This was determined by esterification of benzoic acid with methanol in which a heavy oxygen isotope had been placed. The ester obtained in the reaction

$$R-C\overset{\displaystyle O}{\underset{\displaystyle OH}{\big\langle}} + R'OH \rightleftarrows R-C\overset{\displaystyle O}{\underset{\displaystyle OR'}{\big\langle}} + H_2O \qquad\qquad [14\text{-}52]$$

contained the heavy oxygen. Hence the benzoic acid must have furnished the OH group for the water. From this fact and from the

$$\text{⬡-CO } \boxed{(\text{OH} + \text{H})} \text{ O}^{18}\text{CH}_3 \overset{\text{H}^+}{\rightleftharpoons} \text{⬡-COO}^{18}\text{CH}_3 + \text{HOH} \qquad [14\text{-}53]$$

fact that H⁺ is a necessary catalyst if the esterification is to be run in a practical time, the mechanism of Expression 14-54 has been proposed. Each step is reversible.

$$
\text{R-C} \overset{O}{\underset{OH}{\big\langle}} + H^+ \rightleftharpoons \text{R-C} \overset{\overset{\oplus}{OH}}{\underset{OH}{\big\langle}} \longleftrightarrow \text{R-C} \overset{\oplus\ OH}{\underset{OH}{\big\langle}} \overset{R'OH}{\rightleftharpoons}
$$

$$
\text{R-}\underset{\boxed{OH\ \ H}}{\overset{OH\ \ R'}{\underset{|}{C}} :O^\oplus} \rightleftharpoons \text{R-C} \overset{O}{\underset{OR'}{\big\langle}} + H_2O + H^+ \qquad [14\text{-}54]
$$

2. From Acyl Halides or Acid Anhydrides. Both acyl halides and acid anhydrides react with alcohols to produce esters and liberate a molecule of an acid. Mechanisms closely related to the

$$
\text{R-C} \overset{O}{\underset{Cl}{\big\langle}} + R'OH \longrightarrow \text{R-C} \overset{O}{\underset{OR'}{\big\langle}} + HCl \qquad [14\text{-}55]
$$

$$
\text{R-C} \overset{O}{\underset{O}{\big\langle}} \overset{O}{\underset{R}{C\big\langle}} + R'OH \longrightarrow \text{R-C} \overset{O}{\underset{OR'}{\big\langle}} + R\text{-COOH} \qquad [14\text{-}56]
$$

one written for direct esterification account for the formation of esters in these two reactions. The carbon in an acyl halide will certainly be electron-deficient with two such strongly electron-attracting atoms as Cl and O attached to it. (See p. 285.) The al-

$$
\text{R-C} \overset{O}{\underset{Cl}{\big\langle}} \longleftrightarrow \text{R-C} \overset{\oplus\ \overset{.\ \ominus}{O}:}{\underset{Cl}{\big\langle}} \overset{R'OH}{\rightleftharpoons} \text{R-C}\overset{\overset{..\ominus}{O}\ R'}{\underset{Cl\ H}{:O^\oplus}} \longrightarrow \text{R-COOR}' + HCl \qquad [14\text{-}57]
$$

cohol may thus form a new coordinate covalent bond with this carbon. Loss of HCl then gives the new ester.

A similar mechanism is rational for the acylation of alcohols with acid anhydrides.

14-13. Reactions of Esters

1. Hydrolysis. (See p. 289.) The reversal of direct esterification may be accomplished by either acidic or basic hydrolysis. Alkaline hydrolysis is frequently a faster method of accomplishing the desired result.

$$R-C\underset{OR'}{\overset{O}{\diagdown}} + OH^- \longrightarrow R-C\underset{O^-}{\overset{O}{\diagdown}} + R'OH \qquad [14\text{-}58]$$

$$R-C\underset{OR'}{\overset{O}{\diagdown}} + H_3O^+ \longrightarrow R-C\underset{OH}{\overset{O}{\diagdown}} + R'OH + H^+ \qquad [14\text{-}59]$$

Exercise 14-5. Show the stepwise hydrolysis of an ester, following a pattern analogous to the hydrolysis of an amide (p. 199).

2. Reduction. (See also p. 276.) The reduction of an ester to two alcohols is accomplished (both in the laboratory and commercially) by catalytic hydrogenation with nickel and in the laboratory by sodium and alcohol. Since the products are both alcohols, the reaction is most useful if the boiling points of the two products are far enough apart so that they can be readily separated by distillation. With glyceryl esters of long-chain acids, the alcohol may be separated from glycerol by steam distillation.

$$R-C\underset{OR'}{\overset{O}{\diagdown}} + 2H_2 \xrightarrow{\text{Ni}} R-CH_2OH + R'OH \qquad [14\text{-}60]$$

$$\begin{array}{l} R-COO-CH_2 \\ R-COO-CH \\ R-COO-CH_2 \end{array} + 6H_2 \xrightarrow{\text{Ni}} 3RCH_2OH + \begin{array}{l} CH_2OH \\ CHOH \\ CH_2OH \end{array} \qquad [14\text{-}61]$$

3. Ammonolysis. Ammonolysis of an ester in the presence of NH_4Cl as an acid catalyst is, as we have seen, an excellent way to prepare many amides. The reaction goes with short-chain esters

$$R-C{\overset{O}{\underset{OR'}{\lessgtr}}} + NH_3 \xrightarrow[\text{liq.}]{NH_4^+} R-C{\overset{O}{\underset{NH_2}{\lessgtr}}} + R'OH \qquad [14\text{-}62]$$

and aqueous ammonia, but with esters of higher molecular weight liquid ammonia gives better yields.

4. Grignard Reaction. Grignard reagents add to the carbonyl group of an ester in the same manner as with aldehydes or ketones. The addition intermediate contains a carbon bearing two single bonds to oxygen, which configuration, we have found from previ-

$$R-C{\overset{O}{\underset{OR'}{\lessgtr}}} + \overset{\delta-\ \delta+}{R''\ MgX} \longrightarrow \left[R-\overset{OMgX}{\underset{OR'}{C}}-R'' \right] \longrightarrow R-C{\overset{O}{\underset{R''}{\lessgtr}}} + R'-OMgX \qquad [14\text{-}63]$$

ous experience, is likely to be unstable. (Oxygen has a great tendency to form a double bond in such circumstances.) Here the instability is relieved by loss of R'OMgX, leaving a ketone. Since the ketones readily add Grignard reagents, one would not expect the reaction to stop at this stage, especially if an excess of the Grignard is used. The second addition intermediate may be decomposed to a tertiary alcohol. Ketones are not often successfully prepared by this method, but the method is good for tertiary alcohols containing two like groups.

$$R-\overset{O}{\overset{\|}{C}}-R'' + R''MgX \longrightarrow R-\overset{OMgX}{\underset{R''}{C}}-R'' \xrightarrow{HOH} R-\overset{R''}{\underset{R''}{C}}-OH + Mg{\overset{OH}{\underset{X}{\diagdown}}} \qquad [14\text{-}64]$$

Exercise 14-6. Show by equations the effect of treating ethyl benzoate with an excess of phenylmagnesium bromide.

14-14. Occurrence of Esters

Esters occur widely in nature and give their odors to fruits and flowers. Perfumes and artificial fruit flavors frequently contain synthetic esters. One artificial raspberry flavor is a mixture of nine esters, two organic acids, acetaldehyde, glycerol, and ethyl alcohol.

Waxes are largely high-molecular-weight esters of long-chain fatty acids and long-chain alcohols. Beeswax, for example, contains myricyl palmitate, $CH_3(CH_2)_{14}COOC_{31}H_{63}$, along with the free acid and alcohol and esters of cholesterol. Carnauba wax, containing mostly myricyl cerotate, $CH_3(CH_2)_{24}COOC_{31}H_{63}$, is used in making electrical insulation for wire and cable.

Oils and fats are glyceryl esters of long-chain fatty acids. Fats are solid at room temperature while oils are liquid.

14-15. Oils and Fats

The general formula for an oil or fat is

$$
\begin{array}{l}
\overset{\displaystyle O}{\overset{\|}{\underset{}{}}} \\[-2pt]
R-C-O-CH_2 \\
\overset{\displaystyle O}{\overset{\|}{\underset{}{}}}\quad | \\[-2pt]
R'-C-O-CH \\
\overset{\displaystyle O}{\overset{\|}{\underset{}{}}}\quad | \\[-2pt]
R''-C-O-CH_2
\end{array}
$$

The three acid groups may be the same or different and may be saturated or unsaturated. Oils are likely to contain a larger fraction of unsaturated acids than fats since the esters of unsaturated acids generally exhibit lower melting points. The three esterified acids found most frequently in a number of oils and fats are palmitic, stearic, and oleic. The formulas for the glyceryl esters of these acids are given here. The esters are representative of the compounds found in naturally occurring substances. Their common names are tripalmitin, tristearin, and triolein, respectively.

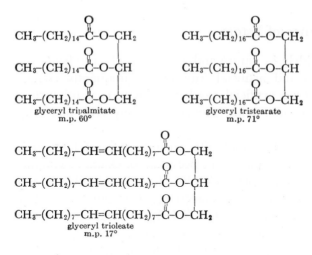

$$CH_3-(CH_2)_{14}-\overset{\overset{\displaystyle O}{\|}}{C}-O-CH_2$$
$$CH_3-(CH_2)_{14}-\overset{\overset{\displaystyle O}{\|}}{C}-O-CH$$
$$CH_3-(CH_2)_{14}-\overset{\overset{\displaystyle O}{\|}}{C}-O-CH_2$$

glyceryl tripalmitate
m.p. 60°

$$CH_3-(CH_2)_{16}-\overset{\overset{\displaystyle O}{\|}}{C}-O-CH_2$$
$$CH_3-(CH_2)_{16}-\overset{\overset{\displaystyle O}{\|}}{C}-O-CH$$
$$CH_3-(CH_2)_{16}-\overset{\overset{\displaystyle O}{\|}}{C}-O-CH_2$$

glyceryl tristearate
m.p. 71°

$$CH_3-(CH_2)_7-CH=CH(CH_2)_7-\overset{\overset{\displaystyle O}{\|}}{C}-O-CH_2$$
$$CH_3-(CH_2)_7-CH=CH(CH_2)_7-\overset{\overset{\displaystyle O}{\|}}{C}-O-CH$$
$$CH_3-(CH_2)_7-CH=CH(CH_2)_7-\overset{\overset{\displaystyle O}{\|}}{C}-O-CH_2$$

glyceryl trioleate
m.p. 17°

A. Uses of Fats and Oils

Fats and oils are used as foodstuffs, in the manufacture of soap, and in the compounding of paints.

1. Foodstuffs. Butter, lard, and tallow are the fats most commonly used in foods. Olive oil, coconut oil, corn oil, and soybean oil have attained some importance for the same purpose. Since housewives in this country are averse to using oils for cooking, a large industry has grown out of the discovery that one can change oils to fats by removing some of the double bonds in the unsaturated acid groups. Pure cottonseed oil or coconut oil can be hydrogenated catalytically to solid fats, which are sold as oleomargarine or under such trade names as Crisco and Spry. As an example of the "hardening of oils," as it is called, triolein can be

$$CH_3-(CH_2)_7-CH=CH-(CH_2)_7-\overset{\overset{\displaystyle O}{\|}}{C}-O-CH_2$$
$$CH_3-(CH_2)_7-CH=CH-(CH_2)_7-\overset{\overset{\displaystyle O}{\|}}{C}-O-CH \;+\; 3H_2 \;\overset{Ni}{\longrightarrow}\;$$
$$CH_3-(CH_2)_7-CH=CH-(CH_2)_7-\overset{\overset{\displaystyle O}{\|}}{C}-O-CH_2$$

$$CH_3-(CH_2)_{16}-\overset{\overset{\displaystyle O}{\|}}{C}-O-CH_2$$
$$CH_3-(CH_2)_{16}\;\overset{\overset{\displaystyle O}{\|}}{C}-O-CH$$
$$CH_3-(CH_2)_{16}-\overset{\overset{\displaystyle O}{\|}}{C}-O-CH_2$$

[14-65]

changed to tristearin (Exp. 14-65). Note the result of more drastic reduction of an ester on page 273.

Exercise 14-7. What would you expect to get by catalytic reduction of glyceryl stearate at elevated temperature and high hydrogen pressure?

2. Saponification. A soap is a metal salt of a long-chain fatty acid. A detergent is any substance that will aid in removing dirt (oil, grease, etc.) with water. Some soaps are detergents, and some detergents are soaps. Detergent soaps are water-soluble soaps; the common ones are sodium or potassium salts of fatty acids. The sodium soaps are generally hard, the potassium soaps soft.

Saponification is the term originally applied to the process of making soap by heating a fat or oil with a base. Now the term is applied more generally to the alkaline hydrolysis of any ester. Saponification is illustrated by the reaction of sodium hydroxide

$$
\begin{array}{c}
\underset{\substack{\text{O}}}{\text{CH}_3(\text{CH}_2)_{14}\text{-C-O—CH}_2} \\
\underset{\substack{\text{O}}}{\text{CH}_3\text{-(CH}_2)_{14}\text{-C-O-CH}} \quad + 3\text{NaOH} \longrightarrow \quad 3\text{CH}_3(\text{CH}_2)_{14}\text{-C}\overset{\text{O}}{\underset{\text{ONa}}{\diagup}} \quad + \quad \underset{\substack{\text{CH}_2\text{OH}}}{\overset{\text{CH}_2\text{OH}}{\text{CHOH}}} \\
\underset{\substack{\text{O}}}{\text{CH}_3\text{-(CH}_2)_{14}\text{-C-O-CH}_2} \\
\text{glyceryl tripalmitate} \qquad\qquad\qquad \text{sodium palmitate} \qquad \text{glycerol}
\end{array}
$$

[14-66]

and glyceryl palmitate. Glycerol is always a by-product of soap manufacture and is itself an important article of commerce.

3. Paints. The unsaturated acids that are present in the drying oils of paints make tough films when exposed to air. The process of drying a paint does not involve the removal of water at all, but instead is due to cross-linking of the unsaturated bonds in the acid constituents of the oils. It is a polymerization rather than a dehydration. Linseed oil, from flax seed, is the most widely used drying oil, but some synthetic drying oils (S.D.O.) are available.

Exercise 14-8. Write a reaction between glyceryl palmitate and liquid ammonia under pressure in the presence of an ammonium salt.

B. *Metabolism of Fats*

Fats are practically untouched in the digestive tract until they reach the intestine. The lipases of the pancreatic juices in the alkaline medium of the intestine catalyze the hydrolysis of fats to glycerol and acids. The acids pass through the walls of the intestine into the lymphatic system. There at least 60% of them is immediately synthesized back to fats and transported into the blood stream. The remaining acids, along with the glycerol, go into the portal vein and directly to the liver.

In the liver, through the agency of phosphorus compounds, the glycerol may be changed to glycogen (see p. 415) in a cycle part of which is congruent with that of carbohydrates. The acids are oxidized in a process that involves, in one stage (perhaps not the first), the carbon atom that is β to the carboxyl group. Biochemists do not agree whether or not one of the alternative paths shown in Expression 14-67 as ways of getting to the β-hydroxy acid is correct as a step in the metabolic process.

$$
\begin{array}{l}
\text{R–CH}_2\text{–CH}_2\text{–COOH} \qquad\qquad \text{R–CH–CH}_2\text{–COOH} \longrightarrow \\
\qquad\qquad \longrightarrow \text{R–CH=CH–COOH} \longrightarrow\ \overset{|}{\text{OH}} \\[2mm]
\text{R–C–CH}_2\text{–COOH} \longrightarrow \text{R–COOH} + \text{CH}_3\text{–COOH} \qquad [14\text{-}67] \\
\ \overset{\|}{\text{O}} \qquad\qquad\qquad\qquad\qquad \longrightarrow \text{CO}_2 + \text{H}_2\text{O}
\end{array}
$$

In any case, β-oxidation as one step is well established by some experiments of Knoop (1904). When he fed a number of phenyl-substituted acids of the series

$\langle\text{--}\rangle\text{–(CH}_2)_n\text{–COOH}$

to dogs, the end product excreted by the dogs was either benzoic acid (when $n = 2$ or 4) or phenylacetic acid

⬡-CH₂COOH

(when $n = 1$ or 3). These results suggested to Knoop that the carbon chain was lost two carbon atoms at a time during metabolism, by β-oxidation.

Since the acids occurring in natural fats have an even number of carbon atoms, one suggestion for the biosynthesis of fats also involves addition of two carbons at a time. Some kind of aldol condensation (with acids, not aldehydes) is suggested as one possible path, but the details of the biosynthesis of fats are also uncertain.

Nitriles, or Cyanides

Hydrogen cyanide is an extremely poisonous, colorless liquid boiling near room temperature (26°C). It forms a very stable compound with the hemoglobin in the blood in much the same way as carbon monoxide. Because of its lethal nature, beginning students in chemistry are always warned against its preparation.

When mixed with water under pressure, hydrogen cyanide is hydrolyzed to formic acid.

$$H-C{\equiv}N + 2H_2O \xrightarrow{190°} H-COOH + NH_3 \qquad\qquad [14\text{-}68]$$

14-16. Nomenclature of Nitriles

Expression 14-68 shows the relation of cyanides to acids and is the basis for the preferred method of naming these compounds. A compound is named as the nitrile of the acid to which it may be hydrolyzed. HC≡N is named formonitrile by this system (*formic* − *ic* + *onitrile*).

An alternative method of naming nitriles is to name the group attached to the C≡N group and follow it with the word *cyanide*. By this method CH_3–C≡N is named methyl cyanide. Some members of the series are named by both methods in Table 14-2.

TABLE

14-2   *Nitriles, or Cyanides*

FORMULA	NAMES		B.P.°C
H–C≡N	formonitrile	hydrogen cyanide	26
CH₃–C≡N	acetonitrile	methyl cyanide	81
CH₃–CH₂–C≡N	propionitrile	ethyl cyanide	98
$\overset{\displaystyle CH_3}{\underset{\displaystyle CH_3}{\diagdown}}$ CH–CH₂–C≡N	isovaleronitrile	isobutyl cyanide	130
C₆H₅C≡N	benzonitrile	phenyl cyanide	191
CH₃(CH₂)₇CH=CH(CH₂)₇C≡N	oleonitrile		

14-17. Preparation of Nitriles

1. From Alkyl Halides. Primary alkyl halides can be heated with aqueous alkali cyanides to produce nitriles. The hydrolysis of the alkyl halide is a competing reaction and is so important with secondary and tertiary alkyl halides that nitriles cannot be prepared profitably from them. Aryl halides will not react. (For the hydrolysis of KCN see p. 187.)

$$CH_3–CH_2–CH_2–Br + KCN \xrightarrow[86\%]{CuCN} CH_3–CH_2–CH_2–C≡N + KBr \qquad [14-69]$$

Exercise 14-9. Show how an alcohol could result from treatment of an alkyl halide with aqueous potassium cyanide.

2. Dehydration of an Amide. Preparation of a nitrile by the dehydration of an amide has already been described as a reaction of the amides (p. 269). The amides are prepared from the corre-

$$3CH_3–CONH_2 + P_2O_5 \xrightarrow[72\%]{} 3CH_3–C≡N + 2H_3PO_4 \qquad [14-70]$$

$$3\langle \rangle - C \begin{matrix} O \\ \diagdown \\ NH_2 \end{matrix} + P_2O_5 \xrightarrow[95\%]{} 3\langle \rangle - C \equiv N + 2H_3PO_4 \qquad \textbf{[14-71]}$$

benzamide   benzonitrile

sponding acids by heating the ammonium salt until a molecule of water is driven off (Exp. 14-39). At elevated temperature, in the presence of ThO₂, a one-step process of attaining the transformation of an acid into a nitrile has been accomplished. This reaction is carried out commercially on long-chain acids.

$$CH_3(CH_2)_{14}COOH + NH_3 \xrightarrow[ThO_2]{500°} CH_3(CH_2)_{14}CN + 2H_2O \qquad \textbf{[14-72]}$$

palmitic acid   palmitonitrile

3. From a Sulfonic Acid.   The nitrile group is commonly placed on a benzene ring by a fusion of solid sodium cyanide and a sodium aryl sulfonate.

$$CH_3-\langle \rangle-SO_3Na + NaCN \longrightarrow CH_3-\langle \rangle-C\equiv N + Na_2SO_3 \qquad \textbf{[14-73]}$$

4. Hydrogen Cyanide.   The simplest of the nitriles can easily be prepared by simply treating a metal cyanide with a strong mineral acid such as sulfuric. The choice of hydrochloric acid for this

$$Na^+CN^- + H^+OSO_2OH^- \longrightarrow Na^+OSO_2OH^- + HCN \qquad \textbf{[14-74]}$$

reaction would be a poor one since HCl is a gas at room temperature.

Sodium cyanide is made commercially by passing ammonia over melted sodium mixed with finely divided carbon. Two reactions take place: sodium amide is first formed, and it then reacts with carbon to yield sodium cyanide.

$$Na + NH_3 \xrightarrow[100\%]{350°} NaNH_2 + \tfrac{1}{2}H_2 \qquad \textbf{[14-75]}$$

$$NaNH_2 + C \xrightarrow[90\%]{700°} NaCN + H_2 \qquad \textbf{[14-76]}$$

Since potassium metal cannot be handled so readily as sodium, potassium cyanide is made by the fusion of potassium carbonate with ground coke and by the passage of ammonia into the hot mixture.

$$K_2CO_3 + 4C + 2NH_3 \xrightarrow{900°} 2KCN + 3CO + 3H_2 \qquad [14\text{-}77]$$

A high-temperature reaction that leads directly to hydrogen cyanide is now being used commercially. Methane is the source of carbon, and ammonia of nitrogen. Controlled oxidation of this mixture at 750°C gives hydrogen cyanide in good yield.

$$2CH_4 + 3O_2 + 2NH_3 \xrightarrow{750°} 2HCN + 6H_2O \qquad [14\text{-}78]$$

Exercise 14-10. What changes in oxidation numbers are involved in Expression 14-78?

5. Addition of Hydrogen Cyanide to an Alkene. Recently found catalysts allow the direct addition of hydrogen cyanide under pressure to alkenes. Yields are 60–75% on the lower members of the series (2–4 carbons). At a temperature of 130° and under

$$CH_3\text{-}CH{=}CH_2 + HCN \xrightarrow[75\%]{Co_2(CO)_8} CH_3\text{-}\overset{\displaystyle CH_3}{\underset{}{CH}}\text{-}C{\equiv}N \qquad [14\text{-}79]$$

100 atmospheres pressure, in the presence of dicobalt octacarbonyl, propylene will add HCN to give isobutyronitrile in 75% yield. (Is this addition contrary to Markownikoff's rule?)

14-18. Reactions of Nitriles

1. Hydrolysis. (See p. 267.) The hydrolysis of a nitrile can be accomplished fairly readily in either alkaline or acid solution. The reaction is slow in neutral solution except for nitriles that are soluble in water (1–4 carbon atoms). Acid hydrolysis involves first the formation of an amide, which can be isolated if desired (Exp.

$$R\text{-}C{\equiv}N + OH^- + H_2O \longrightarrow R\text{-}C{\overset{\displaystyle O}{\underset{\displaystyle O^-}{\big<}}} + NH_3 \qquad [14\text{-}80]$$

$$R-C{\equiv}N + H_3O^+ + H_2O \longrightarrow R-COOH + NH_4^+ \hspace{2cm} [14-81]$$

14-40). The amide stage represents a partial hydrolysis, since amides themselves can be hydrolyzed to carboxylic acids (p. 268).

2. Reduction. Nitriles may be reduced to primary amines by chemical reducing agents such as sodium plus ethanol or by hydrogen in the presence of a catalyst. In the process using Raney nickel* as the catalyst, the formation of the primary amine is accompanied by the formation of a secondary amine. (See p. 298.)

$$\langle \_ \rangle{-}C{\equiv}N + 4Na + 4C_2H_5OH \longrightarrow \langle \_ \rangle{-}CH_2NH_2 + 4NaOC_2H_5 \hspace{1cm} [14-82]$$

$$2R-C{\equiv}N + 4H_2 \xrightarrow{\text{Ni}} R-CH_2-NH-CH_2R + NH_3 \hspace{1.5cm} [14-83]$$

3. With Grignard Reagents. Grignard reagents add to the carbon-nitrogen triple bond to yield the products expected on the assumption that the nitrile molecule has at least the two contributing structures

$$R-C{\equiv}N \longleftrightarrow \overset{\delta+\ \ \delta-}{R-C{=}N}$$

The addition product of the nitrile and the Grignard reagent can be hydrolyzed to a ketone and a magnesium salt. The salt shown

$$\overset{\delta+\ \ \delta-}{R'-C{=}N} + R-Mg-X \longrightarrow R'-\underset{R}{\underset{|}{C}}{=}N-MgX \xrightarrow{H_2O} \underset{R}{\overset{R'}{>}}C{=}O + \begin{bmatrix} & NH_2 \\ Mg & \\ & X \end{bmatrix}$$

$$\Big\downarrow {\scriptstyle H_2O}$$

$$NH_3 + Mg{<}\underset{X}{\overset{OH}{}}$$

$$[14-84]$$

---

* Raney nickel, an alloy patented by Murray Raney (U.S. Patent 1,628,190, May 1927), contains nickel and aluminum in a 1:1 ratio. The catalyst is prepared by dissolving the aluminum away from the alloy with sodium hydroxide. The finely divided nickel obtained in this way makes an excellent hydrogenation catalyst for high-pressure reactions and some low-pressure reductions.

in brackets cannot be isolated but is itself hydrolyzed to a basic magnesium halide. This is the best method of utilizing a Grignard to prepare a ketone, but most ketones are more readily prepared by other methods.

14-19. Comparison of Activity of Acid Derivatives in Terms of Electronic Structures

The acid derivatives discussed in this chapter, except the nitriles,

R–C≡N, have had the general structure $R-\overset{\displaystyle O}{\overset{\displaystyle \|}{C}}-A$. In the study of ketones and aldehydes, the behavior of the carbonyl group was shown to be that of an electron-seeking group, and the reactions were readily interpreted on the assumption of contributions from the two resonance structures.

$$R-\overset{\displaystyle O}{\overset{\displaystyle \|}{C}}-R \longleftrightarrow R-\overset{\displaystyle :\overset{..}{O}:^{\delta-}}{\underset{\delta+}{C}}-R$$

When one R group is replaced by A, the modification of properties depends on the character of A.

1. **R–C–OH**. $R-\overset{\displaystyle \mathbf{O}}{\overset{\displaystyle \|}{C}}-\mathbf{OH}$ The OH group in the acid is an electron-attracting group; so there is some tendency to increase the formal positive charge on the carbon by pulling electrons away from it. This tendency is opposed by the proclivity of the oxygen for increasing its covalence to three—that is, for sharing one of its unshared pairs. The most important contributing structures of the acid are the two shown below. The importance of structure *b* is deduced from

$$\underset{(a)}{R-\overset{\displaystyle \overset{\delta-..}{:\overset{}{O}:}}{\underset{\delta+\;..}{C}}-\overset{..}{O}H} \longleftrightarrow \underset{(b)}{R-\overset{\displaystyle \overset{..\delta-}{:\overset{}{O}:}}{\underset{\delta+}{C}}=\overset{..}{O}H}$$

the poor acylating ability of an acid in comparison with an acyl halide or an acid anhydride, for example, in the acylation of NH_3, ROH, or an aromatic hydrocarbon.

Canonical structure *b* must aid in the ionization of the acid because a positively charged oxygen would repel a proton. The carboxylate ion is somewhat more stable than the acid because of the exact equivalence of the ionic resonance structures *c* and *d*. (See p. 120.) Canonical structure *e* also probably makes some contribution to the ion.

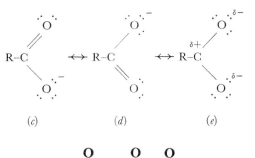

2. R–C⟍X
   $$\overset{O}{R-C}\diagup$$
   and R–C–O–C–R.   In the earlier sections of this

chapter all the reactions that went with acyl halides also went with acid anhydrides. There is, however, quite a difference in vigor and speed of reaction between the two acylating agents. Water reacts violently with acetyl chloride in a test tube and evolves so much heat that it cannot be held in the hand. Acetic anhydride hydrolyzes only after several hours at room temperature and must

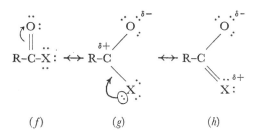

be heated on the steam bath to produce a rapid reaction. This qualitative difference is encompassed in the electronic structures that can be drawn. The lawful structure *g* is much more important than *h* in the total picture of an acyl halide.

The propensity of oxygen and nitrogen to increase their covalences (form multiple bonds) is very much greater than that of any monovalent atom such as a halogen. Among the halogens, chlorine shows a greater tendency toward double-bond formation than bromine, and iodine shows none. In the acid anhydride, the structure *k* corresponding to *h* makes a greater contribution to the total structure. Hence the carbonyl carbon in acetic anhydride is not so

$$
\underset{(i)}{\text{R–C–O–C–R}} \quad \longleftrightarrow \quad \underset{(j)}{\text{R–C–O–C–R}} \quad \longleftrightarrow \quad \underset{(k)}{\text{R–C=O–C–R}}
$$

positive as the carbonyl carbon in an acyl halide. The acyl halide is therefore a more reactive acylating agent (electron-seeking) than an acid anhydride.

$$
\overset{O}{\underset{}{\|}} \qquad \overset{O}{\underset{}{\|}}
$$

3. R–C–NH₂, R–C–OR′, and R–C≡N.   Amides, esters, and nitriles are much poorer acylating agents than the compounds containing the two functions just discussed. The structures *n* and *q* make even more important contributions to the total structure of amides and esters, respectively, than *k* does to acid anhydrides.

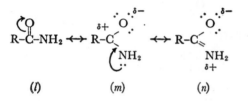

(l) (m) (n)

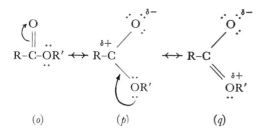

(o) (p) (q)

Consequently, amides and esters are poor acylating agents. The inclination of the carbonyl carbon to react with electron-donating agents is not nearly so marked as in the acyl halides and acid anhydrides. Whereas the latter two react with water (a weak base), it takes hydroxyl ions (a strong electron donor) to accomplish hydrolysis (even then not rapid) of an ester or amide. As might be expected for neutral substances, protons will also catalyze the hydrolysis. The presence of an acid would discourage the contributions of structures n and q (and h and k).

There is also some evidence that amides tautomerize (see p. 334) in the following way:

$$R-C\underset{NH_2}{\overset{O}{\lessgtr}} \rightleftarrows R-C\underset{NH}{\overset{OH}{\lessgtr}}$$

This appears to be more important in amides in which an R' group is substituted on nitrogen than in $RCONH_2$.

The triple bond in nitriles makes it impossible to draw a complete analogy to esters and amides. Nitrogen cannot exhibit its

$$R-C\equiv N: \overset{\delta+ \quad ..\delta-}{\longleftrightarrow} R-C=N:$$
$$(r) \qquad\qquad (s)$$

tendency to increase its covalence to four here because carbon cannot expand its octet to ten electrons. The only influence that can be exerted, therefore, is the slight edge that nitrogen has over carbon in negative character. Canonical structure *s* actually makes a rather large contribution to the structure of a nitrile, enough to promote the hydrolysis reaction in acidic or basic solution.

14-20. Correlation of Hydrolytic Reactions of Acyl Halides, Acid Anhydrides, Esters, Amides, and Nitriles

The hydrolyses of various acid derivatives have been treated separately in this chapter. These reactions, however, have many features in common, a fact that makes a single treatment profitable. In the preceding paragraph it was suggested that the carbon in the acid function of these various derivatives is electron-deficient to varying degrees but electron-deficient in every case. All of these compounds should therefore have some tendency to form a covalent bond with an electron donor (a nucleus-seeking reagent). The electron deficiency in the carbon atom of an acyl halide is so pronounced that only a weak base, H_2O, is necessary to hydrolyze it. Of course, the molecular weight of the compound will still determine how rapidly the hydrolysis proceeds.

$$\text{R-C}\overset{\text{O}}{\underset{\text{Cl}}{\diagdown}} \longrightarrow \text{R-C}\overset{\overset{\delta+}{\overset{..}{\text{O}}}{}^{\delta-}}{\underset{\text{Cl}}{\diagdown}} + H_2O \rightleftarrows \left[\text{R-C}\overset{\overset{\ominus}{\text{O}}\ \text{H}}{\underset{\text{Cl H}}{:\text{O}^{\oplus}}} \right] \longrightarrow \text{R-C}\overset{\text{O}}{\underset{\text{OH}}{\diagdown}} + HCl$$

[14-85]

The hydrolysis of acid anhydrides follows a similar pattern.

For most of the neutral derivatives—esters, amides, and nitriles—appreciable rates of reaction are obtained only in acid or alkaline media. The three kinds of derivative may be treated as one. An ester is the chosen example. The hydrolysis in acid solution (Exp. 14-86) will, of course, be the reverse of esterification (see Exp. 14-54). In basic solution the mechanism of hydrolysis of an ester

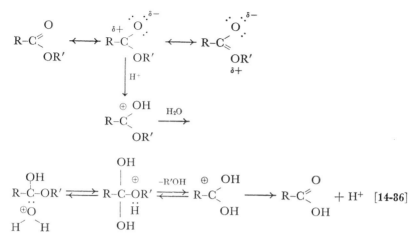

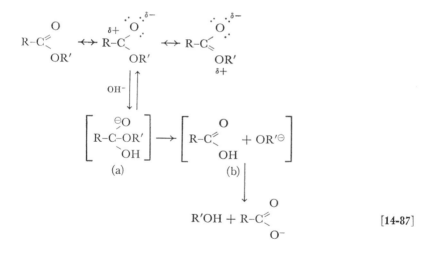

again centers on the positive carbon of the carbonyl group. The complex *a* may lose either an $OH^{\ominus}$ or an $OR'^{\ominus}$. The first is sim-

ply a reversal to the ester, but the second is a step forward. This reaction will go to an end because the alkoxide ion in *b* can take a proton from RCOOH to give an alcohol and a carboxylate ion.

Similar mechanisms may be written for the ammonolysis of acyl halides, acid anhydrides, and esters, since ammonia, like water, alcohols, and hydroxide ions, has unshared pairs of electrons available for the formation of coordinate covalent bonds.

SUMMARY

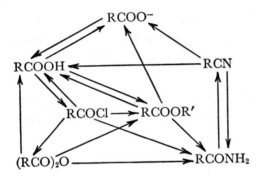

In the accompanying diagram the acid derivatives discussed in this chapter are interrelated. The diagram, of course, cannot be complete without being more cluttered. The following additional preparations and reactions are recalled to mind.

1. Salts of aliphatic acids may be converted to aldehydes (Chap. 11) and ketones (Chap. 12). A salt of an organic acid may be converted to its anhydride, generally readily.

2. Acyl halides and acid anhydrides undergo Friedel-Crafts reactions to form aromatic and aromatic-aliphatic ketones. The chemical properties of acyl halides and acid anhydrides are qualitatively the same.

3. Amides may also be prepared by heating ammonium salts of carboxylic acids. Amides may be reduced catalytically.

4. Esters occur widely in nature as the fragrant constituents of fruits and flowers, as fats, oils, and waxes. Esters can be reduced catalytically to alcohols and take part in Grignard reactions.

5. Alkyl halides can be used to prepare aliphatic nitriles from solution, but it is necessary to fuse aryl sulfonates with alkali cyanides to prepare aromatic nitriles. Nitriles can be reduced to amines, and their reactions with Grignard reagents can frequently be stopped at the ketone stage.

A theoretical discussion of the relative activities of acylating agents is given in terms of electronic concepts.

REFERENCES

1. Schuette, "Oleomargarine," J. Chem. Education, 5, 1621 (1928).
2. von Loesecke, "The Banana—a Challenge to Chemical Investigation," J. Chem. Education, 7, 1541 (1930).
3. Crocker, "The Chemistry of Flavor," J. Chem. Education, 22, 567 (1945).
4. Saul, "Odor and the Organic Chemist," J. Chem. Education, 23, 296 (1946).
5. Fieser and Fieser, *Organic Chemistry*, 3rd ed. (D. C. Heath & Co., Boston, 1956), Chaps. 15, 18.
6. Snell, "Soap and Glycerol," J. Chem. Education, 19, 172 (1942).
7. Bossert, "The Metallic Soaps," J. Chem. Education, 27, 10 (1950).

EXERCISES

(Exercises 1–10 will be found within the text.)

11. Write electronic structures for acetyl chloride, sodium propionate, acetic anhydride, ethyl n-butyrate, formamide, thionyl chloride, phosphorus trichloride, ethyl cyanide, chloral, phenyl magnesium iodide.

12. Write structural formulas for propionyl chloride, n-butyric anhydride, isopropyl n-butyrate, n-octyl isovalerate, potassium propionate, 3-ethyl-4-methyl-pentanoic acid, ethyl trichloroacetate, trichloroethyl acetate.

13. Write structural formulas for all isomers having the formula $C_5H_{10}O_2$. Name them.

14. Name the following compounds:

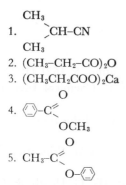

1. CH_3–CH–CN with CH_3
2. $(CH_3$–CH_2–$CO)_2O$
3. $(CH_3CH_2COO)_2Ca$
4. (phenyl)–C $\overset{O}{\underset{OCH_3}{}}$
5. CH_3–C $\overset{O}{\underset{O-\text{(phenyl)}}{}}$

6. $C_{12}H_{25}OSO_3H$
7. CH_3–CH_2–CH_2–$COBr$
8.  Br–(phenyl)–$COOCH(CH_3)_2$
9. CH_2=CH–$CONH_2$
10. CH_3–$(CH_2)_{16}$–$COONH_4$
11. CH_3–CH_2–CH_2–$COOCH_2$
    CH_3–CH_2–CH_2–$COOCH$
    CH_3–CH_2–CH_2–$COOCH_2$
12. 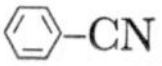 (phenyl)–CN

15. Write equations for any reactions that will occur between the following reagents and compounds:

REAGENTS
1. benzoyl chloride
2. acetic anhydride

COMPOUNDS
1. hexane
2. 1-propanol
3. ethyl ether

16. Write equations for three methods of preparing n-propyl acetate.

17. Prepare

1. n-butyric acid from n-butyl alcohol
2. propionic acid from ethyl alcohol
3. isobutyric acid from propylene
4. n-butyric acid from propionic acid
5. propionic anhydride
6. *p*-methylbenzamide from *p*-methylbenzonitrile

18. How much acetic acid is required to esterify 40 g of tetramethyleneglycol?

19. What is the percentage yield of ethyl benzoate if 100 g of benzoic acid yields 100 g of the ester?

20. What would you guess were the relative rates of esterification of acetic acid with 1-butanol, 2-butanol, and 2-methyl-2-propanol?

21. What are the differences between an oil, a fat, and a wax?

22. Why is sodium lauryl sulfate called a "soapless soap"?

23. What happens to a fat in the human digestive system?

24. When an ester is hydrolyzed with $HO^{18}H$, the O^{18} appears in the acid. Write for the hydrolysis a mechanism that will account for this observation.

25. A compound, $C_9H_{10}O_2$, was hydrolyzed with sodium hydroxide solution, and part of the solution was distilled. The distillate contained water and a compound that reacted positively to the iodoform test. The residue was acidified, and a white precipitate formed. The precipitate was found to have a molecular weight of 122. What was the original compound? Write equations for the various reactions described.

26. From compound A, having the formula C_8H_9NO, ammonia was liberated by hot alkali. Acidification of the cold hydrolyzed solution precipitated compound B, a substance of formula $C_8H_8O_2$. Vigorous oxidation of either A or B gave an acid of formula $C_8H_6O_4$. Write structural formulas for compounds A, B, and C and equations to show how each was obtained.

Properties of the Covalent Bond of Carbon with Nitrogen

15 Ammonia and Amines

15-1. The Nitrogen System

The world functions, in large part, on an oxygen system of compounds. What would it be like if it were based on a nitrogen system instead? Respiration would be carried on in nitrogen instead of oxygen, and the abundant oxides, water, carbon dioxide, and sand (SiO_2), would be replaced by their respective nitrides, ammonia, cyanogen $(CN)_2$, and silicon nitride (Si_3N_4). The analogies can be carried to ridiculous lengths, but some comparisons will be useful, if they serve no other purpose, in helping us to remember properties.

It is to be noted that the NH_2 group has the same electronic

$$\overset{..}{:}\!N\!:\!H \qquad \overset{..}{:}\!\overset{..}{O}\!:\!H$$
$$\underset{H}{\overset{..}{}}$$

structure as the OH group. A general rule may be stated: Radicals having the same electronic structures have many properties in common.

TABLE

15-1   *The Oxygen and Nitrogen Systems of Compounds*

O SYSTEM	N SYSTEM	MIXED SYSTEM
H–OH, water	H–NH$_2$, ammonia	
NaOH, sodium hydroxide	NaNH$_2$, sodium amide	
R–OH, an alcohol	R–NH$_2$, a prim-amine	
R–O–R, an ether	R–NH–R, a sec-amine	

	R–N–R, a tert-amine
	  |
	  R	

HO–SO$_2$–OH, sulfuric acid	H$_2$N–SO$_2$–NH$_2$, sulfamide	H$_2$N–SO$_2$–OH, sulfamic acid
HO–OH, hydrogen peroxide	H$_2$N–NH$_2$, hydrazine	H$_2$N–OH, hydroxylamine

RO–OH, an alkyl
  hydroperoxide
	RNH–NH$_2$, an alkyl
	  hydrazine
		RNH–OH,
		  N-alkylhydroxylamine*
		H$_2$N–OR,
		  O-alkylhydroxylamine*
		R–NH–OR,
		  N, O-dialkylhydroxyl-
		  amine*

RO–OR, an alkyl peroxide		R–N–OR, a
		  |
		  R
		trialkylhydroxylamine

* Alkyl groups may be substituted on C, S, N, O, and other elements in various compounds. When there may be confusion in the location of an alkyl group, a capital letter may be used to show its position, thus:

CH$_3$–NH–OH,   N-methylhydroxylamine
H$_2$N–O–CH$_3$,   O-methylhydroxylamine

15-2. Oxidation Numbers of Nitrogen

Nitrogen forms the oxides shown (with their most probable structures) in Table 15-2.

The oxidation state of the nitrogen in each of these compounds can be found by applying the rules given in Chapter 13 for carbon

TABLE

15-2 | *Oxides of Nitrogen*

FORMULA	NAME	PROBABLE STRUCTURE
N_2O	nitrous oxide	$N\equiv N \rightarrow O$
NO	nitric oxide	$N=O$
N_2O_3	nitrogen sesquioxide	$O=N-O-N=O$
$\left\{\begin{array}{l}NO_2 \\ N_2O_4\end{array}\right\}$	$\left\{\begin{array}{l}\text{nitrogen dioxide} \\ \text{nitrogen tetroxide}\end{array}\right\}$	$\left\{\begin{array}{c}O\leftarrow N=O \\ \\ O \qquad\qquad O \\ \diagdown \qquad \diagup \\ N-N \\ \diagup \qquad \diagdown \\ O \qquad\qquad O\end{array}\right\}$
N_2O_5	nitrogen pentoxide	$O=N-O-N=O$ $\quad\downarrow\qquad\downarrow$ $\quad O\qquad O$

compounds. Bonds joining like atoms are assumed to be equally shared, and hence a zero oxidation state exists between the two. Nitrogen is more electropositive than oxygen, and so any N-O bond gives an oxidation state of $+1$ for nitrogen and -1 for oxygen. One new idea enters into the picture in these compounds. Wherever a formal charge exists on an atom, that must be added to the number obtained by counting the atom-to-atom bonds in the molecule. Thus, in $: \overset{\cdot\cdot}{N} \overset{x}{\underset{x}{:}} N : \overset{xx}{\underset{xx}{O}} \overset{x}{\underset{x}{:}}$, atom 1 has an oxidation number of zero; atom 2 has one bond to oxygen and a formal charge of $+1$, so its oxidation number is $+2$; atom 3 has one bond to nitrogen and a formal charge of -1, so its oxidation number is -2. If the structure of nitrous oxide is written as $N\overset{\leftharpoonup}{=}N=O$, then the first nitrogen has an oxidation number of -1 since it has such a formal charge. The central nitrogen has an oxidation number of $+3$, and the oxygen -2.

In $:\overset{\cdot \cdot}{N}\overset{\times \times}{\underset{\times}{:O}}\overset{\times}{:}$, the oxidation state of nitrogen is $+2$ and of oxygen -2.

Exercise 15-1. What are the oxidation numbers of nitrogen and oxygen in N_2O_3, NO_2, N_2O_4, N_2O_5, HNO_2, and HNO_3?

Besides these positive oxidation states, nitrogen, like carbon, is electronegative with respect to hydrogen and hence shows some negative oxidation numbers. In this series of oxides and hydrides, then, nitrogen exhibits the entire spectrum of oxidation states from $+5$ to -3.

TABLE

15-3 | *Other Compounds of Nitrogen*

FORMULA	OXIDATION STATE	NAME
N≡N	0	nitrogen
H $\|$ H–N–OH	−1	hydroxylamine
H H $\|$ $\|$ H–N–N–H	−2	hydrazine
NH_3	−3	ammonia

In all hydrides and oxides, with the exception of nitric oxide, nitrogen forms three bonds with other atoms or, if a coordinate covalent bond is present, a maximum of four bonds. Nitrogen has not been observed to expand its shell beyond eight electrons, and it is suggested that it is impossible for any of the first-row elements to do so.

15-3. Nomenclature of Amines

We form the common names of amines by adding *amine* to the

TABLE

15-4 | *Amines*

FORMULA	COMMON NAME	IUC NAME	TYPE
CH_3-NH_2	methylamine	methanamine	primary
$CH_3-CH_2-NH_2$	ethylamine	ethanamine	primary

$CH_3-CH-CH_3$
    |
   NH_2
	isopropylamine	2-propanamine	primary

$CH_3-NH-CH_2-CH_3$	methylethylamine	——	secondary

$CH_3-CH_2-N-CH-CH_3$
      / |
   CH_3 CH_3
	methylethylisopropylamine	——	tertiary

$CH_3-NH-CH-CH_2-CH_3$
      |
    CH_3
	methyl sec-butylamine	——	secondary

$\langle \rangle -NH_2$	aniline	aniline	primary
$\langle \rangle -NH-CH_3$	methylaniline	methylaniline	secondary
$\langle \rangle -NH-\langle \rangle$	diphenylamine	diphenylamine	secondary
$\langle \rangle -N\overset{C_2H_5}{\underset{C_2H_5}{}}$	diethylaniline	diethylaniline	tertiary

names of the radicals attached to nitrogen. If those radicals are different, the shorter or shortest is commonly named first.

The degree of an amine—primary, secondary, or tertiary—is determined by the number of hydrogen atoms on the nitrogen and not by the nature of the groups attached to it. A primary amine has two hydrogen atoms on nitrogen, a secondary amine has one hydrogen on nitrogen, and a tertiary amine has no hydrogen on nitrogen. The alkyl group may be secondary while the amine is still primary; sec-butyl amine, for example, is a primary amine.

$$CH_3-CH_2-CH-NH_2$$
$$\underset{}{|}$$
$$CH_3$$

sec-butyl amine

Many aromatic amines have common names that do not reveal their nature; aniline is aminobenzene. Mixed aliphatic-aromatic amines are frequently named as substituted anilines; $C_6H_5-N(CH_3)_2$ is called dimethylaniline. (For others, see Table 15-4.) We name aromatic amines such as $C_6H_5NHC_6H_5$ in the same way as aliphatic amines—that is, by naming each radical attached to nitrogen; the example given is named diphenylamine.

Exercise 15-2. Write structural formulas for tert-butyl amine, ethyl-aniline, *p*-nitroaniline, isopropyl di-n-propylamine, methylethylani-line, and *p*-ethylaniline.

15-4. Preparation of Amines

1. Ammonolysis of Halides. Amines were first prepared by A. W. von Hofmann in 1850 by heating an alkyl halide with ammonia. Although a mixture of primary, secondary, and tertiary amines is obtained, the method is practiced commercially since the products can be separated by differences in boiling points. A large excess of ammonia is used so that the reaction shown in Ex-

$$R:Cl + :NH_3 \xrightarrow{\Delta} [RNH_3]^+Cl^- \qquad [15\text{-}1]$$

$$[RNH_3]^+Cl^- + :NH_3 \underset{\rightleftarrows}{\overset{\Delta}{\rightleftarrows}} RNH_2 + NH_4^+Cl^- \qquad [15\text{-}2]$$

pression 15-2 goes to some extent to decompose the substituted ammonium chloride first formed. This primary amine, RNH_2, may then undergo corresponding reactions until a tertiary amine is the final product. The method is unsatisfactory for the preparation of pure amines.

$$RNH_2 + RCl \longrightarrow [R_2NH_2]^+Cl^- \qquad [15\text{-}3]$$

$$[R_2NH_2]^+Cl^- + NH_3 \rightleftarrows R_2NH + NH_4^+Cl^- \qquad [15\text{-}4]$$

$$R_2NH + RCl \longrightarrow [R_3NH]^+Cl^- \qquad [15\text{-}5]$$

$$[R_3NH]^+Cl^- + NH_3 \rightleftarrows R_3N + NH_4^+Cl^- \qquad [15\text{-}6]$$

The ammonolysis of a halogen on a benzene ring is much more difficult than that of an alkyl halide. Aniline, however, is prepared commercially in large quantities by this method. The reac-

$$\text{\char"27E8}\text{\char"27E9}-Cl + 2NH_3 \xrightarrow{CuO} \text{\char"27E8}\text{\char"27E9}-NH_2 + NH_4Cl \qquad [15\text{-}7]$$

tion is carried out at about 300° and at pressures as high as 300 atmospheres. In contrast, many alkyl halides are ammonolyzed at room temperature and atmospheric pressure within a few hours.

2. Reduction of an Oxime, Amide, Nitrile, or Nitro Compound. Catalytic reduction by hydrogen of an oxime, amide, nitrile, or nitro compound (Chap. 16) yields a primary amine. The reactions may also be carried out with other chemical reducing agents. In acid solution (HCl), nitro compounds, both aliphatic and aromatic, are readily reduced by tin, zinc, or iron. Oximes and nitriles are frequently reduced by sodium and alcohol, which form a strongly basic medium.

Aniline is prepared commercially by the action of iron and steam, in the presence of small amounts of hydrochloric acid, on nitro-

$$\text{\char"27E8}\text{\char"27E9}-NO_2 + Fe + H_2O \xrightarrow[95\%]{\substack{\text{trace of}\\ \text{HCl}}} \text{\char"27E8}\text{\char"27E9}-NH_2 + Fe_3O_4 \qquad [15\text{-}8]$$

benzene. Sodium amalgam has been used with water in alcohol

$$CH_3-(CH_2)_5-\overset{H}{\underset{}{C}}=NOH + Na(Hg) + H_2O \xrightarrow[73\%]{C_2H_5OH}$$

$$CH_3-(CH_2)_5-CH_2NH_2 + NaOH + Hg \qquad [15\text{-}9]$$

to reduce n-heptaldoxime to n-heptylamine.

Nitriles in alcoholic solution are frequently reduced by sodium.

$$CH_3-CH_2-CH_2-CH_2-C\equiv N + Na + C_2H_5OH \xrightarrow[68\%]{}$$

$$CH_3CH_2CH_2CH_2\underset{NH_2}{CH_2} + NaOC_2H_5 \qquad [15\text{-}10]$$

Catalytic reduction of an amide is frequently used in the preparation of long-chain amines (p. 268).

$$CH_3(CH_2)_{10}CONH_2 + H_2 \xrightarrow[\substack{250° \\ 48\%}]{CuCr_2O_4} \underset{\text{n-dodecylamine}}{CH_3(CH_2)_{10}CH_2NH_2} + H_2O \qquad [15\text{-}11]$$

Exercise 15-3.   Find the changes in the oxidation numbers of the various atoms involved in the reductions in Expressions 15-8, 15-9, 15-10, and 15-11, and balance the equations.

3. Hofmann Hypobromite Reaction.   Short-chain aliphatic amines (1–7 carbons) are conveniently prepared in the laboratory by the treatment of an amide with sodium hypobromite solution. It is to be noted that the product has one less carbon than the

$$\underset{\substack{\| \\ O}}{R\text{-}C\text{-}NH_2} + NaOBr + 2NaOH \longrightarrow RNH_2 + Na_2CO_3 + NaBr + H_2O$$
$$[15\text{-}12]$$

starting amide. This is the only method mentioned so far of shortening a carbon chain by one carbon. Aromatic acid amides also undergo the degradation.

There are probably several steps in this reaction, and some of the intermediates can be isolated. The first of such intermediates is an N-bromo-substituted amide, which loses HBr, probably in two steps. The substance isolated after the N-bromoamide is an

$$\underset{\text{n-butyramide}}{CH_3\text{-}CH_2\text{-}CH_2\overset{\overset{\displaystyle O}{\|}}{\text{-}C}\text{-}NH_2} + NaOBr \longrightarrow$$

$$\underset{\text{N-bromo-n-butyramide}}{CH_3\text{-}CH_2\text{-}CH_2\overset{\overset{\displaystyle O}{\|}}{\text{-}C}\overset{\overset{\displaystyle H}{\,}}{\text{-}N}\text{-}Br} + NaOH \xrightarrow{OH^-}$$

$$\left[CH_3\text{-}CH_2\text{-}CH_2\overset{\overset{\displaystyle O}{\|}}{\text{-}C}\text{-}N \right] \xrightarrow{\text{rearranges}} \underset{\text{n-propyl isocyanate}}{CH_3\text{-}CH_2\text{-}CH_2\text{-}N=C=O} \xrightarrow{HOH}$$

$$\left[CH_3\text{-}CH_2\text{-}CH_2\overset{\overset{\displaystyle H}{\,}}{\text{-}N}\overset{\overset{\displaystyle O}{\|}}{\text{-}C}\text{-}OH \right] \xrightarrow[90\%]{-CO_2} \underset{\text{n-propyl amine}}{CH_3\text{-}CH_2\text{-}CH_2\text{-}NH_2} \qquad [15\text{-}13]$$

isocyanate. The last reaction is well known. An isocyanate is easily hydrolyzed to an amine and carbon dioxide. This reaction is pictured as an addition of water and a subsequent loss of carbon dioxide.

4. Gabriel Synthesis of Pure Aliphatic Primary Amines.
A nitrogen-containing compound called phthalimide is used for the preparation of pure aliphatic primary amines. Since phthalimide

$$\text{phthalimide} \quad NH + KOH \longrightarrow \quad N\text{-}K + H_2O \qquad [15\text{-}14]$$

is a slightly acidic substance, it will dissolve in strong bases like KOH to form a salt—in this case, potassium phthalimide. Alkyl halides will react with the isolated dry salt, on being heated, to yield an N-alkyl substituted phthalimide, which can be hydrolyzed in acid solution to an amine. The amines prepared by the Gabriel

$$N\text{-}K + C_2H_5\text{-}Br \longrightarrow N\text{-}C_2H_5 \xrightarrow[H^+]{2HOH} \begin{array}{c} COOH \\ \\ COOH \end{array} + C_2H_5NH_2 \atop \text{ethyl amine}$$

$$[15\text{-}15]$$

synthesis are free of products of side reactions; hence it is a good method for synthesizing pure primary amines.

Exercise 15-4.   Show how to prepare $HO\text{-}CH_2\text{-}CH_2\text{-}NH_2$ by the Gabriel synthesis.

15-5. Reactions of Amines

A. *Reactions as Bases*

1. With Water.   When an R group is substituted for a hydrogen in water to give an alcohol, the modification in the chemical properties of water is not marked for many reactions. By analogy one might expect a similar situation in the nitrogen system. An R group substituted for hydrogen in ammonia actually results in a slightly stronger base when the R group is short (1–4 carbons). When a benzene ring is substituted for a hydrogen, the resulting amine is a much weaker base than ammonia. Hydrazine and hydroxylamine

$$NH_3 + H_2O \rightleftarrows NH_4^+ + OH^- \tag{15-16}$$

$$RNH_2 + H_2O \rightleftarrows RNH_3^+ + OH^- \tag{15-17}$$

$$R_2NH + H_2O \rightleftarrows R_2NH_2^+ + OH^- \tag{15-18}$$

$$R_3N + H_2O \rightleftarrows R_3NH^+ + OH^- \tag{15-19}$$

$$\bigcirc\text{-}NH_2 + H_2O \rightleftarrows \bigcirc\text{-}NH_3^+ + OH^- \tag{15-20}$$

$$H_2N\text{-}NH_2 + H_2O \rightleftarrows H_2N\text{-}NH_3^+ + OH^- \tag{15-21}$$

$$H_2N\text{-}OH + H_2O \rightleftarrows HO\text{-}NH_3^+ + OH^- \tag{15-22}$$

have basic strengths comparable to that of ammonia. Expressions 15-16–15-22 show the behavior of the various types of amines as bases.

2. Salt Formation.   If the amines can act as bases in the ways just described, they should form salts with acids. The salts formed with sulfuric, hydrochloric, and picric acids are the only ones ordinarily prepared.

Sulfuric acid can form two kinds of salts with amines since it has two ionizable hydrogens. If one ionizable hydrogen is replaced,

$$CH_3NH_2 + HOSO_3H \longrightarrow [CH_3NH_3]^+[OSO_3H]^- \tag{15-23}$$
$$\text{methylammonium hydrogen sulfate}$$

$$2CH_3NH_2 + HOSO_3H \longrightarrow [CH_3NH_3]_2^{++}[OSO_3]^= \tag{15-24}$$
$$\text{methylammonium sulfate}$$

the salt is called an alkylammonium hydrogen sulfate; if both are replaced, the salt is an alkylammonium sulfate.

The salts of hydrochloric acid, on the other hand, are more frequently called amine hydrochlorides and are written with a dot between the formula for the amine and HCl: $RNH_2 \cdot HCl$.

$$\text{◯-NHCH}_3 + \text{HCl} \longrightarrow \overset{\text{CH}_3}{\underset{\substack{\text{methylaniline}\\ \text{hydrochloride}}}{\text{◯-NH}}} \cdot \text{HCl} \qquad \text{[15-25]}$$

Phenylhydrazine forms a hydrochloride, which, being less unstable in air than the free base, gives a convenient way to handle the compound.

$$\text{◯-NH–NH}_2 + \text{HCl} \longrightarrow \underset{\substack{\text{phenylhydrazine}\\ \text{hydrochloride}}}{\text{◯-NH–NH}_2} \cdot \text{HCl} \qquad \text{[15-26]}$$

Hydroxylamine also forms comparable salts with strong acids.

$$\text{NH}_2\text{OH} + \text{HOSO}_3\text{H} \longrightarrow \underset{\substack{\text{hydroxylamine hydrogen}\\ \text{sulfate}}}{[\text{HONH}_3]^+[\text{OSO}_3\text{H}]^-} \qquad \text{[15-27]}$$

$$2\text{NH}_2\text{OH} + \text{HOSO}_3\text{H} \longrightarrow \underset{\text{hydroxylamine sulfate}}{[\text{HONH}_3]_2^+[\text{OSO}_3]^=} \qquad \text{[15-28]}$$

The picric acid salts of amines are frequently used to identify amines, particularly tertiary amines. These picrates are yellow crystalline salts, easily recrystallized, and many of them have sharp characteristic melting points.

[15-29]

When two aryl groups are substituted in ammonia, the basic strength is reduced so far that the amine will no longer dissolve in hydrochloric acid.

$$\underset{\text{diphenyl amine}}{\overset{\text{H}}{\text{◯-N-◯}}} + \text{HCl} \longrightarrow \text{no reaction} \qquad \text{[15-30]}$$

The amines are readily liberated from their salts by treatment with strong inorganic bases such as KOH, NaOH, and $Ca(OH)_2$.

$$\begin{array}{l} C_2H_5 \\ \quad \diagdown N \cdot HCl + NaOH \longrightarrow (C_2H_5)_3N + NaCl + H_2O \qquad\qquad \textbf{[15-31]} \\ C_2H_5 \; C_2H_5 \end{array}$$

B. Isocyanide Reaction (Test for a Primary Amine)

Both aliphatic and aromatic primary amines react with chloroform in the presence of potassium hydroxide to produce an isocyanide that has a very nauseating odor. This test is so sensitive that it can be used to detect the presence of primary amines as impurities in secondary and tertiary amines. The student should

$$\diagdown\!\!\!\bigcirc\!\!-NH_2 + CHCl_3 + 3KOH \xrightarrow[40\%]{} \diagdown\!\!\!\bigcirc\!\!-N\!\!\equiv\!\!C + 3KCl + 3H_2O \qquad\qquad \textbf{[15-32]}$$
$$\text{phenyl isocyanide}$$

compare the electronic structures of the cyanide and isocyanide groups.

The most important chemical property of an isocyanide is the ease with which it can be rearranged into a cyanide. Warming to $250°$ will accomplish this.

$$R\!-\!N\!\!\equiv\!\!C \xrightarrow{250°} R\!-\!C\!\!\equiv\!\!N \qquad\qquad \textbf{[15-33]}$$

C. Acylation

Just as ammonia reacts with acyl halides or acid anhydrides to form acid amides, so primary and secondary amines react to yield substituted amides. The acyl halides of low molecular weight react violently with ammonia or amines; those of higher molecular weight undergo the same reaction readily if the reactants are shaken together in aqueous alkali.

$$CH_3-\overset{\overset{O}{\|}}{C}-Cl + 2CH_3CH_2CH_2NH_2 \longrightarrow CH_3-\overset{\overset{O}{\|}}{C}-\overset{\overset{H}{|}}{N}-C_3H_7 + C_3H_7NH_2 \cdot HCl$$
N-n-propylacetamide

$$[15\text{-}34]$$

Acetyl chloride reacts violently with n-propyl amine, half of the amine being used up in the formation of the amine hydrochloride. The same compound can be prepared with a slower evolution of heat by the use of acetic anhydride.

$$\begin{matrix} CH_3-C\overset{\nearrow O}{\underset{\searrow}{}} \\ O + CH_3CH_2CH_2NH_2 \longrightarrow CH_3-\overset{\overset{O}{\|}}{C}-NHC_3H_7 + CH_3COO^- \ \overset{+}{H_3N}-C_3H_7 \\ CH_3-C\overset{\nwarrow}{\underset{\searrow O}{}} \end{matrix}$$

$$[15\text{-}35]$$

When benzoyl chloride is shaken with an amine in basic solution, the reaction, known as the Schotten-Baumann reaction, is a test for primary or secondary amines. This reaction, it will be noted, is a replacement of a hydrogen on the nitrogen by an acyl

$$\langle\text{O}\rangle\text{-}C\overset{\nearrow O}{\underset{\searrow Cl}{}} + \langle\text{O}\rangle\text{-NH}_2 + NaOH \longrightarrow \langle\text{O}\rangle\text{---}\overset{\overset{O}{\|}}{C}\text{-}\overset{\overset{H}{|}}{N}\text{---}\langle\text{O}\rangle + NaCl + H_2O$$

$$[15\text{-}36]$$

$$\langle\text{O}\rangle\text{-}C\overset{\nearrow O}{\underset{\searrow Cl}{}} + \langle\text{O}\rangle\text{-}\overset{\overset{H}{|}}{N}\text{-CH}_3 + NaOH \underset{90\%}{\longrightarrow} \langle\text{O}\rangle\text{-}\overset{\overset{O}{\|}}{C}\text{-}\overset{\overset{CH_3}{|}}{N}\text{---}\langle\text{O}\rangle + NaCl + H_2O$$

$$[15\text{-}37]$$

group. Hence it is evident that tertiary amines cannot give such a reaction. Note that this can be run in OH^-, in contrast to CH_3COCl, which reacts violently even with H_2O. (See pp. 259–260.)

Exercise 15-5. Write a method of preparing acetanilide.

D. *Nitrous Acid*

Nitrous acid can be used to distinguish primary, secondary, and tertiary amines in both the aromatic and the aliphatic series. The reactions are illustrated in Expressions 15-38, 15-39, and 15-40.

$$RNH_2 + HONO \longrightarrow ROH + N_2 + H_2O \text{ (gas bubbles off)} \qquad [15\text{-}38]$$

$$\begin{matrix} R \\ \diagdown \\ NH + HONO \longrightarrow \\ \diagup \\ R' \end{matrix} \quad \begin{matrix} R \\ \diagdown \\ N\text{--}N\text{=}O + H_2O \text{ (yellow oil separates)} \\ \diagup \\ R' \end{matrix} \qquad [15\text{-}39]$$

a dialkylnitrosoamine

$$\begin{matrix} R \\ \diagdown \\ N\text{--}R'' + HONO \longrightarrow \\ \diagup \\ R' \end{matrix} \quad \text{(no visible reaction although the amine} \\ \text{dissolves in the acid to form a salt)} \qquad [15\text{-}40]$$

Nitrous acid cannot be isolated, for it decomposes spontaneously to nitric oxide and nitrogen dioxide. Consequently it is prepared

$$2HONO \longrightarrow NO + NO_2 + H_2O \qquad [15\text{-}41]$$

within the reaction mixture at the time of use by addition of HCl to a solution of a nitrite.

$$Na^+NO_2^- + H^+Cl^- \longrightarrow HONO + Na^+Cl^- \qquad [15\text{-}42]$$

The evolution of nitrogen from a primary aliphatic amine is analogous to the preparation of nitrogen from ammonium nitrite. Nitrogen is frequently prepared in the laboratory by warming a mixture of ammonium chloride and sodium nitrite. An examination of the character of the nitrogens in these two substances will reveal that they are in the same oxidation states as the nitrogens in an amine and nitrous acid, respectively.

$$\overset{-3}{N}H_4^+Cl^- + Na^+\overset{+3}{O}NO^- \longrightarrow \overset{0}{N_2} + 2H_2O + Na^+Cl^- \qquad [15\text{-}43]$$

$$RNH_3^+Cl^- + HONO \longrightarrow N_2 + H_2O + ROH + H^+Cl^- \qquad [15\text{-}44]$$

The reactions with aliphatic amines can be run at room temperature. With aromatic amines, if side reactions are to be avoided, it is best to proceed at $0°C$. At this temperature the reactions of the three types take the courses shown in Expressions 15-45, 15-46, and 15-47. The first of these reactions results in a clear solution;

$$\text{⬡-NH}_2 + \text{HONO} \xrightarrow[0°C]{\text{HCl}} \left[\text{⬡-N}_{\parallel}^{+}\text{N}\right]^+\text{Cl}^- + 2\text{H}_2\text{O} \qquad \text{[15-45]}$$

benzene diazonium
chloride

$$\text{⬡-}\overset{\text{H}}{\underset{\mid}{\text{N}}}\text{-R} + \text{HONO} \xrightarrow[0°C]{\text{HCl}} \text{⬡-}\overset{\text{R}}{\underset{\mid}{\text{N}}}\text{-N=O} + \text{H}_2\text{O} \qquad \text{[15-46]}$$

N-nitrosoalkyl-
aniline

$$\text{⬡-}\overset{\text{R}}{\underset{\mid}{\text{N}}}\text{-R}' + \text{HONO} \xrightarrow[0°C]{\text{HCl}} \text{O=N-⬡-}\overset{\text{R}}{\underset{\mid}{\text{N}}}\text{-R}' + \text{H}_2\text{O} \qquad \text{[15-47]}$$

p-nitrosodialkylaniline

to distinguish between the latter two requires further reactions.

15-6. Reactions Subsequent to Diazotization

The reaction in which a diazonium salt is formed was discovered by Peter Griess (1858) and is called a diazotization (from *azote*, French for nitrogen). These diazonium compounds are interesting in themselves and useful industrially, especially in the manufacture of dyes. Most of them decompose when dry, some with explosive violence. For that reason they are prepared in solution and used for some subsequent reaction at once without isolation. Some can be stabilized, generally by forming a double salt with $ZnCl_2$, and may then be handled dry.

The reactions of these compounds are best accounted for by two structures rather than one, for the compounds appear to behave in some reactions as if they had one structure and in other reactions as if they had another. The two types of reactions that the diazo compounds undergo are those in which nitrogen is lost and those in which nitrogen is not lost. Reactions involving the loss of nitrogen occur in acid solution and probably involve the

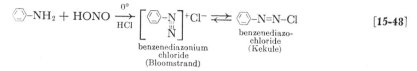

$$[15\text{-}48]$$

diazonium chloride structure (proposed by Bloomstrand), which is favored in such solutions. Reactions involving no loss of nitrogen occur in basic solution and probably involve the diazo structure (proposed by Kekule), which is more stable in alkaline solution.

A. *Reactions Involving Loss of Nitrogen*

1. Reaction with Water. When a solution of a diazonium salt formed at 0°C is warmed on a steam bath, nitrogen is evolved and a phenol is formed. A diazonium hydrogen sulfate generally gives a better yield from this reaction than the corresponding chloride.

$$CH_3\text{-}\langle\rangle\text{-}NH_2 + HONO \xrightarrow[0°C]{H_2SO_4} \left[CH_3\text{-}\langle\rangle\text{-}\underset{N}{\overset{N}{N}} \right]^+OSO_3H^- \xrightarrow[100°C]{H_2O}$$

$$CH_3\text{-}\langle\rangle\text{-}OH + N_2 + H_2SO_4 \qquad [15\text{-}49]$$
$$55\%$$

2. Reduction with Hypophosphorous Acid. The diazonium group may be replaced by hydrogen upon reacting with aqueous hypophosphorous acid. The reaction is an oxidation-reduction.

$$\left[CH_3\text{-}\langle\rangle\text{-}\underset{N}{\overset{N}{N}}^{,NO_2} \right]^+ Cl^- + H_3PO_2 + H_2O \xrightarrow{80\%} CH_3\text{-}\langle\rangle^{,NO_2} + N_2 + HCl + H_3PO_3$$

$$[15\text{-}50]$$

Exercise 15-6. What atoms change oxidation numbers in Expression 15-50?

3. Sandmeyer Reactions. A diazonium compound will form with copper I chloride a double salt that can be decomposed to the corresponding chloro compound. In like manner a nitrile group may be made to replace an amine on an aromatic ring by diazotiza-

$$\left[\underset{NO_2'}{\bigcirc}\text{-N}\overset{|||}{\underset{N}{\equiv}}\right]^{+}Cl^{-} \xrightarrow[95\%]{CuCl} \underset{NO_2'}{\bigcirc}\text{-Cl} + N_2 \qquad [15\text{-}51]$$

tion and subsequent treatment with copper I cyanide. These reactions embodying complex copper salts are known as Sandmeyer reactions.

$$\left[\bigcirc\text{-N}\overset{|||}{\underset{N}{\equiv}}\right]^{+}Cl^{-} + \text{Cu(CN)} \xrightarrow[72\%]{} \bigcirc\text{-CN} + N_2 + CuCl \qquad [15\text{-}52]$$

4. Gattermann Reactions. Reactions similar to the Sandmeyer reactions may be accomplished by the use of copper "bronze" (copper powder) as a catalyst for decomposing the diazonium salt. This catalytic replacement of the diazonium group was discovered by Gattermann.

$$\left[\bigcirc\text{-N}\overset{|||}{\underset{N}{\equiv}}\right]^{+}Cl^{-} \xrightarrow[75\%]{Cu\ powder} \bigcirc\text{-Cl} + N_2 \qquad [15\text{-}53]$$

$$\left[\bigcirc\text{-N}\overset{|||}{\underset{N}{\equiv}}\right]^{+}Cl^{-} + \text{NaCN} \xrightarrow[\text{powder}]{Cu} \bigcirc\text{-CN} + Na^{+}Cl^{-} + N_2 \qquad [15\text{-}54]$$

B. Reactions Without Loss of Nitrogen

1. Preparation of Phenylhydrazine. Mild reducing agents such as sulfur dioxide and sodium sulfite will reduce a diazo compound to a hydrazine. Phenylhydrazine, which is used extensively for the identification of carbonyl groups, is prepared from benzenediazochloride in this way. It is released from its hydrochloride by

$$\bigcirc\text{-N=N-Cl} + 2Na_2SO_3 + 2H_2O \xrightarrow[84\%]{} \underset{\substack{\text{phenylhydrazine}\\\text{hydrochloride}}}{\bigcirc\text{-}\overset{H}{\underset{|}{N}}\text{-NH}_2 \cdot HCl} + 2Na_2SO_4 \qquad [15\text{-}55]$$

bases, and it must be distilled at reduced pressure because it decomposes at its boiling point under atmospheric pressure. What are the changes in oxidation numbers in this reduction?

2. Coupling Reactions with Phenols and Tertiary Amines. Phenols and tertiary amines react with diazochlorides to give azo compounds, usually highly colored. The strongly acid solution in which the diazotization takes place must be neutralized by sodium acetate before the phenol or amine is added. The azo group is –N=N–, in which both nitrogens are attached to carbon atoms. When one of the nitrogens is attached to another element, the group is called the diazo group.

$$CH_3-\langle\bigcirc\rangle-N=N-Cl + \langle\bigcirc\rangle\begin{smallmatrix}OH\\CH_3\end{smallmatrix} \xrightarrow{CH_3-COONa}$$
p-toluenediazo-
chloride

$$CH_3-\langle\bigcirc\rangle-N=N-\langle\bigcirc\rangle\begin{smallmatrix}-OH\\CH_3\end{smallmatrix} + CH_3COOH + Na^+Cl^- \qquad [15\text{-}56]$$
4-hydroxy-3,4'-dimethylazobenzene

$$\langle\bigcirc\rangle-N=N-Cl + \langle\bigcirc\rangle-N\begin{smallmatrix}C_2H_5\\C_2H_5\end{smallmatrix} \xrightarrow[95\%]{CH_3COONa}$$

$$\langle\bigcirc\rangle-N=N-\langle\bigcirc\rangle-N\begin{smallmatrix}C_2H_5\\C_2H_5\end{smallmatrix} + CH_3COOH + Na^+Cl^- \qquad [15\text{-}57]$$
p-diethylaminoazobenzene

The indicator methyl orange is an azo compound made from sulfanilic acid and dimethyl aniline by the reactions shown in Expression 15-58.

$$HO_3S-\langle\bigcirc\rangle-NH_2 \xrightarrow[\substack{0°\\HCl}]{HONO} HO_3S-\langle\bigcirc\rangle-N=N-Cl + \langle\bigcirc\rangle-N\begin{smallmatrix}CH_3\\CH_3\end{smallmatrix} \xrightarrow[85\%]{NaO-\overset{O}{\overset{\|}{C}}-CH_3}$$
sulfanilic acid

$$HO_3S-\langle\bigcirc\rangle-N=N-\langle\bigcirc\rangle-N\begin{smallmatrix}CH_3\\CH_3\end{smallmatrix} \begin{cases} \xrightarrow{OH^-} ^-O_3S-\langle\bigcirc\rangle-N=N-\langle\bigcirc\rangle-N\begin{smallmatrix}CH_3\\CH_3\end{smallmatrix} \quad \text{yellow} \\ \text{methyl orange} \qquad\qquad [15\text{-}58] \\ \xrightarrow{H^+} HO_3S-\langle\bigcirc\rangle-\overset{H}{\underset{|}{N}}-N=\langle\bigcirc\rangle=\overset{+}{N}\begin{smallmatrix}CH_3\\CH_3\end{smallmatrix} \quad \text{red} \end{cases}$$

15-7. Orientation

A primary, secondary, or tertiary amine group on a benzene ring is a strong *o,p*-directing group and generally dominates the orientation. However, amine groups are also readily oxidized, and some protection of the group is frequently desirable. This is often accomplished for primary and secondary amines by making the N-acyl derivative and subsequently removing it by hydrolysis after the desired substitution on the ring has been run. For a pertinent example, see the nitration of aniline on page 388.

SUMMARY

In general, amines can be prepared by reduction from any compounds containing C-N linkages in which the nitrogen has a higher oxidation state than −3. Gabriel's synthesis is excellent for preparing pure primary amines. The Hofmann hypobromite method is important for shortening a carbon chain by one carbon.

Aliphatic amines of six carbons or fewer are weak bases comparable to ammonia. Long-chain aliphatic amines and aryl amines are much weaker bases than ammonia. Two aryl groups on nitrogen make the amine neutral. Two acyl groups on nitrogen (an imide) result in a weak acid.

The classes of amines may be distinguished by the isocyanide reaction, the Schotten-Baumann reaction, and diazotization reactions. Reactions subsequent to diazotization of primary aromatic amines allow for replacement of the amine group by H, OH, X, CN and others (important in synthetic chemistry) and for coupling reactions (the basis of an important segment of the dye industry).

REFERENCE

Audrieth, "A Classification of Compounds of H and N," J. Chem. Education, 7, 2055 (1930).

EXERCISES

(Exercises 1–6 will be found within the text.)

7. What are the oxidation numbers of nitrogen in the following compounds or ions?

1. NO_2^-
2. NH_2OH
3. CH_3-NHOH
4. H_2N-SO_3H
5. $(CH_3)_2N-C-C_6H_5$
   $\quad\quad\quad\quad\;\; \overset{\|}{O}$
6. C_2H_5-ONO
7. NO_3^-

8. $(CH_3)_3NH^+$
9. $C_6H_5-N\overline{\overline{\underline{\equiv}}}C$
10. $\left[C_6H_5-N\atop{\overset{|||}{N}}\right]^+Cl^-$
11. $C_6H_5-N=N-Cl$ (each nitrogen)
12. dimethylanilinium picrate (p. 304) (each nitrogen)
13. $CH_3-C\equiv N$

8. What change in the oxidation number (if any) of nitrogen takes place in the following transformations?

1. $R-CONH_2 \longrightarrow R-C\equiv N \longrightarrow R-COO^- + NH_3$
2. $R-NO_2 \longrightarrow R-NH_2$

9. Write structures for the following:

1. 2-aminoethanol
2. 3-pentanamine
3. methyl isopropyl amine
4. ethylaniline
5. tetramethyl ammonium hydroxide
6. *p*-toluenediazonium chloride
7. 2,4-dimethylaniline
8. *m*-nitroaniline
9. aniline hydrochloride

10. Write structures for the eight isomers of $C_4H_{11}N$, and name them. Give the class of each.

11. Identify by equations, using specific compounds as examples:

1. Hofmann hypobromite reaction
2. Gabriel synthesis
3. Schotten-Baumann reaction
4. Hofmann ammonolysis reaction
5. isocyanide reaction
6. diazotization
7. acylation
8. Sandmeyer reaction
9. Gattermann reaction
10. coupling reaction

12. Complete and balance the following reactions:

    1. ⬡–CH_2CONH_2 + NaOBr ⟶

          CH_2–CH_2

    2. CH_2⟨   ⟩$CHNH_2$ + $CHCl_3$ + NaOH ⟶

          CH_2–CH_2

    3. diazotization of ethylaniline

    4. Schotten-Baumann reaction on *o*-chloroaniline

    5. action of nitrous acid on sec-butyl amine.

13. How could you most readily decide whether a given liquid is an alcohol, aldehyde, ketone, acid, amine, or ether?

14. Indicate how to prepare

    1. isopropyl amine from ethanol

    2. n-propyl amine from butyric acid

    3. ethanol from propionic acid

    4. 1,2-dibromopropane from 1-propanol

    5. methylacetylene from n-propyl alcohol

    6. pure isobutylamine

    7. propionic acid from n-butyric acid

15. What product would you expect to be formed if ethylene oxide were allowed to react with ammonia? with a primary amine? with a secondary amine? with ammonia, the ethylene oxide being in excess?

16. In what reactions of amines is a C-N bond broken?

17. What volume of nitrogen would be evolved if 7.3 g of n-butylamine were treated with nitrous acid at 0° and at one atmosphere of pressure?

18. Write an equation for the preparation of 1,4-butanediamine by the Hofmann hypobromite reaction.

19. In liquid ammonia, sodium amide, $NaNH_2$, will react with methanol. Write the equation for the reaction. Is this an acid-base reaction in the Arrhenius sense? in the Brönsted sense? in the Lewis sense?

16 Nitro Compounds

The nitro group in a carbon compound has the same electronic structure as the NO_2 group in nitric acid. Two equivalent structures may be written for each of the nitro compounds, and these substances are believed to be stabilized by this resonance.

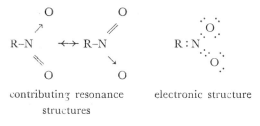

contributing resonance   electronic structure
structures

The nitroalkanes that are available on the market, together with a few aromatic nitro compounds, are shown in Table 16-1.

TABLE
16-1   *Commercially Available Nitro Compounds*

FORMULA	NAME	B.P. °C	M.P. °C
CH_3-NO_2	nitromethane	101	
$CH_3-CH_2-NO_2$	nitroethane	114	
$CH_3-CH_2-CH_2-NO_2$	1-nitropropane	131	
$CH_3-CH-CH_3$ NO_2	2-nitropropane	120	
⬡$-NO_2$	nitrobenzene	209	
NO_2 ⬡$-NO_2$	*m*-dinitrobenzene		90
NO_2 CH_3-⬡$-NO_2$ NO_2	2,4,6-trinitrotoluene (TNT)		82

315

16-1. Preparation of Nitro Compounds

1. Direct Nitration. Both nitroalkanes and aromatic nitro compounds are prepared by direct nitration. As we have seen, the nitration of an aromatic hydrocarbon is easier than that of an alkane. The latter reaction takes place in the gaseous state at $400°$ in stainless steel under pressure. Most nitrations of aromatic hydrocarbons can be carried out near room temperature or on a steam bath at normal pressure.

The nitro group is a *meta*-directing group when it is attached to the aromatic ring. Since it is a strong electron-attracting group, it tends to pull electrons away from the ring. This means that electrons are less available for forming new bonds with electron-seeking groups such as halogens, nitric acid, fuming sulfuric acid, and the reagents used in the Friedel-Crafts reaction. The presence

electron drift

of a nitro group, then, deactivates the ring so that it is more diffi-cult to introduce a second substituent than it was the first. Whereas nitrobenzene is made by nitrating benzene at about $60°$, the second nitro group goes in only when the temperature reaches $100°$ and requires a much longer reaction time.

$$\text{⟨⟩-NO}_2 + \text{HONO}_2 \xrightarrow[75\%]{\text{H}_2\text{SO}_4,\ 100°} \underset{\textit{m-dinitrobenzene}}{\text{⟨⟩-NO}_2} + \text{H}_2\text{O} \qquad [16\text{-}1]$$

To account for the formation of *m*-dinitrobenzene in terms of electronic structures, consider again the structure of the nitro group in nitric acid. There are three strongly electronegative oxygen atoms pulling electrons away from the nitrogen. An extreme picture of this tendency to pull electrons from nitrogen is shown on p. 317, the nitrogen being left with only a sextet of electrons in its outer shell, and a formal charge of $+2$. Though this extreme

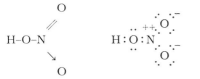

structure may not make an important contribution to the total picture of the nitric acid molecule, there may be a tendency in the direction indicated, and that tendency may be accentuated under the proper circumstances. The presence of sulfuric acid is one proper circumstance, doubtless furnishing an $\overset{\oplus}{N}O_2$ fragment for the attacking agent (Exp. 6-11).

When the nitro group is stationed on an aromatic ring, there is a source of relief for the pull of electrons from nitrogen by the oxygens—namely, the ring itself. Electrons in the ring are mobile and are available for the electron-deficient nitrogen. That is to say, the canonical structures *b* and *c* in Expression 16-2 make an im-

[16-2]

portant contribution to the hieroglyphic that represents nitrobenzene. (One more structure equivalent to *b* can be written.) The contributions made by these lawful structures leave the two *meta* positions with the greatest electron density. Consequently, when the nitro group is the directing group, electron-seeking agents such as $\overset{\oplus}{N}O_2$ (from nitric acid), as well as halogens and fuming sulfuric acid, substitute at the *meta* position.

2. Replacement of a Halide.   Aliphatic nitroalkanes in which the alkyl group is a straight chain may be successfully produced in small quantities in a pure state by the treatment of the alkyl bromide or iodide (but not the chloride) at room temperature with silver nitrite. At elevated temperatures a mixture of alkyl nitrite and other products is obtained along with the nitroalkane.

$$CH_3\text{-}(CH_2)_3\overset{+}{B}\overset{-}{r} + AgNO_2 \xrightarrow[73\%]{} CH_3\text{-}(CH_2)_3NO_2 + \underline{AgBr} \qquad [16\text{-}3]$$

3. Replacement of Amine Group on Aromatic Nucleus. Some aromatic nitro compounds not easily made by direct nitration can be made by replacement subsequent to diazotization of an amine. (See p. 605 for an example.)

16-2. Reactions of Nitro Compounds

Nitro compounds may be classified in the same way as alcohols. A primary nitro compound is one in which the carbon carrying the nitro group is bonded to two hydrogens and one carbon (or three hydrogens in the case of nitromethane). In a secondary nitro compound, the carbon carrying the nitro group is bonded to one hydrogen and two carbons. A tertiary nitro compound has no hydrogens on the carbon carrying the nitro group. Some of the properties of these compounds depend on the class of the carbon bearing the nitro group. Of necessity all aromatic nitro compounds are tertiary.

1. As Weak Acids. Primary and secondary nitro compounds are weakly acidic. They dissolve slowly in sodium hydroxide but not in sodium hydrogen carbonate. It has been shown that the primary and secondary nitroalkanes have two possible structures, called the nitro form and the aci form (the word *aci* comes from

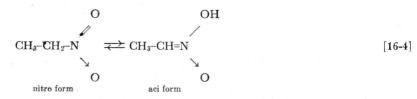

$$[16\text{-}4]$$

acid). Since nitroethane is probably an equilibrium mixture of the two forms, if we add a reagent that reacts with one particular form, all the original compound will eventually react as that form. A base comes near to being such a reagent. Though a strong base will react with either form of a nitro compound, it reacts much more rapidly with the aci form.

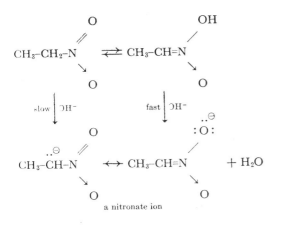

[16-5]

Secondary nitro compounds behave in a similar fashion. Tertiary nitro compounds have no hydrogen available on the α carbon,

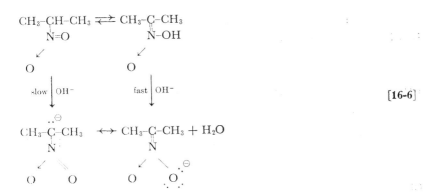

[16-6]

cannot undergo the shift to an aci form, and cannot dissolve in bases. Examples:

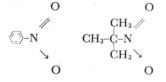

2. Aldol Condensation.

The lability of the hydrogen atom on the α carbon is demonstrated by the fact that primary and secondary nitro compounds will undergo the aldol condensation with aldehydes. The hydrogen on the α carbon of the nitro compound is more easily removed (labile) than that of the aldehyde, and this determines that the final product is a nitro alcohol.

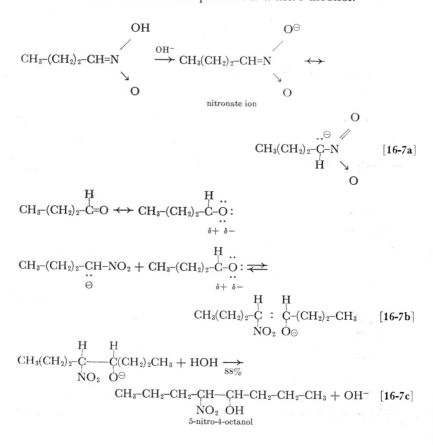

The reaction, like the aldol condensation of aldehydes, may be written in three steps (see p. 215). The nitronate ion is more commonly written with the negative charge located on one of the oxygen atoms of the aci form of the nitro compound. But it seems likely that the electronic structure written with the pair associated with the α carbon atom may make an important contribution to the structure of the nitronate ion. The resulting product of the aldol condensation is more simply accounted for if this is so.

In the second step, the nitronate ion adds to the polar form of the carbonyl group as shown. This ionic product then extracts a proton from a water molecule, regenerating the catalytic hydroxyl ion. Dilute alkaline solutions, such as NaOH, K_2CO_3, and $Ca(OH)_2$, catalyze this aldol condensation of a nitro compound and an aldehyde.

The condensation goes readily with aldehydes that are somewhat soluble in water, and with formaldehyde further condensation will take place as long as there is a hydrogen atom on the α carbon of the nitro compound.

$$CH_3\text{-}CH_2\text{-}NO_2 + H\text{-}CHO \xrightarrow{OH^-} CH_3\text{-}\underset{\underset{NO_2}{|}}{CH}\text{-}CH_2OH + H\text{-}CHO \xrightarrow[98\%]{OH^-}$$

$$CH_3\text{-}\underset{\underset{CH_2OH}{|}}{\overset{\overset{CH_2OH}{|}}{C}}\text{-}NO_2 \qquad [16\text{-}8]$$

2-nitro-2-methyl-
1,3-propanediol

Exercise 16-1. Write a three-step mechanism to show the formation of 2-nitro-1-propanol. (See Exp. 16-7.)

Exercise 16-2. Write equations to show the reactions of 3-nitro-2-pentanol with the following reagents: hydrogen and catalyst, acetic anhydride, PCl_5, dilute NaOH, P_2O_5.

3. Hydrolysis in Acid Solution. The hydrolysis products of primary and secondary nitro compounds in acid solution are readily understood if the mechanism is assumed to be addition of the acid to the aci form. The intermediate, in which the functional

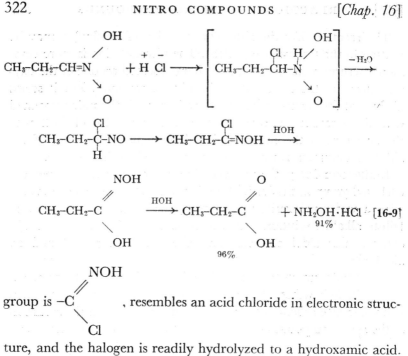

$$CH_3-CH_2-\overset{\overset{\displaystyle NOH}{\parallel}}{\underset{\underset{\displaystyle OH}{\diagdown}}{C}} \quad \xrightarrow{\text{HOH}} \quad CH_3-CH_2-\overset{\overset{\displaystyle O}{\parallel}}{\underset{\underset{\displaystyle OH}{\diagdown}}{C}} \quad + NH_2OH \cdot HCl \quad [16\text{-}9]$$

$$\underset{96\%}{} \qquad\qquad\qquad\qquad 91\%$$

group is $-\overset{\overset{\displaystyle NOH}{\parallel}}{\underset{\underset{\displaystyle Cl}{\diagdown}}{C}}$, resembles an acid chloride in electronic struc-

ture, and the halogen is readily hydrolyzed to a hydroxamic acid.

The functional group in a hydroxamic acid, $-\overset{\overset{\displaystyle NOH}{\parallel}}{\underset{\underset{\displaystyle OH}{\diagdown}}{C}}$, also has

the properties of a carboxylic acid, to which it can be further hydrolyzed. The final products of the hydrolysis, then, are a carboxylic acid and a hydroxylamine salt—a hydrochloride in the example cited.

When a secondary nitro compound is hydrolyzed, the mechanism can be written analogously until a chloro-nitroso compound is formed. Since no hydrogen remains on the carbon carrying the chlorine atom and the nitroso group, the tautomeric (see p. 334) oximino group (–C=NOH) cannot form. Both groups hydrolyze, however, and acetone, hyponitrous acid, and HCl are formed. Hyponitrous acid is only a hypothetical product; it is too unstable to be isolated, and it decomposes by various simultaneous reactions.

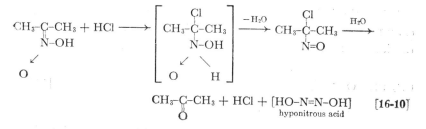

$$CH_3\text{-}C\text{-}CH_3 + HCl + [HO\text{-}N\text{=}N\text{-}OH] \qquad [16\text{-}10]$$
hyponitrous acid

Tertiary nitro compounds, including all aromatic members of the series, cannot be hydrolyzed. This is the predicted result if the mechanism shown above for primary and secondary nitro compounds is correct.

4. Action of Halogens. In the presence of a small amount of alkali, primary and secondary nitro compounds react in such a manner that the halogen is substituted on the carbon that carries the nitro group. This may come about in the way shown in Expression 16-11. The reaction can be controlled so that the first

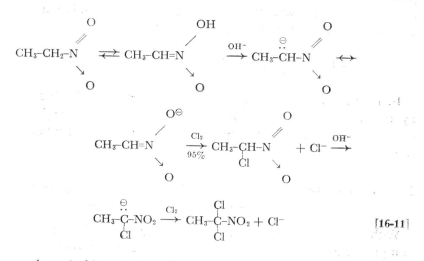

product, 1-chloro-1-nitroethane, can be isolated. The second compound, 1,1-dichloro-1-nitroethane, is a good insecticide for orchard insects.

When the halogenation is run under anhydrous conditions (for example, in the presence of P_2O_5), the attack by the halogen is random along the chain; so a mixture is obtained.

$$CH_3-CH_2NO_2 + Cl_2 \xrightarrow{P_2O_5} CH_3-\underset{Cl}{CHNO_2} + \underset{Cl}{CH_2}-CH_2-NO_2 + HCl \qquad [\textbf{16-12}]$$

In an aromatic nitro compound, the orientation of the nitro group to halogenation and all other ring substitutions is *meta* (as explained on p. 316), and such substitutions are difficult because the group deactivates the ring. Only a 50% yield is obtained when nitrobenzene is chlorinated.

$$\langle\underline{}\rangle\text{-NO}_2 + Cl_2 \xrightarrow[50\%]{FeCl_3} \underset{Cl}{\langle\underline{}\rangle}\text{-NO}_2 + HCl \qquad [\textbf{16-13}]$$

5. Reduction. The nitro group in any compound is comparatively easy to reduce, either by chemical reducing agents in acid solution or by catalytic hydrogenation in neutral solution, to a primary amine, generally giving a good yield.

$$\langle\underline{}\rangle\text{-NO}_2 + Sn + HCl \xrightarrow[95\%]{} \langle\underline{}\rangle\text{-NH}_2 + SnCl_2 + SnCl_4 + H_2O \qquad [\textbf{16-14}]$$

In neutral or nearly neutral solution the reduction leads to a hydroxylamine. Zinc dust in the presence of NH_4Cl is capable of reducing nitrobenzene to phenylhydroxylamine at room temperature.

$$\langle\underline{}\rangle\text{-NO}_2 + Zn + H_2O \xrightarrow[68\%]{NH_4Cl} \langle\underline{}\rangle\text{-}\overset{H}{\underset{phenylhydroxylamine}{N}}\text{-OH} + Zn(OH)_2 \qquad [\textbf{16-15}]$$

With aryl nitro compounds in alkaline solution, a series of reduction products intermediate in oxidation number between the nitro group and the amine group may be formed, the particular product depending on the choice of reducing agent. With sodium methoxide in methyl alcohol, a yellow crystalline substance called azoxy-

benzene is obtained as a bimolecular reduction product of nitrobenzene.

$$\text{⟨⟩-NO}_2 + \text{NaOCH}_3 \xrightarrow[100\%]{} \text{⟨⟩-N=N-⟨⟩} + \text{H-C} {\stackrel{O}{_{ONa}}} + \text{H}_2\text{O} \qquad [\textbf{16-16}]$$
$$\downarrow$$
$$\text{O}$$
azoxybenzene

Azobenzene is a red crystalline compound obtained by mild
reduction of azoxybenzene or by reduction of nitrobenzene itself
with zinc dust and NaOH, methyl alcohol being the solvent. If

$$\text{⟨⟩-NO}_2 + \text{Zn} + \text{OH}^- \xrightarrow[86\%]{\text{CH}_3\text{OH}} \text{⟨⟩-N=N-⟨⟩} + \text{ZnO}_2^= + \text{H}_2\text{O} \qquad [\textbf{16-17}]$$
azobenzene

more concentrated NaOH is used in water solution, the reduction
goes one step further, to hydrazobenzene, a colorless solid melting
at 131°.

$$\text{Zn} + \text{⟨⟩-NO}_2 + \text{OH}^- \xrightarrow[80\%]{} \text{⟨⟩-}\overset{\text{H}}{\underset{}{\text{N}}}\text{-}\overset{\text{H}}{\underset{}{\text{N}}}\text{-⟨⟩} + \text{ZnO}_2^= + \text{H}_2\text{O} \qquad [\textbf{16-18}]$$
hydrazobenzene

Another intermediate reduction product of nitrobenzene, nitrosobenzene, is best obtained by oxidation of phenylhydroxylamine
with a dichromate in acid solution. Nitrosobenzene is a colorless
solid when pure but is green as usually prepared.

$$\text{⟨⟩-NHOH} + \text{Cr}_2\text{O}_7^= + \text{H}^+ \xrightarrow[70\%]{} \text{⟨⟩-N=O} + \text{Cr}^{+3} + \text{H}_2\text{O} \qquad [\textbf{16-19}]$$
nitrosobenzene

In $\text{H-O-N} \overset{\nearrow O}{\underset{\searrow O}{}}$, nitrogen has four bonds to oxygen; so the value

of the oxidation number is $+4$ from this source. In addition, there
is a formal charge of $+1$ on the nitrogen, due to one coordinate
covalent bond to oxygen. The oxidation number of nitrogen in
nitric acid is, therefore, $+5$.

In a nitro compound, $-\overset{|}{\underset{|}{C}}-N\overset{\nearrow O}{\underset{\searrow O}{}}$, there are three bonds to oxy-

gen ($+3$), one bond to carbon (-1), and a formal charge ($+1$) on nitrogen; so its oxidation number is $+3$.

The oxidation states of nitrogen in the successive reduction products of nitrobenzene will be found with the formulas below.

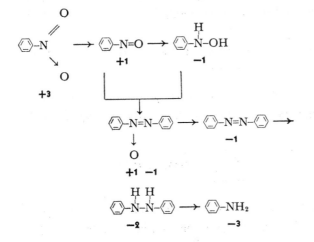

Exercise 16-3. Balance the equations given in Expressions 16-14–16-19. (To obtain the oxidation numbers of N, see pp. 295–296.)

SUMMARY

Drastic conditions are needed for the preparation of nitroalkanes by direct nitration; aromatic nitro compounds are easier to synthesize.

Primary and secondary aliphatic nitro compounds are weak acids; tertiary nitro compounds (of necessity, aromatic nitro compounds are tertiary) are neutral.

Primary and secondary aliphatic nitroalkanes can take part in aldol condensations with aldehydes, can be hydrolyzed in acid solution, and can be halogenated to give products that have recently attained industrial importance. Nitroalkanes are readily reduced to amines catalytically. Aromatic nitro compounds reduce to amines in acid solution, in neutral solution yield hydroxylamines, and in basic solution yield a series of bimolecular reduction products depending on the reducing agent.

EXERCISES

(Exercises 1–3 will be found within the text.)

4. Write structural formulas for 2-nitrobutane, p-nitrotoluene, 2-nitro-1-butanol, azobenzene, 1,1-dichloro-1-nitroethane, p-nitrophenol.

5. Name the following compounds:

1. $CH_3-C(NO_2)=CH_2$

2. 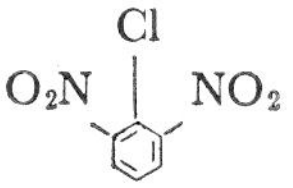 $O_2N \mid NO_2$

3. $CH_3-CHNO_2-CHOH-CH_3$

4. $CH_3-\underset{\underset{CH_2OH}{|}}{\overset{\overset{CH_2OH}{|}}{C}}-NH_2$

5. $CH_3-\underset{\underset{NO_2}{|}}{CH}-CH_2OOC-CH_3$

6. $CH_3-CH_2-COOCH_2-CH_2NO_2$

6. Write the aldol condensation stepwise for the following reactions:

1. $CH_3-CH_2-CHO + OH^-$
2. $CH_3-CO-CH_3 + OH^-$
3. $CH_3CH_2CH_2NO_2 + CH_3CH_2CHO + OH^-$

7. Write equations for the reaction of nitromethane, of nitroethane, and of 2-nitropropane with excess formaldehyde in basic solution.

8. What products would you expect to be formed in a high-temperature reaction of isobutane and nitric acid?

9. Why are no aromatic mononitro compounds soluble in alkali?

10. Show how hydroxylamine hydrogen sulfate could be obtained from 1-nitropropane.

11. Write equations for the reaction of chlorine with 2-nitropropane

1. under anhydrous conditions
2. in the presence of sodium hydroxide solution

12. Synthesize the following compounds:

1. *m*-dinitrobenzene
2. *o*-nitrophenol
3. *m*-bromonitrobenzene
4. 2-amino-2-ethyl-1,3-propanediol
5. 2-nitropropyl propionate
6. 2-nitro-2-butene
7. 3-nitro-2-chloropentane

REVIEW PROBLEMS FOR CHAPTERS 1–16

1. List all the methods you can find for increasing and decreasing the number of carbon atoms attached by carbon-to-carbon bonds in a molecule. Indicate which methods are limited to a change of one carbon atom.

2. Indicate a chemical means of distinguishing between

1. ethanol and acetaldehyde
2. acetyl chloride and acetic anhydride
3. acetamide and sodium acetate
4. acetamide and tripalmitin
5. a mineral oil and a vegetable oil (glyceride)

3. By suitable equations, indicate the various substances with which methyl magnesium bromide will react.

4. With the organic compound given and any inorganic chemicals needed, devise methods for carrying out the following conversions. More than one step will usually be required.

1. $C_2H_5OH \longrightarrow C_2H_5COOH$
2. $CH_3C{\equiv}CH \longrightarrow CH_3COCH_3$
3. $CH_3CH_2CH_2OH \longrightarrow CH_3CH_2CH_2\text{-}O\text{-}CH_2CH_2CH_3$
4. $CH_3CH{=}CHCH_3 \longrightarrow CH_3C{\equiv}CCH_3$
5. $CH_3CH(OH)CH_3 \longrightarrow CH_3CH_2CH_3$
6. $CH_3CH_2CH_2OH \longrightarrow CH_3CH_2CH_2CH_2CH_2CH_3$
7. $(CH_3)_2CHCH_2OH \longrightarrow (CH_3)_3C\text{-}OH$

5. Compound A, of formula $C_5H_{12}O$, reacts with CH_3MgI to liberate one mole of methane. On oxidation, A is converted to compound B, which forms an oxime. When B is treated with $NaOH + I_2$, iodoform and sodium isobutyrate are formed. What is the structure of A?

6. An alcohol has the molecular formula $C_6H_{14}O$. When treated with a solution of zinc chloride in hydrochloric acid, it gives a clear solution that becomes cloudy in a few minutes. When the alcohol is dehydrated over alumina, it gives a product, C_6H_{12}, that is a pure substance not containing any isomers. This material reacts rapidly with bromine in the dark to give $C_6H_{12}Br_2$. Treatment of $C_6H_{12}Br_2$ with alcoholic potassium hydroxide gives C_6H_{10}, which gives a precipitate with ammoniacal silver nitrate reagent. Write structures for these compounds and an equation for the reaction that occurs in each step.

7. Compound A has the molecular formula C_8H_9Br. The bromine cannot be removed by treatment with silver nitrate or hot sodium hydroxide. When compound A is refluxed with alkaline permanganate, an acid having seven carbon atoms and still containing bromine is formed. If A is treated with magnesium in ether and the reagent poured over Dry Ice, an acid having nine carbon atoms and no bromine is obtained. Oxidation of the nine-carbon acid with alkaline permanganate gives a diacid containing eight carbon atoms. This diacid is nitrated, and careful investigation of the mononitro product shows that only one isomer is obtained. Write equations for all reactions described and determine the structure of compound A.

8. Compound A was a pleasant-smelling liquid having the formula $C_{10}H_{12}O_2$. It was not soluble in cold base, but when refluxed with sodium hydroxide solution it gradually dissolved. The resulting solution was cooled and acidified, whereupon a solid, B, having the composition $C_8H_8O_2$, precipitated. When B was heated with soda lime, toluene was obtained. Compound A was next treated with an excess of the Grignard reagent prepared from bromobenzene, and compound C ($C_{20}H_{18}O$) was obtained. C lost water readily to give compound D ($C_{20}H_{16}$). D, on being oxidized with permanganate, gave a ketone ($C_{13}H_{10}O$) and compound E, an acid ($C_7H_6O_2$). Heating E with soda lime gave benzene. Write structural formulas for A, B, C, D, and E.

9. An unknown, A, of formula $C_5H_{10}O_2$, gave a monophenylhydrazone and positive responses to tests with Fehling's solution and Tollen's reagent. It reacted with acetyl chloride to give an ester, liberated CH_4 from CH_3MgI, and gave a positive response to the iodoform test. Mild oxidation of A produced B, an acid whose formula was $C_5H_8O_3$. B gave a monophenylhydrazone and a positive response to the iodoform test, but negative responses to Fehling's and Tollen's tests. Write structures for A and B.

10. The alkaloid nicotine has been synthesized by the following series of reactions. [See Noller, *Chemistry of Organic Compounds* (W. B. Saunders Co., Philadelphia), p. 610.] State the conditions necessary for each step, and name each process—for example, "hydrolysis of an ester," "reduction of a ketone." Each arrow may involve more than one step.

$$RCN + Br(CH_2)_3OC_2H_5 \longrightarrow R\overset{\overset{\displaystyle O}{\|}}{C}(CH_2)_3OC_2H_5 \longrightarrow$$

$$\underset{1}{} \qquad \underset{2}{}$$

$$\underset{\underset{NH_2}{|}}{R}CH-(CH_2)_3OC_2H_5 \longrightarrow \underset{\underset{NH_2}{|}}{R}CH-(CH_2)_3Br \longrightarrow$$

$$\underset{3}{} \qquad \underset{4}{}$$

$$\begin{matrix} R-CH\!\!-\!\!-\!\!-\!\!CH_2 \\ \underset{}{NH}\diagdown \quad \diagup CH_2 \\ CH_2 \end{matrix} \longrightarrow \begin{matrix} R-CH\!\!-\!\!-\!\!-\!\!CH_2 \\ CH_3-N\diagdown \quad \diagup CH_2 \\ CH_2 \end{matrix}$$

$$\underset{5}{} \qquad \qquad \underset{6}{}$$

PART THREE

FOUNDATION FOR THE FRAMEWORK

---

17 Isomerism

Part Two of this text was devoted to a rather detailed study of the properties of important functional groups in carbon compounds as a basis for an elementary picture of the chemistry of the covalent bond. The whole treatment was based on two fundamental assumptions: (1) that carbon is tetravalent and tetrahedral; (2) that carbon atoms may be joined to other carbon atoms in single, double, and triple bonds. It is high time that these assumptions be justified. Much of this chapter may appear to be geometry, but it is good chemistry too. First, more detailed geometrical implications (topology) of the tetrahedron and joined tetrahedra will be deduced, and then, in the next chapter, the historical and theoretical basis for these two assumptions will be carefully laid.

17-1. Types of Isomerism

Isomers are compounds having the same molecular formula but different structural formulas. The difference in structural formula will be apparent in some cases if we write the structure in two dimensions; in other cases it is necessary to look at the structure in three dimensions.

The various types of isomerism will be classified in the following way:

A. Structural isomers
   1. Positional
   2. Functional (including tautomers)
B. Spatial isomers, or stereoisomers
   1. Optical
   2. Geometric

17-2. Structural Isomerism

The two classes of structural isomerism have already been discussed. Examples are given in the following list:

Positional isomers:
  butane, 2-methylpropane
  pentane, 2-methylbutane, 2,2-dimethylpropane
  o-xylene, *m*-xylene, *p*-xylene
  1-butanol, 2-butanol
  2-pentanone, 3-pentanone

Functional isomers:
  ethanol, methyl ether
  propanal, propanone
  propionic acid, methyl acetate
  1-butyne, 1,3-butadiene

17-3. Tautomerism

Tautomerism is a special case of functional isomerism. If two isomeric forms of the same compound are directly and readily interconvertible, the two forms are called tautomers and the phenomenon is called tautomerism. This definition was first clearly stated by Butlerov (1876) though the name was given by Laar (1885).

1. Nitro-aci Tautomerism. It was suggested in the last chapter that some of the chemistry of the primary and secondary nitro compounds is best accounted for on the assumption that two readily interconvertible structures exist, the nitro form and the aci

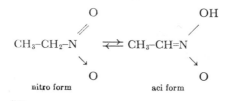

$$\underset{\text{nitro form}}{CH_3-CH_2-N \overset{O}{\underset{O}{\nwarrow}}} \quad \rightleftarrows \quad \underset{\text{aci form}}{CH_3-CH=N \overset{OH}{\underset{O}{\nwarrow}}}$$

[17-1]

form. It is to be noted that the two forms differ only by the position of a hydrogen atom. Since the bonds involved in the two structures have nearly equal energy, the two forms might be expected to have similar stability. Lability of the hydrogen on a carbon that is α to a functional group is a necessity for tautomerism.

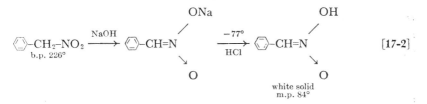

Both of the possible forms of at least one nitro compound have been isolated. If phenylnitromethane is dissolved in NaOH and then regenerated in the cold at Dry Ice temperatures with HCl, a white crystalline solid separates. This solid slowly reverts to the liquid nitro form on standing at room temperature.

The fact that the two forms have slightly different properties has led to the designation of the liquid as the nitro form and the white solid as the aci form. The aci form dissolves very quickly in bases, gives a red-brown color with Fe^{+3}, and reacts rapidly with bromine (see p. 323). The nitro form dissolves slowly in bases, gives no color with Fe^{+3}, and does not react with bromine until sodium hydroxide is added.

Exercise 17-1. Write equations to show the reactions of the aci form of phenylnitromethane with each of the following reagents: dilute NaOH, Fe^{+3}, Br_2.

2. Keto-enol Tautomerism. (See also Chap. 23.) Diketones in which the second keto group is in the β position with respect to the first also undergo some reactions that suggest tautomerism. The β-keto esters behave in like manner.

Benzoylacetone:

$$CH_3\text{-}\overset{O}{\overset{\|}{C}}\text{-}CH_2\text{-}\overset{O}{\overset{\|}{C}}\text{-}\langle\rangle \;\rightleftharpoons\; CH_3\text{-}\overset{O}{\overset{\|}{C}}\text{-}CH=\overset{OH}{\overset{|}{C}}\text{-}\langle\rangle \qquad\qquad [17\text{-}3]$$

keto form, 1%   enol form, 99%

Acetoacetic ester:

$$CH_3-\overset{O}{\overset{\|}{C}}-CH_2-\overset{O}{\overset{\|}{C}}-OC_2H_5 \rightleftarrows CH_3-\overset{OH}{\overset{|}{C}}=CH-\overset{O}{\overset{\|}{C}}-OC_2H_5 \qquad [17\text{-}4]$$

keto form, 92%   enol form, 8%

Ethyl acetoacetate, commonly called acetoacetic ester, gives reactions that are characteristic of a ketone, an ester, an alcohol, and an alkene. These may all be accounted for if the equilibrium shown in Expression 17-4 obtains. (Exp. 17-7 does not give strong

$$CH_3-\overset{O}{\overset{\|}{C}}-CH_2-\overset{O}{\overset{\|}{C}}-OC_2H_5 + NaHSO_3 \longrightarrow CH_3-\overset{OH}{\overset{|}{\underset{SO_3Na}{C}}}-CH_2-\overset{O}{C\underset{OC_2H_5}{\diagdown}} \qquad [17\text{-}5]$$

a methyl ketone

$$CH_3-\overset{O}{\overset{\|}{C}}-CH_2-\overset{O}{\overset{\|}{C}}-OC_2H_5 + H_2O \rightleftarrows CH_3-\overset{O}{\overset{\|}{C}}-CH_2-COOH + C_2H_5OH \quad [17\text{-}6]$$

an ester

$$CH_3-\overset{OH}{\overset{|}{C}}=CH-\overset{O}{\overset{\|}{C}}-OC_2H_5 + Na \longrightarrow \left[CH_3-\overset{:\overset{..}{O}:}{\overset{|}{C}}=CH-C\overset{O}{\underset{OC_2H_5}{\diagup}} \right]^- Na^+ + \tfrac{1}{2}H_2$$

an alcohol   [17-7]

support to the argument, for it is now considered probable that sodium can displace hydrogen directly from the keto form as well as from the enol form.)

$$CH_3-C\overset{\diagup OH}{\underset{\diagdown CH-\overset{O}{\overset{\|}{C}}-OC_2H_5}{}} + Br_2 \longrightarrow CH_3-\overset{OH}{\overset{|}{C}}-\overset{}{\underset{Br}{C}H}-\overset{O}{\overset{\|}{C}}-OC_2H_5 \longrightarrow$$

an alkene

$$CH_3-\overset{O}{\overset{\|}{C}}-\underset{Br}{\overset{}{C}H}-\overset{O}{\overset{\|}{C}}-OC_2H_5 + HBr \qquad [17\text{-}8]$$

The enol form is an alkene (ene) as well as an alcohol (ol). The estimated percentages of the two forms in equilibrium come from a rapid titration with bromine. The validity of the values given depends on the assumption that the enol (alkene) reacts more rapidly with bromine than the keto form. The first disappearance of the red color of bromine is taken as the end-point. Upon standing, an equilibrium involving more enol is again established.

A few enol forms have been isolated in the pure state by repeated distillation in all-quartz apparatus. The enol form readily reverts to the equilibrium mixture of the two forms in a short time when stored in glass. There is evidently enough alkali on the surface of the glass to catalyze this conversion. (For tautomerism in sugars see p. 407.)

17-4. Optical Isomerism

Spatial isomers, or stereoisomers, are isomers having the same structural formulas but differing in the spatial relations of the atoms. Optical isomers have identical physical properties, with the exception of one, and the same gross chemical properties, even though they may differ slightly in chemical activity toward some reagents. The one physical property that differs is the ability of the isomers to rotate plane-polarized light.

Three different kinds of lactic acid (2-hydroxypropionic or α-hydroxypropionic acid) have properties dependent on the source of the acid, as shown in Table 17-1. Though all three kinds have the same index of refraction and the same density, the racemic

TABLE
17-1 | *Physical Properties of Lactic Acids*

SOURCE	NAME	M.P. °C	INDEX OF REFRACTION	DENSITY AT 20°	SPECIFIC ROTATION OF PLANE-POLARIZED LIGHT
sour milk	racemic	17	1.44	1.248	0°
muscle	dextro	53	1.44	1.248	+2.24°
fermentation of sugar	levo	53	1.44	1.248	−2.24°

acid (defined on p. 342) has a lower melting point than the dextro and levo (the optically active) forms. It is usual to find some such difference in physical properties between the optically active and the other forms of these isomers. Another example is available in the tartaric acids (see Table 21-8, p. 471). The melting point and the solubility in water and alcohol of the dextro and levo forms of tartaric acid are exactly the same but differ from those of racemic and meso tartaric acids.

17-5. Plane-polarized Light

Many of the properties of light are accounted for if we describe it as a transverse wave motion like that propagated when a pebble is dropped into a quiet pool of water. The water moves up and down while the wave appears to move toward the edge of the pool; that is, the line of propagation is perpendicular to the motion of the particles. A ray of light is propagated as a vibration in all possible directions perpendicular to the direction of propagation. A ray of light that has passed through a Nicol prism (cut from clear calcite, $CaCO_3$, and put together in a special way) vibrates in only one plane and is said to be plane-polarized. Likewise, if a beam of light from a sodium vapor lamp, for example, as at F in the figure (p. 339), is passed through a Nicol prism, E, the emergent beam vibrates in parallel planes.

The Nicol prism is analogous to the pages of a book (p. 340), which will allow passage of a knife blade through them if the plane of the blade is parallel to the plane of the pages, but will not allow passage of the blade if it is held perpendicular to the pages or at an angle to them. The Nicol prism takes out all the light from a beam of light except that vibrating in a particular single plane (or in parallel planes). When one of a certain group of liquid organic compounds (or a solution of a compound) is placed in a beam of plane-polarized light, the plane in which the light is polarized will be rotated. Depending upon the nature of the compound, the plane will be rotated either clockwise (as we look toward the source of the light) or counterclockwise. For example, ordinary sugar solutions (sucrose) rotate the plane of polarized light clockwise; nicotine turns the plane counterclockwise. The magnitude of this rotation depends upon the nature and concentration (in the case of solutions) of the compound, the wavelength of the light, and the length of the tube. This rotation is observed by means of a second Nicol prism, B, which is rotated until the variation in the intensity of the light passing through it indicates the orientation of the plane of polarization. When the analyzing prism is at right angles to the plane of the light beam, no light will pass through; when it is parallel to the plane of the light, all of the light will pass through. When all the light passes through the

Diagram of a polarimeter. Approximate positions of the following parts are lettered at the top: A, the observer; B, a Nicol prism called the analyzer, mounted in a circular scale measured in degrees, which can be turned clockwise or counterclockwise; C, a tube containing the liquid (or solution) to be tested; D, a piece of quartz covering a half moon in the circular field of vision and dividing the field for accurate comparison of brightness; E, a Nicol prism cut and mounted to transmit light vibrating in parallel planes—called the polarizer; F, a sodium vapor lamp, the light source.

polarimeter, both halves of the divided field are equally bright, and the rotation can be read on the circular scale. By analogy, two books with their pages perpendicular will not permit passage of a knife through both of them.

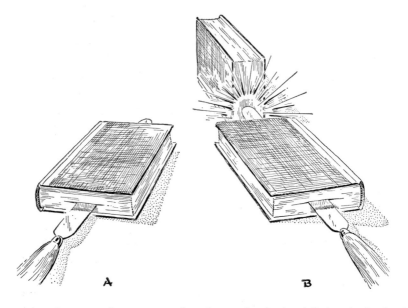

A   B

A substance that rotates the plane of polarized light clockwise is called dextro, and one that rotates it counterclockwise is called levo. As may be noted in Table 17-1, the specific rotation of dextro lactic acid is $+2.24°$ and that of levo lactic acid is equal but in the opposite direction, or $-2.24°$. In Table 21-8 (p. 471) the specific rotation of dextro tartaric acid is $+14°$ and that of the levo isomer is $-14°$.

17-6. The Asymmetric Carbon Atom

What is the nature of the substances that cause the rotation of the plane of polarized light? The tetrahedral nature of the carbon atom will account for the existence of two different compounds having the same structural formula if four different groups are

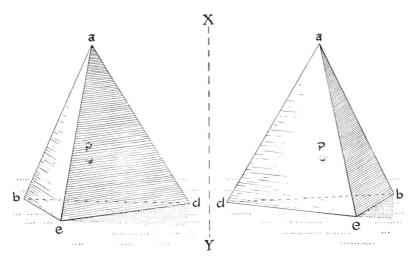

attached to a single carbon atom. In the adjacent figure four groups, *a*, *b*, *d*, and *e*, are shown attached to a central carbon atom (point *P*) in two different ways. Such a statement by itself is generally unconvincing. It is suggested that you make the ball-and-stick models of these two arrangements to test the validity of the statement.

If an imaginary mirror is placed perpendicular to this page on the dotted line *XY* in the accompanying figure, it will be seen that

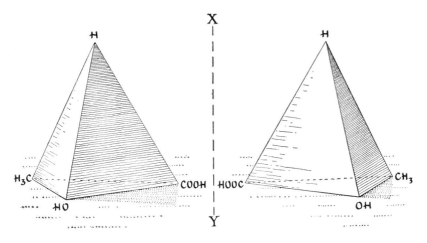

the dextro and levo lactic acids are mirror images of each other. They are as different as the individual gloves of a pair. Each tetrahedron is asymmetric (not symmetrical); that is, it is impossible to pass a plane anywhere through the molecule so that exactly similar parts are obtained.

It is not important to remember which individual structure of a pair has a dextro rotation when the compound is tested for this property. Indeed, the absolute configuration of no single substance was known before 1951. For present purposes one structure may arbitrarily be assigned the dextro configuration, and its mirror image is then called the levo form. (For a convention used in sugar chemistry, see p. 401.)

17-7. Racemic Modification

When a carbon becomes asymmetric during a synthesis by any of the methods so far discussed in this course, equal mixtures of *d* and *l* forms of the resulting compound are always obtained. The reason for this will be apparent if we consider the three-dimensional structures of the compounds involved in the synthesis shown in Expression 17-9. (The carbon that is to become asymmetric is

$$CH_3\text{-}CHO + HCN \longrightarrow \underset{\underset{CN}{|}}{\overset{\overset{H}{|}}{CH_3\text{-}C\text{-}OH}} \xrightarrow[H^+]{H_2O} \underset{\underset{COOH}{|}}{\overset{\overset{H}{|}}{CH_3\text{-}C\text{-}OH}} \qquad [17\text{-}9]$$

not shown in the accompanying diagram but is understood to lie

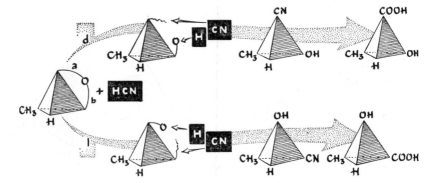

at the center of the tetrahedron.) When one of the double carbon-oxygen bonds breaks to add HCN, there is a 50-50 chance (it is assumed) that the bond at *a* is the one that breaks. Breaking at *a* will yield the cyanohydrin labeled *d* (dextro); breaking at *b* will yield the *l* (levo) cyanohydrin. Hydrolysis of each of these compounds yields the corresponding lactic acid. The assumption is correct, since 50-50 mixtures of *d* and *l* forms of asymmetric compounds are found in all laboratory syntheses not involving an asymmetric carbon in the starting substances. The resulting equimolecular mixture of *d* and *l* forms of a compound is optically inactive and is called a racemic modification.

The simple process of mixing equivalent amounts of the *d* and *l* forms of a compound will also, of course, result in the racemic modification.

Exercise 17-2. Write a synthesis for 2-pentanol by a Grignard method. In which reaction is the asymmetric carbon introduced? How many forms of the product exist? If the synthetic 2-pentanol is now converted to 2-chloropentane, how many forms of the new compound are there?

17-8. Racemization

A third process accomplishes the same result. If an optically active substance is subjected to the action of heat, light, or certain solvents or chemical reagents, half of it may change to the opposite configuration, giving a racemic modification. This process is called racemization. An example is the racemization of a *d* tartaric acid salt by heating of its aqueous solution, during which the optical rotation changes from that of the original tartaric acid salt to zero.

Very wide variations in the stability of optically active compounds are known. Some compounds are not racemized short of complete destruction; a few racemize on standing at room temperature in the solid state. A discussion of the mechanism by which compounds racemize is beyond the scope of this book.

17-9. Compounds Containing Two or More Asymmetric Carbons

If two asymmetric carbon atoms occur in the same molecule, there are four different forms of the compound, as shown in the adjacent figure. The four forms are characterized as two pairs of

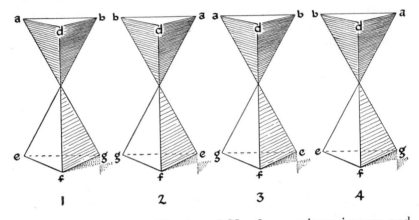

racemic modifications: No. 1 and No. 2 are mirror images and form one *dl* pair; No. 3 and No. 4 are mirror images and form a second *dl* pair. Such a compound is 3-amino-2-pentanol: there are four forms, and they are characterized as two *dl* pairs (that is, their *character* is that of two *dl* pairs).

$$
\begin{array}{c}
\text{H} \\
\text{CH}_3-\text{*}\text{C}-\text{OH} \\
\text{C}_2\text{H}_5-\text{*}\text{C}-\text{NH}_2 \\
\text{H}
\end{array}
$$

2 asym. *C's
2 *dl* pairs

If the two asymmetric carbons each have the same four groups attached, the number of forms is reduced to three. Forms 5a and 6a are a *dl* pair; form 7a is found to be optically inactive and is called a meso form. The meso form is defined in the following way: whenever a plane of symmetry can be passed through a molecule that is capable of optical activity, that form is optically inactive

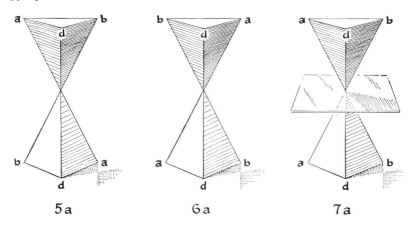

5a   6a   7a

and is called a meso form. Such a plane, parallel to the planes *abd*, can be passed through form 7a. For convenience, on plane surfaces, structures 5a, 6a, and 7a can be pictured in projection, as shown in the accompanying figure, in which a carbon is under-

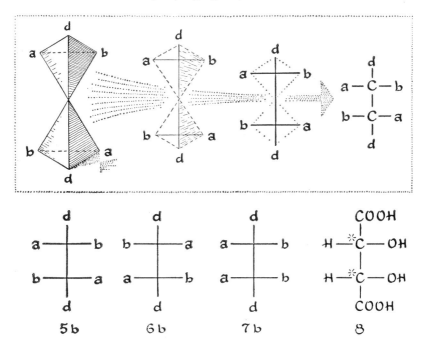

5b   6b   7b   8

stood to be at each intersection of lines. An example of the compound with three forms is tartaric acid, the meso form of which is shown as structure 8.

A racemic modification and a meso form are both optically inactive; the difference is that the first is optically inactive by external compensation (compensation by a second molecule) and the second by internal compensation (compensation within the same molecule). In meso-tartaric acid, one asymmetric carbon is dextro, the other is levo, and the two compensate each other.

To decide whether two structures of a compound are identical when projected as above (and when three-dimensional models are not available) we need only one rule. If, by rotating one figure in the plane of the paper or blackboard, we can superimpose it on the second, the two structures are identical. If structure 5b is rotated in the plane of the paper through 180°, it will not be superimposable on 6b and is therefore a different structure.

Table 17-2 shows the number and character of stereoisomers to be expected with any number of asymmetric carbons. Beyond 5 asymmetric carbons, the number of forms is scarcely of practical importance since the difficulties of isolating and identifying the large number of forms are insurmountable by present chemical or physical methods.

| TABLE 17-2 | *Number of Isomers Possible with One or More Asymmetric Carbons* |

NO. OF ASYMMETRIC CARBONS	NO. OF FORMS	CHARACTER
1	2	1 *dl* pair
2	4	2 *dl* pairs
3	8	4 *dl* pairs
4	16	8 *dl* pairs
5	32	16 *dl* pairs
n	2^n	2^{n-1} *dl* pairs
2 asym. C's alike	3	1 *dl* pair and 1 meso

17-10. Resolution of Optical Isomers

The first separation of optical isomers into d and l forms was accomplished by Louis Pasteur for sodium ammonium tartrate when he mechanically separated crystals that were themselves mirror images. This method, using a magnifying glass, besides being tedious, cannot be applied to most substances since the d and l forms do not necessarily crystallize in separate crystals. (For an interesting account of Pasteur's work see p. 378.) Many substances that occur in nature are already optically active and need no resolution. In the manufacture of wine, for example, d tartaric acid is left in the bottom of the casks as a by-product of the fermentation of sugar.

The most frequently used method of resolution can best be described with an example. The racemic modification of an optically active base, such as 1,2-diaminopropane, reacts with one isomer of an optically active acid, such as tartaric, to form salts in water

$$
\begin{array}{c}
\text{H } \text{ H} \\
CH_3\text{-}\overset{|}{C}*\text{---}\overset{|}{C}\text{-H} \\
\overset{|}{N}H_2 \overset{|}{N}H_2
\end{array}
+
\begin{array}{c}
COOH \\
H\text{-}\overset{|}{C}*\text{-OH} \\
HO\text{-}\overset{|}{C}*\text{-H} \\
COOH
\end{array}
\longrightarrow
\begin{bmatrix}
CH_3 \\
\overset{|}{C}HNH_3 \\
CH_2NH_3
\end{bmatrix}^{++}
\begin{bmatrix}
COO \\
\overset{|}{C}HOH \\
\overset{|}{C}HOH \\
COO
\end{bmatrix}^{=}
\qquad [17\text{-}10]
$$

dl 1,2-diaminopropane d tartaric acid dl propylenediamine d tartrate
        or
dl propylenediamine

solution. If dl propylenediamine is used with d tartaric acid, two compounds will form in equal quantity. The two products, however, are no longer complete mirror images and hence do not necessarily have the same physical properties. In the example chosen, l propylenediamine d tartrate is less soluble in water than the other salt. By repeated crystallizations (15–25 are required in

dl propylenediamine
        + $\longrightarrow$ d propylenediamine d tartrate
2 d tartaric acid l propylenediamine d tartrate   [17-11]

some cases) it can be separated from the d propylenediamine *d* tartrate. A strong base such as NaOH can be used to liberate the resolved *l* or *d* propylenediamine from the appropriate salt.

17-11. Optical Activity Without Asymmetric Atoms

In the definition of optical activity (p. 337) no mention was made of the necessity for an asymmetric atom in compounds exhibiting the phenomenon. The only criterion for optical activity in a compound is the nonexistence of a plane of symmetry in the molecule.

Some compounds without an asymmetric atom exhibit optical activity because the entire molecule, rather than a single atom, is asymmetrical. Examples are the allenes, the spiranes, and inositol.

1. Allenes. An optically active allene is an alkadiene having the general formula shown in the diagram, in which *a* and *b* are

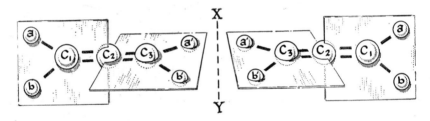

any two different groups. The asymmetry is due to the two double bonds on the same carbon, whereby, in the left half of the diagram, the atoms to the left of C_2 are in a plane perpendicular to the plane of the atoms to the right of C_2, and to the impossibility of rotation about a double bond. The number of forms of an allene is two, and their character is that of one *dl* pair.

The first allene to be resolved into its *d* and *l* forms was the compound

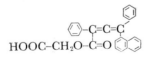

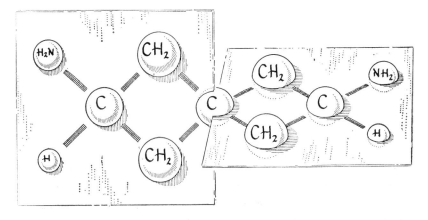

2. Spiranes.   A spirane may be considered an allene in which one or both double bonds are replaced by rings. The two rings are in planes perpendicular to each other and hence give rise to optical isomerism without an asymmetric atom. Spiranes also give rise to just one *dl* pair of isomers. The diaminospirocycloheptane shown here has been resolved.

3. Inositol.   The inositol type of compound is any 1,2,3,4,5,6-hexa-substituted cyclohexane. The only one in which the optically active forms are known is inositol itself, which is hexahydroxy-

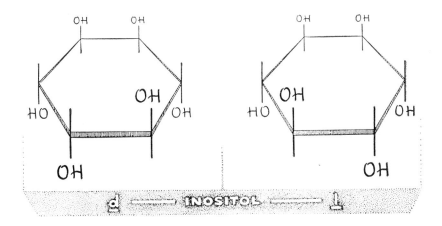

cyclohexane. Besides the *dl* pair illustrated, there are seven *cis-trans* isomers (p. 352) through each of which at least one plane of symmetry may be passed. Inositol occurs widely in nature, usually associated with fats, in both plants and animals. Soybean oil contains a rather high concentration of inositol.

17-12. The Walden Inversion

In a number of reactions of optically active compounds the direction of rotation of the plane of polarized light is inverted. Although this does not necessitate an inversion of configuration, a number of reactions are known in which the cycle may be completed; that is, the original compound with opposite rotation may be obtained by a second reaction. If *l* malic acid is treated with

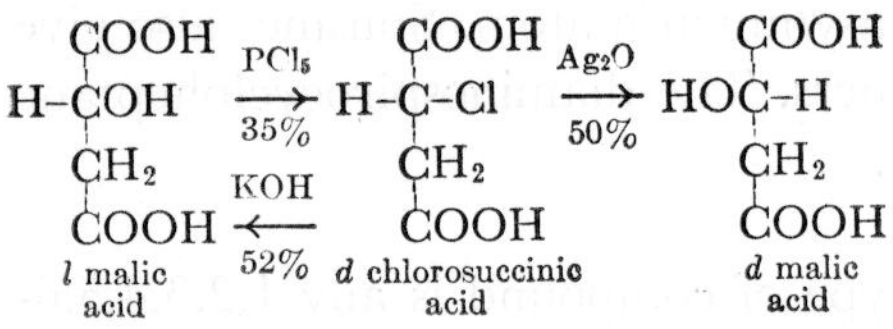

$$
\begin{array}{ccccc}
\text{COOH} & \xrightarrow{PCl_5} & \text{COOH} & \xrightarrow{Ag_2O} & \text{COOH} \\
\text{H-COH} & \xrightarrow{35\%} & \text{H-C-Cl} & \xrightarrow{50\%} & \text{HOC-H} \\
\text{CH}_2 & \text{KOH} & \text{CH}_2 & & \text{CH}_2 \\
\text{COOH} & \xleftarrow{52\%} & \text{COOH} & & \text{COOH} \\
l \text{ malic} & & d \text{ chlorosuccinic} & & d \text{ malic} \\
\text{acid} & & \text{acid} & & \text{acid}
\end{array}
$$

[17-12]

PCl_5, *d* chlorosuccinic acid is obtained, and Ag_2O, acting on this compound, will yield *d* malic acid. Of necessity, an inversion in configuration has taken place in one of the two reactions. The first such inversion was observed by Paul Walden in 1893.

Many Walden inversions are known in molecules of octahedral configuration, in which cobalt, for example, rather than carbon, is the central atom. At least one inversion is known in which no carbon atoms appear at all; so the phenomenon is not restricted to carbon chemistry. Accomplishment of a Walden inversion aided in the acceptance of the tetrahedral theory (see p. 381).

The mechanism now accepted for a Walden inversion is illustrated in the accompanying sketch. If we assume that the inversion (Exp. 17-12) occurs in the first step, then, according to the theory, PCl_5 must have approached the *l* malic acid molecule at the face

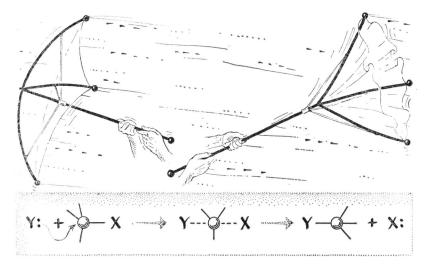

opposite the OH group. Then the molecule turned inside out
(inverted), like an umbrella in a wind, to give the new compound,
d chlorosuccinic acid. At some point in the reaction, the molecule
must have assumed a planar structure (transition state). This
theory requires the revision of the picture of a rigid molecule that
a ball-and-stick model suggests.

17-13. Asymmetry in Nature

Man's synthetic methods of producing organic compounds always result in racemic modifications (p. 342) when optically inactive starting compounds are used. In nature, in contrast, synthesis of compounds capable of optical activity usually results in
optically active forms. Little progress has been made in discovering
a reason for this or how the synthesis takes place. Very often in
metabolic processes one form of an optically active pair is destroyed
while the other is untouched; frequently one form is much more
active biologically than the other. Enzyme systems are often so
specific in their catalytic activity that it takes different enzymes
to act on the d and l forms.

The importance of stereochemistry will be evident from the following facts: The sugars (Chap. 19) that occur in nature, as well as starch and cellulose, are optically active substances. The amino acids (Chap. 20) found in proteins all occur as optically active forms except the one that contains no asymmetric atom. The study of stereochemistry is essential to an understanding of the chemistry of vitamins such as ascorbic acid, hormones such as thyroxine and cortisone, alkaloids such as nicotine and strychnine, and antibiotics such as penicillin.

17-14. Geometric Isomerism

If it is assumed, for the compound 1,2-dibromoethane, that there is free rotation about the single bond joining the carbons, then

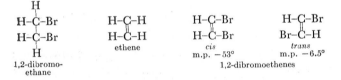

there is only one compound having that name. If a mechanical stick-and-spring model is made of 1,2-dibromoethene, the double bond gives rigidity to the structure, and it is not possible to rotate the groups about the double bond. Two different models can be made for the compound. In the laboratory, two compounds hav-

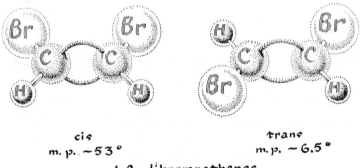

cis
m. p. ~53°

trans
m.p. ~6.5°

1, 2-dibromoethenes

ing this same formula but differing in several properties can be isolated. This example and others have given rise to the belief that there is restricted rotation about the double bond and that the spring-and-stick model gives an accurate picture, in this regard, of this type of isomerism.

A compound must have two features before it can exhibit geometric isomerism: (1) a double bond in the molecule to restrict rotation at a carbon-carbon or carbon-nitrogen bond or some other structural feature that will restrict rotation about a bond; (2) two different groups on each of the atoms at the restriction. This means, for example, that there is only one form of ethene, or of isobutylene, since each carbon carries like groups, or of 1-butene, since one of the carbons carries like groups, but two forms of 2-butene.

$$CH_3-\overset{|}{\underset{|}{C}}-CH_3$$
$$H-C-H$$
isobutylene

$$CH_3-CH_2-\overset{|}{\underset{|}{C}}-H$$
$$H-C-H$$
1-butene

$$CH_3-\overset{|}{C}-H$$
$$CH_3-C-H$$
cis

$$CH_3-\overset{|}{C}-H$$
$$H-C-CH_3$$
trans

2-butene

The phrase "some other structural feature that will restrict rotation," in the definition of geometric isomerism, commonly

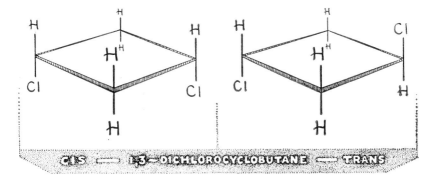

refers to a ring structure, which, of course, restricts complete rotation about *single bonds*. For example, seven of the isomers of inositol (p. 349) are not optical isomers but are geometric isomers. In 1,3-dichlorocyclobutane, two geometric isomers are possible.

As the definitions show, geometric isomerism and optical isomerism are not in any way related though both may occur in the same molecule.

Geometric isomers are distinguished by the names *cis* and *trans*. For a pair of geometric isomers that have at least one like group on the double-bonded carbons, the compound called the *cis* form has the like groups near each other. When the like groups are across from each other, the compound is called the *trans* form. If all four groups on the two carbons are different, it is necessary to indicate which groups are *cis* or *trans* to each other.

17-15. Properties of Geometric Isomers

In general, the *cis* isomers have lower melting points, are less stable, and, if a functional group is present, are more soluble in water, than the *trans*. This difference in physical properties is shown in Table 17-3 for two pairs of *cis-trans* isomers.

TABLE

17-3 | *Comparison of* Cis *and* Trans *Isomerism*

	CH_3-C-H $\parallel$ $HOOC-C-H$ *cis*	CH_3-C-H $\parallel$ $H-C-COOH$ *trans*	$H-C-COOH$ $\parallel$ $H-C-COOH$ *cis*	$H-C-COOH$ $\parallel$ $HOOC-C-H$ *trans*
Name	isocrotonic acid	crotonic acid	maleic acid	fumaric acid
M.P. °C	15	72	130	200 (subl.)
Sol. in 100 g 　H_2O at 25°	40	8.3	79	0.7
Conc. of H^+ 　in 0.1M 　solution	——	.001	0.034	0.009

The *cis* form of a compound can frequently be changed into the more stable *trans* form merely by the application of heat. The reverse reaction, which requires absorption of energy, has been accomplished in a few cases by ultraviolet radiation.

$$\begin{array}{c} \text{H–C–COOH} \\ | \\ \text{H–C–COOH} \end{array} \underset{\text{ultraviolet}}{\overset{250°}{\rightleftarrows}} \begin{array}{c} \text{H–C–COOH} \\ | \\ \text{HOOC–C–H} \end{array} \qquad [17\text{-}13]$$

17-16. Determination of Configuration

No absolute method of determining *cis* and *trans* configuration in a pair of such isomers can be applied in every case. Two chemical methods have been used. Other methods depend on relating such physical properties as melting points, solubility, dipole moments, interatomic distances, and densities to configuration.

1. Cyclization. Van't Hoff (1875) first distinguished maleic and fumaric acids by the greater ease of converting the former into an anhydride. Maleic acid loses a molecule of water to give an anhydride at 100° and reduced pressure, but it takes a temperature of 250° and a strong dehydrating agent (P_2O_5) to convert fumaric acid into the same anhydride. The much milder condition

$$\begin{array}{c} \text{H–C–COOH} \\ | \\ \text{H–C–COOH} \end{array} \underset{\text{cold H}_2\text{O}}{\overset{100°\ \text{red. pressure}}{\rightleftarrows}} \begin{array}{c} \text{H–C–C} \\ \ \ \ \ \ \ \| \ \ \ \ O \\ \text{H–C–C} \end{array}^{O} \overset{P_2O_5}{\longleftarrow} \begin{array}{c} \text{HOOC–C–H} \\ | \\ \text{H–C–COOH} \end{array} \qquad [17\text{-}14]$$

for the conversion of the maleic acid suggested that it was the *cis* isomer. It is rational to assume that the *trans* isomer would require the more drastic conditions since a rearrangement is necessary. The slow hydrolysis of the anhydride with *cold* water back to maleic acid is also convincing evidence since a rearrangement under these conditions is extremely unlikely.

Going from a cyclic compound to a *cis* isomer is sometimes possible. The catalytic oxidation of benzene is the commercial method of making maleic acid.

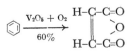

$$\frac{V_2O_5 + O_2}{60\%}$$ H–C–C=O
                                    \O   [17-15]
                                 H–C–C=O

2. Conversion into Known Structures. Once the configuration of a particular compound has been established, isomers of unknown structure may be converted into it by paths known to result in no rearrangement. Trichlorocrotonic acid can be hydrolyzed easily to fumaric acid and hence bears the *trans* configuration. By reduction of *trans*-trichlorocrotonic acid with sodium amalgam in neutral solution, *trans*-crotonic acid, of melting point 72°C, is obtained. Liquid crotonic acid (called isocrotonic acid), of melting point 15°C and boiling point 163°C, is therefore designated as the *cis* isomer.

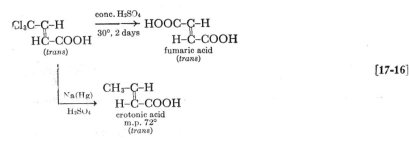

Cl₃C–C–H   conc. H₂SO₄   HOOC–C–H
  HC–COOH   ————————→   H–C–COOH
  (trans)   30°, 2 days   fumaric acid
                                   (trans)
                                                            [17-16]
       |
       | Na(Hg)   CH₃–C–H
       |————————→   H–C–COOH
       H₂SO₄   crotonic acid
                      m.p. 72°
                      (trans)

Though the configurations of the compounds cited in the last two paragraphs are reasonably certain, one cannot look upon the evidence with the same satisfaction as upon the Körner absolute method (p. 375). There is no certain proof that the reactions cited are free of rearrangements. One may say that the conditions under which the reactions are carried out have not been accompanied by rearrangements in cases where that possibility could be tested. Hence it is likely (but not certain) that in these cases, also, rearrangements do not occur.

SUMMARY

Extension of the subject of isomerism to closer inspection in three dimensions requires a number of definitions of new terms: optical isomerism, dextro form, meso form, racemic mixture, asymmetric atom, asymmetric molecule, racemization, resolution, and geometric isomerism. The phenomena and significance of tautomerism and the Walden inversion are also discussed. Finally, with the study of isomerism, we can begin to lay the foundation for the fundamental assumptions that we made at the beginning of the course—the tetravalent and tetrahedral nature of carbon—and for the evolution of ideas leading to the hypothesis that atoms are joined together in molecules.

REFERENCES

1. Bent, "Aspects of Isomerism and Mesomerism," J. Chem. Education, 30, 220, 284, 328 (1953).
2. Senior, "An Evaluation of the Structural Theory of Organic Chemistry," J. Chem. Education, 12, 409, 465 (1935).
3. Senior, "On Certain Chemical Definitions," J. Chem. Education, 13, 508 (1936).
4. Senior, "On Certain Relations Between Chemistry and Geometry," J. Chem. Education, 15, 464 (1938).
5. Goldfarb and Smorgousky, "Introduction to Isomerism and Structural Theory," J. Chem. Education, 13, 22 (1936).

EXERCISES

(Exercises 1–2 will be found within the text.)

3. Which of the following may exist in stereoisomeric forms? How many stereoisomeric forms of each would you expect?

   1. $CH_3-CHOH-CH_2-CH_3$

2. $CH_3-CH_2-CHBr-CH_2-CH_3$
3. $CH_3-CHBr-CHBr-CH_3$
4. $CH_3-CHCl-CHBr-COOH$
5. $(CH_3)_2CBr-COOH$
6. $CH_3-CH_2-CH(CH_3)-CH(CH_3)_2$
7. $CH_3-CHOH-CHOH-CHOH-COOH$

4. Write a formula for a five-carbon primary amine that is optically active and a formula for one that is not.

5. Write formulas for all isomeric pentanols that are optically active.

6. Write the formula for a monochlorocarboxylic acid that can exist in two stereoisomeric forms.

7. Devise a means of resolving the optically active amine shown below and of recovering both the dextro and the levo form.

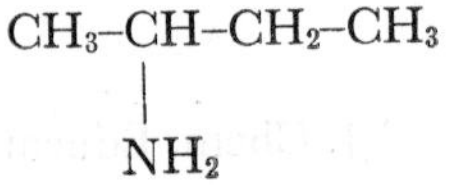

$$CH_3-CH-CH_2-CH_3$$
$$|$$
$$NH_2$$

8. From only the following information, what conclusion can be drawn regarding the number of times Walden inversion occurs and at what point?

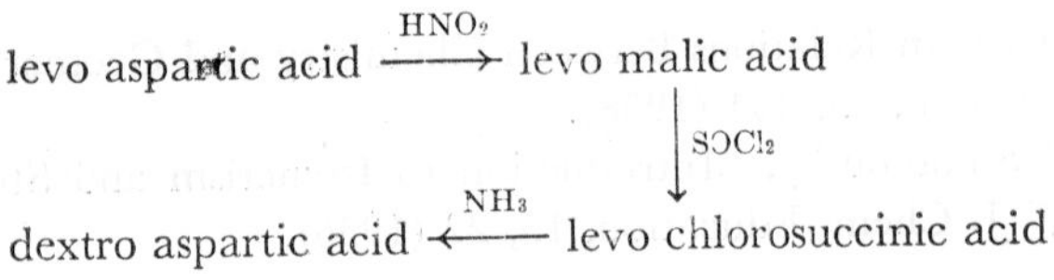

levo aspartic acid $\xrightarrow{\text{HNO}_2}$ levo malic acid

$\downarrow \text{SOCl}_2$

dextro aspartic acid $\xleftarrow{\text{NH}_3}$ levo chlorosuccinic acid

9. Levo lactic acid is esterified to form dextro ethyl lactate. Hydrolysis produces the levo acid. Is this evidence for a Walden inversion?

10. Which of the following may exist in stereoisomeric forms? How many stereoisomeric forms of each would you expect?

1. $CH_2=CH-CH_2-CH_3$ 4. $CH_3-CH=CH-COOH$
2. $CH_3-CH=CH-CH_3$ 5. $CH_3-CBr=CBr-CH_3$
3. $CH_3-CH=CH-CH_2-CH_3$ 6. $CH_3-CH=CBr_2$

11. Name the type of isomerism (if any) to be expected of the following substances, and write formulas for the isomeric forms:

1. $C_6H_5-\underset{\underset{O}{\|}}{C}-CH_2-\underset{\underset{O}{\|}}{C}-C_6H_5$

2. $C_6H_5-CHOH-CH_2-C_6H_5$
3. 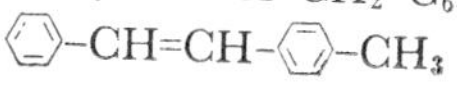 ⬡$-CH=CH-$⬡$-CH_3$

12. Write structural formulas for two substances of formula $C_6H_{10}O$ that are

1. functional isomers
2. optical isomers
3. geometric isomers

13. Write formulas to indicate the configuration of all stereoisomeric forms of

$$C_6H_5-CH=CH-\underset{\underset{CH_3}{|}}{CH}-COOH$$

14. What are the number and the character of stereoisomers of the following compounds?

1. CH_3-CHNH_2-COOH
2. $CH_3-CHOH-CHOH-CH_3$
3. $CH_3-CHOH-CHOH-CH_2-CH_3$
4. $HO-\underset{\underset{CH_2-COOH}{|}}{\overset{\overset{CH_2-COOH}{|}}{C}}-COOH$
5. cyclohexanol
6. $CH_3-CH=CH-CH_2-CHOH-COOH$
7. $CH_2OH-CHOH-CHOH-CH_2OH$
8. $CHO-CHOH-CHOH-CHOH-CH_2OH$
9. $CH_3-\underset{\underset{N-OH}{\|}}{C}-C_6H_5$
10. $C_6H_5-CHNH_2-COOH$

18 Evolution of the Covalent Bond

On the radio program called Twenty Questions, all things are classified as animal, mineral, or vegetable. Guessing games of various kinds are based on the same classification. This is a carry-over from the seventeenth century, when, in textbooks in general science, all chemical substances were treated under three headings. The mineral world was inanimate, and the belief was prevalent that only a vital force could synthesize any substance existing in the other two worlds. Consequently, chemists busied themselves with minerals, which, they felt, were unchanging and could be relied on to behave in a predictable way.

The mental barriers erected on this classification and on the "vital force" theory undoubtedly delayed work in science through the early nineteenth century. But men have always erected walls that more courageous men have been willing to climb over, to walk around, or to burrow through. Lavoisier, in France, was one of these venturesome and courageous men. He demonstrated that the products of all life processes are composed mainly of carbon, hydrogen, and oxygen and less frequently contain nitrogen, sulfur, phosphorus, and other elements. He developed methods of analyzing organic compounds so that their composition could be fairly well established, a task that previously had been considered futile. The work of Berzelius on accurate determinations of atomic weights helped immeasurably. It is well to recall that the first suggestion that substances might well be represented by formulas (Berzelius) was made about 1810. The atomic theory and Avogadro's hypothesis were not generally accepted until about 1860.

A large number of organic compounds were known before 1800, however, and some of their properties had been studied: alcohol (12th century), ether (1544), methyl alcohol and acetone (1661),

360

benzoic acid (1608), glycerol (1783), fats, esters, hydrogen cyanide, and oxalic, tartaric, and other acids.

18-1. Birth of Organic Chemistry (1828)

Lavoisier accounted for the nature of the acidic substances among organic compounds by supposing that oxygen (the acid-forming element) was combined with an organic "compound radical." This idea grew in popularity until organic chemistry was called the chemistry of compound radicals. Gay-Lussac showed that cyanogen behaves in many ways like the element chlorine; oxalic acid was looked upon as containing the compound radicals CO and CO_2; alcohol was ethylene and water. All of these led Berzelius to say that organic substances are composed in the same way as inorganic except that radicals take the place of elements.

The search for compound radicals in organic substances led to the discovery of isomerism, first looked upon as an error since it seemed incredible that two substances could have the same analysis. Indeed, the very existence of the analytical method was based on the postulate that every substance had its own unique percentage composition. Wöhler, in 1822, reported an analysis for silver cyanate; in 1823, when Liebig found the same analysis for silver fulminate, he pointed out that Wöhler must have made an error. Liebig then repeated both analyses and found them both correct. In 1825 Faraday found two hydrocarbons (now called α-butylene and ethylene) that had the same composition, CH_2. The only explanation was that the elements had different arrangements in

			% C	% H
ethylene:	$CH_2=CH_2$	C_2H_4	85.7	14.3
α-butylene:	$CH_3-CH_2-CH=CH_2$	C_4H_8	85.7	14.3

each of those two pairs of substances. (It is now known also that there are different numbers of carbons and hydrogens in the two alkenes.) Berzelius did not immediately accept the facts or the explanation but later himself found identical compositions for racemic tartaric acid and dextro tartaric acid and suggested the name "isomerism" for this phenomenon.

The barrier between the animal and vegetable worlds began to fall when Chevreul discovered that some acids and fats can be isolated from either source. At this time the known organic substances were divided into three classes: acids, which contained more oxygen than would form water with the hydrogen present; carbohydrates, which contained just enough oxygen to form water with the hydrogen present; and oils, resins, or alcoholic substances, which contained less oxygen than would combine with the hydrogen. Hydrocarbons were ignored in the classification, and in other respects also the classification is not acceptable at the present time.

The last mental barrier to be removed before all substances could be considered as obeying the laws of one chemistry was destroyed by Wöhler's fortuitous synthesis of urea, a substance believed to come only from life processes. By warming potassium cyanate with ammonium sulfate (purely inorganic substances), Wöhler expected to get ammonium cyanate upon evaporation of the solution but instead obtained urea and potassium sulfate. A synthesis for shattering the doctrine of vital force could hardly have been better chosen, for urea, containing the waste nitrogen from metabolic processes, was one of the best-known of animal products.

18-2. The Radical Theory

In the year of Wöhler's synthesis, Dumas suggested that ethylene was a radical in the same sense with which the term was used in inorganic chemistry. He pointed out advantageous analogies between ammonia and its compounds and ethylene and its addition products. Some formulas illustrating the analogies are given in Table 18-1. (The original formulas of Dumas, using atomic weights C = 6 and O = 16, are translated into corrected formulas.) The radicals were thought to be capable of independent existence and, indeed, to exist as such in each of the compounds cited. Ethylene was considered a base that formed hydrates (alcohol) and salts (ethers and esters). Dumas at one time said, "Ethylene would turn litmus blue if it were only soluble in water"!

TABLE
18-1   *Dumas Formulas*

Substance	Modern Formula	Dumas Formula (C = 12, O = 16)	NH₃ Analogue
ethylene	$CH_2=CH_2$	C_2H_4	NH_3
ethyl bromide	CH_3-CH_2Br	$C_2H_4 \cdot HBr$	NH_4Br
alcohol	CH_3-CH_2OH	$C_2H_4 \cdot H_2O$	NH_4OH
ether	$C_2H_5-O-C_2H_5$	$2C_2H_4 \cdot H_2O$	[ammonium oxide]
ethyl acetate	$CH_3-COO-C_2H_5$	$2C_2H_4 \cdot C_4H_6O_3 \cdot H_2O$	CH_3COONH_4
ethyl oxalate	$(COOC_2H_5)_2$	$2C_2H_4 \cdot 2CO_2 \cdot H_2O$	$(COONH_4)_2$
ethylsulfuric acid	$C_2H_5OSO_3H$	$C_2H_4 \cdot SO_3 \cdot H_2O$	NH_4OSO_3H

Soon after Dumas' announcement, Liebig and Wöhler (1832) published a paper on the chemistry of oil of bitter almonds (benzaldehyde). They were able to show that the benzoyl radical remained intact in a number of transformations: benzaldehyde = benzoyl + hydrogen; benzoic acid = benzoyl + oxygen; benzoyl chloride = benzoyl + chlorine; benzamide = benzoyl + ammonia; ethyl benzoate = benzoyl + alcohol. (Benzoyl was written $C_{14}H_{10}O_2$, double the present formula, C_6H_5-CO-.)

Berzelius was at first pleased with this bold achievement, but he believed that no radical could contain oxygen. Lavoisier had defined a radical as a group of elements that behaves like a single element and unites with oxygen to form an acid. Berzelius reconciled the new facts with the old definition by regarding benzoyl as an oxide of the true radical, C_7H_5, and benzoic acid as a higher oxide.

By 1838, the radical theory had been modified far enough for Liebig to list two essential characteristics of a compound radical: (1) it is an unchanging constituent in a series of compounds; (2) it can be replaced by other simple bodies in these compounds.

18-3. Dualism and Substitution

Throughout this period of the radical theory, Berzelius maintained that the formation of a compound was due to the uniting of an electropositive substance (ordinarily a metal or a base) with an electronegative substance (oxygen or an acid). This dualistic nature of substances was extended to the following situation: A metal might combine with oxygen to form an oxide, but this would not necessarily completely overcome the electropositive character of the metal. This oxide could then react further with an acid to form a salt. Likewise, sulfur, for example, was electropositive and could unite with oxygen, the most electronegative of all the elements, but the acid formed by the two (sulfuric) was still capable of reacting with metals or oxides of metals. Berzelius used his theory also in accounting for the behavior of organic compounds and was so fond of it that he clung to it even after Dumas had struck it telling blows.

The death blow came to dualism with the discovery that chlorine (a strongly electronegative substance) can substitute for hydrogen (an electropositive element) without drastically changing the character of the product. The discovery is linked with royalty by interesting circumstances. Wax candles used in the palace of the Tuileries gave off an offensive odor during a royal function (1834), and Dumas as a chemist was asked to find out why. The wax had been bleached by chlorine gas, and he found that some had combined with the wax. Further investigation by Dumas revealed that chlorine would replace hydrogen volume for volume in a number of substances. Acetic acid was chlorinated to trichloroacetic, and the product was found still to have many of the properties of vinegar. This astounded Berzelius, who never quite gave up the dualistic theory. His theory, of course, held that an element could have only one character, either electronegative or electropositive. Dumas was somewhat carried away by his theory of substitution and went so far as to suggest that every element (even carbon) might be replaced by some other without alteration in the properties of the compound. Liebig violently opposed this view, and Wöhler laughed it to scorn. In a letter to Berzelius [published in Liebig's *Annalen*, 33, 308 (1840)]* Wöhler wrote a skit in French ridiculing Dumas' theory and using the pseudonym S. C. H. Windler (swindler). Wöhler later wrote to Liebig that he had no idea the skit might be printed and that the least Liebig could have done, as editor of the journal, was to substitute a French pseudonym such as Ch. Arlatan.

18-4. Theory of Types

By 1840, Dumas regarded the dualistic theory as harmful to the development of organic chemistry and abandoned it entirely. He introduced the theory of types, which was developed by Laurent and Gerhardt. Molecules were no longer regarded as being composed always of two parts; instead, the entire compound was a unit, and its character depended primarily on the arrangement of atoms.

By 1850, Gerhardt recognized four types of carbon compounds as being analogous to simple inorganic molecules. Examples of each type are listed on the next page.

The type theory was given credibility by the work of Wurtz, Hofmann, and Williamson. Wurtz discovered the amines in 1848, and Hofmann prepared primary, secondary, and tertiary amines shortly thereafter by heating alkyl halides with ammonia. The amines resembled ammonia quite closely in properties, and this strengthened the type theory.

Williamson's work on ethers (1850) brings up the type of experimental evidence that may be called chemical proof. At this time most chemists were writing alcohol, sodium ethoxide, and ether with the following formulas ($C = 6$, $O = 16$): (C_4H_5O,HO), (C_4H_5O,NaO), (C_4H_5O). Using formulas obtained from vapor

* Liebig's *Annalen der Chemie* was one of the first journals publishing the results of chemical research.

HYDROGEN TYPE	HYDROCHLORIC ACID TYPE	WATER TYPE	AMMONIA TYPE
$\left.\begin{array}{l} H \\ H \end{array}\right\}$	$\left.\begin{array}{l} H \\ Cl \end{array}\right\}$	$\left.\begin{array}{l} H \\ H \end{array}\right\} O$	$\left.\begin{array}{l} H \\ H \\ H \end{array}\right\} N$
$\left.\begin{array}{l} C_2H_5 \\ H \end{array}\right\}$	$\left.\begin{array}{l} C_2H_5 \\ Cl \end{array}\right\}$	$\left.\begin{array}{l} C_2H_5 \\ H \end{array}\right\} O$	$\left.\begin{array}{l} C_2H_5 \\ H \\ H \end{array}\right\} N$
$\left.\begin{array}{l} C_2H_5 \\ C_2H_5 \end{array}\right\}$	——	$\left.\begin{array}{l} C_2H_5 \\ C_2H_5 \end{array}\right\} O$	$\left.\begin{array}{l} C_2H_5 \\ C_2H_5 \\ H \end{array}\right\} N$
			$\left.\begin{array}{l} C_2H_5 \\ C_2H_5 \\ C_2H_5 \end{array}\right\} N$
$\left.\begin{array}{l} C_2H_3O \\ H \end{array}\right\}$	$\left.\begin{array}{l} C_2H_3O \\ Cl \end{array}\right\}$	$\left.\begin{array}{l} C_2H_3O \\ H \end{array}\right\} O$	$\left.\begin{array}{l} C_2H_3O \\ H \\ H \end{array}\right\} N$
acetaldehyde	acetyl chloride	acetic acid	acetamide

density measurements, Laurent and Gerhardt wrote these three compounds by the type formulas as follows:

$$\left.\begin{array}{l} C_2H_5 \\ H \end{array}\right\} O \qquad \left.\begin{array}{l} C_2H_5 \\ Na \end{array}\right\} O \qquad \left.\begin{array}{l} C_2H_5 \\ C_2H_5 \end{array}\right\} O$$

Guided by the experiments of Hofmann, who had obtained substituted ammonias by treating alkyl halides with ammonia, Williamson expected to prepare a substituted alcohol by treating potassium ethoxide with ethyl iodide. Instead, he obtained ether, and he saw that Gerhardt's formulas would account for the result:

$$\left.\begin{array}{l} C_2H_5 \\ K \end{array}\right\} O + \left.\begin{array}{l} C_2H_5 \\ I \end{array}\right\} \longrightarrow KI + \left.\begin{array}{l} C_2H_5 \\ C_2H_5 \end{array}\right\} O \qquad [18\text{-}1]$$

On the basis of the old formulas (atomic weights: $C = 6$, $O = 16$) it was still possible to give an explanation if it was assumed that potassium ethoxide first decomposed to potassium

$$C_4H_5O,KO \longrightarrow KO + C_4H_5O \qquad\qquad [18\text{-}2]$$

$$C_4H_5I + KO \longrightarrow KI + C_4H_5O \qquad\qquad [18\text{-}3]$$

oxide and ether and that the oxide then reacted with ethyl iodide to give a second molecule of ether. Williamson clinched his argument by using methyl iodide rather than ethyl iodide. The result was the mixed ether, methyl ethyl ether, rather than an equal mixture of dimethyl and diethyl ethers. The result could readily

$$\left.\begin{matrix}C_2H_5\\K\end{matrix}\right\}O + \left.\begin{matrix}CH_3\\I\end{matrix}\right\} \longrightarrow \left.\begin{matrix}C_2H_5\\CH_3\end{matrix}\right\}O + KI \qquad\qquad [18\text{-}4]$$

be accounted for with Gerhardt's formulation. This is a proof based on experiment, and it gave great impetus to the acceptance of the type theory.

So far nothing has been said about the arrangement of atoms in the organic molecule. Gerhardt thought that this problem could never be solved. Formulas, for him, represented methods of formation or decomposition and could not be refined further. He even considered that a compound could have several different formulas to show its behavior in different reactions. (Consider the present view of tautomerism, p. 334.) The theory of types, however, did much to classify the large number of organic substances into a few categories.

18-5. Bonds Within Molecules

It was not until 1858 that any chemist considered that atoms were bonded to one another in a molecule. This original and startling idea (independently proposed by Kekule and Couper) cleared the way for the general acceptance of Avogadro's hypothesis and finally made the distinction between an atom and a molecule possible. Although the theory may appear to have come from nowhere, a close examination of the literature will reveal germs of the idea in several places before the theory actually appeared. It often happens, when the time is right, that a scientific theory is proposed in more than one laboratory.* This is not to detract from the man who first put the new theory into words. [For a longer account and an appreciation of Kekule, see F. R. Japp, "Kekule Memorial Lecture," J. Chem. Soc., 73, 98–138 (1898).]

Kekule was influenced in his thinking by the work of Liebig, Dumas, Gerhardt, and Williamson. Kolbe and Frankland also made important contributions at this time; Kekule probably did not consider them, but they nevertheless fit into the total picture.

Kekule's idea that carbon was tetravalent came at the conclusion of a study of methane and related hydrocarbons. In going through the series from methane to carbon tetrachloride, Kekule

$$CH_4 \quad CH_3Cl \quad CH_2Cl_2 \quad CHCl_3 \quad CCl_4$$

found that the hydrocarbon residue, the group that is left after each displacement, would combine with one more equivalent of chlorine. After considering the other radicals of the methane homologues that have the general formula C_nH_{2n+1}, he boldly suggested that, when several carbon atoms occur in a compound radical, *they are connected to one another*. This was one of the first properties mentioned in this course as characteristic of carbon atoms and is now considered axiomatic.

Kolbe's work on the electrolysis of acids and Frankland's work on organometallic compounds had preceded this theory. Kolbe electrolyzed solutions of salts of carboxylic acids and obtained carbon dioxide and aliphatic hydrocarbons at the positive pole. This could be interpreted as indicating some bonding between the two parts of the original acid although Kolbe did not so interpret it. Kolbe and Frankland both hydrolyzed acetonitrile to acetic acid. Gerhardt wrote the product as a molecule of the water type, $\left.\begin{array}{c} CH_3CO \\ H \end{array}\right\}O.$

Frankland found that a metal attached to an alkyl group would

---

* For this reason there is no assurance that scientific secrets ("atomic secrets," for example) can be kept very long by one country.

take up only half as much oxygen as the metal alone would take up. For example, magnesium forms the oxide MgO, but, if R is attached to Mg, two R-Mg radicals will be present in a corresponding oxide, (R-Mg)$_2$O. He was also impressed by the symmetry of construction of inorganic compounds [*Annalen*, 85, 368 (1853)]. He wrote: "When the formulae of inorganic chemical compounds are considered, even a superficial observer is struck with the general symmetry of their construction; the compounds of nitrogen, phosphorus, antimony, and arsenic especially exhibit the tendency of these elements to form compounds containing 3 or 5 equivalents of other elements, and it is in these proportions that their affinities are best satisfied; thus in the 1:3 ratio, we have NO$_3$, NH$_3$, NS$_3$; PO$_3$, PH$_3$, PCl$_3$; SbO$_3$, SbH$_3$, SbCl$_3$; AsO$_3$, AsH$_3$, AsCl$_3$, etc., and in the 1:5 ratio NO$_5$, NH$_4$O, NH$_4$I; PO$_5$, PH$_4$I, etc. Without offering any hypothesis regarding the cause of this symmetrical grouping of atoms it is sufficiently evident from the examples just given, that such a law or tendency prevails, and that, no matter what the character of the uniting atoms may be, the *combining power* of the attracting element is always satisfied by the same number of atoms."

The concept of graphic formulas followed immediately after the postulation of joined atoms. Kekule wrote ethyl alcohol in his text-

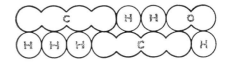

KEKULE FORMULA FOR ETHYL
ALCOHOL (1861)

book (1861) with circles to represent bonds for the atoms: ④ for C, ② for O, and ① for H. Couper, who was not inhibited by the type theory, as Kekule was, in 1858 had pictured compounds by structural formulas that are essentially the same as those now in use.

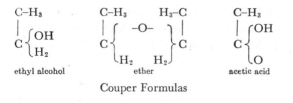

ethyl alcohol   ether   acetic acid

Couper Formulas

Simultaneously Butlerov in Russia was making contributions in the same direction as Kekule and Couper. Butlerov apparently was the first to use the term "chemical structure," and he suggested that a single chemical formula could show how atoms were joined together. His is the first suggestion that the arrangement of atoms in a molecule determined the properties of the compound. He furthermore suggested that the arrangement of atoms in a molecule (other than valence bonds) might be important in relating chemical properties to structure. One such relationship already discussed is Markownikoff's rule. (Markownikoff was a student of Butlerov.)

Probably Kekule, Couper, and Butlerov should share about equally the credit for the theory of bonding within molecules. They may even have influenced one another's thinking, for Butlerov knew Kekule (1857–58) and saw Couper in Paris (1857) in Wurtz's laboratory.

It must not be supposed that the theory of bonds within molecules gained immediate acceptance. As late as 1877, Kolbe published a paper decrying the use of formulas [J. prakt. Chem. (2), 15, 475 (1877)], and a little later he wrote [J. prakt. Chem. (2), 24, 418 (1881)]: "If anything is fitted to show what confusion structural chemistry has caused in the minds of chemists it is van't Hoff's recent publication, 'On the Arrangement of Atoms in Space.' I should not have mentioned this and the unbelievable chemical nonsense in it, if it had not originated under the auspices of Kekule."

18-6. Structure of Benzene

By 1860, then (if we forget Kolbe), a satisfactory graphic formula could be written for the known aliphatic compounds, but the aromatic compounds, which had a high ratio of carbon to hydrogen atoms, would not fit into the picture. Any theory of the structure of the aromatic compounds must account for three facts: (1) they never contain fewer than six carbons; (2) any decomposition short of complete destruction ends with the carbon content at six carbons; (3) homology exists in aromatic compounds as well as in the aliphatic series.

It was these facts for which Kekule sought an explanation in his "Structurtheorie" of the benzene ring. How the ideas of the tetravalent carbon atom, the aliphatic carbon chain, and the benzene ring came to him he related to the German Chemical Society in 1890 at a celebration called the "Kekulefeier." The following quotation is from the Journal of the Chemical Society, 73, 100 (1898). Compare Benfey, J. Chem. Education, 35, 21 (1958).

"During my stay in London I resided for a considerable time in Clapham Road in the neighbourhood of the Common. I frequently, however, spent my evenings with my friend Hugo Müller at Islington, at the opposite end of the giant town. We talked of many things, but oftenest of our beloved chemistry. One fine summer evening I was returning by the last omnibus, 'outside' as usual through the deserted streets of the metropolis, which are at other times so full of life. I fell into a reverie (*Träumerei*), and lo, the atoms were gambolling before my eyes! Whenever, hitherto, these diminutive beings had appeared to me, they had always been in motion; but up to that time I had never been able to discern the nature of their motion. Now, however, I saw how, frequently, two smaller atoms united to form a pair; how a larger one embraced two smaller ones; how still larger ones kept hold of three or even four of the smaller; whilst the whole kept whirling in a giddy dance. I saw how the larger ones formed a chain, dragging the smaller ones after them, but only at the ends of the chain. I saw what our Past Master, Kopp, my highly honoured teacher and friend, has depicted with such charm in his 'Molekularwelt'; but I saw it long before him. The cry of the conductor: 'Clapham Road,' awakened me from my dreaming; but I spent a part of the night in putting on paper at least sketches of these dream forms. This was the origin of the *Structurtheorie*."

Then he relates a similar experience that he had when the idea of the benzene ring occurred to him. This refers to a later period, when Kekule was a professor in Ghent, but may be quoted here in connection with the previous passage. He describes how he was at work one evening:

"I was sitting, writing at my text-book; but the work did not progress; my thoughts were elsewhere. I turned my chair to the fire and dozed. Again the atoms were gambolling before my eyes. This time the smaller groups kept modestly in the background. My mental eye, rendered more acute by repeated visions of the kind, could now distinguish larger structures, of manifold conformation: long rows, sometimes more closely fitted together; all twining and twisting in snake-like motion. But look: What was that? One of the snakes had seized hold of its own tail, and the form whirled mockingly before my eyes. As if by a flash of lightning I awoke; and this time also I spent the rest of the night in working out the consequences of the hypothesis.

"Let us learn to dream, gentlemen," adds Kekule; "then perhaps we shall find the truth; . . . but let us beware of publishing our dreams before they have been put to the proof by the waking understanding."

The paper describing the consequences of Kekule's thinking on the structure of benzene is also worth reading and will be found in the following references: Bull. soc. chim, 3, 98–110 (1865), and Ann., 137, 129 (1866). The benzene ring is written thus in that first paper on the subject:

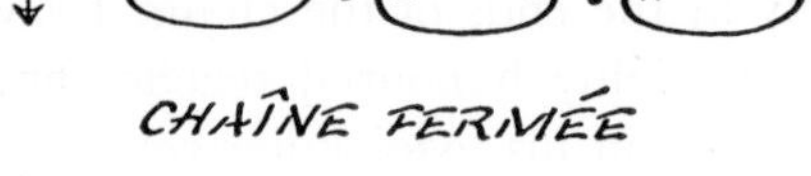

CHAÎNE FERMÉE

Kekule also had some fragmentary evidence from which he deduced that benzene formed one monosubstitution product and three disubstitution products. None of the evidence accumulated since that time disputes his conclusion. The six-membered ring fits the newer data and also the fact that there are just three tri-substitution products.

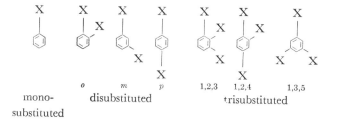

mono-substituted   disubstituted   trisubstituted

18-7. Ozonolysis as a Tool for Proof of Structure

It was mentioned on page 122 that benzene, in sunlight, can be made to add chlorine and give hexachlorocyclohexane. Ozone also adds to double bonds to give an ozonide that can be hydrolyzed to products allowing identification of the fragments. Ethylene, for example, gives an ozonide that decomposes into two molecules of

$$CH_2{=}CH_2 + O_3 \longrightarrow \underset{\substack{\text{ethylene}\\\text{ozonide}}}{CH_2{-}CH_2} \xrightarrow[SO_2]{H_2O} 2H{-}CHO + SO_3 \qquad [18\text{-}5]$$

formaldehyde when a mild reducing agent such as SO_2 is used during the hydrolysis to prevent further oxidation.

With benzene a triozonide can be prepared; this decomposes into three molecules of ethanedial (glyoxal).

$$\bigcirc + O_3 \longrightarrow \underset{O_3}{\overset{O_3}{\bigcirc}}O_3 \xrightarrow[SO_2]{H_2O} \begin{matrix} OCH \\ OCH \end{matrix} \quad \begin{matrix} CHO{\smallsetminus}CHO \\ CHO{\diagup}CHO \end{matrix} \qquad [18\text{-}6]$$

The ozonization and the addition of chlorine indicate the presence of three double bonds in the benzene molecule, but an early

objection to Kekule's structure was that there should be two di-
substituticn products, as shown in the diagram, although no more

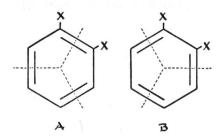

than one had ever been found. The suggestion was made that a
rapid oscillation of the double and single bonds between the two
possible structures would account for this. It was only in 1932 that
this was substantially supported by the identification of three different products from the ozonization of *o*-xylene. From formula B
one would expect two molecules of ethanedial and one of 2,3-
butanediene; from molecule A one would expect two molecules of
propanal-2-one and one of ethanedial. If A and B are present in
equal parts, the ratio ethanedial : propanal-2-one : 2,3-butanediene
should be 3:2:1. This was verified quantitatively in 1941.

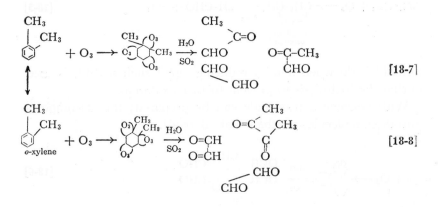

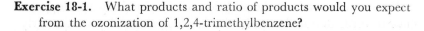

Exercise 18-1.   What products and ratio of products would you expect
from the ozonization of 1,2,4-trimethylbenzene?

18-8. Identification of *o*, *m*, *p* Isomers: Körner's Method

After three isomers of disubstitution products of benzene are found, how is it possible to determine which is which?

dibromobenzenes

It was one of Kekule's students, Körner, who saw how Kekule's theory could be subjected to experiment: by an attempt to identify the three possible disubstitution products of a hexagonal benzene structure. If each of the three compounds shown above is nitrated to a mononitro derivative, the *ortho*-dibromobenzene will yield two possible products; the *meta* compound, three products; and the *para* compound, only one. By careful manipulation Körner was able to identify six nitrodibromobenzenes from the three starting compounds. This enabled him to designate the structures of the three starting dibromobenzenes in an absolute manner.

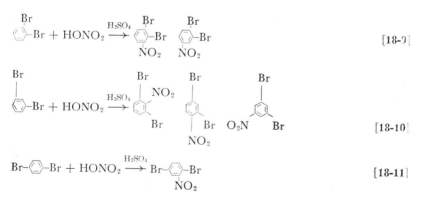

18-9. The Benzene Problem

Of the twelve structures listed on p. 376, the last seven (those illustrated) have been at one time or another considered seriously as possible answers to what may be called the benzene problem.

Exercise 18-2. Assuming that carbon always carries four bonds and the validity of the following observed data, eliminate structures 1–5 as possible competitors for the structure of benzene:

1. Only one compound C_6H_5Br has been prepared from benzene.
2. Only three compounds $C_6H_4Br_2$ have been prepared from benzene.

1. $CH_3–C≡C–C≡C–CH_3$
2. $CH_2=CH–CH=CH–C≡CH$
3. $CH_2=CH–C≡C–CH=CH_2$
4. $CH_3–CH_2–C≡C–C≡C–H$

5. CH———CH
   $\|$ $\|$
   CH CH
   $\diagdown \diagup$
   C
   $\|$
   CH_2

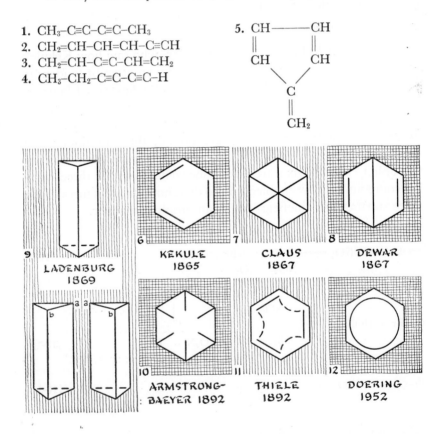

The Claus and Dewar structures require, respectively, three bonds and one bond that are double the length of the other carbon-to-carbon bonds in the molecule. If such bonds existed, they would of necessity be very much weaker than the other bonds. Since X-ray measurements have become available for research in carbon compounds, no C-C distances of comparable length have been found in benzene or any other aromatic compound.

The Ladenburg formula can be eliminated on stereochemical grounds. Disubstitution products in which the substituents are different and on adjacent carbons would be asymmetric, and hence there would be two optical isomers. No such pairs have ever been found. Finally, and decisively, X-ray diagrams indicate a planar configuration for benzene (all atoms lying in a single plane).

The Armstrong-Baeyer model suggested that the fourth bond of carbon was directed toward the center of the ring. Bonds are pairs of electrons, and it is now believed that there is no force to hold the pair at the center.

Thiele's theory of partial valence brought a new idea to the structure of benzene. A 1,3-conjugated system of double bonds adds 1,4 and leaves a double bond at the 2,3 position. Thiele suggested that all four of the carbons in a conjugated system had some tendency to form a double bond, or, in other words, that the double bond at 2,3 was always just ready to form. This idea he carried over to the completely conjugated system in benzene and suggested that each C-to-C bond carried a partial double-bond character. The resonance structures for the benzene ring given in phenol, phenoxide ion, and aniline (p. 384) are only a slight mod-

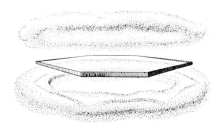

ification of the Kekule and Thiele formulas. Rather than having alternate single and double bonds in the ring, benzene is now believed to have six equivalent bonds. Neither Kekule formula is the structure of benzene, but each makes a contribution to the total structure, which cannot be written in a completely satisfactory manner with classical bonds. Doering has suggested formula 12, which is intended to depict six nonlocalized electrons in a circle superimposed on the hexagon. Formula 12 is a two-dimensional projection of the three-dimensional picture shown on page 377. The six electrons, in two clouds, make a sandwich of the hexagon. It is implied that the highest probability of finding an electron is in the clouds above and below the plane of the ring rather than inside the ring.

18-10. The Tetrahedral Carbon (1874)

The concept of the tetrahedral nature of carbon was introduced at the beginning of this course with dogmatic statements of fact but without proof. Now the contention that carbon is tetrahedral can be rigorously defended in a manner that can be called a chemical proof of structure.

At the time that Pasteur did his classical work on tartaric acid salts, it was, of course, known that carbon was tetravalent. Biot had observed that some quartz crystals turned the plane of polarized light to the right while others turned the plane to the left. The two types of crystals could also be distinguished by crystal faces that had the relations of right and left gloves. Pasteur found that sodium ammonium tartrate also crystallized out in large enough specimens so that under a microscope two types could be distinguished by a difference in hemihedral faces. Upon separating the two types of crystals, he found that their solutions rotated plane-polarized light in opposite directions. This suggested not only that the crystal was asymmetric but that the molecules in the solution must also be asymmetric. This was so startling that Biot, an experienced worker in the field, requested Pasteur to repeat the experiment before his eyes. Pasteur's own account of this part of the story is interesting enough to repeat:

"The announcement of the above facts naturally placed me in communication with Biot, who was not without doubts concerning their accuracy. Being charged with giving an account of them to

the Academy, he made me come to him and repeat before his eyes
the decisive experiment. He handed over to me some racemic
acid which he had himself previously studied with particular care,
and which he had found to be perfectly indifferent to polarized
light. I prepared the double salt in his presence, with soda and
ammonia which he had likewise desired to provide. The liquid was
set aside for slow evaporation in one of his rooms. When it had
furnished about 30 to 40 grams of crystals, he asked me to call at
the *Collège de France* in order to collect them and isolate before
him, by recognition of their crystallographic character, the right
and left crystals, requesting me to state once more whether I really
affirmed that the crystals which I should place at his right would
deviate to the right, and the others to the left. This done, he told
me that he would undertake the rest. He prepared the solution
with carefully measured quantities, and when ready to examine
them in the polarizing apparatus, he once more invited me to
come into his room. He first placed in the apparatus the more
interesting solution, that which ought to deviate to the left. Without even making a measurement, he saw by the appearance of the
tints of the two images, ordinary and extraordinary, in the analyser,
that there was a strong deviation to the left. Then, very visibly
affected, the illustrious old man took me by the arm and said, 'My
dear child, I have loved science so much all my life that this makes
my heart throb.' ''

Van't Hoff and Le Bel independently suggested that the four
equivalent bonds of carbon could be interpreted in a strictly mechanical way only if they were symmetrically arranged on the
surface of the carbon atoms. They both suggested a tetrahedral
structure, and the theory still stands with the slight modification
that the tetrahedron is not necessarily completely symmetrical in
all organic compounds. This theory is considered to be correct
because no new facts that cannot be explained by it have been
brought to light, and because it has been necessary to discard
other possible structures for one or more reasons. These are always
the criteria for the retention or rejection of any chemical theory.

Let us examine the number of possible isomers for the most

likely arrangements of four atoms about a central carbon atom: the tetrahedron with the carbon at the center, the square with the carbon in the center (a planar structure), and the rectangular pyramid with the carbon at the apex. If a, b, c, and d are different

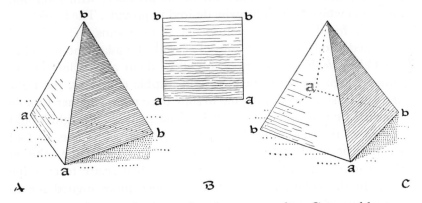

groups attached to the central carbon atom, then Ca_4 would represent compounds similar to CH_4 and CCl_4; Ca_3b would represent CH_3Cl, $CHCl_3$, etc.; Ca_2b_2, CH_2Cl_2, etc.; Ca_2bc, $CH_2Cl–COOH$, etc.; $Cabcd$, lactic acid, etc. Table 18-2 contrasts the number of isomers of such compounds possible for each of the three configurations mentioned with the actual number known to exist. The tetrahedral configuration is the only one that accounts for the number observed.

TABLE 18-2	Number of Possible Isomers for Various Configurations			
TYPE OF COMPOUND	TETRAHEDRON	SQUARE	PYRAMID	NUMBER OBSERVED
Ca_4	1	1	1	1
Ca_3b	1	1	1	1
Ca_2b_2	1	2	2	1
Ca_2bc	1	2	3	1
$Cabcd$	2	3	6	2

The fact that no stereoisomerism is observed in alkynes is accounted for by the tetrahedral configuration. With a planar or pyramidal structure the formation of an alkyne with a triple bond between the carbons is impossible.

All of the above structures will account for the existence of *cis* and *trans* isomers in alkenes of type 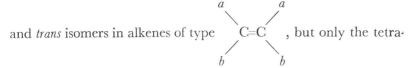 , but only the tetra-hedral configuration gives nonsuperimposable optically active isomers for alkadienes of the type . These are called al-lenes (p. 348), and a number of pairs of them have been resolved. In a planar or pyramidal structure these isomers would be only geometric and not optical. The optical isomerism in spiranes and inositol (p. 349) is accounted for with tetrahedral carbons.

The chemistry of carbon-ring compounds is consistent with a tetrahedral arrangement. Planar or pyramidal carbons would involve enormous strains in the cyclic molecules so formed. It is impossible to make the models of many complex cyclic compounds —for example, hexamethylenetetramine—except with tetrahedral carbon. Though the models cannot reveal all the chemical properties of a molecule, they have been useful in numerous cases for revealing possible difficulties in synthesis.

The accomplishment of a Walden inversion (1893; see p. 350) strengthened the tetrahedral theory (1874) but also modified the early idea that the structure was a rigid, unchangeable tetrahedron. If one tetrahedron is inverted into its mirror image during a chemical reaction, of necessity it cannot preserve a rigid structure throughout the inversion. The tetrahedral nature of carbon has not been seriously questioned since Fischer (1914) showed that interchanging two groups on an asymmetric carbon would result in a configuration opposite to that of the starting compound.

18-11. Evidence for Resonance in Benzene

What reason is there for saying that the Kekule structure is not an accurate picture of the benzene molecule and hence must be modified slightly to account for the latest observations? (Review pp. 119–120.)

1. Interatomic Distances. As determined by X-ray measurements and electron diffraction patterns, the C-C bond distance in organic molecules varies only slightly from one substance to another. In alkanes the distance is 1.54 A; in alkenes it is 1.34 A. If the Kekule structures were completely correct for benzene, one would expect to find both distances revealed in benzene bonds. Instead, only one C-C bond distance is found by X-ray measurements, 1.39 A. This is not that of a single or a double bond, nor an average of the two (1.44 A), and it gives support to the view that all is not well with the Kekule structures.

2. Heats of Hydrogenation. When a mole of cyclohexene is reduced in the presence of nickel, a large amount of heat is liberated; when a mole of benzene is hydrogenated under the same conditions, much less than three times this amount of heat is liberated. For cyclohexene, 28.6 kilocalories of heat is liberated when

$$\bigcirc + H_2 \xrightarrow{Ni} \bigcirc + 28.6 \text{ kcal} \qquad [18\text{-}12]$$

$$\textcircled{} + 3H_2 \longrightarrow \bigcirc + 49.8 \text{ kcal} \qquad [18\text{-}13]$$

one mole is reduced. For three cyclohexene double bonds, then, $3 \times 28.6 = 85.8$ kcal should be liberated. The difference, $85.8 - 49.8 = 36.0$ kcal, may be considered a measure of the stability that a benzene ring possesses in excess of that expected for three ordinary double bonds. This is called resonance energy.

The question may now arise whether the evidence for the Kekule structures of benzene and *o*-xylene in the ozonolysis reactions discussed on pages 373–374 contradicts the evidence given here for the resonance picture. The question will be left open for your thinking and future reading.

18-12. The Phenomenon of Resonance

The phenomenon of resonance is not restricted to benzene, aromatic compounds in general, or alkadienes (p. 104) though these are the examples that have been discussed. Stabilization by resonance is prominent in many functional groups.

From the postulates (p. 120) accruing in a study of the resonance phenomenon only structures involving displacement of electrons need be considered when we decide whether a given canonical structure will make any contribution to the total structure of the molecule or ion. The complete equivalence of the three structures in Expression 18-14 for the nitrate ion, in which all atoms have

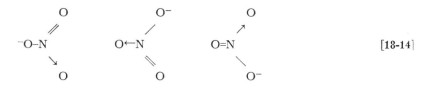

$$[18\text{-}14]$$

electron octets, portends that resonance is very important in this ion. Other structures not equivalent to these, in which octets do not exist for each of the four atoms, can be written, but they make a less important contribution to the structure of the ion.

In the canonical structures for some other ions, molecules, and functional groups, given in Expressions 18-15–18-24, the individual

Carbonate ion:

$$^-O\text{-}\overset{O}{\underset{}{C}}\text{-}O^- \quad O=\overset{O^-}{\underset{}{C}}\text{-}O^- \quad ^-O\text{-}\overset{O^-}{\underset{}{C}}=O \quad ^-O\text{-}\overset{O^-}{\underset{+}{C}}\text{-}O^- \qquad [18\text{-}15]$$

Carbon dioxide:

$$O=C=O \quad \overset{-}{O}\text{-}C\overset{+}{\equiv}O \quad \overset{+}{O}\equiv C\text{-}\overset{-}{O} \quad \overset{+}{O}=C\text{-}\overset{-}{O} \quad \overset{-}{O}\text{-}C=\overset{+}{O} \qquad [18\text{-}16]$$

Carbon monoxide:

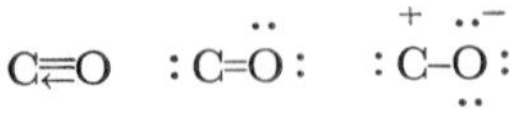

$$C \equiv O \quad : \overset{..}{C} = \overset{..}{O} : \quad : \overset{+}{C} - \overset{..}{\overset{..}{O}} :^{-}$$ [18-17]

Carbonyl group:

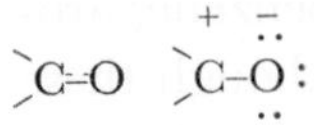

$$\diagdown C{-}O \quad \diagdown \overset{+}{C}{-}\overset{..}{\overset{..}{O}} :^{-}$$ [18-18]

Nitro group:

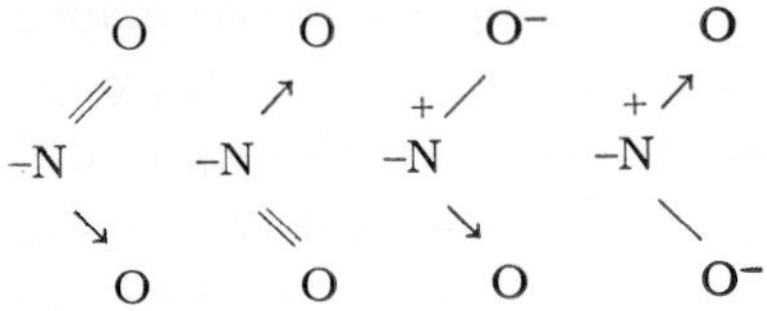

[18-19]

Carboxylate ion:

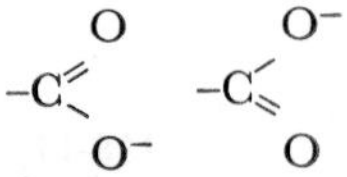

[18-20]

Carboxyl group:

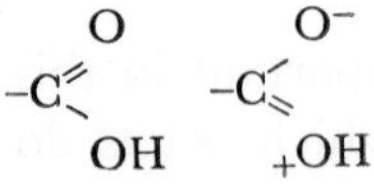

[18-21]

Phenol:

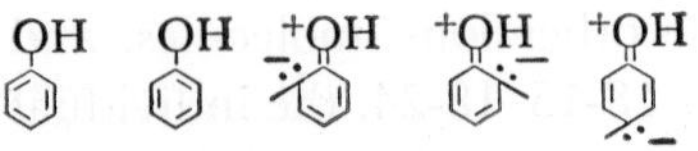

[18-22]

Phenoxide ion:

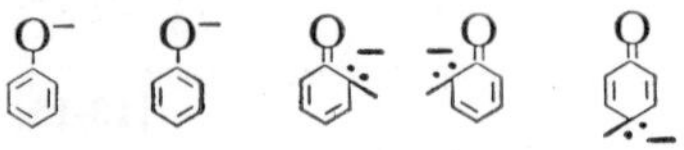

[18-23]

Aniline:

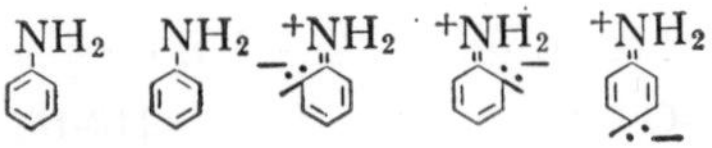

[18-24]

structures do not necessarily contribute equally unless they are exactly equivalent. *It must be remembered that none of these disparate structures has any actual existence.* Writing them is justified only by the fact that we do not have any better method of representing resonance structures on paper.

18-13. Orientation in the Benzene Ring

The mechanism of substitution in the benzene ring has already been discussed for a few substances—for example, phenol and toluene. Now the general subject of orientation will be discussed in terms of an electronic picture. In phenol and aniline the electronic displacements shown in Expressions 18-22 and 18-24 make important contributions to the molecule. This means that there is a greater electron density at the two *ortho* positions and the *para* position than at the two *meta* positions. Consequently, an electron-seeking agent is more likely to attack the *ortho* and *para* positions. The common reagents for substitution in a benzene ring are electron-seeking agents—for example, $^{\oplus}NO_2$ from $HONO_2$, $Br^{\oplus}$ from Br_2, $R^{\oplus}$ from $[R]^{\oplus}[AlCl_4]^{\ominus}$ in the Friedel-Crafts reaction, and $^{\oplus}SO_3H$ from fuming sulfuric acid.

Electronic displacements similar to those shown for the –OH and –NH$_2$ groups on the benzene ring are also exhibited by –OR, –NHR$_2$, and –NH–CO–R. All of these groups are *o,p*-directing groups.

A halogen on a benzene ring affects the orientation in a way that is not apparent in the cases just cited, although the effect is present there also. A halogen, being an electron-attracting atom, attracts electrons from the benzene ring, thus making them less available for the entire ring. This is called the inductive effect (see pp. 458ff.). The resonance effect, of which the principal contributing structures are shown in Expression 18-25, is in the opposite direction.

[18-25]

Though the resonance effect is the more influential, a halogen is a weaker *o,p*-directing group than the hydroxy or amino group.

The halogen deactivates the ring (inductive effect) so that substitution is more difficult than in benzene itself.

Exercise 18-3. Write electronic mechanisms for the chlorination, nitration, sulfonation, and Friedel-Crafts reactions of chlorobenzene. (See also pp. 121–124.)

An alkyl group is also a weak *o,p*-directing group, having a weak inductive effect in the direction opposite to that of a halogen; that is, an alkyl group in an alkyl benzene tends to release a pair of electrons to the ring. The positions of greatest electron density will be the *o,p*-positions (Exp. 18-26), and electron-seeking agents will consequently orient *o,p*.

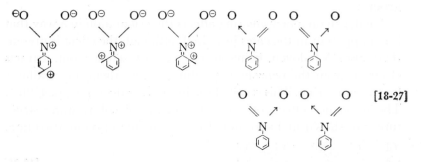

$$[18\text{-}26]$$

Exercise 18-4. Write electronic mechanisms for the nitration and sulfonation of ethylbenzene. (See pp. 122–123.)

The nitro group is a *m*-directing group. It is an electron-attracting group of about the same electronegativity as a halogen. The benzene ring is therefore deactivated by it, and substitution in nitrobenzene is more difficult than in benzene itself. It is to be noted that the *o,p* positions in Expression 18-27 have an electro-

$$[18\text{-}27]$$

positive character. This leaves the two *m* positions with the greater electron density by default, and hence the nitro group is a *m*-directing group. Similarly, the $-C\equiv N$, $-COOH$, $-CHO$, $-CO-R$, and

–SO₃H groups, which are observed to be *m*-directing, are accounted for by the same interpretation, and they all deactivate the ring.

Summary: *o,p*-directors: –OH, –OR, –X, –NH₂, –NHR, –NR₂, –NHCOR, and –R; *m*-directors: –NO₂, –N̄R₃, –CO–R, –CHO, –C≡N, –SO₃H, and –COOH.

Only major products are predicted by the foregoing rules. Small amounts of other products are frequently obtained (see p. 389).

When a benzene derivative contains two substituents, the orientation of a third substituent depends on the character of those already present. If both groups direct the entering group to the same position, a single product will predominate; but, if both groups direct *meta*, the ring will be strongly deactivated, and the third group will enter with difficulty. If an *o,p*-director and a *m*-director are both present, the *o,p*-director will always dominate. When two *o,p*-directors point to different positions, mixtures will be obtained. If one of the two directors is hydroxy or amino, it will determine the orientation of the product since these are the strongest directing groups.

A number of examples in which these rules apply will be given. Since secondary products such as H₂O and CH₃COOH are omitted, these reactions are not written as equations.

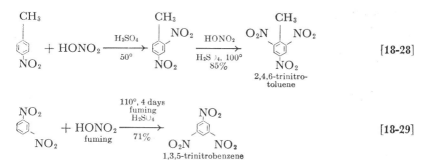

A good yield of 2,4,6-trinitrotoluene is obtained by treatment of toluene with a mixture of nitric and sulfuric acids at 100°. On the other hand, a good yield of 1,3,5-trinitrobenzene is obtained only by vigorous treatment of benzene with a mixture of *fuming* nitric and *fuming* sulfuric acids. This is a graphic example of the activating influence of even a weak *o,p*-director in the presence of a *m*-director.

Substances that are easily oxidized cannot be nitrated without special precautions (generally low temperatures or protection of the easily oxidized group). Phenol can be nitrated at low temperature to give good yields of *o*- and *p*-nitrophenol.

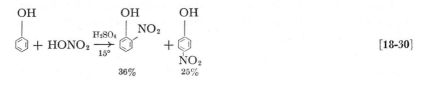

$$\text{[18-30]}$$

Aniline is an easily oxidizable substance and a strong *o,p*-directing group. The directing power can be attenuated and the group protected from oxidation by acetylation. The amine group can later be freed from the amide by hydrolysis.

$$\text{[18-31]}$$

The nitration of acetanilide results in a 90% yield of *p*-nitro-acetanilide. The *ortho* isomer is easily removed by recrystallization, and *p*-nitroaniline can then be recovered by acid hydrolysis (Exp. 18-32).

$$O_2N-\langle\rangle-NHCOCH_3 \xrightarrow[H_2O]{H_2SO_4} O_2N-\langle\rangle-NH_2 \qquad \text{[18-32]}$$

Nitration of the anilinium ion (anilinium sulfate, for example) results in *meta* substitution (Exp. 18-33).

$$\langle\rangle-\overset{+}{N}H_3 \xrightarrow{HONO_2} \overset{NO_2}{\langle\rangle}-\overset{+}{N}H_3 \xrightarrow{OH^-} \overset{NO_2}{\langle\rangle}-NH_2 \qquad \text{[18-33]}$$

Bromination of *o*-cresol results in the formation of 4,6-dibromo-*o*-cresol since the hydroxy group is a much stronger *o,p*-directing group than either an alkyl group or a halogen.

[18-34]

18-14. Relative Amounts of *o, m, p* Isomers

Since the rules of orientation suggest only the predominant products, and small amounts of other isomers are generally obtained, some idea of these relative values is given in Table 18-3.

TABLE
18-3
| *Relative Isomeric Product Composition of Aromatic Substitution Reactions*

COMPOUND	REACTION	% COMPOSITION OF PRODUCTS		
		ORTHO	META	PARA
toluene	nitration −30°	57	3.5	39
	0°	58	4	37
	60°	60	5	35
chlorobenzene	nitration	30		70
bromobenzene	nitration	38		62
phenol	nitration	40		60
acetanilide	nitration	5		95
benzoic acid	nitration	19	80	1
nitrobenzene	nitration	6	93	<1
toluene	bromination	40		60
phenol	bromination	10		90
chlorobenzene	chlorination	39	6	55
bromobenzene	chlorination	41.5	6	52.5
chlorobenzene	sulfonation			100
toluene	sulfonation 0°	43	4	53
	100°	13	8	79

Very little difference in the percentage composition of the product is observed in nitrations at various temperatures, but sulfonations are strongly influenced by the temperature at which they are carried out.

It is not to be thought that the total yield of isomers is 100% in any of the reactions listed in the table. The percentages given show the relative yields of the three isomers.

SUMMARY

Highlights in the evolution of our present ideas on the covalent bond are summarized in Table 18-4.

The rules of orientation in aromatic compounds, appearing on pages 385–387, are rationalized in terms of the resonance and inductive effects.

REFERENCES

This chapter has referred to original sources for information on many of the developments outlined in Table 18-4. Since many students will not yet be prepared to read French or German, or the journal cited may not be available in the library, a collection containing some of the original passages (translated) has been suggested as the first reference, with page references to some of the developments.

1. Leicester and Klickstein, *A Source Book in Chemistry, 1400–1900* (McGraw-Hill Book Co., New York, 1952).

Berzelius	pp. 258–261	dualism theory
Chevreul	287–298	fats, saponification
Wöhler	309–312	urea
Dumas	321–328	esters, chloroacetic acid
Gerhardt & Laurent	345–353	theory of types
Hofmann	364–368	amines
Kolbe	369–373	organic radicals
Pasteur	374–379	asymmetry
Williamson	379–384	theory of etherification
Frankland	384–392	organic radicals

TABLE
18-4

Contributions to the Theory of the Covalent Bond

SCIENTIST	NATIONALITY	DATES	CONTRIBUTION
Antoine Lavoisier	French	1743–1794	Analysis of carbon compounds
Michel Chevreul	French	1786–1889	Fats and fatty acids
J. L. Gay-Lussac	French	1778–1850	Organic analysis, cyanogen
J. J. Berzelius	Swedish	1779–1848	Atomic weights, dualism theory
Friedrich Wöhler	German	1800–1882	Synthesis of urea, benzoyl radical
Justus von Liebig	German	1803–1873	First laboratory instruction, benzoyl, acetyl radicals
J. B. A. Dumas	French	1800–1884	Nitrogen analysis, theory of substitution
Auguste Laurent	French	1807–1853 ⎱	Theory of types
Charles Gerhardt	French	1816–1856 ⎰	
A. W. Williamson	English	1824–1904	Ether synthesis
C. A. Wurtz	French	1817–1884	Amines, glycols, Wurtz reaction
A. W. von Hofmann	German	1818–1892	Synthesis of amines
Hermann Kolbe	German	1818–1884	Synthesis by electrolysis
Edward Frankland	English	1825–1899	Suggestion of combining power, organometallic compounds
F. A. Kekule	German	1829–1896	Quadrivalent carbon, carbon chain, structure of benzene
A. M. Butlerov	Russian	1828–1886	Structural organic theory
Stanislao Cannizzaro	Italian	1826–1910	Exposition of theoretical chemistry
R. W. Bunsen	German	1811–1899	Gas analysis
P. E. M. Berthelot	French	1827–1907	Organic synthesis, esters
Louis Pasteur	French	1822–1895	Fermentation, tartaric acid
J. H. van't Hoff	Dutch	1852–1911 ⎱	Tetrahedral carbon
J. A. Le Bel	French	1847–1930 ⎰	
J. J. Thomson	English	1856–1940	Discovery of the electron
Johannes Stark	German	1874– ⎱	Detailed structural theory of valence based upon electrons
Walter Kossel	German	1888– ⎰	
H. G.-J. Moseley	English	1887–1915	Determination of atomic numbers
Niels Bohr	Danish	1885–	Quantum theory of the atom
G. N. Lewis	American	1875–1946	Octet theory, covalent bonds

Kekule	417–425	aromatic compounds
Körner	425–427	absolute method
van't Hoff	445–453	asymmetric carbon
Le Bel	459–462	asymmetric carbon

2. Hunsberger, "Theoretical Chemistry in Russia," J. Chem. Education, 31, 504 (1954).
3. Mackle, "Evolution of Valence Theory and Bond Symbolism," J. Chem. Education, 31, 618 (1954).
4. Finegold, "The Liebig-Pasteur Controversy," J. Chem. Education, 31, 403 (1954).

EXERCISES

(Exercises 1–4 will be found within the text.)

5. Show how the following conversions could be effected:

1. nitrobenzene to 1,3-diaminobenzene
2. chlorobenzene to 3-nitrobenzoic acid
3. benzene to *m*-chloronitrobenzene
4. benzene to *p*-chloronitrobenzene
5. benzene to 2,4,6-trinitrotoluene
6. benzene to *m*-nitroacetophenone
7. *o*-aminotoluene to *o*-chlorobenzoic acid
8. *o*-aminotoluene to 2-methyl-4-hydroxyazobenzene

6. Synthesize the following:

1. *p*-cresol
2. *p*-bromoacetophenone
3. 2,4,6-tribromoaniline from benzene
4. *p*-nitroaniline
5. 2,4,6-tribromo-3-methylphenol from *m*-cresol
6. *p*-aminobenzenesulfonic acid
7. acetanilide
8. N-methylbenzamide
9. *p*-nitrobenzoic acid
10. 3,5-dinitrobenzoic acid
11. *m*-nitrotoluene (start with *p*-methylaniline)

NEIGHBORS ON JOINED TETRAHEDRA

PART FOUR

NEIGHBORS ON JOINED
TETRAHEDRA

19 Carbohydrates

Up to this point we have emphasized the properties of covalent compounds containing a single functional group. We have mentioned a few compounds, such as the glycols. glycerol, and the nitro alcohols, in which a second functional group has little influence on the chemical properties of the first. Now, in the next five chapters, we shall delineate the chemistry of several polyfunctional compounds. In the first two classes, carbohydrates and amino acids, the presence of the second functional group has a profound effect on the properties of the compound, sometimes modifying the properties of each functional group and sometimes giving the compound new properties not exhibited at all by the functional groups separately. Carbohydrates and amino acids are important in life processes (both plant and animal) and so are important to the biologist, biochemist, doctor, nurse, biophysicist, agriculturist, and others, as well as to the chemist.

19-1. Photosynthesis

The synthesis of carbohydrates from simple starting materials, carbon dioxide and water, takes place continuously in plants. Chlorophyll, the green coloring matter in plants, is the catalyst for the synthesis. Man has so far not been able to duplicate the process in the laboratory. The remarkable part of the synthesis is that the products contain more energy than the starting materials. The extra energy comes from the sunlight absorbed during the reduction of the carbon dioxide. The equation given here only

$$6CO_2 + 6H_2O \xrightarrow[\text{chlorophyll}]{h\nu} C_6H_{12}O_6 + 6O_2 \qquad [19\text{-}1]$$

suggests the beginning and the end of the process. There are many complex intermediate steps, some of which are known with certainty.

The fact that energy can be obtained in this way from the sun has always intrigued man and led him to speculate on the possibility of harnessing the sun's energy for himself as a direct source of power.

19-2. Definition of Carbohydrate

The name "carbohydrate" suggests a combination of carbon and water, and the early work on these compounds led to their representation as $C_n(H_2O)_n$. A large number of carbohydrates have this molecular formula, but many other compounds that have the properties usually associated with carbohydrates do not. The present definition is based on a knowledge of the functional groups present in a carbohydrate. A carbohydrate is a polyhydroxy aldehyde, ketone, or hemiacetal or any substance that can be hydrolyzed to a polyhydroxy aldehyde, ketone, or hemiacetal.

The carbohydrates whose chemistry will be discussed are the sugars, cellulose, starch, and some of their derivatives.

19-3. Classification of Carbohydrates

The carbohydrates may be classified in several ways, one of which is to divide the group into mono-, di-, tri-, and poly-saccharides. A disaccharide will hydrolyze to two molecules of a simple monosaccharide. A monosaccharide is a polyhydroxy aldehyde or ketone. The other terms are obvious from this statement. The characteristic ending of the name of a sugar is -*ose*; a three-oxygen monosaccharide containing an aldehyde group is an aldotriose, and the corresponding ketone is a ketotriose.

$$\begin{array}{ll}
\text{CHO} & \text{CH}_2\text{OH} \\
\text{H–C–OH} & \text{C=O} \\
\text{CH}_2\text{OH} & \text{CH}_2\text{OH} \\
\text{glyceraldehyde,} & \text{dihydroxyacetone,} \\
\text{an aldotriose} & \text{a ketotriose}
\end{array}$$

The most important sugars in the plant world are aldo- and keto-pentoses and -hexoses. In the animal world the aldo- and keto-hexoses are the predominant ones. Glucose is an aldohexose, and fructose is a ketohexose. Besides glucose, only two other aldo-hexoses occur in nature, galactose and mannose. (For structures, see p. 403.)

The three important disaccharides are maltose, lactose, and sucrose.

Though the mono- and di-saccharides are sweet-tasting, the polysaccharides are tasteless. Starch, cellulose, and glycogen, the most important polysaccharides, are substances of high molecular

$$(C_6H_{10}O_5)_n + nH_2O \xrightarrow{H^+} nC_6H_{12}O_6 \qquad\qquad [19\text{-}2]$$
$$\text{cellulose} \qquad\qquad\quad \text{glucose}$$

weight that can be hydrolyzed to glucose. The three polysaccharides are considered to be made up of some pattern of glucose units from which a molecule of water is lost for each two units that are joined together. The value of n may range from 1,000 upward for cellulose.

19-4. Glucose

The present accepted structure of glucose slowly evolved from the work of chemists, over a period of about fifty years, by a process of scientific discovery similar to that described for benzene in the last chapter. Some of the important chemical reactions leading to this structure will be discussed here, though not in historical order.

Glucose is a white crystalline substance whose hydrate melts at 90°. Anhydrous glucose, of melting point 146°C, can be obtained by crystallization from alcoholic solution. Analysis yields the empirical formula CH_2O, and a determination of the molecular weight indicates the molecular formula $C_6H_{12}O_6$.

Glucose reduces Fehling's solution and Tollens' reagent, the implication being that an aldehyde group is present. Glucose can

$$(C_5H_{11}O_5)\text{--CHO} \xrightarrow[\text{reagent}]{\text{Tollen's}} (C_5H_{11}O_5)\text{--COOH} \qquad\qquad [19\text{-}3]$$
$$\text{glucose} \qquad\qquad\qquad\quad \text{gluconic acid}$$

also be oxidized by other reagents, such as bromine water, to gluconic acid. Oxidation of a ketone should give a product with at least one less carbon than the starting compound, whereas gluconic acid contains the same number of carbons as glucose.

Glucose can be reduced to sorbitol, a hexahydroxy alcohol, by catalytic hydrogenation. The gluconic acid obtained by oxidation of glucose can be reduced with HI (a strong reducing agent) to caproic acid, a known compound having the structure $CH_3-CH_2-CH_2-CH_2-CH_2-COOH$. This indicates a six-carbon chain in the original compound.

If glucose is acetylated with acetic anhydride in the presence of anhydrous potassium acetate, a penta-acetate is formed. Since a carbon carrying two hydroxy groups gives an unstable configuration, the five hydroxy groups forming the penta ester are probably on different carbon atoms.

From these observations the logical structure to write for glucose would be $CH_2OH-CHOH-CHOH-CHOH-CHOH-CHO$. But this does not go far enough, for there are four asymmetric carbons in this structure and hence sixteen possible compounds. Still more information is needed.

The glucose we have described occurs in grape sugar and is more precisely named D(+)glucose, the plus sign in parenthesis referring to a dextro rotation. (For the meaning of the capital D see section 19-6 below.) This optically active sugar has a specific rotation of +52°. A freshly prepared solution of D(+)glucose has a rotation of +113°, which gradually changes to +52°. Upon recrystallization from alcohol plus water, D(+)glucose gives the initial rotation of +113° again. A second form of D(+)glucose can be obtained by crystallization from concentrated solution at 110°C; this has an initial rotation of +19°, which also changes to the equilibrium value of +52° on standing. These observations indicate that D(+)glucose must exist in some easily interconvertible forms that readily change back to some equilibrium state.

To understand this last phenomenon, which is called mutarotation, let us recall one more reaction of an aldehyde, hemiacetal formation. Since D(+)glucose is both an aldehyde and an alcohol, there is opportunity for hemiacetal formation within the molecule itself. The resulting ring structures, if the closure takes place to form a six-membered ring, are shown in Exp. 19-4. The six-mem-

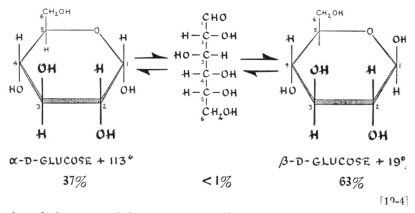

α-D-GLUCOSE + 113°   β-D-GLUCOSE + 19°

37%   <1%   63%

[19-4]

bered rings containing oxygen are shown in planes perpendicular to the vertical lines. An OH group that appears on the right in the straight-chain aldehyde structure is written below the plane of the ring in the α and β forms. It will be noted that the formation of the ring introduces a new asymmetric atom at carbon No. 1 so that two forms of D(+)glucose are possible in the ring structure. (The numbers of carbons 1–5 show them to be at the intersections of lines in the figure.) These two forms are referred to as α and β glucose.

Exercise 19-1. Show by projection formulas (as in Exp. 19-4) the two hemiacetals formed by D(+)glucose and containing five-membered rings.

Since D(+)glucose exhibits reactions that are characteristic of both the straight-chain aldehyde-alcohol structure and the ring structures (two other hemiacetals of five-membered rings formed at the No. 4 carbon instead of No. 5 can be prepared under proper conditions), an equilibrium between all the tautomeric structures is assumed to take place. The predominant forms are the α and β forms shown. Mutarotation is the process by which any form of an optically active substance that can exist in tautomeric forms comes to an equilibrium rotation (not zero). Sugars are the common examples. Any sugar that exists in a hemiacetal structure is capable of mutarotation.

It is conventional, in writing chain structures for sugars, to place the No. 1 carbon atom at the top. The glucose structure is written with the hydroxyl on the third carbon at the left and all the others at the right. The story of how these final refinements in the structure of D(+)glucose were determined will be found in advanced treatises such as Gilman's *Organic Chemistry*. Some of the story is given in the remainder of this chapter. This, of course, was the most difficult part of the determination of structure.

19-5. D(−)fructose

D(−)fructose (Exp. 19-5) is a ketohexose, found in honey and

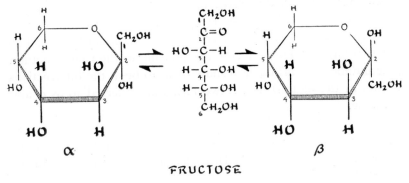

FRUCTOSE

[19-5]

many fruits, and is the only important keto sugar. It also exhibits mutarotation. At equilibrium its rotation is −92°.

Exercise 19-2. Write the projection formulas for two hemiacetal structures for D(−)fructose containing five-membered rings.

19-6. Conventions in Configuration

Since the absolute configuration was not known for any substance before 1949, it was helpful in sugar chemistry to relate all sugars to a particular starting compound whose configuration was arbitrarily assigned. The reference compound is the glyceraldehyde that has a dextro rotation, arbitrarily called D(+)glyceraldehyde. In the three-dimensional Exp. 19-6, the unbroken lines are to be considered

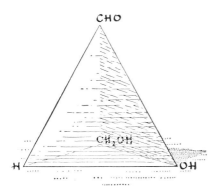

[19-6]

in the plane of the paper and the broken lines behind the plane of the paper. When the four groups attached to the asymmetric carbon in glyceraldehyde are placed as shown here, the compound is assigned the D configuration. The projection of this structure into two dimensions (Exp. 19-7), then, is also D(+)glyceraldehyde.

$$\begin{array}{c} \text{1CHO} \\ | \\ \text{H–C–OH} \\ \text{2} | \\ \text{3CH}_2\text{OH} \end{array}$$

[19-7]

D(+)glyceraldehyde

When a four-carbon sugar is made from D(+)glyceraldehyde, the new carbon forms a bond with the carbon of the aldehyde group. Since the configuration of the original carbon 2 (carbon 3 in the new compound) does not change during this reaction, the two sugars resulting from the increase in the carbon chain have the D configuration with respect to D(+)glyceraldehyde (by definition). Both of these sugars, D(−)erythrose and D(−)threose, have levo rotation, and this is designated by the added notation shown in Expression 19-8. A capital D relates the configuration

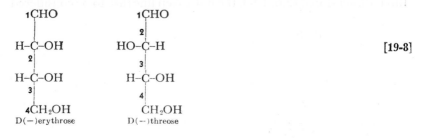

$$[19\text{-}8]$$

of a sugar to D(+)glyceraldehyde; the + or − in parenthesis gives the sign of rotation. Earlier in this text (Chap. 17) a small italic *d* or *l* designated rotation. Thus grape sugar, or dextrose, may be denoted as D(*d*)glucose or as D(+)glucose. It is necessary to recognize both notations in order to read the literature. The configurations of the L series of sugars are mirror images of those of the corresponding D compounds.

Exercise 19-3. Write the structure of L(−)mannose.

In Table 19-1, which shows the D series of aldoses from D(+)glyceraldehyde to the six-carbon sugars, a horizontal bar represents an OH group on a carbon in the chain, and H is not shown for that carbon.

19-7. Properties of Sugars

Though some properties of sugars are those that would be expected of an aldehyde and an alcohol if one functional group had

TABLE
19-1 | *D Aldoses*

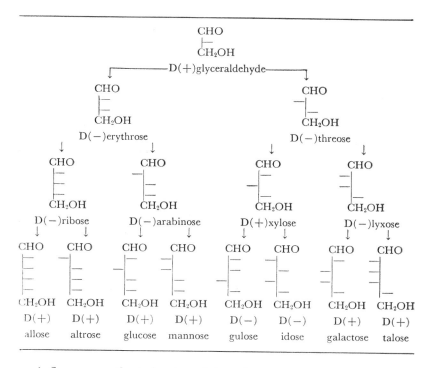

no influence on the behavior of the other, many reactions are due to the influence of the hydroxy groups on the aldehyde (or keto) group.

The many alcohol groups present in sugars make them very soluble in water. Sugars are difficult to crystallize because they are so soluble, and they frequently form syrups. Molasses is a concentrated solution of impure sucrose.

The naturally occurring aldohexoses, $D(+)$glucose, $D(+)$mannose, and $D(+)$galactose, and the ketohexose $D(-)$fructose, undergo fermentation with the proper yeast to yield ethyl alcohol and carbon dioxide.

1. Oxidation. Fehling's solution, Benedict's solution, and Tollens' reagent are all used as tests for the group of sugars called reducing sugars. A reducing sugar is a polyhydroxy aldehyde, ketone, or hemiacetal. Sucrose (see p. 414) is excluded since it has an acetal, rather than a hemiacetal, linkage. Glucose and fructose are both reducing sugars. The oxidation of glucose to gluconic acid can also be accomplished by other mild oxidizing agents such

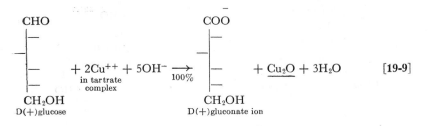

as NaOBr. Nitric acid oxidizes the sugars to dicarboxylic acids; the product of aldehyde sugars contains the same number of carbons, but in that of keto sugars the chain is shortened. $D(-)$fructose, for example, is oxidized to meso tartaric acid and glycolic acid (p. 463) while $D(+)$glucose is converted to saccharic acid in low yield (Exp. 19-10).

2. Reduction. Mild reduction of an aldehyde or keto sugar gives the corresponding alcohol. A very strong reducing agent, such as HI, will reduce a sugar to the corresponding hydrocarbon.

$$\begin{array}{ccc} \text{CHO} & \xrightarrow[30\%]{\substack{\text{H}_2\text{O}\\ \text{Na(Hg)}}} & \text{CH}_2\text{OH} \\ (\text{CHOH})_4 & & (\text{CHOH})_4 \\ \text{CH}_2\text{OH} & & \text{CH}_2\text{OH} \\ \text{glucose} & & \text{mannitol} \end{array} \qquad [19\text{-}11]$$

3. Acetal Formation: Glycosides. Since D(+)glucose exists in equilibrium with hemiacetal forms, it would be logical to suppose that an acetal could be formed from it under the proper conditions. Dry HCl catalyzes the formation of acetals with aldehyde sugars. D(+)glucose, with methyl alcohol, for example, forms a methyl acetal. In sugar chemistry the acetals are called

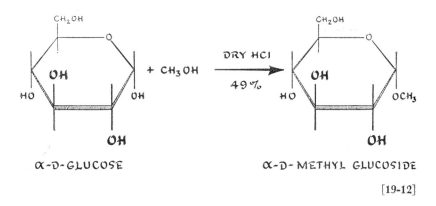

α-D-GLUCOSE   α-D-METHYL GLUCOSIDE

[19-12]

glycosides, and, if glucose is the sugar, the acetal is a glucoside (Exp. 19-12). Sugars frequently occur in plants as glycosides.

Exercise 19-4. Write the structure of β-D-methyl mannoside.

4. Ether Formation. To ensure the formation of the six-membered ring in glucose to the exclusion of the open-chain compound or the five-membered ring, preparation of a polyether is frequently employed. In the Williamson synthesis a sodium alkoxide reacts with an alkyl halide to give an ether. In sugar chemistry the ethers may be made by grinding a glycoside with Ag_2O and then treating the reaction mixture with methyl iodide. Acetals are formed in acid and easily decomposed in acid solution, but they are stable to bases. In the last reaction shown in Exp. 19-13, the 2,3,4,6-tetramethylglucose is easily obtained by treatment of the glucoside with acid, and carbons 1 and 5 may now be worked on in basic solution with assurance that carbons 2, 3, 4, and 6 are protected.

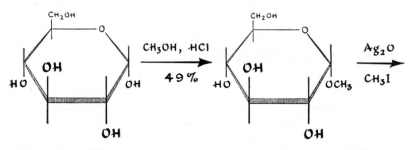

α-D-GLUCOSE   α-D-METHYL GLUCOSIDE

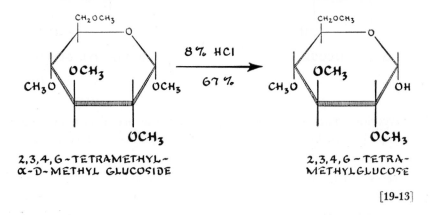

2,3,4,6-TETRAMETHYL-   2,3,4,6-TETRA-
α-D-METHYL GLUCOSIDE   METHYLGLUCOSE

[19-13]

Exercise 19-5. Write a structure for 2,3,4,6-tetramethyl-α-D-methyl galactoside.

5. Effect of Alkali. Allowing a sugar to stand in solution with a trace of alkali establishes rather quickly an equilibrium among the tautomeric forms of the sugar. Only carbons 1 and 2 are affected, and the equilibrium is assumed to be established through an intermediate ene-diol. Any one of the three sugars D(+)glucose, D(−)fructose, and D(+)mannose soon changes to an equilibrium mixture of all three. The formation of the ene-diol results in

the temporary loss of asymmetry in carbon 2. When the hydrogen
from the OH group on carbon 1 shifts back to carbon 2, there is a

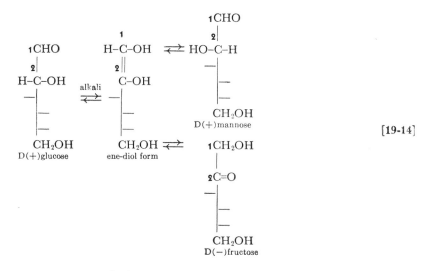

[19-14]

chance that D(+)mannose will form rather than D(+)glucose.
The shift of hydrogen may occur from the OH group on carbon 2,
however, and that gives D(−)fructose.

Note that this process is not a racemization (p. 343); only
carbon 2 changes its configuration. Though carbon 2 may be said
to have undergone a racemization, the configurations of carbons 3,
4, and 5 remain unchanged. Hence the resulting products are
still optically active, as is the mixture, in solution. Note also the
difference between this reaction and mutarotation.

In strong alkali, sugars turn brown, an evidence of decomposition. The nature of the decomposition products has not been established completely.

This behavior in alkaline solution is mentioned particularly to
show one of the difficulties encountered in establishing the configurations of sugars. Very often a reagent that gives a known reaction
with the carbonyl or alcohol group in monofunctional compounds
induces tautomeric changes of the kind mentioned here in a carbohydrate. Since this was not always understood in the early work

on sugars, much of that work has been invalidated, and it is difficult to interpret the literature of 1890–1910 on the stereochemistry
of sugars.

6. Acetylation. Esters of polyalcohols can be prepared with
any assurance of purity only with short-chain acids. In glycerol,
for example, it is not easy to get complete esterification of all three
OH groups with benzoyl chloride, largely because of steric effects.
One might expect no alleviation of this difficulty in the sugars, and
there is none. Acetic anhydride is the best esterifying agent for the
sugars. With this reagent D(+)glucose yields β-D-glucosepentace-

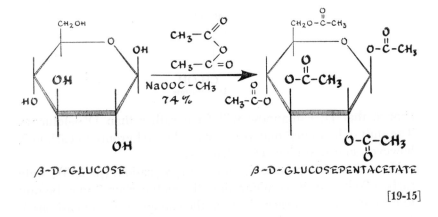

β-D-GLUCOSE   β-D-GLUCOSEPENTACETATE

[19-15]

tate when freshly fused sodium acetate is the catalyst (Exp. 19-15).
The α form is obtained in greater yield with $ZnCl_2$ as catalyst.

Exercise 19-6. Write a structure for α-D-galactosepentacetate.

7. Osazones. Aldo and keto sugars react with phenylhydrazine
to form phenylhydrazones, but the reaction goes further than with
an ordinary carbonyl group, and a diphenylhydrazone is formed.
The reaction involves an oxidation as well as two condensations.
With glucose and fructose the reactions proceed as in Expressions 19-16 and 19-17.

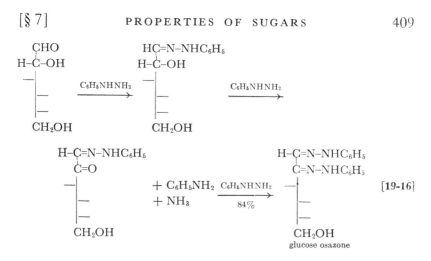

Exercise 19-7. Write the oxidation-reduction reaction in the second step of Expression 19-16 to show the changes in oxidation numbers.

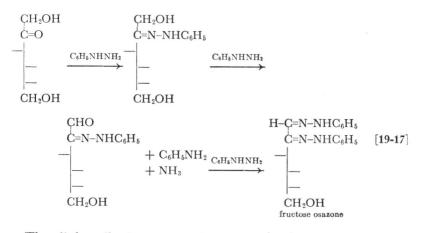

The diphenylhydra*zones* of the sugars (*oses*) were named *osazones* by Emil Fischer. Glucose and fructose osazones have the same melting points and are actually the same compound. This establishes the fact that the configurations of carbons 3, 4, and 5 in glucose and fructose are the same and was one of the important links in the final acceptance of the present structures of the two compounds. The osazones are beautiful yellow crystalline compounds that look like flower petals under the microscope.

The oxidation of the sugar chain by phenylhydrazine does not proceed further than the second carbon, probably because of the stabilizing influence of the conjugated system in the osazone and hydrogen bridging (p. 617) between nitrogens to give a chelate ring.

8. Addition of Hydrogen Cyanide (Kiliani Synthesis). One method of lengthening a carbon chain in a sugar involves the addition of HCN to an aldose as the first step, a reaction described already (p. 209) as a property of aldehydes and ketones. The cyanohydrin can be hydrolyzed to the corresponding acid, which spontaneously forms a γ lactone ring.

$$HO-CH_2CH_2CH_2COOH \xrightarrow[\text{distill}]{-H_2O} \underset{\text{a }\gamma\text{ lactone}}{\overset{\displaystyle CH_2-CH_2-C=O}{\underset{\displaystyle CH_2\text{————}O}{|\qquad\qquad|}}} \qquad [19\text{-}18]$$

a γ hydroxy acid

A γ lactone is formed from a γ hydroxy acid by the loss of a molecule of water between the two functional groups. Lactones are

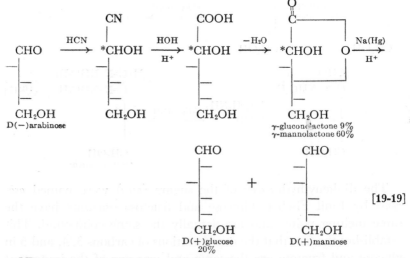

* The carbon marked with the asterisk may have either of two configurations, and hence a mixture of D(+)mannose and D(+)glucose is obtained from D(−)arabinose, as indicated in the last step. The calcium salt of gluconic acid can be separated before the reduction is run, and the yield (20%) of D(+)glucose is based on calcium gluconate as starting compound.

intramolecular esters. The lactones of sugar acids are easily reduced to aldehyde-alcohols by sodium amalgam in dilute sulfuric acid solution. This method of lengthening the carbon chain, when applied to a sugar, is known as the Kiliani synthesis.

9. Degradation: the Wohl-Zemplén Method. The process of degrading a sugar by one carbon (converse of the reaction described in paragraph 8 above) may begin with an oxime derivative of an aldose. An oxime is dehydrated to a nitrile by a dehydrating agent such as acetic anhydride or phosphorus pentoxide. When acetic anhydride is used on D(+)glucose oxime, a pentacetate of the nitrile is formed. The Wohl-Zemplén method, using sodium acetate with acetic anhydride, requires an elevated temperature for the nitrile formation. The nitrile pentacetate is converted to the next lower sugar in one step from this stage by sodium methoxide in chloroform.

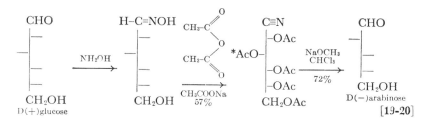

19-8. The Disaccharides†

Of the three important disaccharides, lactose, maltose, and sucrose, only the last is produced in large quantities. It is a leading

* The symbol Ac stands for CH_3-C-.
$$\overset{O}{\underset{\|}{}}$$

† In the following projection formulas for di- and poly-saccharides, the glucose unit has been arbitrarily pictured in the same sense (–CH_2OH group at the top in cyclic structures and –CHO at the top in open-chain structures) in every formula for quick recognition. This means that α and β linkages are sometimes shown in distorted long bonds. It has seemed better pedagogically to do this rather than to invert some monosaccharide units, a procedure that allows for quick recognition by most students only when they stand on their heads.

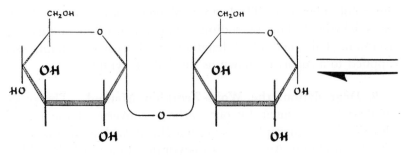

α - MALTOSE

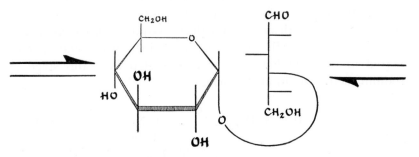

ALDEHYDE FORM

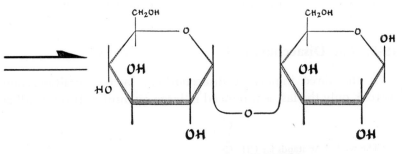

β - MALTOSE

[19-21]

organic chemical, on a tonnage basis, and is produced in almost 100% purity.

1. Maltose. Maltose may be obtained from starch by partial hydrolysis catalyzed by the enzyme diastase.* Further hydrolysis of maltose yields two molecules of glucose. The two glucose units are joined in an acetal linkage at the No. 4 carbon of one unit. Maltose is a reducing sugar and so is considered to have an aldehyde group in equilibrium with two possible hemiacetal structures, shown in Exp. 19-21 as the α and β forms. A very small amount of the aldehyde form is sufficient to give the responses characteristic of a reducing sugar.

Exercise 19-8. Write a structure for maltose osazone.

2. Lactose. Lactose is called milk sugar because of its source. Lactose is an acetal, or glycoside, of galactose with glucose, at

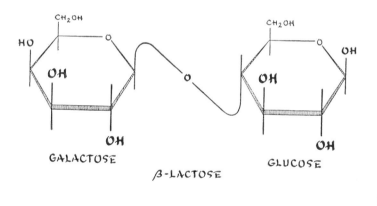

[19-22]

carbon 4 of the glucose unit (Exp. 19-22). It is a reducing sugar and forms an osazone, a fact that indicates a free carbonyl group. The β form of the lactose molecule is the one occurring in milk.

* An enzyme (modern name, biocatalyst) is a catalyst for a reaction taking place in living matter. See also page 417.

3. Sucrose.   Sucrose, in contrast to the other two disaccharides, is a nonreducing sugar. This is accounted for in the structure, which has been demonstrated to have two monosaccharides joined in two acetal linkages. Hence sucrose cannot exhibit the tautomerism with a straight-chain carbonyl structure that accounts for the reducing property of maltose and lactose. The two sugar units in sucrose are glucose and fructose, joined as shown in Exp. 19-23.

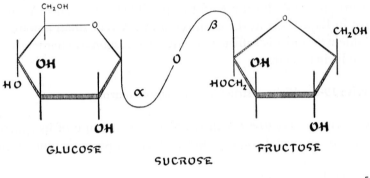

[19-23]

When sucrose is hydrolyzed, a change in the sign of rotation takes place. Sucrose has a specific rotation of +66°, glucose of +52°, and fructose of −92°. The mixture of fructose and glucose obtained from this hydrolysis is therefore called invert sugar, and the change is called inversion. The rotation of invert sugar is −40°, the difference between glucose and fructose.

If sucrose is given a value of 100 on a sweetness scale, some other sugars have the values given in Table 19-2. It is to be noted that invert sugar is sweeter than sucrose. Bakeries and candy kitchens, therefore, profitably hydrolyze sucrose before using it.

19-9. Sweetening Agents

Sweetening agents are by no means limited to polyhydroxy aldehydes, and there are still no reliable rules for predicting whether a particular compound will be sweet or not. The wide variety of functions that may produce sweetness is hinted at in Table 19-2.

TABLE
19-2 | *Relative Sweetness of Various Substances*

D(+)glucose	74	invert sugar	127
D(−)fructose	173	sodium saccharin	66,700
maltose	32	Dulcin	34,600
sucrose, 2% solution	100	2-propoxy-5-nitroaniline	400,000
lactose	16	sodium cyclohexylsulfamate	5,100

Saccharin is probably the best-known sweetening agent besides the sugars. It is 667 times as sweet as sucrose (2% solution), but it has a bitter after-taste for some people. Dulcin, whose formula is

$$
\begin{array}{c} SO_2 \\ \diagdown N^- \ Na^+ \\ C \\ \| \\ O \end{array}
$$

sodium saccharin

C_2H_5–O–⟨ ⟩–$NHCONH_2$, has recently been found to produce cancer in rats and is no longer in use. The sweetest substance known, 2-propoxy-5-nitroaniline, is toxic. Sodium cyclohexylsulfamate, $C_6H_{11}NHSO_3Na$, a new sweetening agent (Sucaryl), has no after-taste and may compete favorably with saccharin.

19-10. The Polysaccharides

The polysaccharides cellulose, starch, and glycogen are all polymers of glucose, and complete hydrolysis of any of them yields D(+)glucose. One of the principal differences among the three

$$(C_6H_{10}O_5)_n + nH_2O \xrightarrow{H^+} nC_6H_{12}O_6 \qquad [19\text{-}24]$$

a polysaccharide   glucose

substances is the value of *n*. The value varies considerably with the method of determining it, but for starch the average is > 500 glucose units, for glycogen it is $> 1,000$, and for cellulose $n = 1,000–3,000$ glucose units. Another important difference is that starch and glycogen are made up of α-glucose units whereas cellulose has β-glucose linkages.

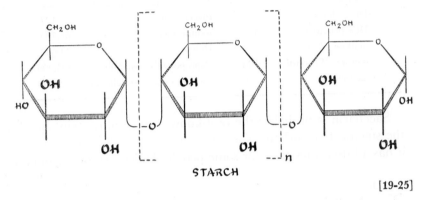

STARCH

[19-25]

Starch is the medium of carbohydrate storage for many plants, especially potatoes and the grains rice, wheat, rye, barley, oats, and corn, and glycogen occupies a like position in the animal world. Cellulose in a fairly pure state (90–100%) occurs only in cotton. Straw is about 30% cellulose, corn cobs 37%, and wood pulp 50–60%.

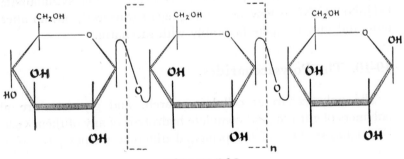

CELLULOSE

[19-26]

Starch is divided into two fractions if dissolved in hot water. The soluble fraction appears to be a long-chain molecule, as shown in Exp. 19-25. The water-insoluble part has been designated as a branched-chain molecule.

An oversimplified picture of the cellulose molecule is shown in Exp. 19-26.

19-11. Metabolism of Carbohydrates

The digestion of carbohydrates begins in the mouth by the action of ptyalin from the saliva. Ptyalin is an enzyme that catalyzes the conversion of starch to dextrins (shorter chains) and maltose (a disaccharide). Further hydrolysis takes place in the stomach before the hydrochloric acid present has a chance to deactivate the ptyalin. Most of the hydrolysis of starch and sugars, then, takes place in the intestine under the influence of the enzyme amylase. Dextrins, maltose, and other sugars are hydrolyzed to monosaccharides.

The monosaccharides pass through the walls of the intestine and are carried by the blood to the liver, where they are synthesized to glycogen. Apparently sugars other than glucose (principally fructose and galactose) are isomerized to glucose before the synthesis of glycogen. At any rate, glycogen contains only α-D-glucose units.

The oxidation of glycogen is the principal source of energy for life processes. In muscles the glycogen synthesized from blood glucose is converted through a phosphoric acid moiety to a glucose or fructose phosphate and then to a fructose diphosphate, which is split into three-carbon fragments by a complex process probably involving several intermediates. The three-carbon fragments are lactic acid and its immediate precursor, pyruvic acid, CH_3–CO–COOH. The final oxidation of lactic acid to carbon dioxide and water is accompanied by the release of considerable energy, some

$$CH_3\text{–}CHOH\text{–}COOH + 3O_2 \longrightarrow 3CO_2 + 3H_2O \qquad [\textbf{19-27}]$$

of which is used up to convert about four-fifths of the lactic acid back to glycogen. One-fifth of the lactic acid is used for muscle contraction to do work. If strenuous exercise causes an accumulation of lactic acid in the muscles, rest may be needed to allow the blood to transport enough oxygen to consume it. A tired feeling and stiffness in muscles may result from this accumulation of lactic acid.

19-12. Cellulose Derivatives

1. Cellulose Acetate. Cellulose acetate is prepared commercially by treatment of cellulose from wood with acetic anhydride and sulfuric acid (Exp. 19-28). If all the hydroxyl groups are acety-

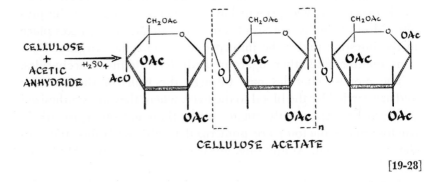

CELLULOSE ACETATE

[19-28]

lated, the product is still, on the average, only a triacetate. The product found to be most useful commercially contains an intermediate number of acetate groups between a tri- and a di-acetate. Such a cellulose acetate makes up the bulk of the products used for safety film, safety glass laminations, lacquers, and fibers such as Celanese.

2. Cellulose Nitrate. A mixture of nitric and sulfuric acids can be used to make a nitrate ester of cellulose. Complete esterification to a trinitrate is difficult, and the stage of esterification is consequently designated by the percentage nitrogen content. The mononitrate has a calculated nitrogen content of 6.7%, the dinitrate (Exp. 19-29) one of 11.1%, and the trinitrate one of 14.1%. When the nitrogen content is above 13%, the cellulose nitrate is classed as guncotton and is used in the manufacture of smokeless powder. It burns rapidly enough to be called explosive. When the

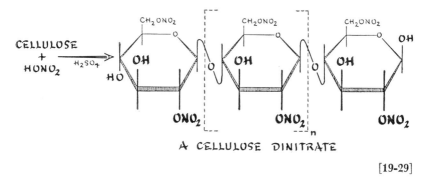

A CELLULOSE DINITRATE

[19-29]

nitrogen content is 11–12%, the cellulose nitrate is soluble in organic solvents, is inflammable but not explosive, and can be used for making plastics and lacquers. This pyroxylin, as it is called, when dissolved in an ether-alcohol mixture, is sold as collodion and used for protecting small cuts on the skin from dirt. It forms a flexible skin of cellulose nitrate after the solvent evaporates. Celluloid, one of the first plastics, was made of pyroxylin softened by camphor. Artificial leather is a fabric coated with a surface of cellulose nitrate.

3. Cellulose Ethers. Ethers of cellulose have been prepared, and the ethyl ether has become commercially important as a plastic material for sizing paper and for making transparent wrapping material, adhesives, and Ethoraon, a new competitor for rayon. Ethyl cellulose is made by letting ethyl chloride react with cellulose that has been treated with NaOH (Exp. 19-30). Two or three hydroxy groups per glucose unit are changed into ethers.

4. Regenerated Cellulose. All methods of making rayon (cellulose acetate and ethyl cellulose are not classed as rayon here) depend on putting cellulose in solution and reprecipitating it as a fiber. Various solvents, including the cuprammonium ion, have been used for the purpose, but more than 90% of the rayon in this country is made by the xanthate process (also called the viscose process).

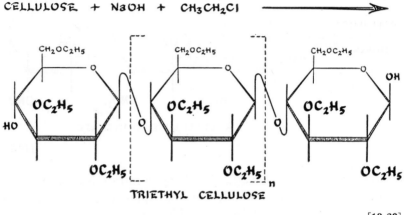

$$\text{CELLULOSE} + \text{NaOH} + \text{CH}_3\text{CH}_2\text{Cl} \longrightarrow$$

TRIETHYL CELLULOSE

[19-30]

This method involves a reaction of alcohols not previously mentioned. An alcohol will react with carbon disulfide in the presence of sodium hydroxide to form a xanthate (a dithiocarbonate), as in Expression 19-31.

$$\text{ROH} + \text{CS}_2 + \text{NaOH} \longrightarrow \text{R–O–}\overset{\overset{\text{S}}{\|}}{\text{C}}\text{–SNa} \qquad [19\text{-}31]$$

sodium alkyl
xanthate

In the xanthate process for making rayon, cellulose is treated with 18% NaOH for two hours and then allowed to age for two days. What happens during this aging process is not well understood, but the alkali cellulose, after this treatment, will react with CS_2 to give a water-soluble sodium cellulose xanthate (Exp. 19-32). A second aging process now gives a viscous (hence the name Viscose) solution from which one can regenerate the cellulose as fibers by squirting the solution through holes into sulfuric acid or as a sheet (cellophane) by running it through rollers into the acid.

The xanthate probably has only one dithiocarbonate linkage per glucose unit, and this is thought to be on carbon 2. Some degradation of the cellulose chain probably accompanies the alkali treat-

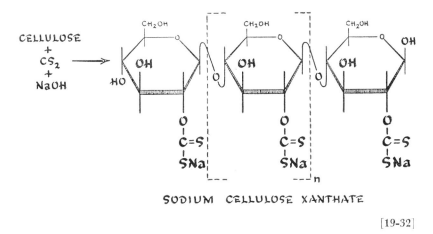

SODIUM CELLULOSE XANTHATE

[19-32]

ment and the regeneration in acid solution so that n is smaller, on the average, in the rayon than it was in the original cellulose molecule.

SUMMARY

One of the most difficult tasks in the history of carbon chemistry was the elucidation of the structures of the six-carbon sugars. Much of the credit goes to the monumental work of Emil Fischer. Only the bare bones of this story are given here. A hint of some of the difficulties encountered in sugar chemistry is given in the account of their reactions and the formation of derivatives from them. One method of lengthening the sugar chain by one carbon (Kiliani) and one method of degrading a sugar (Wohl-Zemplén) are given. A few properties of three disaccharides and of starch and cellulose are delineated.

REFERENCES

1. Fieser and Fieser, *Organic Chemistry*, 3rd ed. (D. C. Heath & Co., Boston, 1956), pp. 472–484, "Carbohydrate Metabolism."
2. Leicester and Klickstein, *A Source Book in Chemistry, 1400–1900* (McGraw-Hill Book Co., New York, 1952), pp. 499–506, Emil Fischer on sugars.
3. Dunn, "Use of Microorganisms in Production of Chemical Products," J. Chem. Education, 19, 387 (1942).
4. Lardy, "Vitamins and Carbohydrate Metabolism," J. Chem. Education, 25, 262 (1948).
5. Calvin, "Path of Carbon in Photosynthesis," J. Chem. Education, 26, 639 (1949).
6. Schoch, "Cellulose, Glycogen, and Starch," J. Chem. Education, 25, 626 (1948).
7. Heines, "John Mercer and Mercerization," J. Chem. Education, 21, 430 (1944).

EXERCISES

(Exercises 1–8 will be found within the text.)

9. Show the configuration of the products formed when $D(-)$fructose reacts with

1. nitric acid
2. sodium amalgam and water
3. phenylhydrazine
4. acetic anhydride
5. a trace of NaOH
6. methanol and dry HCl, followed by methyl iodide and Ag_2O

10. Balance the equation for the reaction occurring in 9-3, which involves oxidation-reduction.

11. Cite two reactions that lead to the conclusion that carbon atoms 3, 4, and 5 have similar configurations in $D(+)$glucose, $D(-)$fructose, and $D(+)$mannose.

12. Show how $D(-)$arabinose can be converted to $D(+)$glucose. Would any other hexose be produced simultaneously?

13. Show the conversion of $D(+)$galactose to

    1. $D(-)$lyxose
    2. α-D-methyl galactoside
    3. 2,3,4,6-tetramethyl-D-galactose
    4. β-D-galactose pentacetate
    5. an osazone

14. The results of the following experiments have been used to indicate the configuration of $D(+)$glucose. Show how this may be done, assuming that no inversion occurs in any step.

    1. $D(+)$glucose $\xrightarrow{Br_2}$ gluconic acid $\xrightarrow[Fe^{++}]{H_2O_2}$ 2,3,4,5-tetrahydroxypentanal $\xrightarrow{Br_2}$ 2,3,4,5-tetrahydroxypentanoic acid $\xrightarrow[Fe^{++}]{H_2O_2}$ 2,3,4-trihydroxy-butanal $\xrightarrow[HNO_3]{Br_2}$ meso tartaric acid

    2. gluconic acid $\xrightarrow{HNO_3}$ glucaric acid $\xrightarrow[Fe^{++}]{H_2O_2}$ 2,3-dihydroxybutan-1,4-dial $\xrightarrow{Br_2}$ l tartaric acid

    3. glucose $\xrightarrow[MnO_4^-]{cold}$ mixture of d and l tartaric acids

    Note: Fenton's reagent $(H_2O_2 + Fe^{++})$ causes loss of CO_2 from the COOH group and formation of an aldehyde at the former carbon atom 2.

15. Write the formula for the product obtained by complete methylation of maltose. Show the products obtained when this is hydrolyzed.

16. Suggest a simple chemical means of distinguishing between maltose and sucrose. Account, in terms of structure, for the difference in behavior between these two sugars.

17. A, B, and C are all aldohexoses. A and B yield the same osazones but on reduction give different hexahydroxy alcohols. B and C yield different osazones but the same alcohol. Show the configurational relationship of A, B, and C.

18. By the Kiliani synthesis, two sugars, D(+)glucose and D(+)mannose, are obtained from D(−)arabinose. Oxidation of 2,3,4,6-tetramethyl-glucose yields meso-2,3,4-trimethoxyglutaric acid. Similar oxidation of 2,3,4,6-tetramethylmannose yields *d* trimethoxyglutaric acid. Methylation of D(−)arabinose to 2,3,4-trimethylarabinose and subsequent oxidation also yield *d* trimethoxyglutaric acid. Assuming that the structure of D(−)arabinose is known, deduce the structure of D(+)mannose and D(+)glucose. Write the sequence of reactions described. (For the structure of glutaric acid see p. 446.)

19. How much Cu_2O is obtained by treatment of 5 g of n-butyraldehyde with excess Fehling's solution?

20. How much propionaldehyde is required to deposit 0.108 g of silver on a mirror from Tollen's reagent?

21. If the percentage of carbon in a compound is 40.0 and of hydrogen is 6.67, and the remainder is oxygen, what is the empirical formula? If the molecular weight is 180, what is the molecular formula?

20 Proteins and Amino Acids

A protein is a complex, organic, nitrogen-containing compound of high molecular weight that yields α amino acids upon hydrolysis. A protein may yield other substances as well in the hydrolytic process.

20-1. Occurrence of Proteins

Proteins occur in plants to an appreciable extent only in the seeds. In animals, on the other hand, proteins occur throughout the body. Hair, nails, feathers, horns, hoofs, skin, muscles, tendons, and nerve tissue are largely protein in character. Some examples of proteins and their estimated molecular weights are given in Table 20-1.

TABLE
20-1 | *Molecular Weights of Some Proteins*

Pepsin (an enzyme)	36,000	Hemoglobin in blood	68,000
Albumin in egg white	43,000	Gelatin	150,000
Casein in milk	98,000	Tobacco mosaic virus	50,000,000

Although proteins obtained from different sources vary in properties, the elementary composition is remarkably constant. The percentage composition is as follows: C, 51–55; H, 6–7; N, 15–18 (average 16); O, 20–24; S, 0–3; P, 0–1; other elements, <1.

20-2. Classification of Proteins

Proteins may be classified into two large groups by the results of their hydrolysis. The simple proteins yield only α amino acids on hydrolysis; the conjugated proteins yield other substances as well. A third class, called derived proteins, are substances produced from complex natural proteins by various reagents. The separation of these classes into various groups depends on solubility in various media and coagulation by various reagents as well as on the results of hydrolysis.

A. Simple Proteins.   The simple proteins are either fibrous (insoluble in neutral solvents) or soluble. The fibrous proteins include fibroin, the protein of silk; elastin, the protein of elastic tissue such as ligaments and arteries; and keratin, the protein in hair, horn, feathers, hoofs, and nails. The soluble proteins include:
  1. albumins, soluble in water and salt solutions: egg albumin, serum albumin from blood
  2. globulins, insoluble in water but soluble in neutral salt solutions such as $MgSO_4$ and $NaCl$: edestin from hemp seed
  3. glutelins, not in group 1 or 2 but soluble in dilute acids or bases: glutenin from wheat
  4. prolamines, insoluble in water or absolute ethanol but soluble in 80% ethanol: zein from corn
  5. histones, strongly basic, soluble in water but insoluble in ammonium hydroxide: globin from hemoglobin
  6. protamines, strongly basic, soluble in water and ammonium hydroxide: salmine from salmon sperm
B. Conjugated Proteins.   The hydrolysis of conjugated proteins yields some nonprotein substances. The foreign group is called a prosthetic group. Conjugated proteins include:
  1. chromoproteins, a colored group united with a simple protein: hematin from hemoglobin
  2. phosphoproteins, yielding phosphoric acid on hydrolysis: casein in milk

3. glycoproteins, having a carbohydrate as the prosthetic group: mucin in saliva
4. nucleoproteins, yielding a nucleic acid on hydrolysis: yeast
5. lecithoproteins, containing lecithin or some other phospholipid: lecithin from egg yolk

C. Derived Proteins. The fragments derived from proteins during hydrolysis in the alimentary canal are examples of derived proteins: proteoses, peptones, polypeptides, and peptides.

The coagulation of proteins can be accomplished by alcohol, bases and acids, and heat. An example of the last is the cooking of an egg white, which coagulates the albumin. The nature of the protein is altered so that it can never be returned to its original form. Proteins may often be coprecipitated with alkaloidal reagents such as morphine, strychnine, and tannic acid. Separations may often be thus effected.

20-3. The Essential Amino Acids

Proteins are polymeric in nature, and their properties have been revealed, in large measure, through their hydrolytic products. All proteins hydrolyze to mixtures of α amino acids, from which thirty different acids have been definitely identified. If one is not too stringent in the interpretation of rules for identification, two more may be added. This is a surprisingly small number of building blocks for the proteins of the great variety of plants and animals in the world.

There are three groups of these amino acids: (1) the neutral amino acids containing one amine group and one carboxylic acid group per molecule; (2) the basic amino acids containing two amine groups and one acid group; (3) the acidic amino acids containing one amine group and two acid groups.

In the following list, the α amino acids marked with an asterisk have been found to be essential to life in the animal body. These amino acids cannot be successfully synthesized from other substances in the body of a rat and hence must be present as such in the diet or present in proteins. The others in the list may be synthesized by the animal body from other substances in the diet. Absence of any of these essential amino acids leads to some form of malnutrition. Three of them, lysine, tryptophane, and **arginine,** have been shown to be essential to man.

1. glycine

H_2NCH_2COOH

2. alanine

$CH_3–CH–COOH$
       NH_2

3. *leucine

CH_3
       $\diagdown$
              $CH–CH_2–CH–COOH$
CH_3 NH_2

4. *isoleucine

$CH_3–CH_2–CH—CH–COOH$
              CH_3 NH_2

5. norleucine

$CH_3–CH_2–CH_2–CH_2–CH–COOH$
                         NH_2

6. *lysine

$H_2N–CH_2–CH_2–CH_2–CH_2–CH–COOH$
                              NH_2

7. *arginine

$H_2N–C–NH–CH_2–CH_2–CH_2–CH–COOH$
       NH NH_2

8. *valine

CH_3
$\quad$ $CH-CH-COOH$
CH_3 $\qquad$ NH_2

9. serine

$HOCH_2-CH-COOH$
$\qquad NH_2$

10. cystine

$HOOC-CH-CH_2-S-S-CH_2-CH-COOH$
$\qquad NH_2 \qquad\qquad\quad NH_2$

11. *threonine

$CH_3-CH-CH-COOH$
$\qquad OH\ NH_2$

12. *methionine

$CH_3-S-CH_2-CH_2-CH-COOH$
$\qquad\qquad\qquad NH_2$

13. aspartic acid

$HOOC-CH_2-CH-COOH$
$\qquad\qquad NH_2$

14. glutamic acid

$HOOC-CH_2-CH_2-CH-COOH$
$\qquad\qquad\qquad NH_2$

15. proline

$CH_2\!\!-\!\!-\!\!CH_2$
$CH_2 \qquad CH-COOH$
$\qquad NH$

16. hydroxyproline

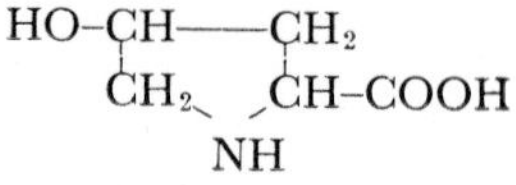

$$HO-CH\!-\!\!-\!CH_2$$
$$CH_2\quad CH\!-\!COOH$$
$$NH$$

17. *tryptophane

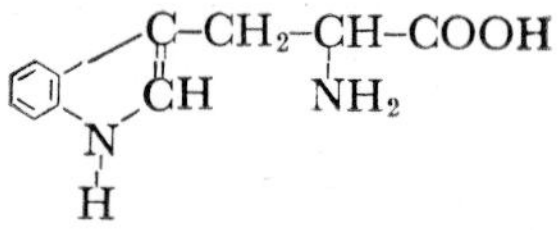

$$C\!-\!CH_2\!-\!CH\!-\!COOH$$
$$CH\qquad NH_2$$
$$N$$
$$H$$

18. *histidine

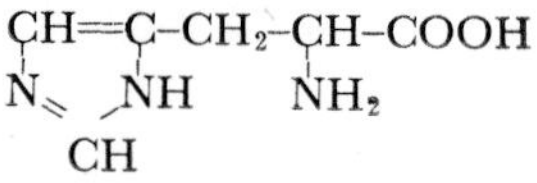

$$CH\!=\!C\!-\!CH_2\!-\!CH\!-\!COOH$$
$$N\quad NH\qquad NH_2$$
$$CH$$

19. *phenylalanine

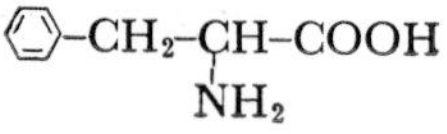

$$CH_2\!-\!CH\!-\!COOH$$
$$NH_2$$

20. tyrosine

$$HO\!-\!\!\!-\!CH_2\!-\!CH\!-\!COOH$$
$$NH_2$$

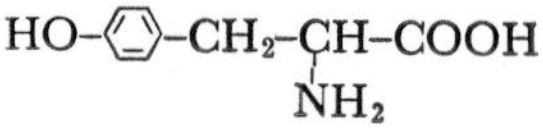

All amino acids that occur in proteins are optically active except glycine, and all these amino acids have the L configuration in relation to $D(+)$glyceraldehyde. Some animals are able to assimilate a D amino acid by changing it into the L compound. If these animals are fed a racemic amino acid, they metabolize the whole. Other animals eliminate the D form from the body.

20-4. Amino Acids in Proteins

The α amino acid content varies widely from one protein to another although the total percentage of nitrogen is nearly constant. The percentage composition of a few proteins containing glycine, alanine, glutamic acid, and proline is shown in Table 20-2.

TABLE 20-2	Percentage Composition of Various Proteins in α Amino Acids			
PROTEIN	GLYCINE	ALANINE	PROLINE	GLUTAMIC ACID
gelatine	25	9	10	11
silk	35	23	1	11
casein	0.5	1.9	8	22
zein	0.0	10	9	31
hair*	4.7	1.5	3.4	0.0

* Contains 19% cystine.

20-5. Metabolism of Proteins

When food containing proteins is taken into the body, the first important action is due to the enzyme pepsin in the acid medium of the stomach. (See p. 413 for enzyme.) Some degradation of the protein takes place at this stage. Further hydrolysis and finally complete hydrolysis proceed in the small intestine in the presence of enzymes, among which trypsin and erepsin are important.

$$\text{protein} \xrightarrow[\substack{H^+ \text{ in} \\ \text{stomach}}]{\text{pepsin}} \text{proteoses} \xrightarrow[\substack{OH^- \text{ in} \\ \text{intestine}}]{\text{trypsin}} \text{peptones} \xrightarrow{\text{erepsin}} \text{polypeptides} \longrightarrow$$

$$\text{simple peptides} \longrightarrow \alpha \text{ amino acids} \qquad [20\text{-}1]$$

Proteoses, peptones, and polypeptides are polymers of α amino acids of decreasing chain length. They constitute one group of derived proteins (p. 426). It is not to be understood that hydrolysis catalyzed by pepsin stops at the proteose stage, but, in general, hydrolysis in the stomach leaves large fragments of the protein molecule. Trypsin splits proteins only to the proteose-peptone stage, but the average degradation is nearer completion than the action catalyzed by pepsin. Erepsin, once thought to be a single enzyme, is now known to be a group of enzymes that act on the derived proteins of lower molecular weight.

After hydrolysis the blood receives the α amino acids through the walls of the intestine and carries them to the cells, where one of three things may happen. (1) The amino acid may be synthesized back to a body protein. (2) Oxidation may take place to provide energy. For alanine, as an example, an over-all picture of what

$$\underset{\substack{\text{L(+)alanine}}}{\underset{\overset{\displaystyle |}{\text{NH}_2}}{\text{CH}_3\text{-CH-COOH}}} \xrightarrow{\text{oxidation}} \underset{\overset{\displaystyle \|}{\text{NH}}}{\text{CH}_3\text{-C-COOH}} \xrightarrow{\text{H}_2\text{O}}$$

$$\underset{\substack{\text{pyruvic acid}}}{\text{NH}_3 + \underset{\overset{\displaystyle \|}{\text{O}}}{\text{CH}_3\text{-C-COOH}}} \xrightarrow{\text{oxidation}} \text{CO}_2 + \text{H}_2\text{O} + \text{energy} \qquad [20\text{-}2]$$

happens is provided by the transformations shown in Expression 20-2. The carbon dioxide and ammonia are eliminated from the body as urea, which is synthesized in the liver. (3) If the diet is

$$2\text{NH}_3 + \text{CO}_2 \xrightarrow{\text{liver}} \underset{\substack{\text{urea}}}{\underset{\displaystyle \diagdown \text{NH}_2}{\overset{\displaystyle \diagup \text{NH}_2}{\text{C=O}}}} + \text{H}_2\text{O} \qquad [20\text{-}3]$$

low in carbohydrates or fats, proteins may be transformed into either of these or used to make hormones, purines, or other body necessities.

20-6. Abnormal Metabolism

If a protein that has not been completely hydrolyzed gets through the walls of the intestine, it acts as a violent poison. Serum sickness is caused by traces of protein carried over from the animal that furnished the serum. Venoms from snakes are poisonous because they introduce proteins directly into the blood. Some allergies to food are ascribed to the ability of proteins in particular persons to escape hydrolysis before entering the blood. Asthma and hay fever may be due to foreign proteins from weed pollen entering the body.

20-7. Preparation of Amino Acids

A. α Amino Acids

1. Strecker Synthesis. Aldehydes add hydrogen cyanide to form cyanohydrins. In the presence of ammonia, the OH group in a cyanohydrin is replaced by the NH_2 group. Subsequent hydrolysis of the amino nitrile yields an amino acid. If sodium cyanide is allowed to react with ammonium chloride, the two reagents HCN and NH_3 are prepared *in situ*, and the cyanohydrin need not be isolated.

$$CH_3\text{-CHO} + HCN \longrightarrow CH_3\text{-}\underset{\underset{CN}{|}}{\overset{\overset{H}{|}}{C}}\text{-OH} \xrightarrow{NH_3} CH_3\text{-}\underset{\underset{NH_2}{|}}{\overset{\overset{H}{|}}{C}}\text{-CN} \xrightarrow[HCl]{2H_2O}$$

$$CH_3\text{-}\underset{\underset{NH_2\cdot HCl}{|}}{CH}\text{-COOH} \xrightarrow[60\%]{Pb(OH)_2} CH_3\text{-}\underset{\underset{NH_2}{|}}{CH}\text{-COOH} \quad [20\text{-}4]$$
$$\underset{dl \text{ alanine hydrochloride}}{} \qquad\qquad \underset{dl \text{ alanine}}{}$$

Serine is now prepared by this method, with ethyl cellosolve as the starting compound. (For another example of the Strecker synthesis, see p. 517.)

$$CH_3\text{-}CH_2\text{-O-}CH_2\text{-}CH_2OH \xrightarrow[35\%]{Cu, 325°} CH_3\text{-}CH_2\text{-O-}CH_2\text{-CHO} \xrightarrow[NH_4Cl]{NaCN}$$
$$\underset{\text{ethyl cellosolve}}{}$$

$$CH_3\text{-}CH_2\text{-O-}CH_2\text{-}\underset{\underset{NH_2}{|}}{\overset{\overset{H}{|}}{C}}\text{-CN} \xrightarrow[\substack{\text{heat \&} \\ \text{pressure} \\ 51\%}]{HBr} HOCH_2\text{-}\underset{\underset{NH_2\cdot HBr}{|}}{CH}\text{-COOH} \quad [20\text{-}5]$$
$$\underset{dl \text{ serine hydrobromide}}{}$$

2. Ammonolysis (Hofmann). The Hell-Volhard-Zelinsky method of preparing α halogen acids is classical (Chap. 21). In the presence of red phosphorus, the halogens will substitute on the α carbon of an acid. The reaction probably proceeds through acid chloride formation (see Chap. 22). Ammonia will then react with the α halogen acid to yield an α amino acid.

$$CH_3\text{-}CH_2\text{-}CH_2\text{-}CH_2\text{-}CH_2\text{-}COOH + Br_2 \xrightarrow{P}$$

$$CH_3\text{-}CH_2\text{-}CH_2\text{-}CH_2\text{-}\underset{Br}{CH}\text{-}COOH \xrightarrow[67\%]{2NH_3} CH_3\text{-}CH_2\text{-}CH_2\text{-}CH_2\text{-}\underset{NH_2}{CH}\text{-}COOH$$

$$dl \text{ norleucine} \qquad [20\text{-}6]$$

3. Gabriel's Synthesis. Gabriel's synthesis of primary amines can be applied generally to the preparation of amino acids. Ethyl α-chloroacetate, for example, reacts with potassium phthalimide

to yield an intermediate that can be hydrolyzed to phthalic acid and glycine hydrochloride. The free amino acid may be obtained by precipitation of the chloride ion with $Pb(OH)_2$ or Ag_2O. How would you remove the phthalic acid from the mixture before isolation of the free amino acid?

B. β Amino Acids

A 1,4 addition of ammonia to an α,β unsaturated acid provides the best general method of obtaining β amino acids. In a closed tube crotonic acid will add NH_3 to give β-aminobutyric acid.

$$CH_3\text{-}CH{=}CH\text{-}COOH + NH_3(aq.) \xrightarrow[61\%]{\substack{\text{closed tube,} \\ 140°}} CH_3\text{-}\underset{NH_2}{CH}\text{-}CH_2\text{-}COOH \qquad [20\text{-}8]$$

The postulated intermediate, if the addition is 1,4, is shown in brackets in Expression 20-9.

$$CH_3-CH=CH-COOH + \overset{\delta+}{H}\ \overset{\delta-}{NH_2} \longrightarrow \left[CH_3-\underset{NH_2}{CH}-CH=C\overset{OH}{\underset{OH}{\diagdown}} \right] \overset{enolic\ H}{\underset{shifts}{\longrightarrow}}$$

$$CH_3-\underset{NH_2}{CH}-CH_2-COOH \qquad [20\text{-}9]$$

C. γ and δ Amino Acids

Acids with an amine group in the γ or δ position are readily prepared by ammonolysis of the corresponding halogen acid.

$$BrCH_2-CH_2-CH_2-COOH + NH_3 \longrightarrow H_2N-CH_2-CH_2-CH_2-COOH \qquad [20\text{-}10]$$
$$\text{γ-bromobutyric acid} \qquad\qquad\qquad \text{γ-aminobutyric acid}$$

20-8. Chemical Properties of Amino Acids

Many of the properties of amino acids are characteristic of the disparate functional groups; other properties involve both functional groups or are modified by the presence of the two groups in the same compound.

1. Effect of Heat. The effect of heat on an amino acid depends on whether the amine group is α, β, or γ to the carboxyl group. In α amino acids, heating causes a loss of water and the formation of a cyclic dimer, called a diketopiperazine.

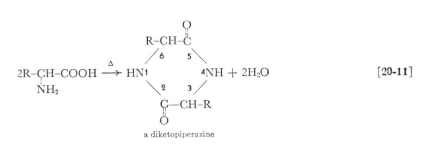

$$2R-\underset{NH_2}{CH}-COOH \xrightarrow{\Delta} \qquad\qquad\qquad [20\text{-}11]$$

a diketopiperazine

The β amino acids lose ammonia readily. This is the reverse of the preparative reaction illustrated in Expression 20-8. Indeed, the addition of NH_3 to an α,β unsaturated acid is reversible. A closed

$$CH_3\text{-}\underset{NH_2}{\underset{|}{CH}}\text{-}CH_2\text{-}COOH \underset{\underset{\text{closed tube}}{\Delta}}{\overset{\Delta}{\rightleftarrows}} CH_3\text{-}CH=CH\text{-}\overset{\overset{O}{\|}}{C}\text{-}OH + NH_3 \qquad [20\text{-}12]$$

system favors the 1,4 addition, but heating in an open system allows the ready loss of NH_3. This is an application of Le Châtelier's principle.

By loss of water a γ amino acid can form a stable five-membered ring within the molecule. This intramolecular dehydration therefore takes precedence over the intermolecular dehydration that occurs with α amino acids. The products are called γ lactams and

$$H_2N\text{-}CH_2\text{-}CH_2\text{-}CH_2\text{-}COOH \longrightarrow \underset{\underset{\text{a }\gamma\text{ lactam}}{CH_2\text{-}NH}}{\overset{CH_2\text{-}CH_2}{\underset{|}{\overset{|}{}}}}\hspace{-0.5em}\diagdown C{=}O + H_2O \qquad [20\text{-}13]$$

are analogous to the γ lactones formed by sugar acids (see p. 410) and γ hydroxy acids (see Chap. 22).

Exercise 20-1. Show by an equation the effect of heat on leucine; on β-alanine; on aspartic acid.

2. Salt Formation. The presence of both an acidic and a basic group endows an amino acid with salt-like properties. The melting points of amino acids are unusually high in comparison with those of other carbon compounds having similar molecular weights. (See Table 20-3.) The structure of glycine may perhaps be better represented as an inner salt having the structure $\overset{+}{H_3N}\text{-}CH_2\text{-}CO\overset{-}{O}$ than by the simple formula H_2NCH_2COOH. The German word *Zwitterion* is descriptive here, and we shall anglicize it hereafter.

The electrostatic forces between ions in crystals of salts such as sodium chloride are so great that considerable heat is necessary to

TABLE
20-3 | *Melting Points of Various Substituted Acetic Acids*

$H_2N–CH_2–COOH$	232°	$HOOC–CH_2–COOH$	135°
$HO–CH_2–COOH$	79°	$H–CH_2–COOH$	16°
$Cl–CH_2–COOH$	63°	$NC–CH_2–COOH$	66°

overcome these forces and melt the crystals. Forces of the same nature act in these amino acids (zwitterions) and the melting points are consequently high, though not as high as those of salts. (Compare NaCl, of melting point 800°C, with glycine, of melting point 232°C.)

Strong acids or bases will release a proton to, or accept one from, the zwitterion, to give a cation or an anion of the amino acid. Amino acids may therefore be called amphoteric in character.

$$\overset{+}{H_3}N–CH_2–\overset{-}{COO} + \overset{+}{H_3O} \longrightarrow \overset{+}{H_3}N–CH_2–COOH + H_2O \qquad [20\text{-}14]$$

$$\overset{+}{H_3}N–CH_2–\overset{-}{COO} + \overset{-}{OH} \longrightarrow H_2N–CH_2–\overset{-}{COO} + H_2O \qquad [20\text{-}15]$$

3. Esterification. The carboxylate group in a zwitterion must be changed to a carboxyl group before the amino acid is easily esterified. We can do this conveniently by forming the amine hydrochloride and esterifying the acid at the same time with alcoholic HCl. Neutralization of the hydrochloride gives the free ester.

$$\overset{+}{H_3}N–CH_2\overset{-}{COO} \xrightarrow{HCl} \underset{NH_2·HCl}{CH_2–COOH} \xrightarrow[95\%]{C_2H_5OH} \underset{NH_2·HCl}{CH_2–COOC_2H_5} \qquad [20\text{-}16]$$

Exercise 20-2. Show by an equation how to prepare diethyl glutamate.

4. Acetylation. At 100°C acetic anhydride will acetylate glycine to N-acetylglycine. Covering up the amine group by making an amide out of it leaves a free carboxyl group, and the compound has the properties of an ordinary carboxylic acid.

$$CH_2\text{-COOH} + CH_3\text{-C}\underset{O}{\overset{O}{\diagdown}} \xrightarrow{81\%} CH_3\text{-}\underset{O}{\overset{O}{C}}\text{-}\overset{H}{N}\text{-CH}_2\text{COOH} + CH_3\text{COOH} \quad [20\text{-}17]$$

under the first reactant: NH_2

under the C: $CH_3\text{-C}\underset{O}{\overset{\diagup}{\diagdown}}$

below the product: N-acetylglycine

The esters of the amino acids are much more readily acetylated than the acids themselves, since the amine group is completely free in these derivatives (no zwitterion formation).

$$H_2NCH_2\text{-C}\underset{OC_2H_5}{\overset{O}{\diagup}} + \quad \begin{matrix} CH_3\text{-C}\overset{O}{\diagup} \\ O \\ CH_3\text{-C}\underset{O}{\diagdown} \end{matrix} \xrightarrow{76\%}$$

$$CH_3\text{-}\underset{O}{\overset{O}{C}}\text{-}\overset{H}{N}\text{-CH}_2\text{-COOC}_2H_5 + CH_3\text{COOH} \quad [20\text{-}18]$$

ethyl N-acetylglycine

Benzoyl chloride reacts with glycine to yield hippuric acid, a substance found in the urine of many animals, particularly the horse. In the laboratory, one may make hippuric acid by shaking

$$\bigcirc\text{-C}\underset{Cl}{\overset{O}{\diagup}} + CH_2\text{-COOH} + 2OH^- \xrightarrow{68\%} \bigcirc\text{-}\underset{O}{\overset{O}{C}}\text{-}\overset{H}{N}\text{-CH}_2\text{-COO}^- + Cl^- + H_2O \quad [20\text{-}19]$$

under the second reactant: NH_2

below the product: hippuric acid anion

benzoyl chloride with glycine in strong basic solution. The acid may be freed from the salt by subsequent acidification with a strong acid. The yield of 68% is based on chloroacetic acid, which is ammonolyzed to glycine in solution but not isolated.

When toxic substances are introduced into the body, externally or by putrefaction, the body may act in several ways to detoxicate the foreign substance. Oxidation, reduction (seldom), or hydrolysis frequently renders the substance nontoxic, but a fourth reaction often occurs. Acetic acid, glycine, or cystine may react with the substance to form a harmless product. Benzoic acid, for example, is converted to hippuric acid by glycine and is eliminated as such in the urine. When sulfanilamide is ingested, it is detoxicated and partially excreted as the acetylated derivative (Exp. 20-20).

$$H_2N-\langle\bigcirc\rangle-SO_2NH_2 + CH_3COOH \longrightarrow CH_3-\overset{\overset{O}{\|}}{C}-\overset{\overset{H}{|}}{N}-\langle\bigcirc\rangle-SO_2NH_2 + H_2O \quad [20\text{-}20]$$

sulfanilamide $\qquad\qquad\qquad\qquad$ p-acetylaminobenzene-
$\qquad\qquad\qquad\qquad\qquad\qquad\qquad$ sulfonamide

5. With Nitrous Acid. Since nitrous acid liberates nitrogen from the amine group in an amino acid in a quantitative manner, the reaction has been applied by Van Slyke as a means of determining primary amine groups in amino acids and proteins.

$$\begin{array}{c}CH_3 \\ {}^{\diagdown}CH-CH-COOH + HONO \xrightarrow[100\%]{} CH_3-CH-CH-COOH + N_2 + H_2O \\ CH_3 \quad NH_2 \qquad\qquad\qquad\qquad CH_3 \quad OH \qquad\qquad [20\text{-}21]\end{array}$$

6. Formation of Peptides. A diketopiperazine can be hydrolyzed in acid solution to a dipeptide—that is, an amide derivative of an amino acid.

$$\begin{array}{c}\overset{O}{\|}\overset{H}{|} \\ C-N \\ CH_2 \diagup\quad\diagdown CH_2 \xrightarrow[H_2O \\ 88\%]{HCl} H_2N-CH_2-\overset{\overset{O}{\|}}{C}-\overset{\overset{H}{|}}{N}-CH_2-COOH \qquad\qquad [20\text{-}22] \\ N-C \qquad\qquad\qquad glycylglycine \\ H\ O \end{array}$$

diketopiperazine

To increase the number of amide linkages until a polypeptide is obtained is a difficult problem, which was first taken up by Emil Fischer about 1900. Among the methods that he used was the action of an α halogen acyl halide on the ethyl ester of an amino acid. Hydrolysis of the ester and ammonolysis of the remaining chlorine then give a tripeptide, diglycylglycine. Although OH⁻ will hydrolyze esters, amides, and alkyl halides, all three of which are present in the case at hand, there is a considerable difference in the ease of hydrolysis. Fischer used 4% NaOH at 25° to hydrolyze chloroacetylglycylglycine ester to chloroacetylglycylglycine and 25% NH_4OH at 100° to effect the ammonolysis.

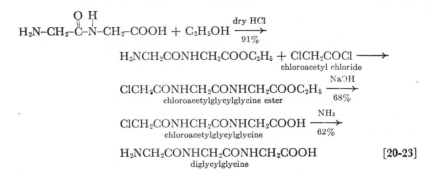

Fischer was able to use this method, along with others, to prepare a polypeptide containing eighteen amino acid units and having a molecular weight of 1,213. The compound he prepared was leucyltriglycylleucyltriglycylleucyloctaglycylglycine. This compound has many of the properties of proteins of low molecular weight and is one of the links in the theory that proteins are polypeptide chains having the general formula

A second observation supporting the peptide theory is that there are very few free amine groups in a protein. In the third place, proteins and polypeptides give the biuret test whereas amino acids do not. The biuret test consists of heating a protein with dilute copper II sulfate and a strong base, which causes a blue-violet color to appear. With shorter-chain peptides, the blue color lightens to pink, but no color is given with amino acids. Biuret, $H_2N-C-NH-C-NH_2$, reacts positively to the test, and its name has been given to this color reaction. The only functional group common to biuret, proteins, and polypeptides is the amide linkage. Hence the biuret test is characteristic of polyamides of this type.

Exercise 20-3. Write structural formulas for all possible tripeptides that can be made from one molecule each of glycine, alanine, and valine.

20-9. Nylon

Silk is largely protein, and nylon is man's imitation of silk. The discovery of nylon was not an accident at all, but the result of a conscious effort by a number of chemists to synthesize a silk-like substance. They even studied the extrusion apparatus of the silkworm in devising a spinneret for nylon.

Whereas a polypeptide is made up of polyamide linkages in which the amine group and the acid group appear in the same starting compound, nylon is made from two molecules, one of which contains two amine groups and the other two carboxyl groups. If we heat hexamethylenediamine and adipic acid together at low

$$\text{HOOC–(CH}_2)_4\text{–COOH} + \text{H}_2\text{N–(CH}_2)_6\text{–NH}_2 \longrightarrow$$

adipic acid   hexamethylenediamine

$$\text{H}_2\text{N–(CH}_2)_6\text{–}\overset{\text{H}}{\underset{}{\text{N}}}\text{–}\left[\overset{\text{O}}{\underset{}{\text{C}}}\text{(CH}_2)_4\text{–}\overset{\text{O}}{\underset{}{\text{C}}}\text{–}\overset{\text{H}}{\underset{}{\text{N}}}\text{–(CH}_2)_6\text{–}\overset{\text{H}}{\underset{}{\text{N}}}\right]_x\text{–}\overset{\text{O}}{\underset{}{\text{C}}}\text{–(CH}_2)_4\text{–}\overset{\text{O}}{\underset{}{\text{C}}}\text{–OH} \quad [20\text{-}24]$$

pressure and elevated temperature over an extended period of time, we obtain a substance of high molecular weight that has the repeating unit shown in Expression 20-24. This polymer can be cold-drawn into elastic threads that are not easy to wet.

The reactions by which the starting compounds are made are already known to the student. In commercial practice the reagents used in the various steps shown in Expression 20-25 may be different.

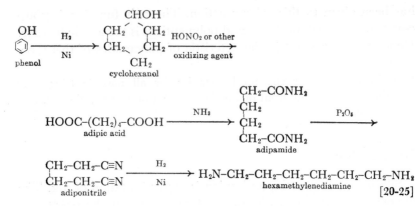

Adiponitrile can now be made from furfural (p. 512) in a process rivaling, commercially, the one just described.

Adiponitrile can now be made from furfural (p. 512) in a process
rivaling, commercially, the one just described.

SUMMARY

The properties of proteins have been deduced, in the main, from the study of their hydrolysis products, the α amino acids. Proteins contain about 16% nitrogen, have high molecular weights, exhibit widely varying solubilities, and are coagulated by various physical and chemical treatments.

Amino acids are neither amines nor acids but rather are salt-like in character. If either functional group is "covered up" by the proper derivative, the other group may function in a more normal way. The salt-like character of amino acids does not make new synthetic methods essential but the methods of isolation of the amino acids after synthesis must be modified with regard to separations since steam distillation and extraction of salts are limited in application.

The properties of amino acids are quite different, depending upon the distance of the amino group from the carboxyl group—α, β, γ, . . . ω.

REFERENCES

1. Merrick, "Plasma Expanders," J. Chem. Education, 30, 368 (1953).
2. Asimov, "Potentialities of Protein Isomerism," J. Chem. Education, 31, 125 (1942).
3. Vickery, "Liebig and Proteins," J. Chem. Education, 19, 73 (1942).
4. Färber, "Development of Protein Chemistry," J. Chem. Education, 15, 434 (1938).
5. Flodin, "Amino Acids and Proteins—Their Place in Human Nutrition Problems," Agricult. and Food Chem., 1, 222 (1953).

EXERCISES

(Exercises 1–3 will be found within the text.)

4. Show how to prepare 2-amino-4-methylpentanoic acid by Strecker's synthesis; by Gabriel's synthesis.

5. Write equations for reactions that take place in the following procedures and combinations:

   1. heating 3-amino-3-phenylpropanoic acid
   2. heating 2-amino-2-phenylethanoic acid
   3. heating 4-amino-4-methylpentanoic acid
   4. valine + nitrous acid
   5. alanine + sodium hydroxide
   6. alanine + hydrochloric acid

6. Show a synthesis of glycylleucylglycylleucine that starts with the amino acids and simple carboxylic acids.

7. Synthesize the following from available starting materials. Show all reagents and intermediate steps.

1. N-acetyvaline
2. 3-aminobutanoic acid
3. ethyl α-aminopropionate
4. 3,6-diethyl-2,5-diketopiperazine
5. ethyl β-aminobutyrate

21 Acids Containing Other Functional Groups

The chemical and physical properties of the mono-carboxylic acids were studied in Chapter 10. The next chapters will be devoted in part to the changes in properties occasioned by the introduction of a second function in a carboxylic acid—for example, a second carboxyl group, a halogen, a hydroxyl group, unsaturation, or a carbonyl group. One such polyfunctional type, the amino acid, has just been discussed.

Dicarboxylic Acids

21-1. Properties of the Dicarboxylic Acids

It has been a general observation that the melting points or boiling points of homologous compounds gradually rise as the number of carbons in the series increases. The solubility of such compounds in water decreases as the molecular weight increases by the addition of methylene groups. The dicarboxylic acids exhibit unexpected properties in these respects, as shown in Table 21-1.

The melting point of a dicarboxylic acid with an odd number of carbons is much lower and its solubility is much higher than that of either of its adjacent homologues. The acid strength of these dicarboxylic acids, however, diminishes in a regular manner as expected. The temperature at the melting point reflects the amount of energy needed to overcome the forces that hold the molecules

445

TABLE
21-1 | *Properties of Dicarboxylic Acids*

FORMULA	COMMON NAME	M.P. °C	g SOL. IN 100 g H_2O AT 20°C	*DISSOC. CONSTANT K_1 (ACID STRENGTH)
HOOC–COOH	oxalic acid	189	9.5	6×10^{-2}
HOOC–CH$_2$–COOH	malonic acid	135	73.5	1.8×10^{-3}
HOOC(CH$_2$)$_2$COOH	succinic acid	185	6.8	6.5×10^{-5}
HOOC(CH$_2$)$_3$COOH	glutaric acid	97	64.	4.5×10^{-5}
HOOC(CH$_2$)$_4$COOH	adipic acid	153	2.	3.9×10^{-5}
HOOC(CH$_2$)$_5$COOH	pimelic acid	103	5.	3.3×10^{-5}
HOOC(CH$_2$)$_6$COOH	suberic acid	140	0.16	3.1×10^{-5}
HOOC(CH$_2$)$_7$COOH	azelaic acid	106	0.24	2.8×10^{-5}
HOOC(CH$_2$)$_8$COOH	sebacic acid	133	0.10	2.6×10^{-5}

* For a definition of dissociation constant, see pp. 458ff.

in a crystal lattice. Why this should vary between dicarboxylic acids of odd- and even-numbered carbon content is not known with certainty.

Cis and *trans* acids show these same differences in melting point and solubility. Maleic acid has a lower melting point and higher solubility than fumaric acid (see Table 21-2). These differences are

TABLE
21-2 | *Properties of Geometric Isomers*

ACID		M.P. °C	SOLUBILITY IN 100 g H_2O AT 20°C
maleic acid	HC–COOH $\parallel$ HC–COOH	130°	60
fumaric acid	HOOC–C–H $\parallel$ H–C–COOH	subl. 200°	0.1

probably connected with the proximity of the carboxyl groups in the *cis* acid, which gives a different type of arrangement in the crystal lattice. The odd-numbered dicarboxylic acids (from the models) are more closely related to the *cis* acid than to the *trans* if it is assumed that there is some tendency for the carbon atoms

$$\begin{array}{ccc} & CH_2 & COOH \\ HOOC & & CH_2 \\ & \text{succinic acid} & \end{array} \qquad \begin{array}{ccc} & CH_2 & CH_2 \\ HOOC & CH_2 & COOH \\ & \text{glutaric acid} & \end{array}$$

to remain in a single plane. If the reason for this anomaly in melting points could be found, the reason for the peculiar solubilities would probably be evident since the two properties are closely related in many known compounds.

21-2. Nomenclature of the Dicarboxylic Acids

The common names of these acids were generally derived from the natural source of the acid and are not systematized. The IUC name is obtained from the parent hydrocarbon of the same number of carbon atoms by the addition of *-dioic*. Malonic acid, for example, is by this system called propanedioic acid.

21-3. Preparation of the Dicarboxylic Acids

1. Hydrolysis of Nitriles. The general methods of preparing carboxylic acids can be applied to the dicarboxylic acids. Glutaric acid is frequently made by hydrolysis of the nitrile with trimethylene glycol as the starting compound.

$$\begin{array}{ccccccc} CH_2OH & & CH_2Br & & CH_2-C\equiv N & & CH_2-COOH \\ CH_2 & \xrightarrow[\substack{H_2SO_4 \\ 95\%}]{HBr} & CH_2 & \xrightarrow[86\%]{NaCN} & CH_2 & \xrightarrow[85\%]{HCl} & CH_2 \\ CH_2OH & & CH_2Br & & CH_2-C\equiv N & & CH_2-COOH \\ & & & & & & \text{glutaric acid} \end{array} \qquad [21\text{-}1]$$

2. Oxidation of Various Compounds. As we saw on page 442, adipic acid is readily obtained by the oxidation of cyclohexanol.

When the ring breaks after the formation of cyclohexanone, a carboxyl group is obtained on each end of the chain. The oxidation can be carried out catalytically on a commercial scale or by nitric acid and other oxidizing agents in the laboratory.

$$
\begin{array}{ccc}
\text{CHOH} & \overset{\text{O}}{\underset{\|}{\text{C}}} & \text{CH}_2\text{–COOH} \\
\underset{\text{CH}_2}{\overset{\text{CH}_2 \quad \text{CH}_2}{\mid \quad \mid}} \xrightarrow{\text{oxidation}} & \underset{\text{CH}_2}{\overset{\text{CH}_2 \quad \text{CH}_2}{\mid \quad \mid}} \xrightarrow[60\%]{\text{HONO}_2} & \underset{\text{CH}_2\text{–COOH}}{\overset{\text{CH}_2}{\underset{\text{CH}_2}{\mid}}}
\end{array}
\qquad [21\text{-}2]
$$

Vigorous oxidation of an alkene will yield an acid. When oleic acid is oxidized by nitric acid, one of the products is a dicarboxylic acid, azelaic.

$$
\text{CH}_3(\text{CH}_2)_7\text{CH=CH}(\text{CH}_2)_7\text{COOH} \xrightarrow{\text{HONO}_2} \text{CH}_3(\text{CH}_2)_7\underset{\text{OH}}{\text{CH}}\text{–}\underset{\text{OH}}{\text{CH}}(\text{CH}_2)_7\text{COOH}
$$

$$
\text{CH}_3(\text{CH}_2)_7\underset{\text{O}}{\overset{}{\text{C}}}\text{-}\underset{\text{O}}{\overset{}{\text{C}}}(\text{CH}_2)_7\text{COOH} \xrightarrow{\text{HONO}_2} \text{CH}_3(\text{CH}_2)_7\text{COOH} + \text{HOOC}(\text{CH}_2)_7\text{COOH}
$$

azelaic acid

$$[21\text{-}3]$$

Exercise 21-1. Balance each of the three oxidation-reduction reactions suggested by the steps in Expression 21-3 on the assumption that the nitric acid is all reduced to NO.

Catalytic oxidation of benzene or naphthalene yields a dicar-

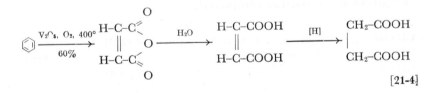

$$[21\text{-}4]$$

boxylic acid anhydride. In the case of benzene, the maleic anhydride obtained can be hydrolyzed and reduced to succinic acid.

$$\text{naphthalene} \xrightarrow[80\%]{V_2O_5,\ O_2,\ 350°} \begin{array}{c} C=O \\ \diagdown O \\ C=O \end{array} \qquad [21\text{-}5]$$

The two isomers of phthalic acid may be obtained by oxidation of the corresponding xylenes.

$$\begin{array}{c} CH_3 \\ \diagdown\text{-}CH_3 \end{array} \xrightarrow[95\%]{MnO_4^-} \begin{array}{c} COOH \\ \diagdown\text{-}COOH \\ \text{isophthalic acid} \end{array} \qquad [21\text{-}6]$$

Exercise 21-2.   Complete and balance the oxidation in Expression 21-6, using MnO_4^- in basic solution as the oxidizing agent.

TABLE
21-3   *Isomers of Phthalic Acid*

FORMULA	COMMON NAME	M.P. °C	SOLUBILITY IN 100 g H_2O AT 25°C
$o\text{-}C_6H_4(COOH)_2$	phthalic acid	206 dec.	0.7
$m\text{-}C_6H_4(COOH)_2$	isophthalic acid	300	0.01
$p\text{-}C_6H_4(COOH)_2$	terephthalic acid	300 subl.	0.001

21-4. Reactions of Dicarboxylic Acids

Dicarboxylic acids form the ordinary acid derivatives in a manner similar to that of the monocarboxylic acids.

Exercise 21-3.   Write reactions for the formation of the following derivatives of succinic acid: amide, salt, ester, nitrile, acid chloride.

A. *Effect of Heat*

1. On Oxalic Acid.   The reaction brought about by the heating of a dicarboxylic acid may take one of several paths, depending on the relative positions of the two acid groups. When oxalic acid is heated, it decomposes into CO, CO_2, and H_2O. The decomposition takes place readily in the presence of H_2SO_4 and is a convenient

$$\begin{array}{l} COOH \\ | \\ COOH \end{array} \xrightarrow{\Delta} CO + CO_2 + H_2O \qquad\qquad\qquad [21\text{-}7]$$

way to prepare CO in the laboratory. The CO_2 is absorbed by bubbling through NaOH solution, in which the CO is not soluble. One may effect the loss of CO_2 only, to obtain formic acid, by heating oxalic acid with glycerol. Probably esterification takes place first, then the loss of CO_2, and then hydrolysis, according to Expression 21-8.

$$\begin{array}{ll} COOH & CH_2OH \\ | & | \\ COOH + CHOH \\ & | \\ & CH_2OH \end{array} \longrightarrow \begin{array}{l} COO\text{-}CH_2\text{-}CH\text{-}CH_2 \\ | \qquad\quad | \quad | \\ COOH \quad\;\; OH\; OH \end{array} \longrightarrow$$

$$CO_2 + H\overset{\displaystyle O}{\overset{\|}{C}}\text{-}O\text{-}CH\text{-}CH\text{-}CH_2 \xrightarrow{H_2O} H\text{-}C\overset{\displaystyle O}{\diagdown} + \underset{OH\;\; OH\; OH}{CH_2\text{-}CH\text{-}CH_2} \quad [21\text{-}8]$$

2. On Malonic Acid.   Malonic acid and alkyl-substituted malonic acids do not form cyclic anhydrides but lose CO_2 quite easily at their melting points. *This is the characteristic behavior of all acids*

$$CH_2\overset{\displaystyle COOH}{\underset{\displaystyle COOH}{\diagup}} \xrightarrow{\Delta} CH_3\text{-}COOH + CO_2 \qquad\qquad [21\text{-}9]$$

with two carboxyl groups on the same carbon and of monocarboxylic acids with a strong negative group on the α *carbon.* When a strong dehydrating agent such as P_2O_5 is present, the anhydride of malonic acid, a very reactive compound called carbon suboxide, is obtained.

$$CH_2\overset{\displaystyle COOH}{\underset{\displaystyle COOH}{\diagup}} \xrightarrow[25\%]{P_2O_5} \underset{\text{carbon suboxide}}{O=C=C=C=O} + 2H_2O \qquad\qquad [21\text{-}10]$$

3. On Succinic and Glutaric Acids. When the dicarboxylic acid can form a cyclic anhydride involving a five- or six-membered ring, that is the preferred path of the reaction induced by heat alone. This possibility exists for the four- and five-carbon dicarboxylic acids. This is the same type of behavior as that exhibited

$$
\begin{array}{ccc}
\text{CH}_2\text{--COOH} & & \text{CH}_2\text{--C} \diagup^{\text{O}} \\
| & \xrightarrow{\Delta} & | \quad\quad\text{O} \qquad\qquad \text{[21-11]} \\
\text{CH}_2\text{--COOH} & & \text{CH}_2\text{--C} \diagdown \\
\text{succinic acid} & & \qquad\quad\text{O} \\
& & \text{succinic anhydride}
\end{array}
$$

by the γ and δ amino acids upon heating and is a corollary of Baeyer's strain theory. The formation of the cyclic anhydrides is greatly aided by the use of dehydrating agents such as acetic anhydride, PCl_3, $POCl_3$, and P_2O_5.

4. On Adipic and Higher Acids. Only traces of a seven-membered ring compound are obtained if adipic acid is heated; instead, the reaction yields a polymeric anhydride of relatively high molecular weight. Dicarboxylic acids containing more than six carbons behave in a similar fashion. If we heat sebacic acid, for example, we obtain after some time a polymer that we can then further transform, by heating for a long time at 200° and low

[21-12]

pressure, into a polymer of much higher molecular weight. The removal of traces of water, by greatly increasing the chain length (y is much larger than x), yields what has been called a super-polyanhydride. This last polymer can be drawn cold into threads, but it has not yet found commercial use, probably because of the effect that alkali (soap) should have on it. During this molecular distillation, as it is called, a trace of a cyclic anhydride with a ring of twenty-two members is also formed.

5. On Phthalic Acid.   Phthalic acid forms a cyclic anhydride at its own melting point. The two isomeric acids, isophthalic and terephthalic, evidently cannot. At least, no compounds are known in which the planar benzene ring takes part in such a caged structure as would be necessary for these compounds. The nonexistent terephthalic anhydride structure is indicated in Expression 21-13.

terephthalic anhydride
(nonexistent)

[21-13]

B. *Formation of Imides*

Amides of dicarboxylic acids are formed by the ammonolysis of esters or acid halides of the acids. Succinamide, glutaramide, and phthalamide, however, readily lose a molecule of NH_3 between the two amide groups to give a cyclic amide which is called an imide.

$$
\begin{array}{ccc}
& \text{O} & \\
\text{CH}_2\text{-C} & & \\
| & \text{NH}_2 & \xrightarrow{\Delta} \\
| & \text{NH}_2 & \\
\text{CH}_2\text{-C} & & \\
& \text{O} & \\
& \text{succinamide} &
\end{array}
\qquad
\begin{array}{c}
\text{O} \\
\text{CH}_2\text{-C} \\
| \qquad \text{NH} + \text{NH}_3 \\
\text{CH}_2\text{-C} \\
\text{O} \\
\text{succinimide}
\end{array}
\qquad [21\text{-}14]
$$

Heating the ammonium salts of any of these three acids also causes ready loss of ammonia.

$$\text{(benzene ring)}\begin{array}{c}\text{COONH}_4\\[4pt]\text{COONH}_4\end{array} \xrightarrow[97\%]{300°} \text{(benzene ring)}\begin{array}{c}\overset{\displaystyle O}{\underset{\displaystyle}{C}}\\[2pt]NH\\[2pt]\underset{\displaystyle O}{\overset{\displaystyle}{C}}\end{array} + NH_3 + 2H_2O \qquad [21\text{-}15]$$

In contrast to amines and amides, which are basic and neutral, respectively, the imides are weakly acidic. An alkyl group releases

$$\underset{\displaystyle ..}{\overset{\displaystyle H}{R:N:H}} \qquad \underset{\displaystyle ..}{\overset{\displaystyle O \; H}{R\text{–}C:N:H}} \qquad \underset{\displaystyle ..}{\overset{\displaystyle O \; H \; O}{R\text{–}C:N:C\text{-}R}}$$

electrons to nitrogen in amines; so short-chain amines are slightly stronger bases than ammonia. The oxygen of the acyl group,

$$R\text{–}\overset{\displaystyle O}{\overset{\displaystyle \|}{C}}\text{–},$$

is a strong electron-attracting atom, and hence there is a strong attraction on the electron pair through the carbonyl group in an amide to pull away from the nitrogen. This *inductive* effect in the acyl group is strong enough to make the amides completely neutral in character. When two acyl groups attract electrons from the same nitrogen, as in an imide (either cyclic or continuous-chain), the effect is great enough to allow the hydrogen remaining on the nitrogen almost to ionize. Imides are strong enough acids to form salts with NaOH but not with NaHCO$_3$. The use of potassium phthalimide (made by dissolving phthalimide in KOH) in the synthesis of primary amines and α amino acids has been mentioned previously.

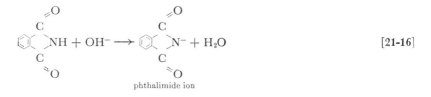

$$\text{(benzene ring)}\begin{array}{c}\overset{\displaystyle O}{\underset{\displaystyle}{C}}\\[2pt]NH\\[2pt]\underset{\displaystyle O}{\overset{\displaystyle}{C}}\end{array} + OH^- \longrightarrow \text{(benzene ring)}\begin{array}{c}\overset{\displaystyle O}{\underset{\displaystyle}{C}}\\[2pt]N^-\\[2pt]\underset{\displaystyle O}{\overset{\displaystyle}{C}}\end{array} + H_2O \qquad [21\text{-}16]$$

phthalimide ion

In other types of reactions—in acid hydrolysis, for example—imides exhibit the properties of amides.

C. *Ketone Formation (Thorpe Reaction)*

It will be recalled that heating a monocarboxylic acid in the vapor phase in the presence of MnO or heating the calcium or barium salt of such an acid yields a ketone. If two carboxyl groups occur in the same molecule, there is the possibility of preparing a

$$(RCOO)_2^- Ca^{++} \longrightarrow R-\overset{O}{\underset{\|}{C}}-R + CaCO_3 \qquad [21\text{-}17]$$

cyclic ketone by the method. Baeyer's strain theory is again an aid in predicting in what compounds this will be likely to happen. When a five- or six-carbon ketone can be formed, the cyclization will go readily; cyclopentanone and cyclohexanone are easily made by this method. The thorium salt frequently gives a better yield than the calcium or barium salt.

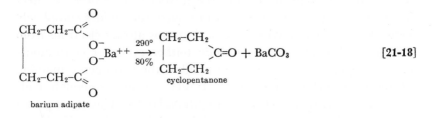

$$[21\text{-}18]$$

Cyclic ketones of this type containing many-membered rings are among the most expensive perfumes. Muscone and civetone (p. 231) are natural ketones of this type, and cyclopentadecanone is a synthetic ketone of similar odor sold on the market as Exaltone. Cyclic ketones of intermediate size (from seven- to ten-membered rings) are difficult to synthesize, but larger rings (from fifteen to eighteen members) are again easier to synthesize. They are probably under no strain.

Some special properties of oxalic acid, malonic acid, and their esters will be discussed in Chapter 22.

Halogen Acids

The most important halogen acids are those in which the halogen is on the carbon adjacent to the carboxyl group. The reason for this is that these are the easiest to make and the most stable.

21-5. Nomenclature of the Halogen Acids

The common name of a halogen acid is derived from the corresponding carboxylic acid, a Greek letter indicating the position of the halogen with respect to the acid group. The Greek letter and the appropriate halogen precede the common name of the carboxylic acid. Thus $Br-CH_2-CH_2-COOH$ is β-bromopropionic acid.

The less commonly used IUC names employ numbers to give the position of the halogen, together with the IUC names of the carboxylic acids. In Table 21-4 are the formulas, common names, and IUC names of a few halogen acids.

TABLE
21-4 | *Halogen Acids*

FORMULA	COMMON NAME	IUC NAME
$ClCH_2-COOH$	chloroacetic acid	chloroethanoic acid
$BrCH_2-CH_2-COOH$	β-bromopropionic acid	3-bromopropanoic acid
CH_3 $\quad\diagdown$ $\qquad C-COOH$ $\quad\diagup\ \ \diagdown$ $CH_3 \qquad Br$	α-bromoisobutyric acid	2-bromo-2-methylpropanoic acid
$I-CH_2-CH_2-CH_2-COOH$	γ-iodobutyric acid	4-iodobutanoic acid

21-6. Preparation of the Halogen Acids

1. Hell-Volhard-Zelinsky Reaction. The commonest method of preparing an α chloro or α bromo acid is direct halogenation in the presence of red phosphorus. The phosphorus halide formed may react with the acid and form an acyl halide, which is then halogenated. The displacement of electrons in the acyl bromide, as shown in Expression 21-19, draws the electrons on the α carbon

$$CH_3-CH_2-COOH \xrightarrow{PBr_3} CH_3-CH_2-C\overset{O}{\underset{Br}{\lessgtr}}$$

$$CH_3-\overset{H}{\underset{H}{\overset{|}{C}}}\rightarrow C\overset{O}{\underset{Br}{\lessgtr}} + \overset{\delta+}{:Br} \overset{\delta-}{:Br:} \longrightarrow CH_3-\overset{Br}{\underset{H}{\overset{|}{C}}}-C\overset{O}{\underset{Br}{\lessgtr}} + HBr \qquad [21\text{-}19]$$

closer to the acyl carbon than they would in the free acid. The α hydrogens may be described as more loosely held than before the acyl bromide was formed. An electron-deficient bromine, $Br^{\oplus}$, then, is more likely to attack the α carbon (displacing a hydrogen) than any farther down the chain, and this accounts for the formation of the α-bromoacyl bromide. Hydrolysis of the acyl

$$CH_3-\overset{}{\underset{Br}{\overset{|}{C}}}H-C\overset{O}{\underset{Br}{\lessgtr}} + HOH \longrightarrow CH_3-\overset{}{\underset{Br}{\overset{|}{C}}}H-COOH + HBr \qquad [21\text{-}20]$$

bromide results in the formation of the α halogen acid. Which halogen is the more easily hydrolyzed in $CH_3-CHBr-COBr$?

Formation of the acyl halide is not really a necessary prelude to the formation of the α halogen acid. The carboxyl group itself draws electrons to it in the same way that the acyl halide group does, though to a lesser degree. The argument just given, then, can be used to account for α halogenation on the acid molecule alone. When PCl_3 is used as a catalyst in this reaction (Exp. 21-19), the yield of α-bromopropionic acid is 85%.

Substituted malonic acids form α halogen acids if there is a hydrogen on the α carbon. The two carboxyl groups have the same type of inductive influence pictured above for the acyl bromide. Hence it is again the α carbon that is attacked. As we have seen

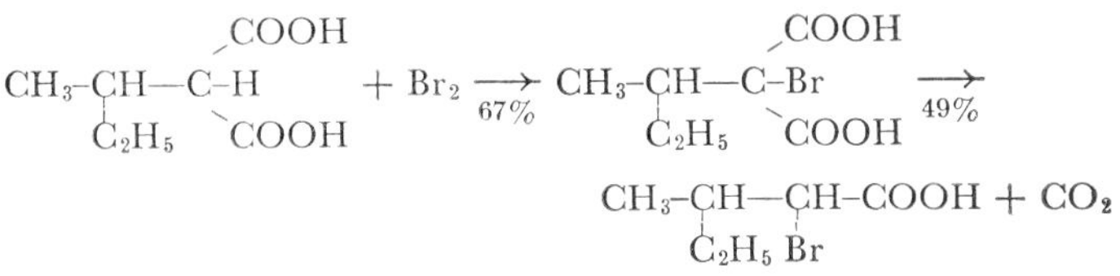

$$\underset{\underset{C_2H_5}{|}}{CH_3-CH} \underset{\diagdown COOH}{\overset{\diagup COOH}{-C-H}} + Br_2 \underset{67\%}{\longrightarrow} \underset{\underset{C_2H_5}{|}}{CH_3-CH} \underset{\diagdown COOH}{\overset{\diagup COOH}{-C-Br}} \underset{49\%}{\longrightarrow}$$

$$\underset{\underset{C_2H_5}{|} \quad \underset{Br}{|}}{CH_3-CH-CH-COOH} + CO_2 \qquad [21\text{-}21]$$

(p. 450), the substituted malonic acids readily lose a molecule of CO_2 when heated.

2. From Hydroxy Acids. The action of PCl_3, PCl_5, or $SOCl_2$ on a hydroxy acid yields a halogen acid. The reaction is not different from what would be expected of the action on an alcohol or an acid alone. The method is not used often since the hydroxy acids are commonly made from the corresponding halogen acid (see p. 464).

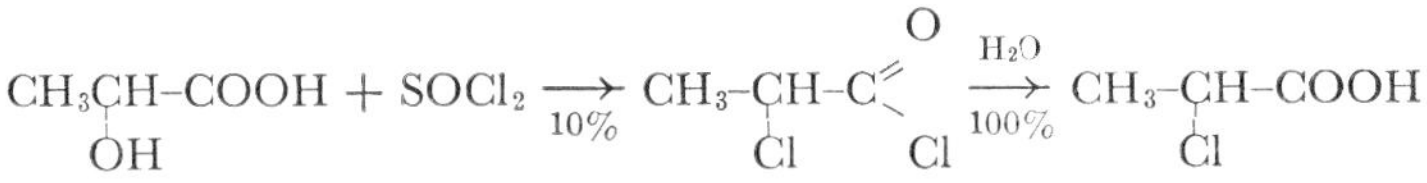

$$\underset{\underset{OH}{|}}{CH_3CH-COOH} + SOCl_2 \underset{10\%}{\longrightarrow} \underset{\underset{Cl}{|} \quad \underset{Cl}{}}{CH_3-CH} \overset{O}{\overset{\diagup}{-C}}_{\diagdown} \underset{100\%}{\overset{H_2O}{\longrightarrow}} \underset{\underset{Cl}{|}}{CH_3-CH-COOH} \quad [21\text{-}22]$$

3. By Oxidation. If a compound containing a halogen also contains another functional group that can be oxidized to a carboxyl group, oxidation is often a good method of preparation.

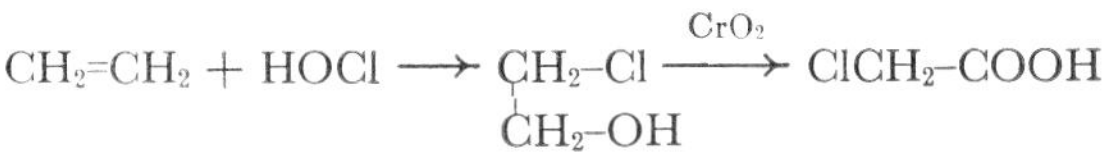

$$CH_2{=}CH_2 + HOCl \longrightarrow \underset{\underset{CH_2-OH}{|}}{CH_2-Cl} \overset{CrO_3}{\longrightarrow} ClCH_2-COOH \qquad [21\text{-}23]$$

$$\underset{(excess)}{HO(CH_2)_5OH} + HBr \overset{H_2SO_4}{\longrightarrow} HO(CH_2)_5Br \overset{HONO_2}{\longrightarrow}$$

$$BrCH_2CH_2CH_2CH_2COOH \qquad [21\text{-}24]$$

If an excess of pentamethylene glycol is treated with a mixture of HBr and H_2SO_4 (Exp. 21-24), the result is a product in which only one alcohol group has been displaced. When this compound is separated from the starting compound, oxidation of the remaining alcohol group will yield δ-bromovaleric acid.

4. By Addition to Unsaturated Acids. Hydrogen halides add 1,4 quite readily to α,β-unsaturated aldehydes or acids. This is the principal method of preparing β halogen acids. Crotonic acid, for example, will add HCl in glacial acetic acid to give β-chlorobutyric acid.

$$CH_3-CH=CH-COOH + HCl \longrightarrow \left[CH_3-\underset{\underset{Cl}{|}}{C}H-CH=C\overset{OH}{\underset{OH}{\diagup}} \right] \overset{H\ shifts}{\underset{to\ carbon}{\longrightarrow}}$$

$$CH_3-\underset{\underset{Cl}{|}}{C}H-CH_2-C\overset{O}{\underset{OH}{\diagup}} \qquad [21\text{-}25]$$

21-7. Reactions of the Halogen Acids

The most important reaction of an α halogen acid is its hydrolysis to an α hydroxy acid. This reaction (see p. 464) and ammonolysis (pp. 433–434) are discussed elsewhere.

The β halogen acids have some properties similar to those of the β hydroxy (p. 466) and β amino acids (p. 436). The most important property is the reversal of the preparative reaction given in Expression 21-25. This reversal is easily accomplished with alcoholic KOH, sometimes merely by warming with NH₃, or even by hydrolysis, which may give more α,β unsaturated acid than β hydroxy acid.

$$CH_3-\underset{\underset{Cl}{|}}{C}H-CH_2-COOH \xrightarrow[100\%]{alc.\ KOH} CH_3-CH=CH-COOH \qquad [21\text{-}26]$$

21-8. Acid Strength; the Inductive Effect

Acid strength in water solution has already been defined as the extent to which a substance is dissociated into ions according to the following equation (HA is any acid, and A^- is its conjugate base):

$$H_2O + HA \rightleftharpoons \overset{+}{H_3O} + A^- \qquad \qquad [21\text{-}27]$$

or

$$HA \rightleftharpoons H^+ + A^- \qquad \qquad [21\text{-}28]$$

Of several ways of expressing acid strength for purposes of comparison, one of the most useful is the number called the dissociation constant of the acid. For different concentrations of acid in water, it has been found that the following relation holds true for dilute solutions:

$$\frac{[\overset{+}{H_3O}][A^-]}{[H_2O][HA]} = K \qquad \qquad [21\text{-}29]$$

Here $[\overset{+}{H_3O}]$ is the formal* or molar concentration of H_3O^+, and the other symbols in brackets have a similar meaning. Since the number of water molecules in any dilute solution of an acid is very large in comparison with the other three species present, it may be considered a constant.† The expression for the dissociation constant is consequently more often written as

$$\frac{[H^+][A^-]}{[HA]} = K_a \qquad \qquad [21\text{-}30]$$

* A formal (F) solution contains one formula weight of a substance per liter of solution. A 0.1F solution of sulfuric acid, then, contains $0.1 \times 98 = 9.8$ g per liter. (See also p. 184.)

† If the following condition exists, there is justification for calling the concentration of water constant. When the concentration of the acid is 0.1F or less, the values $[H_3O^+]$, $[A^-]$, and $[HA]$ are quite small in comparison with $[H_2O]$, which will not vary much from 1,000/18. (A liter of solution will contain nearly 1,000 g of H_2O, the molecular weight of which is 18.) Then $K_a = \frac{1,000}{18} K$ (approx.). A small variation in the number $\frac{1,000}{18} = 55.5$ will have little effect on the values in the equation, and so the simplified equation $\frac{[H^+][A^-]}{HA} = K_a$ may be used. For weak acids with concentrations less than 0.1F, this equation holds within 5–10%.

K_a is constant for a particular acid only at a given temperature and only in quite dilute solutions. Suppose that K_a for acetic acid is 1.8×10^{-5} at $25°$; what is the concentration of hydrogen ions in 0.01F solution?

$$HOOCCH_3 \rightleftharpoons H^+ + CH_3COO^- \qquad\qquad [21\text{-}31]$$

$$K_a = \frac{[H^+][CH_3COO^-]}{[HOOCCH_3]} = 1.8 \times 10^{-5} \qquad\qquad [21\text{-}32]$$

Since each molecule of acetic acid that dissociates yields one hydrogen ion and one acetate ion, of necessity $[H^+] = [CH_3COO^-]$. To solve the equation, we may therefore substitute $[H^+]$ for $[CH_3COO^-]$. Furthermore, in this particular problem $[HOOCH_3]$ $= 0.01 - [H^+]$; so the problem may now be reduced to one unknown in a quadratic equation.

$$\frac{[H^+][H^+]}{0.01 - [H^+]} = 1.8 \times 10^{-5} \qquad\qquad [21\text{-}33]$$

This can be solved by use of the quadratic formula, but an approximate solution may be obtained in a very simple way by an assumption that does not cause an appreciable error in the answer. If $[H^+]$ is very small, compared with 0.01, then $0.01 - [H^+]$ is approximately equal to 0.01. The equation then becomes

$$\frac{[H^+]^2}{0.01} = 1.8 \times 10^{-5} \qquad\qquad [21\text{-}34]$$

Solving, we get

$$[H^+] = 4.2 \times 10^{-4} = 0.00042 \text{ mole/liter} \qquad\qquad [21\text{-}35]$$

The percentage ionization for this concentration of acid may be calculated to be

$$\frac{0.00042}{0.01} \times 100\% = 0.042 \times 100\% = 4.2\% \qquad\qquad [21\text{-}36]$$

Exercise 21-4. Solve the quadratic equation in Expression 21-33 to determine the error involved in the assumption (Exp. 21-34) that

0.01 − [H⁺] = 0.01. This will give a working idea of when the assumption is a valid one to make.

The values of the dissociation constants given in Table 21-1 are for the first hydrogen ion only; for example,

$$\begin{matrix} \text{COOH} \\ | \\ \text{COOH} \end{matrix} \rightleftharpoons H^+ + \begin{matrix} \text{COO}^- \\ | \\ \text{COOH} \end{matrix} \qquad \text{[21-37]}$$

and

$$\frac{[H^+][\text{HOOC–COO}^-]}{\left[\begin{matrix} \text{COOH} \\ | \\ \text{COOH} \end{matrix}\right]} = 6 \times 10^{-2} \qquad \text{[21-38]}$$

The constants for the second stage of the dissociation are much, much smaller than those for the primary ionization.

The inductive effect was discussed in Chapter 18 as one of the influences on the orientation of entering groups in reactions of aromatic compounds. A qualitative measure of the inductive effect is available in the strengths of acids. To facilitate a comparison of certain substituted acetic acids, we have listed their dissociation constants in Table 21-5.

TABLE 21-5 | *Dissociation Constants of Some Substituted Acetic Acids*

ACID	K_a AT 25°	ACID	K_a AT 25°
CH₃–CH₂–CH₂–COOH	1.5×10^{-5}	Br–CH₂–COOH	133×10^{-5}
CH₃–CH₂–COOH	1.34×10^{-5}	I–CH₂–COOH	75×10^{-5}
H–CH₂–COOH	1.82×10^{-5}	HOOC–CH₂–COOH	177×10^{-5}
F–CH₂–COOH	200×10^{-5}	(primary)	
Cl–CH₂–COOH	155×10^{-5}	⬡–CH₂–COOH	5.6×10^{-5}
Cl₂–CH–COOH	$5,000 \times 10^{-5}$	NC–CH₂–COOH	400×10^{-5}
Cl₃–C–COOH	$30,000 \times 10^{-5}$	HO–CH₂–COOH	16×10^{-5}

The only substituents that result in acids weaker than acetic acid are R groups. It has been mentioned previously that R groups tend to release electrons; measurement of these acid strengths is part of the experimental evidence for this statement. If the electron

$$R \longrightarrow CH_2COOH \qquad\qquad [21\text{-}39]$$

drift is in the direction indicated in Expression 21-39, it should be more difficult for the H in the carboxyl group to dissociate. On the other hand, for a group or atom that attracts electrons, the electron drift would be in the opposite direction, and such an acid should be stronger than acetic since the proton may dissociate more readily than in acetic acid itself (Exp. 21-40).

$$Cl \longleftarrow CH_2COOH \qquad\qquad [21\text{-}40]$$

The effect of distance upon the influence of an electron-attracting group may be seen in Table 21-6 for some substituted butyric acids. One must conclude that the inductive effect diminishes very rapidly indeed as the electron-attracting group is substituted on more remote carbons and, in fact, has only a small influence beyond one carbon. A possible exception to this statement is the influence that a group may exert across a benzene ring. An example at least partially free from the taint of other effects is the strength

TABLE 21-6 | *Influence of the Inductive Effect on the Dissociation Constants of Some Acids*

ACID	K_a AT 25°C	ACID	K_a AT 25°C
$CH_3-CH_2-CH_2-COOH$	1.5×10^{-5}	$ClCH_2CH_2CH_2COOH$	3×10^{-5}
$CH_3-CH_2-\underset{\underset{Cl}{\mid}}{C}H-COOH$	139×10^{-5}	$\langle\text{O}\rangle-COOH$	6.9×10^{-5}
		$Cl-\langle\text{O}\rangle-COOH$	10.4×10^{-5}
$CH_3-\underset{\underset{Cl}{\mid}}{C}H-CH_2-COOH$	8.9×10^{-5}	$\underset{Cl}{\langle\text{O}\rangle}-COOH$	15.9×10^{-5}

of *m*- and *p*-chlorobenzoic acids in comparison with benzoic acid (Table 21-6).

Hydroxy Acids

21-9. Nomenclature of the Hydroxy Acids

The important hydroxy acids were given common names derived from the names of the sources in which they were first found. The position of the hydroxy group on the chain may also be designated by means of a Greek letter. The IUC names are only infrequently used. A few important hydroxy acids are listed in Table 21-7.

TABLE
21-7 | *Hydroxy Acids*

FORMULA	COMMON NAMES	
$CH_2OH–COOH$	glycolic acid	hydroxyacetic acid
$CH_3–CHOH–COOH$	lactic acid	α-hydroxypropionic acid
$HOCH_2–CH_2–COOH$	hydracrylic acid	β-hydroxypropionic acid
$HOOC–CHOH–CH_2–COOH$	malic acid	α-hydroxysuccinic acid
$HOOC–CHOH–CHOH–COOH$	tartaric acid	α,α'-dihydroxysuccinic acid
$\begin{array}{c} CH_2–COOH \\ | \\ HO–C–COOH \\ | \\ CH_2–COOH \end{array}$	citric acid	
⬡–CHOH–COOH	mandelic acid	α-hydroxyphenylacetic acid
⬡ OH / COOH	salicylic acid	*o*-hydroxybenzoic acid
$\begin{array}{c} COOH \\ | \\ ⬡ \\ HO \quad | \quad OH \\ OH \end{array}$	gallic acid	3,4,5-trihydroxybenzoic acid

21-10. Preparation of Hydroxy Acids

1. Hydrolysis of an α Halogen Acid. The α halogen acids prepared by the Hell-Volhard-Zelinsky reaction (Exp. 21-19) may be hydrolyzed in boiling water to hydroxy acids.

$$CH_3\text{-}CH_2\text{-}COOH + Br_2 \xrightarrow{P} CH_3\text{-}\underset{Br}{CH}\text{-}COOH \xrightarrow[100°]{H_2O}$$

$$\underset{\underset{78\%}{OH}}{CH_3\text{-}CH\text{-}COOH} + \underset{5\%}{CH_2\text{=}CH\text{-}COOH} \qquad [21\text{-}41]$$

2. Hydrolysis of a Cyanohydrin. The hydrolysis of a cyanohydrin is most frequently carried out in aqueous acid solution.

$$\underset{\substack{\text{acetone cyanohydrin}}}{CH_3\overset{O}{\overset{\|}{C}}\text{-}CH_3 + HCN \longrightarrow \underset{\substack{|\\CH_3}}{CH_3\text{-}\underset{CN}{\overset{OH}{C}}} \xrightarrow[2H_2O]{HCl} \underset{CH_3\quad COOH}{\overset{CH_3\quad OH}{C}} + NH_4Cl} \quad [21\text{-}42]$$

Mandelic acid is readily prepared by heating of the sodium hydrogen sulfite addition product of benzaldehyde with KCN and subsequent hydrolysis of the cyanohydrin.

$$⟨⟩\text{-}CHO + NaHSO_3 \longrightarrow ⟨⟩\text{-}\underset{SO_3Na}{\overset{H}{C}}\text{-}OH \xrightarrow{KCN} ⟨⟩\text{-}\underset{CN}{\overset{H}{C}}\text{-}OH \xrightarrow[52\%]{H_2O + H^+}$$

$$⟨⟩\text{-}\underset{OH}{\overset{H}{C}}\text{-}COOH \qquad [21\text{-}43]$$
$$dl \text{ mandelic acid}$$

3. Reformatsky Reaction. One general method of making a β hydroxy acid involves an organo-metallic compound (Chap. 24). Reformatsky used zinc with an α halogen ester and treated the intermediate organo-zinc compound with an aldehyde or ketone.

It is immediately evident that the organo-zinc compound in the

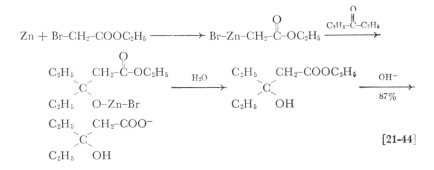

[21-44]

example given is less reactive than Grignard reagents, else it would react with the ester group in the ethyl bromoacetate (and with itself). The Reformatsky reaction is prima-facie evidence that the carbonyl group in a ketone is more reactive than a carbonyl in an ester.

4. Kolbe Synthesis. Aromatic acids cannot carry an OH group on the α carbon, but o-, m-, and p-hydroxybenzoic acids are β, γ, and δ hydroxy acids, respectively. These acids resemble one another more than they resemble the corresponding aliphatic hydroxy acids, and they are synthesized by different methods. Salicylic acid, o-hydroxybenzoic acid, is made commercially by heating the sodium salt of phenol with CO_2 under pressure. The carbonate formed undergoes a rearrangement (see p. 619), in which a new carbon-carbon bond is formed in the o position, resulting in the sodium salt of a β hydroxy acid. The reaction is general for phenols with an *ortho* position unsubstituted. The m- and p-hydroxybenzoic acids are prepared by special methods (see p. 473).

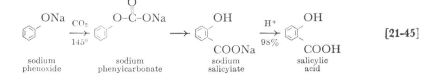

[21-45]

sodium phenoxide   sodium phenylcarbonate   sodium salicylate   salicylic acid

21-11. Reactions of Hydroxy Acids

A. *Effect of Heat*

The hydroxy acids are affected by heat in a way that is analogous to the behavior of the amino acids discussed on pp. 435–436.

1. On α Hydroxy Acids. Two molecules of an α hydroxy acid lose water to form a cyclic ester called a lactide. For some acids the reaction goes so easily that a lactide is formed if the α hydroxy

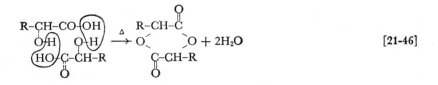

$$[21\text{-}46]$$

acid is placed in a desiccator above sulfuric acid. This cyclic dehydration is competitive with an independent decomposition forming an aldehyde and formic acid.

$$\text{R--CHOH--COOH} \xrightarrow{\Delta} \text{R--CHO} + \text{HCOOH} \qquad [21\text{-}47]$$

2. On β Hydroxy Acids. When the hydroxy group is in the β position, then a water molecule is easily lost to give an α,β-unsaturated acid containing a stable 1,3-conjugated system of double bonds.

$$\underset{\overset{|}{\text{OH}}}{\text{CH}_3\text{--CH--CH}_2\text{--COOH}} \xrightarrow{\Delta} \underset{\text{crotonic acid}}{\text{CH}_3\text{--CH=CH--COOH}} + \text{H}_2\text{O} \qquad [21\text{-}48]$$

3. On γ and δ Hydroxy Acids. The loss of a molecule of water intramolecularly can take place on a γ or δ hydroxy acid since a five- or six-membered ring is obtained in the process. The reaction goes spontaneously in many cases—for example, in the sugar acids, as we saw on page 410. The product is an inner ester, which has been given the name lactone. The γ lactones, in general, are some-

$$R-\underset{\underset{OH}{|}}{C}H-CH_2-CH_2-COOH \longrightarrow R-\underset{\underset{O}{|}}{C}H-CH_2-CH_2-C{=}O$$
a γ lactone
[21-49]

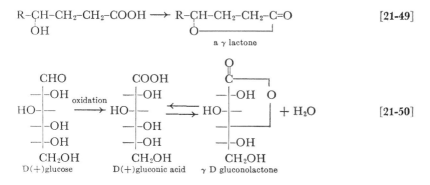

[21-50]

D(+)glucose   D(+)gluconic acid   γ D gluconolactone

what more stable than the δ lactones and are the ones ordinarily formed with sugar acids.

Lactones behave like esters toward ammonia or strong bases. The corresponding amide is formed with ammonia, and the salt of the hydroxy acid with strong bases. Concentrated HBr yields the corresponding γ bromo acid.

$$\underset{\underset{\text{γ-butyrolactone}}{\rule{2.5cm}{0.4pt}O\rule{1cm}{0.4pt}}}{CH_2-CH_2-CH_2-C{=}O} + NH_3 \longrightarrow \underset{\underset{OH}{|}}{CH_2-CH_2-CH_2}-C\overset{O}{\underset{NH_2}{\lessgtr}}$$
γ-hydroxybutyramide
[21-51]

$$\underset{\rule{2cm}{0.4pt}O\rule{1.2cm}{0.4pt}}{CH_2-CH_2-CH_2-C{=}O} + OH^- \xrightarrow[100\%]{Ba(OH)_2} \underset{\underset{OH}{|}}{CH_2-CH_2-CH_2}-COO^-$$
[21-52]

$$\underset{\rule{2cm}{0.4pt}O\rule{1.2cm}{0.4pt}}{CH_2-CH_2-CH_2-C{=}O} + HBr \longrightarrow BrCH_2-CH_2-CH_2-COOH$$
[21-53]

The electronic interpretation of these reactions is similar to the one shown for esters (pp. 286–289). In basic solution the hydrolysis

$$\underset{\rule{2cm}{0.4pt}O\rule{1cm}{0.4pt}}{CH_2-CH_2-CH_2-C{=}O} \longleftrightarrow \underset{\rule{2cm}{0.4pt}O\rule{1cm}{0.4pt}}{CH_2-CH_2-CH_2-\overset{\delta+}{C}-\overset{\delta-}{\underset{\cdot\cdot}{\overset{\cdot\cdot}{O}}}{:}} \xrightarrow{OH^-}$$

$$\underset{\underset{\rule{2cm}{0.4pt}O'\ \ \ OH}{}}{CH_2-CH_2-CH_2-C-O^-} \xrightarrow{HOH} \underset{\underset{OH}{|}}{CH_2-CH_2-CH_2}-C\overset{O}{\underset{O^-}{\lessgtr}}$$
[21-54]

probably takes place as in Expression 21-54. In acid solution the reaction with HBr may go as in Expression 21-55.

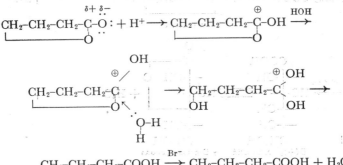

$$CH_2-CH_2-CH_2-COOH \xrightarrow{Br^-} CH_2-CH_2-CH_2-COOH + H_2O \qquad [21-55]$$

4. On ε and ω Hydroxy Acids. Hydroxy acids in which the hydroxy group is in the ε position (or further removed) from the carboxyl do not lose water readily on heating, but, at elevated temperature and with prolonged heating, esterification, first between two and finally among many molecules of the hydroxy acid, results in a polyester. Low-pressure distillation at 200° for a week yields a superpolyester that may be drawn into threads in the same manner as nylon (a superpolyamide) and the superpolyanhydrides. (The last letter of the Greek alphabet, ω (omega), means that the hydroxy group is on the carbon most remote from the carboxyl group.)

$$HO(CH_2)_9-COOH \longrightarrow HO(CH_2)_9-\overset{O}{\underset{\|}{C}}-\left[O(CH_2)_9-\overset{O}{\underset{\|}{C}}\right]_x-OH \qquad [21-56]$$

ω-hydroxydecanoic acid   a superpolyester

5. On Aromatic Hydroxy Acids. The aromatic hydroxy acids lose carbon dioxide slowly when heated to their melting points. When two or three OH groups are present on the benzene ring, the loss of CO_2 goes readily; but, if a nitro group or a halogen is present, the compound is stable to heat. This has led to the following electronic explanation of the reaction: hydroxy groups supply electrons to the benzene ring, the result being that the ring attracts the proton in the carboxyl group; the nitro group and the halogens attract electrons away from the ring.

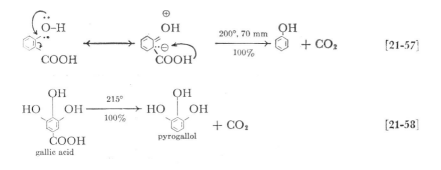

[21-57]

[21-58]

gallic acid   pyrogallol

B. Oxidation of Hydroxy Acids

Oxidation of an α hydroxy acid finally yields an acid with one less carbon than the starting compound. This way of degrading a carbon chain has already been mentioned for the sugars (p. 411). With permanganate in basic solution an aldehyde or ketone would be the intermediate product, of which the former would itself be readily oxidized to an acid.

$$CH_3-CH_2-\underset{\underset{OH}{|}}{CH}-COOH \xrightarrow[OH^-]{MnO_4^-} CH_3-CH_2-CHO + CO_3^= + H_2O$$

$$CH_3-CH_2-COO^- \xleftarrow[\substack{OH^- \\ 66\%}]{MnO_4^-}$$

[21-59]

$$\underset{CH_3}{\overset{CH_3}{>}}\underset{OH}{\overset{}{C}}-COOH \xrightarrow[OH^-]{MnO_4^-} CH_3-\overset{O}{\overset{\|}{C}}-CH_3 + CO_3^= + H_2O$$

[21-60]

Exercise 21-5. Balance equations for the preparation of the three compounds, propionaldehyde, propionate ion, and acetone, indicated in Expressions 21-59 and 21-60.

21-12. Some Important Hydroxy Acids

1. Lactic Acids.   Some of the properties of the three lactic acids (*d*, *l*, and *dl*) were given on page 337. Commercial lactic acid, the racemic mixture, is obtained by fermentation of cane sugar or hydrolyzed starch at a pH of 5.0–5.8 and a temperature of 44° with bacteria called *Lactobacillus delbrueckii*. It is used to some extent as a source of methyl acrylate (for polymers) since the acetate ester loses a molecule of acetic acid upon heating at a high temperature over carborundum, SiC.

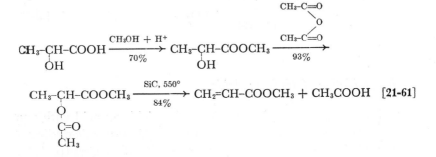

$$CH_3\text{-}\underset{\underset{\displaystyle O}{|}}{\overset{}{CH}}\text{-}COOCH_3 \xrightarrow[84\%]{SiC,\ 550°} CH_2=CH\text{-}COOCH_3 + CH_3COOH \qquad [21\text{-}61]$$

2. Malic Acid.   Green apples contain small amounts of malic acid (*malum*, Latin for apple), which disappear when the apples ripen. Malic acid loses water, as β hydroxy acids do, to give an unsaturated acid: at about 150°, fumaric acid; at higher temperatures, maleic as well.

$$\underset{\substack{\text{malic acid}}}{\overset{\displaystyle COOH}{\underset{\displaystyle COOH}{\overset{|}{\underset{|}{H\text{-}C\text{-}OH}}\,H\text{-}C\text{-}H}}} \xrightarrow{\Delta} \underset{\substack{\text{fumaric acid}}}{\overset{\displaystyle H\text{-}C\text{-}COOH}{HOOC\text{-}C\text{-}H}} \quad or \quad \underset{\substack{\text{maleic acid}}}{\overset{\displaystyle H\text{-}C\text{-}COOH}{H\text{-}C\text{-}COOH}} \qquad [21\text{-}62]$$

3. Tartaric Acids.   There are four kinds of tartaric acid although only three forms of the compound exist. Some of their physical properties are listed in Table 21-8. Note the differences in properties between a *dl* pair and either the *d* or the *l* form.

TABLE
21-8 | *Tartaric Acids*

ACID	M.P. °C	SOL. IN 100 g H_2O AT 25°	SOL. IN 100 g C_2H_5OH AT 25°	K_a AT 25°C	SP. ROTATION $[\alpha]$ AT 20°
d	170	147	43	97×10^{-5}	$+14°$
l	170	147	43	97×10^{-5}	$-14°$
dl (racemic)	206	24	5	97×10^{-5}	0
meso	140	167		60×10^{-5}	0

Dextro tartaric acid is formed during the fermentation of grapes in wine-making and is therefore abundant and cheap. Oxidation of fumaric and maleic acids yields *dl* and meso tartaric acids respectively.

$$
\begin{array}{cc}
\text{H–C–COOH} & \xrightarrow[\;100\%\;]{\text{MnO}_4^-} & \begin{array}{c} \overset{\displaystyle \text{OH}}{\underset{\displaystyle \text{OH}}{\text{H–C–COOH}}} \\ \text{H–C–COOH} \end{array}
\end{array}
\qquad [21\text{-}63]
$$

Exercise 21-6. Complete and balance the equation indicated in Expression 21-63.

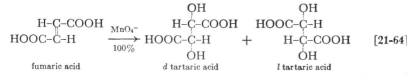

$$
\begin{array}{c}
\text{H–C–COOH} \\
\text{HOOC–C–H} \\
\text{fumaric acid}
\end{array}
\xrightarrow[\;100\%\;]{\text{MnO}_4^-}
\begin{array}{c}
\overset{\text{OH}}{\text{H–C–COOH}} \\
\text{HOOC–C–H} \\
\underset{\text{OH}}{} \\
d \text{ tartaric acid}
\end{array}
\;+\;
\begin{array}{c}
\overset{\text{OH}}{\text{HOOC–C–H}} \\
\text{H–C–COOH} \\
\underset{\text{OH}}{} \\
l \text{ tartaric acid}
\end{array}
\qquad [21\text{-}64]
$$

The acid constituent in about 10% of the baking powder sold in this country is the potassium half salt of tartaric acid; the basic constituent is sodium hydrogen carbonate. In the wet dough, as it warms up, CO_2 is liberated; the dough "rises." The mixed salt,

$$
\begin{array}{c}
\text{COO}^-\text{K}^+ \\
\text{H–C–OH} \\
\text{HO–C–H} \\
\text{COOH} \\
\text{acid}_2
\end{array}
\;+\;
\begin{array}{c}
\quad\quad\;_{\diagup}\text{O}^-\text{Na}^+ \\
\text{C}{=}\text{O} \\
\quad\quad\;^{\diagdown}\text{OH} \\
\text{base}_1
\end{array}
\longrightarrow
\begin{array}{c}
\text{COO}^-\text{K}^+ \\
\text{H–C–OH} \\
\text{HO–C–H} \\
\text{COO}^-\text{Na}^+ \\
\text{base}_2
\end{array}
\;+\;
\begin{array}{c}
\quad\quad\;_{\diagup}\text{OH} \\
\text{C}{=}\text{O} \\
\quad\quad\;^{\diagdown}\text{OH} \\
\text{acid}_1
\end{array}
\xrightarrow{} CO_2 + H_2O
\qquad [21\text{-}65]
$$

sodium potassium tartrate, a by-product of the acid-base reaction, is called "Rochelle salt" (base₂ above). Some free *d* tartaric acid is also included in the baking powder. Flour or starch dilutes the two reactive constituents and absorbs traces of moisture so that the baking powder does not deteriorate while standing.

4. Citric Acid. Citrus fruits, especially the lemon, contain appreciable quantities of citric acid. The sourness in many artificial soft drinks is supplied by this acid. An unusual property, that its calcium salt is less soluble in hot water than in cold, allows it to be isolated from other substances present in lemon juice. The free acid is obtained from the calcium salt by concentration of a water solution after precipitation of calcium as $CaSO_4$.

$$CH_2\text{--}COOH$$
$$HO\text{--}C\text{--}COOH$$
$$CH_2\text{--}COOH$$
citric acid

5. Salicylic Acid. The synthesis of *o*-hydroxybenzoic acid by the Kolbe reaction was mentioned earlier (Exp. 21-45). This acid is used in large quantities for the manufacture of aspirin and two other compounds, oil of wintergreen and salol.

Aspirin is used as an analgesic (to relieve pain) and as an antipyretic (to reduce a fever). In 1939, 5,371,682 pounds was manufactured in the United States, enough for sixty five-grain tablets per person in the entire country. Salol is an intestinal antiseptic. Oil of wintergreen is a flavor and perfume.

Sodium salicylate is used in treating rheumatism and in place of aspirin for the relief of pain.

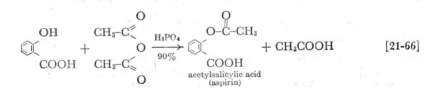

$$[21\text{-}66]$$

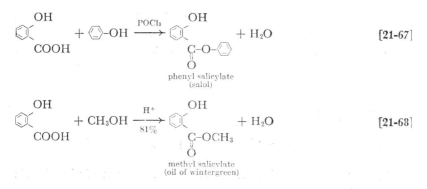

$$\text{phenyl salicylate (salol)} \quad [21\text{-}67]$$

$$\text{methyl salicylate (oil of wintergreen)} \quad [21\text{-}68]$$

6. m- and p-Hydroxybenzoic Acids. Although these two acids are not made in large quantities, the chemistry involved in their syntheses illustrates some useful and interesting reactions. The starting compounds, m- and p-cresol, are available from coal tar. If an ether is made of the cresol to protect the phenolic group from oxidation, the side chain may be oxidized to a carboxyl group. Removal of the ether group gives the desired compound, as is illustrated for m-hydroxybenzoic acid.

$$\langle\!\rangle\text{-ONa} + CH_3I \xrightarrow[97\%]{} \langle\!\rangle\text{-OCH}_3 \xrightarrow[CO_3^-]{MnO_4^-}$$

$$\langle\!\rangle\text{-COO}^- \xrightarrow[\text{aniline}]{HCl} \langle\!\rangle\text{-COOH} + CH_3Cl \quad [21\text{-}69]$$

SUMMARY

A second functional group in a carboxylic acid is most profoundly felt if it reacts with the carboxyl group to form a new function. Four things frequently happen, depending on the distance the new group is from the carboxyl; cyclization, unsaturation, loss of CO_2, or polymerization may occur. Each of these occurrences is possible when the second group is a carboxyl, a halogen, or a hydroxy group in the right place with respect to the original carboxyl group.

The acid strength of a carboxylic acid is affected markedly by the presence of a second group only if the group is on the α carbon. This inductive effect is described for the halogen atom and other groups. K_a is defined.

EXERCISES

(Exercises 1–6 will be found within the text.)

7. 1. Write the expression for the ionization constant of phenylacetic acid.
   2. Find the approximate H^+ concentration in 0.005 formal aqueous solution of this acid, assuming that $K_a = 5.6 \times 10^{-5}$.
   3. Determine a more exact value for the H^+ concentration. What percent error is involved in the assumption made in the approximate calculation?
   4. Find the percent ionization of phenylacetic acid in the above solution.

8. List the formulas of the following acids in the order of decreasing value of K_a:

   $CH_3-CH_2-CHBr-COOH$
   $CH_3-CH_2-CBr_2-COOH$
   $CH_3-CH_2-CCl_2-COOH$
   $CH_3-CHBr-CH_2-COOH$
   $CH_3-CHBr-CHBr-COOH$

9. Write equations to show the effect of heating each of the following substances:

   1. 3-hydroxy-3-methylbutanoic acid
   2. 2,4-dihydroxybenzoic acid
   3. octanedioic acid
   4. calcium pimelate
   5. hydroxysuccinic acid
   6. glutaramide
   7. succinic acid
   8. γ-hydroxybutyric acid

9. α-hydroxybutyric acid
10. ω-hydroxycapric acid

10. Show how to accomplish the following transformations:

1. benzene to bromosuccinic acid
2. ethylene to tartaric acid
3. 2-butenoic acid to β-bromobutyric acid
4. benzenesulfonic acid to salicylic acid
5. cyclohexanol to cyclopentanone
6. ethanol to ethyl malonate (using no other organic substance except KCN)
7. m-cresol to m-hydroxybenzoic acid

11. Indicate simple chemical means of distinguishing between

1. chloroacetic acid and acetyl chloride
2. glycine and acetamide
3. succinimide and ammonium succinate

12. Write balanced equations for

1. methyl chloroacetate + hot aqueous NaOH
2. 3-chlorobutanoyl chloride + hot aqueous NaOH
3. 4-bromobutanamide + hot aqueous NaOH followed by warm H_2SO_4
4. succinic anhydride + excess ethanol + H_2SO_4
5. α-hydroxyisobutyric acid + MnO_4^- (basic solution)
6. $CH_3-(CH_2)_4-CH=CH-CH=CH(CH_2)_7-COOH + HNO_3$

22 Carbonyl Acids and Carbonyl Esters: Acetoacetic Ester and Malonic Ester Syntheses

In the laboratory the Grignard reagent is perhaps the most versatile of all organic reagents for syntheses. Not very far behind, however, are the compounds known as acetoacetic ester and malonic ester. These compounds, $CH_3–CO–CH_2–COOC_2H_5$ and $CH_2(COOC_2H_5)_2$, are so well known that they are seldom referred to by their full names, ethyl β-ketobutyrate and diethyl malonate, respectively. The first of these belongs to the class called β keto esters, which have some special properties, and the second is an ester of a dicarboxylic acid in which the two carboxyl groups are on the same carbon (see p. 450). To get acquainted with the first of these two important compounds, we shall first discuss the general properties of carbonyl acids and esters.

α Carbonyl Acids

22-1. Preparation of α Carbonyl Acids

1. Oxidation of Hydroxy Acids. The α carbonyl acids can be prepared from the corresponding hydroxy acids by oxidation.

$$\begin{array}{l} CH_2OH \\ | \\ COOH \end{array} \xrightarrow[35\%]{H_2O_2\ +\ Fe^{++}} \begin{array}{l} CHO \\ | \\ COOH \end{array} \qquad \text{[22-1]}$$

glycolic acid   glyoxylic acid

476

For example, hydroxyacetic (glycolic) acid is oxidized to glyoxylic acid (also called glyoxalic). In like manner α-hydroxypropionic acid can be oxidized to α-ketopropionic (pyruvic) acid.

$$
\begin{array}{ccc}
CH_3 & & CH_3 \\
CHOH & \xrightarrow{\text{oxidation}} & C{=}O \\
COOH & & COOH \\
\text{lactic acid} & & \text{pyruvic acid}
\end{array}
\qquad [22\text{-}2]
$$

2. Hydrolysis of Halogenated Acids.   More frequently these two acids are prepared by other methods. If acetic acid is treated with bromine until two hydrogens are replaced, the dibromoacetic acid can then be hydrolyzed to glyoxylic acid.

$$
CH_3{-}COOH + Br_2 \xrightarrow[87\%]{} Br_2CHCOOH \xrightarrow[\substack{\text{(low} \\ \text{yield)}}]{HOH} \overset{HO}{\underset{HO}{>}}\!C{-}COOH \xrightarrow{-H_2O} O{=}\overset{H}{\underset{}{C}}{-}COOH
$$

a stable hydrate   glyoxylic acid

$$[22\text{-}3]$$

3. Pyrolysis of Tartaric Acid.   Heating tartaric acid in the presence of a dehydrating agent gives a product that loses CO_2 readily, yielding pyruvic acid. Since tartaric acid is a β hydroxy acid (as well as an α hydroxy acid), the easy loss of H_2O to give an

$$
\begin{array}{ccccc}
COOH & \left[\begin{array}{c}COOH\end{array}\right. & COOH & COOH \\
H{-}C{-}OH & C{-}OH & C{=}O & C{=}O \\
HO{-}C{-}H & \xrightarrow{\substack{\text{dehydrating} \\ \text{agent}}} \quad C{-}H & \longrightarrow \quad CH_2 & \xrightarrow[55\%]{-CO_2} \quad CH_3 \\
COOH & \left.COOH\right] & COOH
\end{array}
\qquad [22\text{-}4]
$$

unsaturated acid is reasonable. The resulting product carries an OH group on a carbon with a double bond, and we should therefore, from previous examples, expect it to rearrange to α-ketosuccinic acid. This is also a β keto acid, and it is characteristic of β keto acids (see Exp. 22-6) to lose CO_2 easily; so the final pyrolytic product of tartaric acid is pyruvic acid.

22-2. Reactions of α Carbonyl Acids

The carbonyl and carboxyl groups present in the α carbonyl acids both display their characteristic reactions. Glyoxylic acid, for example, forms an oxime and a phenylhydrazone, gives Tollen's test and Fehling's test for aldehydes, undergoes the Cannizzaro reaction, and forms salts, esters, and acetals.

Exercise 22-1.   Write equations for the eight specified reactions of glyoxylic acid.

Pyruvic acid adds HCN, forms an oxime, a phenylhydrazone, esters, and an anilide, and can be reduced with sodium amalgam.

Exercise 22-2.   Write equations for the six specified reactions of pyruvic acid.

One important reaction of all α keto acids, which is due to the interaction of the two functional groups, is the loss of carbon monoxide with gentle heating in sulfuric acid. Tagging the carboxyl group in pyruvic acid with a heavy isotope, C^{13}, has shown that the carbon monoxide comes from carbon 1 and not from carbon 2. The advent of nuclear reactions has made a new and powerful tool available for our use in a closer study of mechanisms of reaction.

$$CH_3\text{-}\underset{O}{\overset{}{C}}\text{-}C^{13}OOH \xrightarrow[H_2SO_4]{\Delta} CH_3\text{-}COOH + C^{13}O \qquad [22\text{-}5]$$

β Keto Esters

Acids containing a carbonyl group in the β position are very unstable and lose CO_2 spontaneously. This property makes them

$$CH_3\text{-}\overset{O}{\overset{\|}{C}}\text{-}CH_2\text{-}\overset{O}{\overset{\|}{C}}\text{-}OH \longrightarrow CH_3\text{-}\overset{O}{\overset{\|}{C}}\text{-}CH_3 + CO_2 \qquad [22\text{-}6]$$

of very little importance in themselves, but their intermediate formation in the hydrolysis of esters of the β keto acids is one of the most important reactions in all aliphatic chemistry.

The β keto esters, in comparison with the acids, are stable. The most important one is ethyl β-ketobutyrate, commonly called acetoacetic ester.

22-3. Preparation of β Keto Esters

Acetoacetic ester and some other keto esters are easily prepared in the laboratory by the Claisen condensation, which may be loosely defined as any condensation catalyzed by sodium ethoxide. If we reflux ethyl acetate in absolute alcohol in a solution containing sodium ethoxide, we get an equilibrium mixture in which a large fraction is acetoacetic ester. The mechanism shown in Expression 22-7 is generally accepted at the present time as the most probable.

$$CH_3-\overset{O}{\overset{\|}{C}}-OC_2H_5 + \overset{\ominus}{O}C_2H_5 \rightleftharpoons \overset{\ominus}{C}H_2-C\overset{O}{\underset{OC_2H_5}{\diagdown}} + C_2H_5OH \qquad [22\text{-}7a]$$

$$CH_3-C\overset{\delta+\ :\overset{\delta-\ ..}{O}:}{\underset{OC_2H_5}{\diagup}} + \overset{\ominus}{C}H_2-COOC_2H_5 \rightleftharpoons CH_3-\overset{:\overset{\ominus}{\overset{..}{O}}:}{\underset{CH_2-COOC_2H_5}{\overset{|}{\underset{}{C}}}}-OC_2H_5 \qquad [22\text{-}7b]$$

$$CH_3-\overset{:\overset{\ominus}{\overset{..}{O}}:}{\underset{CH_2-COOC_2H_5}{\overset{|}{\underset{}{C}}}}-OC_2H_5 \underset{76\%}{\rightleftharpoons} CH_3-\overset{O}{\overset{\|}{C}}-CH_2-\overset{O}{\overset{\|}{C}}-OC_2H_5 + \overset{\ominus}{O}C_2H_5 \qquad [22\text{-}7c]$$

This is the three-step mechanism (see Exp. 11-28) that we proposed for the aldol condensation of aldehydes and the reaction of aldehydes with nitroalkanes, both also base-catalyzed. The first step (Exp. 22-7a) is the displacement (induced by the catalyst, a proton acceptor) of a proton from the α position to the functional group. The two resonance forms of the ester (Exp. 22-8) are the important ones; the second contributing form shows a deficiency of electrons

$$CH_3-C\overset{O}{\underset{OC_2H_5}{\diagdown}} \quad\longleftrightarrow\quad CH_3-C\overset{\overset{\delta-}{\cdot\cdot}O\cdot\cdot}{\underset{\delta+\ OC_2H_5}{\diagdown}} \qquad\qquad [22\text{-}8]$$

on the carbon of the ester group. In the second step (Exp. 22-7b) this electron-deficient carbon joins the anion produced in the first step in a new carbon-carbon bond. The resulting unstable ion could conceivably relieve itself in more than one way. Since there are plenty of protons available in the solvent, one possible end of the reaction would be the formation of

$$\left[\begin{array}{c} \text{OH} \\ CH_3-\overset{|}{\underset{|}{C}}-OC_2H_5 \\ CH_2-COOC_2H_5 \end{array}\right]$$

Two single bonds to oxygen from the same carbon have been, in our previous experience, a sign of instability, and we might predict here that this would be an unlikely path for the reaction to take. The path taken is the one shown in step three (Exp. 22-7c), stabilization by loss of an ethoxide ion and the formation of a $\diagup^{\diagdown}C{=}O$ function.

The Claisen condensation proceeds well with other esters if there is a hydrogen on the α carbon. Ethyl propionate, for example, reacts (Exp. 22-9) in like manner to give a β keto ester, ethyl α-propiopropionate. The condensation can be run on a mixture of

$$CH_3CH_2-COOC_2H_5 + \overset{\ominus}{O}C_2H_5 \rightleftharpoons CH_3-\overset{\cdot\cdot\ominus}{C}H-COOC_2H_5 + C_2H_5OH \quad [22\text{-}9a]$$

$$CH_3-CH_2-C\overset{\overset{\cdot\cdot\ \delta-}{\cdot\cdot}O\cdot\cdot}{\underset{\delta+\ OC_2H_5}{\diagdown}} + CH_3-\overset{\cdot\cdot\ominus}{C}H-COOC_2H_5 \rightleftharpoons CH_3-CH_2-\overset{\overset{\ominus}{:\underset{\cdot\cdot}{O}:}}{\underset{\underset{CH_3}{\overset{|}{C}H-COOC_2H_5}}{\overset{|}{C}-OC_2H_5}} \quad [22\text{-}9b]$$

$$CH_3-CH_2-\overset{\overset{\displaystyle ..\ominus}{\displaystyle :O:}}{\underset{\underset{\displaystyle CH_3}{\displaystyle CHCOOC_2H_5}}{C}}-OC_2H_5 \underset{47\%}{\rightleftarrows} CH_3-CH_2-\overset{O}{\overset{\|}{C}}-\underset{\underset{\displaystyle CH_3}{|}}{CH}-\overset{O}{\overset{\|}{C}}-OC_2H_5 + \overset{\ominus}{O}C_2H_5 \qquad [22\text{-}9c]$$

ethyl α-propiopropionate

two esters provided one of them can act only as a source of the carbonyl group and not as a source of α hydrogen. Ethyl oxalate has no α hydrogen and will condense with ethyl acetate, if sodium ethoxide is present, to yield ethyl oxaloacetate.

$$\underset{\underset{\text{ethyl oxalate}}{}}{\overset{\overset{\displaystyle OC_2H_5}{|}}{\underset{\underset{\displaystyle OC_2H_5}{|}}{\overset{\displaystyle C=O}{\underset{\displaystyle C=O}{|}}}}} + CH_3-COOC_2H_5 \xrightarrow[80\%]{NaOC_2H_5}$$

$$C_2H_5O-\overset{O}{\overset{\|}{C}}-\overset{O}{\overset{\|}{C}}-CH_2-\overset{O}{\overset{\|}{C}}-OC_2H_5 + C_2H_5OH \qquad [22\text{-}10]$$

ethyl oxaloacetate

Exercise 22-3. Write a possible three-step mechanism for the reaction shown in Expression 22-10.

22-4. Properties of Acetoacetic Ester

As we suggested on page 336, acetoacetic ester undergoes some reactions characteristic of an ester, others characteristic of a ketone, and still others characteristic of an alcohol or phenol. This behavior is rational if a tautomeric equilibrium between a keto and an enol form is assumed. As an enol, acetoacetic ester gives a deep red

$$CH_3-\overset{O}{\overset{\|}{C}}-CH_2-\overset{O}{\overset{\|}{C}}-OC_2H_5 \underset{\text{keto form}}{\rightleftarrows} CH_3-\overset{OH}{\overset{|}{C}}=CH-\overset{O}{\overset{\|}{C}}-OC_2H_5 \qquad [22\text{-}11]$$

keto form   enol form

color with $FeCl_3$ solution, forms chelates (p. 618) with copper ion, and adds bromine. As a ketone, acetoacetic ester forms a phenyl-hydrazone, adds HCN, and, as a methyl ketone, adds $NaHSO_3$.

Exercise 22-4. Write equations for five of the indicated reactions of acetoacetic ester (omit Cu^{++}).

A. Hydrolysis of Acetoacetic Ester

The hydrolysis of acetoacetic ester takes one of two paths, depending on the concentration of the alkaline solution used to speed up the reaction.

1. Acid Split. In concentrated base, the reaction called the "acid split" yields two acetate ions and one molecule of ethyl alcohol. This hydrolysis is the type of equilibrium reaction that

$$CH_3-\overset{O}{\overset{\|}{C}}-CH_2-\overset{O}{\overset{\|}{C}}-OC_2H_5 + (30\%)OH^- \longrightarrow CH_3-\overset{O}{\overset{\|}{C}}-O^- + CH_3-\overset{O}{\overset{\|}{C}}-OC_2H_5 \overset{OH^-}{\longrightarrow}$$

$$2CH_3-\overset{O}{\overset{\|}{C}}-O^- + C_2H_5OH \qquad [22\text{-}12]$$

occurs in the formation of the ester by the Claisen reaction with an alkaline catalyst. In the present case water is the solvent for the 30% NaOH solution. In such a strong basic solution the shift to acetate ion is favored. In the formation of the ester, an absolute alcohol solution of $NaOC_2H_5$ is the catalyst, and the concentration of the base, $\overline{O}C_2H_5$, is rather small at any time.

2. Ketone Split. When dilute base (10% or less) is used as a catalyst for hydrolyzing acetoacetic ester, the "ketone split" obtains. In dilute solution the more vulnerable carbonyl in the carbethoxy group is attacked; this results first in preliminary hydrolysis of the ester and then in decarboxylation (loss of CO_2). As the con-

$$CH_3-\overset{O}{\overset{\|}{C}}-CH_2-\overset{\overset{\delta-}{\cdots}}{\underset{OC_2H_5}{\overset{\delta+}{C}}\overset{\ddot{O}}{\cdots}} + (10\%)OH^- \longrightarrow CH_3-\overset{O}{\overset{\|}{C}}-CH_2-\overset{O^\ominus}{\underset{OH}{C}}-OC_2H_5 \longrightarrow$$

$$HOC_2H_5 + CH_3-\overset{O}{\overset{\|}{C}}-CH_2-\overset{O}{\overset{\nearrow}{C}}{\ominus} \longrightarrow CH_3-\overset{O}{\overset{\|}{C}}-CH_3 + CO_2 + \overset{\ominus}{O}C_2H_5 \quad [22\text{-}13]$$

centration of base increases, both carbonyl groups are cleaved, and the acid split predominates. It should be emphasized that there is no sharp change in the path of the reaction in going from dilute to concentrated basic hydrolysis.

B. Alkylation

When sodium displaces hydrogen from acetoacetic ester, the hydrogen comes from the methylene group located between the two carbonyl groups. Evidently the attraction of these two groups for electrons loosens the bond between carbon and one of the

$$
\overset{O}{\underset{\|}{CH_3-C}} \longleftarrow CH_2 \longrightarrow \overset{O}{\underset{\|}{C}}-OC_2H_5
$$

hydrogens of the methylene group. The hydrogen is not ionized, but a strong base will displace it. Alternately, it may be that the methylene group first enolizes and that the hydrogen is displaced from oxygen rather than from carbon. The two anions shown in Expression 22-14 are in resonance. A methylene group situated between two such electron-attracting groups is referred to as an

$$
\overset{O}{\underset{\|}{CH_3-C}}-\overset{O}{\underset{\|}{CH_2-C}}-OC_2H_5 + Na \longrightarrow Na^{\oplus}\left(\overset{O^{\ominus}}{\underset{\|}{CH_3-C}}=\overset{O}{\underset{\|}{CH-C}}-OC_2H_5 \right) + \tfrac{1}{2}H_2
$$

$$
\updownarrow
$$

$$
Na^{\oplus}\left(\overset{O}{\underset{\|}{CH_3-C}}-\underset{\underset{\ominus}{\cdot\cdot}}{CH}-\overset{O}{\underset{\|}{C}}-OC_2H_5 \right) \qquad [22\text{-}14]
$$

"active methylene," and the hydrogens are called "active hydrogens." The active hydrogen in acetoacetic ester can be removed by $NaOC_2H_5$ as well as by Na.

$$
\overset{O}{\underset{\|}{CH_3-C}}-\overset{O}{\underset{\|}{CH_2-C}}-OC_2H_5 + NaOC_2H_5 \longrightarrow
$$

$$
Na^{\oplus}\left(\overset{O}{\underset{\|}{CH_3-C}}-\underset{\cdot\cdot}{CH}-\overset{O}{\underset{\|}{C}}-OC_2H_5 \right)^{\ominus} + C_2H_5OH \qquad [22\text{-}15]
$$

The practical value of acetoacetic ester as a starting point for synthetic work lies in the ease with which new carbon-carbon bonds can be formed by treatment of the anion with an alkyl halide. The product is also readily hydrolyzed. If ethyl bromide is used, the ethyl ethylacetoacetate produced can be hydrolyzed in

$$
Na^{\oplus}\left(CH_3-\overset{\overset{\text{O}}{\|}}{C}-\underset{..}{CH}-COOC_2H_5\right)^{\ominus} + C_2H_5-Br \xrightarrow{79\%}
$$

$$
CH_3-\overset{\overset{\text{O}}{\|}}{C}-\underset{\underset{C_2H_5}{|}}{CH}-\overset{\overset{\text{O}}{}}{C}\underset{OC_2H_5}{\diagdown} + Na^+Br^- \qquad [22\text{-}16]
$$

either the ketone split or the acid split. The ketone split gives methyl n-propyl ketone as the principal product, and the acid split gives acetic acid and n-butyric acid.

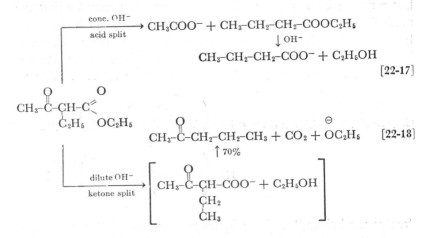

$$
\begin{array}{l}
\xrightarrow[\text{acid split}]{\text{conc. OH}^-} CH_3COO^- + CH_3-CH_2-CH_2-COOC_2H_5 \\
\hspace{4cm} \downarrow OH^- \\
\hspace{2cm} CH_3-CH_2-CH_2-COO^- + C_2H_5OH \\
\hspace{5cm} [22\text{-}17]
\end{array}
$$

$$
CH_3-\overset{\overset{\text{O}}{\|}}{C}-\underset{\underset{C_2H_5}{|}}{CH}-\overset{\overset{\text{O}}{}}{C}\underset{OC_2H_5}{\diagdown}
$$

$$
CH_3-\overset{\overset{\text{O}}{\|}}{C}-CH_2-CH_2-CH_3 + CO_2 + \overset{\ominus}{O}C_2H_5 \qquad [22\text{-}18]
$$

$$
\uparrow 70\%
$$

$$
\xrightarrow[\text{ketone split}]{\text{dilute OH}^-}
\left[
CH_3-\overset{\overset{\text{O}}{\|}}{C}-\underset{\underset{CH_3}{\underset{|}{CH_2}}}{CH}-COO^- + C_2H_5OH
\right]
$$

Isolation of ethyl ethylacetoacetate itself allows for further reaction with the same or a new alkyl halide. In other words, the remaining hydrogen in the methylene group is still active enough to be replaced for a second alkylation. This final alkylation product can likewise be split in two ways. [The second alkylation is ordinarily carried out in toluene at a higher temperature. In alcohol, the base (ethoxide ion) will tend to establish an equilibrium be-

$$CH_3\text{-}\overset{\overset{\text{O}}{\|}}{C}\text{-}\underset{\underset{C_2H_5}{|}}{CH}\text{-}COOC_2H_5 + Na \xrightarrow{\text{toluene}}$$

$$\left(CH_3\text{-}\overset{\overset{\text{O}}{\|}}{C}\text{-}\underset{\underset{C_2H_5}{|}}{\overset{..}{C}}\text{-}COOC_2H_5\right)^{\ominus}Na^{\oplus} + \tfrac{1}{2}H_2$$

$$\underset{CH_3}{\overset{CH_3}{}}\!\!\diagdown CH\text{-}Br$$

$$\longrightarrow CH_3\text{-}\overset{\overset{\text{O}}{\|}}{C}\text{-}\underset{\underset{\underset{CH_3\diagup\diagdown CH_3}{CH}}{|}}{\overset{|}{\underset{}{C}}}\overset{C_2H_5}{}\text{-}COOC_2H_5 + Na^+Br^- \qquad [22\text{-}19]$$

tween the monoalkylated ester and the products expected from a cleavage of this ester, which in this case are ethyl acetate and ethyl n-butyrate.]

$$CH_3\text{-}\overset{\overset{\text{O}}{\|}}{C}\text{-}\underset{\underset{\underset{CH_3\diagup\diagdown CH_3}{CH}}{|}}{\overset{C_2H_5}{C}}\text{-}COOC_2H_5 \xrightarrow[OH^-]{\text{conc.}} CH_3\text{-}COO^- + \underset{\underset{CH_3}{CH_3\text{-}CH}}{\overset{CH_3\text{-}CH_2}{}}\!\!\diagdown CH\text{-}COO^- + C_2H_5OH \qquad [22\text{-}20]$$

$$\bigg\downarrow \substack{\text{dilute} \\ OH^-}$$

$$C_2H_5OH + CH_3\text{-}\overset{\overset{\text{O}}{\|}}{C}\text{-}\underset{\underset{\underset{CH_3\diagup\diagdown CH_3}{C}}{|}}{\overset{C_2H_5}{C}}\text{-}COO^- \xrightarrow{H^+} CH_3\text{-}\overset{\overset{\text{O}}{\|}}{C}\text{-}\underset{\underset{\underset{CH_3\diagup\diagdown CH_3}{CH}}{|}}{CH}\text{-}CH_2\text{-}CH_3 + CO_2 \qquad [22\text{-}21]$$

The following four general formulas indicate the types of products that can be obtained from acetoacetic ester by alkylation:

$$CH_3\text{-}\overset{\overset{\text{O}}{\|}}{C}\text{-}CH_2R \qquad R\text{-}CH_2\text{-}COOH$$

$$CH_3\text{-}\overset{\overset{\text{O}}{\|}}{C}\text{-}\underset{\underset{R'}{|}}{CH}\text{-}R \qquad R\text{-}\underset{\underset{R'}{|}}{CH}\text{-}COOH$$

Malonic Ester

22-5. Preparation of Malonic Ester

The name "malonic ester," without further specification, refers to diethyl malonate since this is the ester commonly used for synthetic purposes. Synthesis of malonic ester involves reactions already familiar to the student. Chloroacetic acid, obtained by direct chlorination of acetic acid, reacts easily with NaCN to give cyanoacetic acid, which can be esterified directly in the presence of sulfuric acid. The nitrile is hydrolyzed and esterified in the same reaction.

$$\text{ClCH}_2\text{-COO}^- \xrightarrow{\text{NaCN}} \text{N}{\equiv}\text{C-CH}_2\text{-COO}^- \xrightarrow[\substack{\text{H}_2\text{SO}_4 \\ 88\%}]{\text{C}_2\text{H}_5\text{OH}} \text{CH}_2\underset{\text{COOC}_2\text{H}_5}{\overset{\text{COOC}_2\text{H}_5}{\diagup\diagdown}} \qquad [22\text{-}22]$$

22-6. Properties of Malonic Ester

Malonic ester is a colorless liquid boiling at 198°C but commonly distilled at reduced pressure because it decomposes to some extent at the elevated temperature. Like other esters, it will react with NH_3 (to form malonamide) and can be reduced (to trimethylene glycol and ethyl alcohol). Hydrolysis of malonic ester in basic solution gives the malonate ion, which, when neutralized with a strong inorganic acid, loses CO_2 readily (Exp. 21-9).

$$\text{CH}_2\underset{\text{COOC}_2\text{H}_5}{\overset{\text{COOC}_2\text{H}_5}{\diagup\diagdown}} \xrightarrow{\text{OH}^-} \underset{\text{COO}^-}{\overset{\text{COO}^-}{\text{CH}_2}} \xrightarrow{\text{H}^+} \underset{\text{COOH}}{\overset{\text{COOH}}{\text{CH}_2}} \xrightarrow{\Delta} \text{CH}_3\text{COOH} + \text{CO}_2 \qquad [22\text{-}23]$$

A. The Active Methylene Group

The position of the $-\text{CH}_2-$ group between two carbethoxy groups ($-\text{COOC}_2\text{H}_5$) endows it with the activity that is characteristic of this group in acetoacetic ester. One of the active hydrogens may be displaced by sodium or by sodium ethoxide. The anion

$$
CH_2 \genfrac{}{}{0pt}{}{COOC_2H_5}{COOC_2H_5} + Na \longrightarrow Na^{\oplus} \left(:CH \genfrac{}{}{0pt}{}{COOC_2H_5}{COOC_2H_5} \right)^{\ominus} + \tfrac{1}{2}H_2 \qquad [22\text{-}24]
$$

$$
CH_2 \genfrac{}{}{0pt}{}{COOC_2H_5}{COOC_2H_5} + NaOC_2H_5 \rightleftharpoons Na^{\oplus} \left(:CH \genfrac{}{}{0pt}{}{COOC_2H_5}{COOC_2H_5} \right)^{\ominus} + C_2H_5OH \qquad [22\text{-}25]
$$

formed in the displacement is stabilized by the various resonance forms, the more important ones being shown in Expression 22-26.

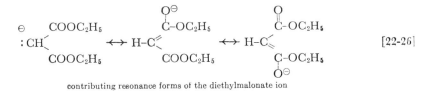

contributing resonance forms of the diethylmalonate ion

B. Alkylation

New carbon-carbon bonds can be made with malonic ester by a base-catalyzed reaction analogous to the C-alkylation of acetoacetic ester. Mono- and di-substituted malonic esters are possible by consecutive reactions, and both types of ester hydrolyze to substituted malonic acids that lose CO_2 readily. Consequently, RCH_2COOH and $RR'CHCOOH$ are the general types of substances that finally result as useful products from the alkylation of malonic ester.

1. Monosubstitution

$$
CH_2 \genfrac{}{}{0pt}{}{COOC_2H_5}{COOC_2H_5} + \overset{\ominus}{O}C_2H_5 \longrightarrow \overset{\ominus}{:}CH \genfrac{}{}{0pt}{}{COOC_2H_5}{COOC_2H_5} + C_2H_5OH \qquad [22\text{-}27]
$$

$$
\overset{\ominus}{:}CH \genfrac{}{}{0pt}{}{COOC_2H_5}{COOC_2H_5} + \overset{\delta+}{C_2H_5}\overset{\delta-}{I} \xrightarrow{81\%} C_2H_5\text{-}CH \genfrac{}{}{0pt}{}{COOC_2H_5}{COOC_2H_5} + I^- \qquad [22\text{-}28]
$$

$$C_2H_5-CH{\overset{\displaystyle COOC_2H_5}{\underset{\displaystyle COOC_2H_5}{}}} \; + \; OH^- \longrightarrow C_2H_5CH{\overset{\displaystyle COO^-}{\underset{\displaystyle COO^-}{}}} \; \xrightarrow[\Delta]{H^+} \; C_3H_7COOH + CO_2$$

$$[22\text{-}29]$$

2. Disubstitution

$$C_2H_5-CH{\overset{\displaystyle COOC_2H_5}{\underset{\displaystyle COOC_2H_5}{}}} \; + \; \overset{\ominus}{O}C_2H_5 \longrightarrow C_2H_5-\overset{\ominus}{\underset{..}{C}}{\overset{\displaystyle COOC_2H_5}{\underset{\displaystyle COOC_2H_5}{}}} \; + \; C_2H_5OH \quad [22\text{-}30]$$

$$C_2H_5\overset{\ominus}{\underset{..}{C}}{\overset{\displaystyle COOC_2H_5}{\underset{\displaystyle COOC_2H_5}{}}} \; + \; \overset{\delta+\;\delta-}{CH_3I} \xrightarrow{89\%} C_2H_5-\underset{\displaystyle COOC_2H_5}{\overset{\displaystyle COOC_2H_5}{C}}-CH_3 \; + \; I^- \qquad [22\text{-}31]$$

$$C_2H_5-\underset{\displaystyle COOC_2H_5}{\overset{\displaystyle COOC_2H_5}{C}}-CH_3 \; + \; OH^- \longrightarrow C_2H_5-\underset{\displaystyle COO^-}{\overset{\displaystyle COO^-}{C}}-CH_3 \; \xrightarrow[84\%]{H^+} \; \underset{\displaystyle CH_3 \quad COOH}{\overset{\displaystyle C_2H_5 \quad H}{C}} \; + \; CO_2$$

$$[22\text{-}32]$$

C. *Dicarboxylic Acids from Malonic Ester*

By treating the diethyl malonate anion with I_2, CH_2I_2, or other dihalogens, dicarboxylic acids may be obtained upon hydrolysis of the intermediate.

$$CH_2I_2 + 2Na^{\oplus}\left(\overset{\ominus}{CH}{\overset{\displaystyle COOC_2H_5}{\underset{\displaystyle COOC_2H_5}{}}}\right) \longrightarrow CH_2{\overset{\displaystyle CH(COOC_2H_5)_2}{\underset{\displaystyle CH(COOC_2H_5)_2}{}}} \; + \; 2Na^+I^-$$

$$2CO_2 + {\overset{\displaystyle COOH}{\underset{\displaystyle COOH}{\overset{|}{\underset{|}{\overset{\displaystyle CH_2}{\underset{\displaystyle CH_2}{\overset{\displaystyle CH_2}{}}}}}}}} \quad \xleftarrow{H^+ + \Delta} \quad CH_2{\overset{\displaystyle CH-COO^-}{\underset{\displaystyle CH-COO^-}{}}}{\overset{\displaystyle COO^-}{\underset{\displaystyle COO^-}{}}} \;\Big|\, OH^- \qquad [22\text{-}33]$$

Exercise 22-5. Could the reaction in Expression 22-33 be applied to acetoacetic ester? Write an equation to show the reaction between ICH_2CH_2I and the sodium salt of malonic ester; between iodine and the sodium salt of acetoacetic ester.

D. *Knoevenagel Reaction*

In the presence of a weak organic base as a catalyst, malonic ester will condense with aldehydes and a few ketones. The organic base is a proton acceptor, and this encourages the first step in the reaction (Exp. 22-34a). The intermediate ion, carrying a negative

$$CH_2\begin{smallmatrix}COOC_2H_5\\COOC_2H_5\end{smallmatrix} + (C_2H_5)_2NH \rightleftarrows (C_2H_5)_2\overset{\oplus}{N}H_2 + :\overset{\ominus}{C}H\begin{smallmatrix}COOC_2H_5\\COOC_2H_5\end{smallmatrix} \qquad [22\text{-}34a]$$

$$\underset{\delta+\ \delta-}{\langle\rangle\text{-}\overset{H}{\underset{}{C}}\text{-}\overset{..}{\underset{..}{O}}:} + :CH\begin{smallmatrix}COOC_2H_5\\COOC_2H_5\end{smallmatrix} \rightleftarrows \langle\rangle\text{-}\overset{H}{\underset{}{C}}\begin{smallmatrix}O^{\ominus}\\COOC_2H_5\\CH\begin{smallmatrix}\\COOC_2H_5\end{smallmatrix}\end{smallmatrix} \qquad [22\text{-}34b]$$

$$(C_2H_5)_2\overset{\oplus}{N}H_2 + \langle\rangle\text{-}\overset{H}{\underset{}{C}}\begin{smallmatrix}O^{\ominus}\\COOC_2H_5\\CH\begin{smallmatrix}\\COOC_2H_5\end{smallmatrix}\end{smallmatrix} \rightleftarrows \langle\rangle\begin{smallmatrix}H\ OH\\C\ \ \ COOC_2H_5\\C\\COOC_2H_5\end{smallmatrix} + (C_2H_5)_2\overset{\oplus}{N}H_2$$

$$\langle\rangle\text{-}\overset{H}{\underset{}{C}}=C(COOC_2H_5)_2 + OH^- \xleftarrow{70\%} \qquad [22\text{-}34c]$$

charge on the oxygen, is capable of the tautomeric equilibrium shown in Exp. 22-34c. Stabilization is established by loss of an OH⁻, which results in an unsaturated ester. Hydrolysis of this product yields cinnamic acid. The Knoevenagel reaction is, there-

$$\langle\rangle\text{-}CH=C\begin{smallmatrix}COOC_2H_5\\COOC_2H_5\end{smallmatrix} \xrightarrow{OH^-} \langle\rangle\text{-}CH=C\begin{smallmatrix}COO^-\\COO^-\end{smallmatrix} \xrightarrow{\Delta\ +\ H^+}$$

$$\langle\rangle\text{-}CH=CH\text{-}COOH + CO_2 \qquad [22\text{-}35]$$
$$\text{cinnamic acid}$$

fore, useful in synthesizing α,β-unsaturated acids. (See also the Perkin reaction, p. 521.) Formaldehyde also undergoes the Knoevenagel reaction.

Exercise 22-6.   Write the Knoevenagel reaction with formaldehyde stepwise.

E. *Preparation of α Halogen Acids*

Alkyl malonic acids may be brominated in the α position to yield α halogen malonic acids, which are unstable to heat just as

$$\begin{matrix} C_2H_5 & COOH \\ & \diagdown CH\text{--}CH \diagup \\ CH_3 & COOH \end{matrix} + Br_2 \longrightarrow C_4H_9\text{--}C\begin{matrix} \diagup COOH \\ \diagdown COOH \end{matrix} \xrightarrow[72\%]{\Delta}$$

$$C_4H_9\text{--}CHBr\text{--}COOH + CO_2 \qquad [22\text{-}36]$$

other acids carrying two carboxyl groups on the same carbon are. They readily lose carbon dioxide to produce α bromo acids. Chloro compounds may be made in the same way.

SUMMARY

Most keto acids are too unstable to be important items of commerce, but they play an important role in metabolic processes (pyruvic acid, in particular), and their esters (especially acetoacetic ester) are quite reactive and very useful synthetically. The active methylene group in acetoacetic ester may be alkylated, and the following families of ketones and acids may be obtained by hydrolytic action on the products: RCH_2COOH, $RR'CHCOOH$, RCH_2COCH_3, and $RR'CHCOCH_3$. The two series of acids may also be obtained by similar procedures on the chemically closely related malonic ester.

EXERCISES

(Exercises 1–6 will be found within the text.)

7. Write a three-step mechanism for the Claisen condensation of ethyl propionate.

8. Using malonic ester, show how to synthesize

    1. hexanoic acid
    2. 2-methylpentanoic acid
    3. β-phenylpropionic acid
    4. 2-butenoic acid
    5. pimelic acid
    6. 3-methylbutanoic acid

9. Show how to prepare as many of the substances in Exercise 8 as can be made from acetoacetic ester.

23 Unsaturated Compounds Containing Functional Groups

In an unsaturated compound containing a functional group, the position of the unsaturation with respect to the functional group determines the compound's properties in some reactions and has no influence in others. Changing the position of the unsaturation along the chain is likely to make a profound difference one or two carbons away from the functional group and to have very little effect five or six carbons away. These modifications in properties will be noted for each of several types of compounds.

Unsaturated Alcohols

23-1. Properties of Unsaturated Alcohols

Some unsaturated alcohols are listed with names and boiling points in Table 23-1.

1. α,β. Vinyl alcohol has never been prepared; all methods that should lead to its synthesis result in acetaldehyde instead. For example, the hydrolysis of either vinyl chloride or acetylene yields acetaldehyde. The tautomerism shown in Expression 23-1 probably favors the aldehyde structure so much that the compound shows no properties of an alcohol.

$$CH_2{=}CH \underset{\longleftarrow}{\overline{\hspace{1.5cm}}} CH_3{-}\underset{H}{\overset{|}{C}}{=}O \qquad \text{[23-1]}$$
$$\underset{OH}{|}$$

TABLE
23-1 | *Unsaturated Alcohols*

FORMULA	COMMON NAME	IUC NAME	B.P.
$CH_2=CHOH$ (does not exist)	vinyl alcohol		
$CH_2=CH-CH_2-OH$	allyl alcohol	2-propen-1-ol	97°
$CH_3-CH=CH-CH_2OH$	crotonyl alcohol	2-buten-1-ol	118°
$CH_2=CH-CH_2-CH_2OH$		3-buten-1-ol	114°
$CH_2=CH-CH-CH_3$ $\overset{\mid}{OH}$		1-buten-3-ol	97°
⬡-$CH=CH-CH_2OH$	cinnamyl alcohol		258°
$CH_2=C-CH_2-CH_2-CH_2-CH-CH_2-CH_2OH$ $\overset{\mid}{CH_3}$ $\overset{\mid}{CH_3}$	citronellol	2,6-dimethyl-1-octen-8-ol	222°
$HC≡C-CH_2OH$	propargyl alcohol	2-propyn-1-ol	115°

Acetone, in contrast to acetaldehyde, does show at least one of the properties of an enol: it will react with sodium, as rapidly as

$$CH_3-\underset{\underset{O}{\|}}{C}-CH_3 \xrightleftharpoons{\quad\quad} CH_2=\underset{\underset{OH}{\mid}}{C}-CH_3 \xrightarrow{Na} CH_2=\underset{\underset{ONa}{\mid}}{C}-CH_3 + \tfrac{1}{2}H_2 \qquad [23\text{-}2]$$

ethanol does, to liberate hydrogen. The equilibrium in this tautomerism is at least slightly greater in favor of the enol than in acetaldehyde.

In contrast to the instability of the aliphatic compounds of the vinyl alcohol type, the aromatic homologues are quite stable in the tautomeric enol form. The complete conjugation of the ring

enol form   keto form   [23-3]
phenol

and the resulting high resonance energy account for this difference in stability. With more OH groups on the benzene ring, the substance becomes more aliphatic; that is, the keto form becomes more important. Phloroglucinol, 1,3,5-trihydroxybenzene, for example, can be made to behave like a phenol or like a ketone. Treatment with CH_3I may yield either a trimethyl ether of the phenol (Exp.

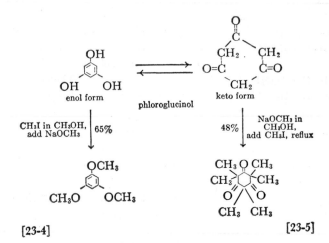

[23-4]   [23-5]

23-4) or a hexamethylated* compound of the triketo form (Exp. 23-5). In the keto form of phloroglucinol the methylene groups are all active because they are located between two carbonyl groups (compare acetoacetic ester). Phloroglucinol also forms a trioxime with hydroxylamine; other properties (color with $FeCl_3$, etc.) are those of a phenol.

2. β,γ. Tautomerism has not been observed in β,γ-unsaturated alcohols. The double bond in this position, however, does have a profound effect on the ease of replacement of the OH group. It will be recalled that the three types of saturated alcohols show rather large differences in activity toward HCl. With $ZnCl_2 + HCl$, for example, a primary alcohol does not react at room temperature,

---

* Both reactions given here for phloroglucinol are referred to as alkylations; the first, in which an ether is formed, is an O-alkylation (oxygen alkylation); the second is a C-alkylation (carbon alkylation). This use of terms is current.

a secondary alcohol reacts within five minutes, and a tertiary alcohol reacts within one minute (and nearly as rapidly with concentrated HCl alone). A β,γ-unsaturated alcohol behaves toward these two reagents as if it were a tertiary alcohol. (For a possible explanation see p. 502.)

⬡–CH₂OH
benzyl alcohol

CH₂=CH–CH₂OH
allyl alcohol

CH–CH
‖　‖
CH　C–CH₂OH
　＼O／
furfuryl alcohol

The similarity in behavior among allyl alcohol, benzyl alcohol, and furfuryl alcohol, with respect to concentrated HCl, has been accounted for on the assumption that all three have the same essential grouping, a double bond one carbon away from the alcohol group:

–C=C–CH₂OH

3. γ,δ. When the double bond of an unsaturated alcohol is moved one more carbon, to the γ,δ position, the activity of the OH group with respect to replacement is approximately equal to that of the OH group of the corresponding saturated alcohol. For example, 1-pentanol and 3-penten-1-ol are similar in the reactivity of the alcohol group.

As would be expected, removal of the double bond to positions still more remote from the OH group has no added significance.

23-2. Preparation of Unsaturated Alcohols

A. *From Acetylene*

As a result of the pioneering work of a German chemist, Reppe, in the 1940's, it is now commercially feasible to manufacture a long list of aliphatic compounds from acetylene. Among them are a series of unsaturated alcohols. Under three atmospheres pressure at 100° an excess of acetylene will react with a 30% aqueous solution of formaldehyde in the presence of copper I acetylide to give a good yield of propargyl alcohol. With an excess of formaldehyde, butynediol is obtained. Controlled hydrogenation of these acety-

$$HC\equiv CH + HCHO \xrightarrow[100°]{CuC\equiv CCu} HC\equiv C-CH_2OH \qquad\qquad [23\text{-}6]$$
excess   propargyl alcohol

$$HC\equiv CH + HCHO \xrightarrow[98\%]{\substack{3\ atm.,\ 100°\\ CuC\equiv CCu}} HOCH_2-C\equiv C-CH_2OH \qquad [23\text{-}7]$$
excess   butynediol

lenic alcohols leads to olefinic alcohols, and hydrogenation at 200 atmospheres with Raney nickel gives the saturated alcohols.

$$HC\equiv C-CH_2OH + H_2 \xrightarrow[pH > 7]{Pd} CH_2=CH-CH_2OH \qquad [23\text{-}8]$$
allyl alcohol

$$HOCH_2-C\equiv C-CH_2OH + H_2 \xrightarrow[200\ atm.]{Ni} HOCH_2-CH=CH-CH_2OH \xrightarrow[H_2]{Ni}$$
butenediol

$$HOCH_2CH_2CH_2CH_2OH \xrightarrow[275°]{NaH_2PO_4} CH_2=CH-CH_2=CH_2 \qquad [23\text{-}9]$$
1,4-butanediol   1,3-butadiene

1,4-Butanediol can be dehydrated at 275° with phosphate catalysts to 1,3-butadiene, a process that may compete with the preparation from petroleum sources in countries where petroleum is less abundant than in the United States.

Reppe was able to condense acetaldehyde with acetylene in a similar series of reactions.

Parallel with Reppe's development of acetylene chemistry in aqueous solvents, Weizmann, in Israel, condensed ketones with acetylene, using potassium hydroxide as the reagent in solvents such as the ethers of glycols (Ex. 11.4, p. 167), where water was altogether absent or present in only small amounts. Acetone, for

$$CH_3COCH_3 + HC\equiv CH \xrightarrow[90\%]{KOH,\ -10°} \underset{\substack{| \\ CH_3\ OH}}{\overset{CH_3}{\diagdown}}C-C\equiv CH + H_2 \xrightarrow{Pd\ on\ BaSO_4}$$

$$\underset{\substack{| \\ CH_3\ OH}}{\overset{CH_3}{\diagdown}}C-CH=CH_2 \xrightarrow[63\%]{-H_2O} CH_2=\underset{\substack{| \\ CH_3}}{C}-CH=CH_2 \qquad [23\text{-}10]$$
isoprene

example, condenses at $-10°$ with acetylene to yield an acetylenic alcohol that can be reduced to an olefinic alcohol. The last compound offers for the first time a good synthetic path to isoprene (Exp. 23-10). With excess acetone, unsaturated glycols are possible, as shown in Expression 23-11.

Exercise 23-1.   Show how the following compounds could be made from acetylene: 1-butyn-3-ol, 1-butyn-3-ene, 2-butanol-3-one, 2,3-butane-diol, 3-hexyn-2,5-diol, 2,4-hexanediol.

B. From Alkenes

High-temperature chlorination of a short-chain alkene proceeds without addition. The allylic product is readily hydrolyzed to the corresponding alcohol. Allyl and methallyl alcohols have been made in this way.

$$CH_3 \atop CH_3 \Big\rangle C{=}CH_2 + Cl_2 \xrightarrow[86\%]{} CH_2{=}\underset{CH_3}{C}{-}CH_2Cl \xrightarrow[90\%]{CO_3^-,\ 120°} CH_2{=}\underset{CH_3}{C}{-}CH_2OH \quad [23\text{-}12]$$

isobutylene   methallyl chloride   methallyl alcohol

C. With Grignard Reagents

Unsaturated alcohols in which the double bond is removed two carbons or more from the OH group are best made from Grignard reagents even though side reactions keep the yields from being high. The principal side reaction that prevents a good yield is the

$$CH_2=CH-CH_2Br + Mg \xrightarrow{\text{ether}} CH_2=CH-CH_2-MgBr + HCHO \longrightarrow$$

$$H-C\begin{matrix} H & OMgBr \\ & \\ & CH_2-CH=CH_2 \end{matrix} \xrightarrow[26\%]{\text{HOH}} CH_2=CH-CH_2-CH_2OH \qquad [23\text{-}13]$$

coupling of the Grignard with more of the allyl halide to give an alkadiene. The coupling reaction can be used profitably, however, for synthesis of 1,5-alkadienes by addition of an allyl halide to the corresponding Grignard.

$$CH_2=CH-CH_2-Br + CH_2=CH-CH_2-MgBr \xrightarrow[68\%]{\text{ether}}$$

$$\underset{\text{1,5-hexadiene}}{CH_2=CH-CH_2-CH_2-CH=CH_2} + MgBr_2 \qquad [23\text{-}14]$$

D. By Reduction of an Aldehyde

1. Meerwein-Ponndorf-Verley Method. Aluminum isopropoxide is a specific reagent for reducing aldehyde groups in the presence of double bonds. The reaction mixture results in an equilibrium between a mild reducing agent (the alkoxide) and a mild oxidizing agent (the carbonyl). One can make the reaction of

$$Al + 3CH_3-CHOH-CH_3 \xrightarrow{95\%} Al\left(-OCH\begin{matrix}CH_3\\ \\ CH_3\end{matrix}\right)_3 + \tfrac{3}{2}H_2 \qquad [23\text{-}15]$$

Expression 23-16 go to completion by boiling off the acetone from the mixture. (Boiling points: cinnamaldehyde, 252°; cinnamyl alcohol, 250°; isopropyl alcohol, 83°; acetone, 56°.)

$$3 \langle\bigcirc\rangle\text{-CH=CH-CHO} + Al\left(-OCH\begin{matrix}CH_3\\ \\ CH_2\end{matrix}\right)_3 \underset{68\%}{\overset{CH_3CHOHCH_3}{\rightleftarrows}}$$

$$(\langle\bigcirc\rangle\text{-CH=CH-CH}_2\text{O-})_3Al + 3CH_3-\underset{O}{\overset{}{C}}-CH_3 \qquad [23\text{-}16]$$

Recently it has been shown conclusively by Doering that the Meerwein-Ponndorf-Verley transformation involves a direct transfer of a hydride ion $(H\!:\!^-)$ from one carbon to another. The Cannizzaro reaction proceeds by a similar mechanism.

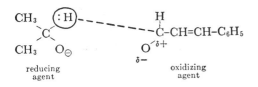

reducing   oxidizing
agent   agent

2. Lithium Aluminum Hydride.

A remarkable new reagent, $LiAlH_4$,* also reduces an aldehyde group without affecting an olefinic bond in the same molecule. Crotonaldehyde can be reduced to crotyl alcohol in 70% yield. In certain cases the double bond can also be reduced by further treatment with more $LiAlH_4$.

$$4CH_3\!-\!CH\!=\!CH\!-\!CHO + LiAlH_4 \overset{+\quad-}{\longrightarrow} (CH_3CH\!=\!CH\!-\!CH_2O)_4AlLi \overset{2HOH}{\underset{70\%}{\longrightarrow}}$$

$$4CH_3CH\!=\!CH\!-\!CH_2OH + Li^+ + AlO_2^- \qquad [23\text{-}17]$$

Cinnamaldehyde, for example, is reduced in 90% yield to cinnamyl alcohol, but this can be reduced in 93% yield to dihydrocinnamyl

$$\langle\underline{}\rangle\!-\!CH\!=\!CH\!-\!CHO \xrightarrow[90\%]{LiAlH_4} \langle\underline{}\rangle\!-\!CH\!=\!CH\!-\!CH_2OH \xrightarrow[93\%]{LiAlH_4} \langle\underline{}\rangle\!-\!CH_2CH_2CH_2OH$$

cinnamyl alcohol   dihydrocinnamyl alcohol

$$[23\text{-}18]$$

alcohol. Lithium aluminum hydride can also be used to reduce ketones, acyl halides, and even acid anhydrides and acids. This is the first convenient laboratory method of reducing an acid directly to an alcohol in one step.

---

* Lithium aluminum hydride is a very convenient reducing agent for many organic reactions since it is soluble in ether. Its preparation depends on this property. Lithium hydride reacts with aluminum chloride in ether solution. The lithium chloride precipitates in ether solution, and a separation is thus effected:

$$4LiH + AlCl_3 \overset{ether}{\longrightarrow} LiAlH_4 + 3LiCl$$

23-3. Reactions of Unsaturated Alcohols

The increased activity of the unsaturated alcohols with respect to replacement of the OH group has been mentioned. Presence of the double bond in the β,γ position does make the alcohol easier to oxidize. Silver oxide does not ordinarily attack a saturated alcohol, but it will oxidize allyl alcohol to acrolein and acrylic acid, depositing a silver mirror.

$$CH_2{=}CH{-}CH_2OH \xrightarrow{Ag_2O} CH_2{=}CH{-}CHO \longrightarrow CH_2{=}CH{-}COOH \qquad [23\text{-}19]$$

In reactions in which hydrogen is displaced or a substance adds to the double bond, unsaturated alcohols exhibit the disparate properties of an alcohol and an alkene.

Exercise 23-2. Write equations for reactions of allyl alcohol with Na, PBr$_3$, Br$_2$, and acetic anhydride.

Unsaturated Halides

23-4. Reactivity of Various Halogen Compounds

When a chemist speaks about a reactive halide in a molecule, he ordinarily refers (unless he states otherwise) to the activity of that halogen with respect to Ag$^+$. Any reagent that reacts in the same manner as Ag$^+$ should show similar activity toward halogens. Ag$^+$ reacts with a halogen directly, and hence the ease of removing the halogen will depend most significantly on the tendency toward ionization that the particular halogen exhibits.

Ionized halides react instantaneously with silver ion. Hence water-soluble halide salts, ammonium salts, and amine hydrohalides are the most reactive substances in this respect. Acyl halides,

$$Ag^+ + X^- \longrightarrow \underline{AgX} \qquad [23\text{-}20]$$

PX$_3$, PX$_5$, SOCl$_2$, NOCl, SO$_2$Cl$_2$, and others that are immediately hydrolyzed by water have been regarded in this text as "almost ionized," hence very reactive toward Ag$^+$. Aqueous silver nitrate will also precipitate silver halides from β,γ-unsaturated halogen compounds and tertiary alkyl halides. These two classes of halogen compounds hydrolyze rather readily in water and hence furnish halogen ions. Other alkyl halides, however, hydrolyze in water only slowly at room temperature. Furthermore, they are not soluble in water; hence, to get any reactions with Ag$^+$, one must use an alcoholic solution (containing some water). Primary alkyl chlorides must be boiled with alcoholic silver nitrate before they give a precipitate; vinyl and aryl halides do not react even under such strenuous conditions. These relative reactivities are summarized in the following list:

Precipitate AgX from aqueous solution:

1. ionized halides
2. acyl halides, SOCl$_2$, PCl$_5$, etc.
3. allyl halides, R$_3$CX

Precipitate AgX from alcoholic solution:

4. alkyl halides (R$_2$CHX, RCH$_2$X)

Do not precipitate AgX after boiling in alcoholic solution:

5. aliphatic polyhalides, such as CHCl$_3$ and CCl$_4$
6. vinyl halides
7. aryl halides

We shall now consider the reason for the difference in activity among the halogens in the various unsaturated compounds. Two effects are at work in all these compounds and cannot always be clearly separated: the inductive effect and the resonance effect. A halogen is an electronegative group; that is, it attracts electrons toward itself. An alkyl group tends to repel electrons. In an alkyl halide these two tendencies reinforce each other and result in a substance of moderate activity toward silver ion. In the unsat-urated compounds (Exp. 23-21) the resonance effect is influential in the resulting activity.

$$CH_2=CH-CH_2-Cl \qquad CH_2=CH-\overset{\oplus}{CH_2} \longleftrightarrow \overset{\oplus}{CH_2}-CH=CH_2 \qquad [23\text{-}21]$$
$$(1) \qquad\qquad\qquad (2) \qquad\qquad\qquad (3)$$

In allyl chloride (1), for example, any tendency for the chlorine atom to form a chloride ion (inductive effect) is greatly aided by the fact that the anticipated ion (2) is exactly equivalent to the ion (3) obtained by the shift of a pair of electrons (resonance effect). A primary postulate of the theory of resonance is that resonance will be important in a molecule or ion when different electronic structures of equal stability can be written for the mole-cule or ion. Since the anticipated ion (2) is stabilized to a large extent by resonance, the tendency to form the ion will be great, and allyl chloride will be very reactive toward reagents that de-pend on this tendency. Structures analogous to the resonance forms of allyl chloride but not exactly equivalent to each other may be written for benzyl chloride and furfuryl chloride. The analogy is justified by the chemical properties, for both these sub-stances give immediate precipitates with silver nitrate solution. The corresponding alcohols also are readily converted to the hal-ides by concentrated HCl, as mentioned previously (p. 495).

$$CH_2=CH-CH_2-Cl \qquad \langle\underset{}{\bigcirc}\rangle-CH_2Cl \qquad \begin{matrix} CH-CH \\ CH\ \ C-CH_2-Cl \\ \diagdown O \diagup \end{matrix}$$
allyl chloride   benzyl chloride

furfuryl chloride

In vinyl chloride (Exp. 23-22) the inductive effect of the halogen (4) is opposed to the tendency of the halogen to assume a double-bond character (5). In structure 5 the halogen has a formal posi-tive charge, which it has only under special circumstances. Indeed,

$$CH_2=CH \rightarrow \overset{..}{\underset{..}{Cl}} : \longleftrightarrow \overset{\ominus}{CH_2}-CH=\overset{\oplus}{\underset{..}{Cl}} : \qquad\qquad [23\text{-}22]$$
$$(4) \qquad\qquad\qquad (5)$$

if we judge by the properties ordinarily associated with chlorine, structure 5 may appear to be extremely unlikely to make a contribution to the structure of vinyl chloride. There is evidence, however, that the chlorine in vinyl chloride has some double-bond character. Electron diffraction measurements have been interpreted to indicate that the carbon-halogen bond is shorter than could be expected for a normal single bond. Double-bond character of the halogen would account for the extremely low activity of the halogen toward silver ion. The closely analogous electronic

TABLE 23-2	Comparison of Bond Lengths in Halides (calculated bond lengths: C=Cl, 1.57 A; C-Cl, 1.76 A)

COMPOUND	OBSERVED BOND LENGTH, C TO Cl
chlorobenzene	1.69 A
vinyl chloride	1.69 A
methyl chloride	1.77 A
tert-butyl chloride	1.78 A

picture that can be drawn for the aryl halides (Exp. 23-23) has been used as a means of tying their properties to those of the vinyl halides (Exp. 23-22).

$$\langle\bigcirc\rangle\text{-}\overset{..}{\underset{..}{Cl}}: \longleftrightarrow \langle\bigcirc\rangle\overset{\ominus}{\underset{}{}}=\overset{\oplus}{\underset{..}{Cl}}: \qquad\qquad [23\text{-}23]$$

When the double bond is separated from the halogen by two or more carbons, the properties of the halogen in unsaturated halogen compounds are not appreciably different from those of the corresponding alkyl halide. If four-carbon compounds are compared, the order of decreasing activity toward silver nitrate solution is

$CH_3\text{-}CH\text{=}CH\text{-}CH_2Cl > CH_2\text{=}CH\text{-}CH_2\text{-}CH_2\text{-}Cl,$

$$CH_3\text{-}CH_2\text{-}CH_2\text{-}CH_2\text{-}Cl >> CH_3\text{-}CH_2\text{-}CH\text{=}\underset{\underset{Cl}{|}}{C}H$$

23-5. Preparation of Unsaturated Halides

1. Vinyl Chloride. Acetylene will add HCl in the presence of mercury II chloride to give vinyl chloride, a gas boiling at $-15°C$. Treatment of ethylene dichloride with NaOH also produces the same compound. In the laboratory an alcoholic solution of NaOH is used, but commercially, since it is economical and feasible to control the conditions very closely, it is possible to use aqueous NaOH to remove HCl from the dichloride without the accompanying hydrolysis. Both methods of making vinyl chloride are commercially feasible.

$$H-C\equiv C-H + HCl \xrightarrow[100\%]{HgCl_2} CH_2=CHCl \qquad [23-24]$$

$$CH_2=CH_2 + Cl_2 \longrightarrow ClCH_2-CH_2-Cl + NaOH \xrightarrow{150°}$$
$$CH_2=CHCl + NaCl + H_2O \qquad [23-25]$$

2. Allyl Chloride. High-temperature halogenation of propene (p. 85) results in substitution without appreciable addition of chlorine to the double bond. In the laboratory all the allyl halides

$$CH_3-CH_2=CH + Cl_2 \xrightarrow[75\%]{600°} ClCH_2-CH=CH_2 + HCl \qquad [23-26]$$

can be easily prepared from the alcohol by boiling with the corresponding constant-boiling hydrohalogen acid (35% HCl, 48% HBr, or 57% HI).

3. 4-Chloro-1-butene. It is better to prepare unsaturated halides in which the double bond is in the γ,δ position by starting with the halogen present and introducing the double bond rather than by working in the reverse order. The γ,δ-unsaturated halides rearrange upon heating, and sometimes during chemical reactions,

$$ClCH_2-CH_2-CH_2-CH_2-Cl + alc.\ NaOH \longrightarrow CH_2=CH-CH_2-CH_2-Cl \qquad [23-27]$$

to form allylic compounds. To state the case explicitly, the conditions for introducing a double bond are not so likely to cause the rearrangement as the conditions for introducing a halogen (boiling acid solution on an alcohol).

$$CH_2=CH-CH_2-CH_2-Cl \xrightarrow{\Delta} CH_3-CH=CH-CH_2-Cl \qquad [23\text{-}28]$$

23-6. Reactions of Unsaturated Halides

The olefinic character of allyl and vinyl halides is evident in reactions with reagents that ordinarily add to double bonds.

Exercise 23-3.   Write equations for the reaction of vinyl chloride with HCl and with Br_2; of allyl chloride with Cl_2 and with HOCl.

1. Silver Nitrate Solution.   The comparative reactivity of various halides was discussed in section 23-4, but equations were not given. Allyl halides react rapidly with aqueous silver nitrate solu-

$$CH_2=CH-CH_2-Cl + Ag^+NO_3^- + H_2O \longrightarrow$$
$$CH_2=CH-CH_2OH + \underline{AgCl} + H^+NO_3^- \qquad [23\text{-}29]$$

tions. In dilute solution the organic product is allyl alcohol. Vinyl halides do not react even with boiling alcoholic silver nitrate.

2. Allylic Rearrangement.   In the past, in the preparation of allyl chlorides, mixtures of products or totally unexpected products have frequently been noted. At present it is generally accepted that the type of rearrangement shown in Expression 23-30 can take

$$R-\underset{\underset{A}{|}}{C}H-CH=CH-R' \xrightarrow{B} R-\underset{\underset{B}{|}}{C}H-CH=CH-R' \qquad [23\text{-}30]$$

$$\xrightarrow[\substack{\text{allylic}\\ \text{rearrangement}}]{B} R-CH=CH-\underset{\underset{B}{|}}{C}H-R'$$

place in these reactions. The two products probably arise in the following way: If B represents a group capable of replacing A in the allylic compound, R–CH–CH=CH–R', the first step may be

$$\underset{A}{\text{R–CH–CH=CH–R'}}$$

loss of A : $^{\ominus}$ to give the canonical incipient carbonium ions (1 and 2 in Exp. 23-31). If new group B : $^{\ominus}$ attacks the positive fragment, both 3 and 4 may result. If R and R' are different, structures 1 and 2 are *not* equivalent but may be *nearly* equivalent in energy. If R and R' are the same group, of course, no apparent rearrangement occurs.

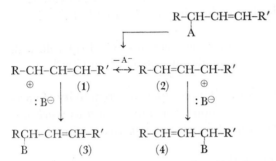

[23-31]

An example of the allylic rearrangement is the reaction of 1-butene-3-ol and PBr₃, of which crotyl bromide is the principal product. Crotyl alcohol gives the same mixture with PBr₃. The best explanation is shown in Expression 23-32. In some cases the rear-

CH₃–CH–CH=CH₂   CH₃–CH=CH–CH₂OH
  | crotyl alcohol
  OH
1-buten-3-ol

OH⁻ ↑ ↓ PBr₃   PBr₃ ↑ ↓ OH⁻

CH₃–CH–CH=CH₂ ↔ CH₃–CH=CH–CH₂
       ⊕   ⊕

OH⁻ ↑ ↓ PBr₃   OH⁻ ↑ ↓ PBr₃   [23-32]

CH₃–CH–CH=CH₂   CH₃–CH=CH–CH₂Br
  | crotyl bromide
  Br   (main product)
(small amount)

rangement is complete in both directions, as in Expression 23-33.

$$\underset{CH_3}{\overset{CH_3}{>}}C=CH-CH_2-Br \underset{PBr_3}{\overset{OH^-}{\rightleftharpoons}} \underset{CH_3 \ OH}{\overset{CH_3}{>}}C-CH=CH_2 \qquad [23\text{-}33]$$

3. Polymerization of Vinyl Chloride.

Vinyl chloride adds to itself to form long chains in the presence of benzoyl peroxide and many other catalysts, such as sunlight, $AlCl_3$, and H_3PO_4. The polymeric product may easily reach a length in which n is 1,000,

$$CH_2=CH \atop Cl \xrightarrow{\left(\bigcirc-\overset{O}{\overset{\|}{C}}-O\right)_2} -CH_2-CH-CH_2-CH-\left[CH_2-CH\right]_n-CH_2-CH- \atop Cl \quad Cl \quad \left[\ Cl\ \right]_n \quad Cl \qquad [23\text{-}34]$$

and the product at this stage is a white powder that may be molded under high pressure into various shapes. The Vinylite polymers, of high molecular weight, are made principally from vinyl chloride, vinyl acetate ($CH_2=CH-OOC-CH_3$), vinylidene chloride ($CH_2=CCl_2$), or mixtures of these monomers. Vinylite resins have been used to make toothbrushes, combs, and a host of miscellaneous toys of various types.

Polyvinyl acetate can be hydrolyzed to polyvinyl alcohol; the polymer of vinyl alcohol exists even though the monomer does not. (For a mechanism of polymerization with benzoyl peroxide as catalyst, see p. 664.)

$$CH_2=CH \atop O \atop C=O \atop CH_3 \longrightarrow \quad -CH_2-CH-\left[CH_2-CH\right]- \atop O \qquad O \atop C=O \qquad C=O \atop CH_3 \qquad \left[\ CH_3\ \right]_x \xrightarrow{OH^-}$$

polyvinylacetate

$$-CH_2-CH-CH_2-CH-\left[CH_2-CH\right]- \atop OH \qquad OH \qquad \left[\ OH\ \right]_y \qquad [23\text{-}35]$$

polyvinyl alcohol

Unsaturated Carbonyl Compounds

23-7. Ketene

The simplest unsaturated carbonyl compound and at the same time the most reactive one is called ketene, $CH_2=C=O$. Heating acetone to a high temperature will drive off a molecule of CH_4, ketene being the second product.

$$CH_3-\overset{O}{\overset{\|}{C}}-CH_3 \xrightarrow[90\%]{700°} CH_2=C=O + CH_4 \qquad [23\text{-}36]$$

Ketene is extremely reactive and may be expected to introduce the acetyl group into any molecule possessing an active hydrogen. The reactions may be thought of as additions to the carbonyl group followed by a tautomeric shift of a hydrogen to a stable acetyl compound. The reactions below show similarities between the behavior of ketene and that of an acid anhydride.

$$\overset{\delta+ \; \delta-}{CH_2}=\overset{\delta+ \; \delta-}{C=O} + H \; OH \longrightarrow \left[CH_2=C\overset{OH}{\underset{OH}{\Big\langle}} \right] \longrightarrow CH_3COOH \qquad [23\text{-}37]$$

$$CH_2=C=O + H-NH_2 \longrightarrow \left[CH_2=C\overset{OH}{\underset{NH_2}{\Big\langle}} \right] \longrightarrow CH_3-C\overset{O}{\underset{NH_2}{\Big\langle}} \quad \text{(also adds } RNH_2\text{)} \qquad [23\text{-}38]$$

$$CH_2=C=O + H-O\overset{O}{\overset{\|}{C}}-CH_3 \longrightarrow \left[\begin{array}{c} CH_2=C\overset{OH}{\underset{O}{\Big\langle}} \\ \underset{CH_3}{\overset{}{C}}=O \end{array} \right] \longrightarrow \begin{array}{c} CH_3-C\overset{O}{\underset{O}{\Big\langle}} \\ CH_3-C\overset{O}{\underset{O}{\Big\langle}} \end{array} \qquad [23\text{-}39]$$

$$CH_2=C=O + H-OC_2H_5 \xrightarrow{H^+} \left[CH_2=C\overset{OH}{\underset{OC_2H_5}{\Big\langle}} \right] \longrightarrow CH_3-C\overset{O}{\underset{OC_2H_5}{\Big\langle}} \qquad [23\text{-}40]$$

$$CH_2=C=O + H-Cl \longrightarrow \left[CH_2=C\begin{smallmatrix} OH \\ \\ Cl \end{smallmatrix} \right] \longrightarrow CH_3-C\begin{smallmatrix} O \\ \\ Cl \end{smallmatrix} \qquad [23\text{-}41]$$

$$CH_2=C=O + X-Mg-R \longrightarrow \left[CH_2=C\begin{smallmatrix} OMgX \\ \\ R \end{smallmatrix} \right] \xrightarrow{HOH} R-\overset{O}{\underset{|}{C}}-CH_3 + MgXOH \qquad [23\text{-}42]$$

In contrast to the additions to the carbonyl group is the addition of Br_2, an electron-seeking agent, to the olefinic double bond in ketene, resulting in α-bromoacetyl bromide. Ketene can be prepared from this product by the action of zinc dust.

$$CH_2=C=O \underset{Zn}{\overset{Br_2}{\rightleftarrows}} Br-CH_2-C\begin{smallmatrix} O \\ \\ Br \end{smallmatrix} \qquad [23\text{-}43]$$

Ketene, which boils at $-41°$, dimerizes readily to diketene, a liquid boiling at $127°$. Ketene cannot be stored for any length of time since it dimerizes so easily, but diketene can be pyrolyzed and so used as a source of ketene. Several structures for the dimer have been proposed, and the current idea from infrared data is that diketene is an equilibrium mixture of at least two structures, the predominating one being vinylaceto-β-lactone.

$$2CH_2=C=O \longrightarrow \underset{\substack{| \quad | \\ O--C=O}}{CH_2=C--CH_2} \rightleftarrows \underset{\substack{| \quad | \\ O-C=O}}{CH_3-C=CH} \qquad [23\text{-}44]$$

vinylaceto-β-lactone β-crotonolactone

Acetoacetic ester is made commercially by the addition of ethyl alcohol to diketene.

$$\underset{\substack{| \\ O-C=O \\ \delta- \; \delta+}}{\overset{\delta+ \; \delta-}{CH_2=C-CH_2}} + H \; OC_2H_5 \longrightarrow \underset{\substack{| \\ OH \\ \text{enol}}}{CH_2=C-CH_2-\overset{O}{\overset{||}{C}}-OC_2H_5} \rightleftarrows$$

$$\underset{\text{keto}}{CH_3-\overset{O}{\overset{||}{C}}-CH_2-\overset{O}{\overset{||}{C}}-OC_2H_5} \qquad [23\text{-}45]$$

Diketene will also add aniline, phenol, ethylene glycol, and other substances bearing active hydrogens. The reaction product from aniline, acetoacetanilide, is an important intermediate in making dyes for color photographic film.

Exercise 23-4.   Write equations for the reaction of diketene with aniline; with phenol; with ethylene glycol.

The reaction in Expression 23-39 is a commercial process for making acetic anhydride.

23-8. α,β-Unsaturated Aldehydes and Ketones

A conjugated system of double bonds,* such as that in α,β-unsaturated aldehydes and ketones, gives a special stability to a molecule. (See alkadienes, p. 104.)

The double bond is somewhat less reactive toward electron-seeking agents, and the carbonyl group adds electron-donating agents more reluctantly than a free carbonyl group. Some of the reactions go by a 1,4-addition mechanism, which is followed by a tautomeric shift to stabilize the enol formed. The important resonance structures contributing to the total picture of acrolein are

$$CH_2=CH-CHO \leftrightarrow \overset{..\delta-}{CH_2}-\overset{\delta+}{CH}-CHO \leftrightarrow CH_2=CH-\overset{H}{\underset{\delta+ \; \delta-}{C-O}} \leftrightarrow CH_2-CH=\overset{\delta+}{CH} \\ \underset{O_{\delta-}}{}$$

$$[23\text{-}46]$$

shown in Expression 23-46. The last one is the most stable and the most important contributing one.

---

* An α,β-unsaturated aldehyde or ketone is referred to as a 1,3-conjugated system in which atom 1 is oxygen. (This is not connected in any way with the naming system.) Addition reactions that involve only the carbonyl group are, then, referred to as 1,2-additions; if the conjugated system takes part, the reaction is called a 1,4-addition.

C=C–C=O
4 3 2 1

TABLE
23-3   *Unsaturated Aldehydes and Ketones*

FORMULA	COMMON NAME	IUC NAME	B.P. °C
$CH_2=CH-CHO$	acrolein	propenal	52
$CH_3-CH=CH-CHO$	crotonaldehyde	2-butenal	103
$\begin{array}{c}CH_3 \quad\quad O \\ \diagdown \quad\quad \| \\ C=CH-C-CH_3 \\ \diagup \\ CH_3\end{array}$	mesityl oxide	4-methyl-3-penten-2-one	129
$\begin{array}{c}CH_3-CH_2-CH=C-CHO \\ \| \\ CH_3\end{array}$		2-methyl-2-pentenal	137
⬡$-CH=CH-CHO$	cinnamaldehyde		252
⬡$-CH=CH-CO-CH_3$	benzalacetone		262
⬡$-CH=CH-CO-$⬡	benzalacetophenone		348
$\begin{array}{c}CH-CH \\ \| \quad\quad \| \\ CH \quad C-CHO \\ \diagdown \diagup \\ O\end{array}$	furfural		162

A. Preparation of α,β-Unsaturated Aldehydes and Ketones

1. Acrolein. Dehydration of glycerol by $KHSO_4$ or H_3PO_4 is the common method of preparing acrolein. Since the OH group from the secondary alcohol should be removed more readily than the primary hydroxyl, the loss of water may occur as in Expression 23-47.

$$\begin{array}{c}CH_2OH \\ CHOH \\ CH_2OH\end{array} \xrightarrow[KHSO_4]{-H_2O} \begin{array}{c}CH-OH \\ CH \\ CH_2OH\end{array} \xrightarrow{\text{tautomerizes}} \begin{array}{c}CHO \\ CH_2 \\ CH_2OH\end{array} \xrightarrow[48\%]{-H_2O} \begin{array}{c}CHO \\ CH \\ CH_2\end{array} \quad [23\text{-}47]$$

Acrolein is also made now by isomerization of propargyl alcohol (p. 496) in acid solution (Exp. 23-48).

$$HC \equiv C-CH_2OH \xrightarrow[\text{isomerizes}]{H^+} \underset{\text{acrolein}}{CH_2=CH-CHO} \qquad \text{[23-48]}$$

2. Furfural. The heterocyclic unsaturated aldehyde known as furfural may be obtained from a number of different sources, such as corn cobs, oat hulls, peanut shells, straw from small grains, and bagasse. Corn cobs yield about 11% furfural by dehydration of their pentosans (glycosides of five-carbon sugars). The production in 1954 was about 100,000,000 pounds from corn cobs. If D(−)arabinose may be considered a representative pentose in the glycoside (pentosan), Expression 23-49 shows the over-all equation.

$$
\begin{array}{c}
\underset{\text{D(−)arabinose}}{
\begin{array}{c}
CHO \\
HO-\!\!\!-H \\
H-\!\!\!-OH \\
H-\!\!\!-OH \\
CH_2OH
\end{array}}
\xrightarrow[70\%]{12\% \ HCl}
\underset{\text{furfural}}{
\begin{array}{c}
CH-\!\!-CH \\
\parallel \quad \parallel \\
CH \quad CCHO \\
\diagdown O \diagup
\end{array}} + 3H_2O \qquad \text{[23-49]}
$$

Three molecules of water are removed per pentose molecule, and the cyclization occurs on oxygen between carbons 2 and 5.

The production of furfural has reached such a large volume because of its selective solvent properties and because it is readily recovered in many processes simply by steam distillation. It exhibits the four following advantageous properties: (1) Furfural dissolves paraffin wax and removes it from the wax oil fraction of petroleum (see p. 49). (2) Furfural is a solvent for the coloring matter in rosin and hence is used in large quantities in the naval stores industries. (3) In the refining of vegetable oils and animal oils and fats, furfural is a selective solvent for the unsaturated compounds. (4) Furfural containing 4% water is a solvent that changes the relative volatilities of isobutane, isobutylene, 1-butene, and 1,3-butadiene. These compounds have boiling points so close together that they are otherwise difficult to separate. Fortunately, in the presence of furfural, the 1,3-butadiene is the least volatile and can be separated from the other three by fractional distillation. The petroleum industry, therefore, uses most of the furfural that is made in this country.

Furfural is also a starting compound for the manufacture of adiponitrile (to make nylon), antimalarial drugs, vitamin B, and *l*-lysine (now being added to bread by one large company).

3. From the Aldol Condensation.   Aldol loses a molecule of water on being distilled in the presence of a trace of iodine. (See Exp. 11-28.)

$$2CH_3-CHO \xrightarrow[85\%]{10\% \ KOH, \ 0°} CH_3-\underset{\underset{\text{aldol}}{OH}}{CH}-CH_2-CHO \xrightarrow[\text{distill}]{I_2} CH_3-CH=CH-CHO$$

[23-50]

One can obtain cinnamaldehyde slowly by shaking benzaldehyde and acetaldehyde with a mild base. The water from the intermediate β-hydroxy-β-phenylpropionaldehyde is lost spontaneously, and the formation of aldol itself (which is reversible) does not interfere, in the end, with the formation of the cinnamaldehyde. The reaction may be called a mixed aldol condensation.

$$\langle \bigcirc \rangle \text{-CHO} + CH_3-CHO \xrightarrow{NaOH} \langle \bigcirc \rangle-\underset{OH}{CH}-CH_2-CHO \xrightarrow{-H_2O} \langle \bigcirc \rangle-\underset{CHO}{CH=CH}$$

$$\underset{OH}{\longrightarrow} CH_3-\underset{OH}{CH}-CH_2-CHO$$

[23-51]

The aldol condensation of acetone to give diacetone alcohol and the latter's easy dehydration to mesityl oxide have already been described (Exp. 12-21).

4. Claisen-Schmidt Condensations.   Benzaldehyde condenses with acetone to benzalacetone in the presence of a mild base or to dibenzalacetone in the presence of excess benzaldehyde. Both products are yellow, low-melting solids and are used in making perfumes. Benzalacetone is said to have the odor of sweet peas.

$$\langle \bigcirc \rangle\text{-CHO} + CH_3-\overset{O}{\underset{||}{C}}-CH_3 \xrightarrow[78\%]{NaOH}$$

$$\langle \bigcirc \rangle\text{-}\underset{\text{benzalacetone}}{CH=CH-\overset{O}{\underset{||}{C}}-CH_3} \xrightarrow[94\%]{NaOH + C_6H_5CHO} \langle \bigcirc \rangle\text{-}\underset{\text{dibenzalacetone}}{CH=CH-\overset{O}{\underset{||}{C}}-CH=CH-}\langle \bigcirc \rangle$$

[23-52]

Below 30° the Claisen-Schmidt reaction can be run on benzaldehyde and acetophenone.

Exercise 23-5. Show the preparation of benzalacetone (Exp. 23-52) as a three-step base-catalyzed condensation of the aldol type.

Exercise 23-6. Write the Claisen-Schmidt reaction for benzaldehyde and acetophenone.

B. *Reactions of* α,β-*Unsaturated Aldehydes and Ketones*

Reagents that react with aldehydes or with alkenes will react with α,β-unsaturated aldehydes and ketones. The products may be different because of 1,4-addition. Only where there is a difference in activity or course of the reaction will they be mentioned specifically here.

1. Reduction and Oxidation. Acrolein and other α,β-unsaturated aldehydes undergo the pinacol reduction in mild alkaline media, whereas ordinary aldehydes are reduced to alcohols. The

$$CH_2{=}CH{-}CHO + H_2O \xrightarrow{\text{Mg(Hg)}} CH_2{=}CH{-}\underset{\underset{OH}{|}}{C}H{-}\underset{\underset{OH}{|}}{C}H{-}CH{=}CH_2 \qquad [23\text{-}53]$$

reaction is important only as an indication that the carbonyl group in a conjugated system is slightly more stable than the aldehyde group in a saturated aldehyde.

Another mildly alkaline reducing agent, aluminum isopropoxide (Meerwein-Ponndorf-Verley method, p. 498), also reduces the carbonyl group without affecting the olefinic double bond.

The corresponding unsaturated acid is obtained from the α,β-unsaturated aldehyde by use of an oxidizing agent that does not affect the olefinic bond (Tollen's reagent, for example).

2. With Hydrogen Chloride. In anhydrous methyl alcohol, hydrogen chloride adds 1,4 to crotonaldehyde. While this 1,4-addition is proceeding, CH_3OH adds 1,2 to the carbonyl group.

The final product, then, is an acetal. In aqueous acid solution the acetal reverts to the aldehyde, which may be oxidized to the β chloro acid.

$$CH_3-CH=CH-CHO \longleftrightarrow CH_3-\overset{\delta+}{CH}-CH=\overset{\delta-}{CH}-O \xrightarrow{HCl}$$

$$\left[CH_3-\underset{Cl}{\overset{|}{CH}}-CH=CH-OH \right] \longrightarrow CH_3-\underset{Cl}{\overset{|}{CH}}-CH_2-CHO \xrightarrow[\text{dry HCl}]{2CH_3OH}$$

$$CH_3-\underset{Cl}{\overset{|}{CH}}-CH_2-\overset{H \ OCH_3}{\underset{OCH_3}{C}} \xrightarrow[\text{HOH}]{H^+} CH_3-CHCl-CH_2-CHO \xrightarrow[\text{agent}]{\text{oxid.}}$$

$$CH_3-\underset{Cl}{\overset{|}{CH}}-CH_2-COOH \qquad\qquad [23\text{-}54]$$

3. With Grignard Reagents. (See also p. 538.) The contrast in activity between the aldehyde and ketone groups is well shown in the α,β-unsaturated compounds. The aldehyde group is reactive enough to yield the normal 1,2-addition compounds characteristic of the carbonyl group; in α,β-unsaturated ketones, 1,4-addition is an important competing reaction.

1,2-addition:

$$CH_3-CH=CH-CHO + C_2H_5MgBr \longrightarrow$$

$$\left[CH_3-CH=CH-\overset{H \ OMgBr}{\underset{C_2H_5}{C}} \right] \xrightarrow[90\%]{H_2O} CH_3-CH=CH-\underset{OH}{\overset{H}{\underset{|}{C}}}-CH_2-CH_3 \quad [23\text{-}55]$$

1,4-addition:

$$\overset{O}{\overset{||}{\bigcirc}\text{-}CH=CH-C-\bigcirc} + \overset{\delta- \ \delta+}{C_2H_5-MgBr} \longrightarrow$$

$$\left[\underset{C_2H_5}{\bigcirc\text{-}\overset{|}{CH}}-CH=\underset{OMgBr}{\overset{|}{C}}\text{-}\bigcirc \xrightarrow{H_2O} \underset{C_2H_5}{\bigcirc\text{-}\overset{|}{CH}}-CH=\underset{OH}{\overset{|}{C}}\text{-}\bigcirc \right] \xrightarrow{58\%}$$

$$\underset{C_2H_5}{\bigcirc\text{-}\overset{|}{CH}}-CH_2-\underset{O}{\overset{|}{C}}\text{-}\bigcirc \qquad\qquad [23\text{-}56]$$

For the specific results of competition between 1,2- and 1,4-addition, see Table 24-1.

4. With Sodium Hydrogen Sulfite.

Hot sodium hydrogen sulfite solution adds to cinnamaldehyde, two moles to one, a ratio that is accounted for only if one molecule adds 1,4. Both moles of $NaHSO_3$ may be removed again by treatment with acid or base.

$$\text{C}_6\text{H}_5\text{-CH=CH-C(H)=O} + NaHSO_3 \longrightarrow \left[\text{C}_6\text{H}_5\text{-CH(SO}_3\text{Na)-CH=CH-OH}\right] \longrightarrow$$

$$\text{C}_6\text{H}_5\text{-CH(SO}_3\text{Na)-CH}_2\text{-C(H)=O} \xrightarrow{\text{NaHSO}_3} \text{C}_6\text{H}_5\text{-CH(SO}_3\text{Na)-CH}_2\text{-C(H)(OH)(SO}_3\text{Na)} \qquad [\textbf{23-57}]$$

5. With Hydroxylamine.

Hydroxylamine condenses with α,β-unsaturated aldehydes to yield oximes, but from the same type of ketone there is competition with 1,4-addition. When benzalacetophenone is the reacting α,β-unsaturated ketone, the compounds of Expressions 23-58 and 23-59 are among the products formed.

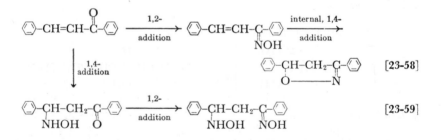

6. With Hydrogen Cyanide.

The greater activity of α,β-unsaturated aldehydes in comparison with α,β-unsaturated ketones results in only 1,2-addition of hydrogen cyanide to the former while the latter yield ketonitriles. Acrolein, for example, adds hydrogen cyanide to give the corresponding unsaturated cyanohydrin, and benzalacetophenone gives a 1,4-addition product.

$$CH_2=CH-CHO + HCN \xrightarrow[96\%]{} CH_2=CH-\underset{\underset{C\equiv N}{|}}{CH}-OH \qquad [23\text{-}60]$$

$$\text{⬡}-CH=CH-CO-\text{⬡} + HCN \xrightarrow[96\%]{} \text{⬡}-\underset{\underset{C\equiv N}{|}}{CH}-CH_2-\overset{\overset{O}{||}}{C}-\text{⬡} \qquad [23\text{-}61]$$
benzalacetophenone

An interesting new use of acrolein is in the preparation of methionine via the Strecker synthesis (p. 433). Methyl mercaptan, CH_3SH, a sulfur analogue of an alcohol, will add 1,4 to acrolein in the presence of HCN, which adds 1,2 in the same reaction

$$CH_2=CH-CHO + CH_3SH + HCN \longrightarrow CH_3-S-CH_2-CH_2-\underset{\underset{C\equiv N}{|}}{CH}-OH \xrightarrow{NH_3}$$

$$CH_3-S-CH_2-CH_2-\underset{\underset{C\equiv N}{|}}{CH}-NH_2 \xrightarrow{HOH} CH_3-S-CH_2-CH_2-\underset{\underset{NH_2}{|}}{CH}-COOH \qquad [23\text{-}62]$$
dl methionine

mixture. Ammonolysis and hydrolysis then give the amino acid methionine (p. 429). Methionine is a useful food supplement for poultry. The 1,4-addition mentioned here is an example of the Michael condensation (next paragraph).

7. Michael Condensation.   Any 1,4-addition, catalyzed by sodium ethoxide, of a substance containing an active hydrogen to an unsaturated compound may be called a Michael condensation. Mesityl oxide will add acetoacetic ester in the presence of $NaOC_2H_5$. The addition product suffers a further internal condensation by loss of an ethyl alcohol molecule to yield a six-membered ring. This internal Claisen condensation was investigated first by Dieckmann and bears his name. An iodoform reaction may then be run on the triketone to give a diketone, called methone or 5,5-dimethyl-1,3-cyclohexanedione. This diketone is in tautomeric equilibrium with 5,5-dimethyldihydroresorcinol. The preparation of methone is given in detail for several reasons: it demonstrates the Michael condensation, a Dieckmann reaction, an acetoacetic ester con-

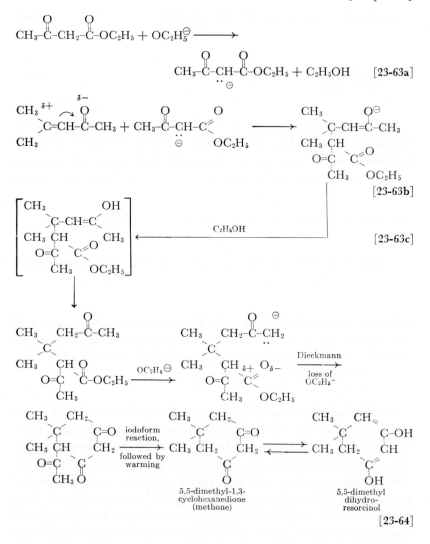

[23-63a]

[23-63b]

[23-63c]

[23-64]

densation, and, more important, the ease of formation of the six-membered ring and its tendency to aromaticity.

Exercise 23-7. Write a series of reactions equivalent to Expressions 23-63 and 23-64, with malonic ester in place of acetoacetic ester.

The Claisen-Schmidt condensation of benzaldehyde and aceto-phenone (p. 513) is followed by a Michael addition if the temperature is raised above 30°.

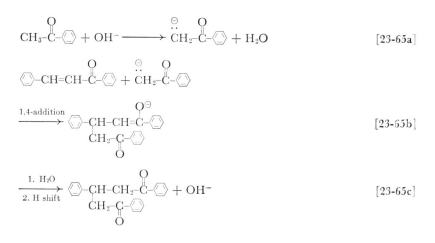

8. Skraup Reaction. An α,β-unsaturated aldehyde will add a primary aromatic amine 1,4; and, if the reaction is allowed to proceed in the presence of a mild oxidizing agent and a strong dehydrating agent, ring closure takes place, and a heterocyclic tertiary amine results. Sulfuric acid is the dehydrating agent employed; the oxidizing agent is ordinarily the nitro compound that would reduce to the amine used. The reaction is quite exothermic, but ferrous sulfate has been found to control the reaction. In the preparation of quinoline, the aldehyde required, acrolein, is ordinarily prepared *in situ* from glycerol, with sulfuric acid as the dehydrating agent. Other mild oxidizing agents, such as arsenic III oxide, may be used in place of the nitro compound.

Ring closure takes place *ortho* to the amine group, of course; and, if both *ortho* positions are open, mixtures may result with *m*-substituted anilines. A variety of quinolines substituted in either ring may be made by the Skraup reaction since other α,β-unsaturated compounds may be used in place of acrolein.

Exercise 23-8. Write equations for the Skraup reaction on methyl vinyl ketone and aniline; on acrolein (from glycerol) and *p*-toluidine (*p*-methylaniline).

Unsaturated Acids and Esters

Table 23-4 lists some important unsaturated acids and esters, with common names, boiling points, and melting points.

23-9. Preparation of Unsaturated Acids and Esters

1. Knoevenagel Reaction. (See p. 489.) *Trans*-crotonic acid is obtained by hydrolysis of the Knoevenagel condensation product of acetaldehyde and malonic ester.

Exercise 23-9. Write the reaction stepwise as a base-catalyzed aldol condensation, using diethyl amine as the base.

$$
\underset{\text{acetone}}{CH_3\text{–}\overset{\displaystyle CH_3}{\underset{}{C}}\text{=}O} + CH_2\!\!\begin{array}{l} COOC_2H_5 \\ COOC_2H_5 \end{array} \xrightarrow[92\%]{\text{piperidine}} CH_3\text{–}\overset{\displaystyle CH_3}{\underset{}{C}}\text{=}C\!\!\begin{array}{l} COOC_2H_5 \\ COOC_2H_5 \end{array} \xrightarrow{\text{NaOH}}
$$

$$
CH_3\text{–}\overset{\displaystyle CH_3}{\underset{}{C}}\text{=}C\!\!\begin{array}{l} COONa \\ COONa \end{array} \xrightarrow[\text{warm}]{H^+} CH_3\text{–}\overset{\displaystyle CH_3}{\underset{}{C}}\text{=}CH\text{–}COOH \qquad [23\text{-}67]
$$

TABLE
23-4 | *Unsaturated Acids and Esters*

FORMULA	COMMON NAME	B.P. °C	M.P. °C
$CH_2=CH-COOH$	acrylic acid	140	
$CH\equiv C-COOH$	propiolic acid	144 dec.	
$CH_3-CH=CH-COOH$	{ *cis*-crotonic acid	165 dec.	15
	trans-crotonic acid	189	72
$HOOC-CH=CH-COOH$	{ maleic acid (*cis*)		130
	fumaric acid (*trans*)		200 subl
⬡-CH=CH-COOH	cinnamic acid (*trans*)		133
⬡-C≡C-COOH	phenylpropiolic acid		135
$CH_3(CH_2)_7CH=CH(CH_2)_7COOH$	oleic acid		16
CH-CH ‖ ‖ CH C-CH=CH-COOH 　 ╲O╱	furylacrylic acid		141
$CH_2=C-COOCH_3$ 　　　 | 　　　 CH_3	methyl methacrylate	102	
$CH_2=CH-COOCH_3$	methyl acrylate	85	
⬡-CH=CH-COOC$_2$H$_5$	ethyl cinnamate	271	
$CH_3-COOCH=CH_2$	vinyl acetate	72	
$CH_3COOCH_2-CH=CH_2$	allyl acetate	103	

One may obtain ethyl cinnamate and cinnamic acid in the same way, using benzaldehyde as the starting compound.

Exercise 23-10. Show the preparation of the unsaturated ester in Expression 23-67 as a base-catalyzed aldol condensation.

Exercise 23-11. Show how ethyl cinnamate and cinnamic acid can be prepared by the Knoevenagel reaction.

2. Perkin Condensation. Aromatic α,β-unsaturated acids may be obtained by an anhydrous reaction between an aromatic aldehyde and an acid anhydride at about 175°C, a salt of the acid being used as a mild basic catalyst. The reaction goes very well

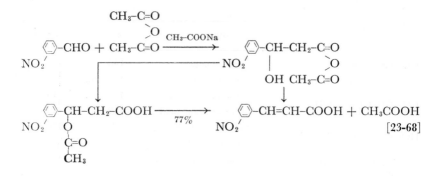

with nitro- or halogen-substituted aromatic aldehydes. The present evidence indicates that an alcohol of the aldol type is an intermediate, and that it may decompose either directly into the substituted cinnamic acid or by first forming an ester of the aldol.

3. Oxidation of Alcohol or Aldehyde.

If the double bond is protected from oxidizing agents before the aldehyde or alcohol group is oxidized, it can frequently be used to advantage in preparing an α,β-unsaturated acid. Bromine adds to allyl alcohol readily, and the resulting dibromide is easily oxidized to the dibromo acid, from which the bromine can subsequently be removed.

$$CH_2=CH-CH_2OH + Br_2 \xrightarrow[91\%]{} \underset{Br \quad Br}{CH_2-CH-CH_2OH} \xrightarrow[\substack{OH^- \\ 76\%}]{MnO_4^-}$$

$$\underset{Br \quad Br}{CH_2-CH-COO^-} \xrightarrow[\substack{then \\ Zn\ dust \\ 91\%}]{H^+} CH_2=CH-COOH \qquad [23\text{-}69]$$

Another route to the same compound involves a 1,4-addition of HCl to acrolein. After oxidation of the product the HCl can be removed by alcoholic KOH.

$$CH_2{=}CH{-}CHO + HCl \xrightarrow[87\%]{} ClCH_2{-}CH_2{-}CHO \xrightarrow[90\%]{HONO_2}$$

$$ClCH_2{-}CH_2{-}COOH \xrightarrow[\substack{KOH \\ 80\%}]{alc.} CH_2{=}CH{-}COO^- \qquad \textbf{[23-70]}$$

Exercise 23-12.   Balance the oxidation-reduction reaction indicated in Expression 23-70, assuming that $HONO_2$ is reduced to NO.

4. Vinylacetic Acid.   Allyl bromide forms a Grignard reagent very readily, and will add to CO_2 in the normal manner. Hydrolysis gives vinylacetic acid. (One complication in this reaction is given on p. 498.) This reaction demonstrates the only good

$$CH_2{=}CH{-}CH_2Br \xrightarrow[\text{ether}]{Mg} CH_2{=}CH{-}CH_2{-}MgBr \xrightarrow{CO_2}$$

$$\underset{\substack{\\ CH_2{-}CH{=}CH_2}}{\overset{\substack{OMgBr \\ \\}}{O{=}C\big\langle}} \xrightarrow[22\%]{H_2O} CH_2{=}CH{-}CH_2{-}C\underset{OH}{\overset{O}{\big\langle}} \qquad \textbf{[23-71]}$$

vinylacetic acid

method of preparing β,γ-unsaturated acids and can be used for making acids in which the double bond is more remote from the carboxyl group. Since the highest temperature reached is the boiling point of ether, the conditions are quite mild, and rearrangement of the double bond to a new position is unlikely.

23-10. Reactions of Unsaturated Acids and Esters

The α,β-unsaturated acids and esters generally add reagents like HBr in the 1,4 manner. The carboxylic acid and ester groups also activate the double bond so that it will add electron-seeking agents, a property not ordinarily associated with an alkene. Acrylic acid, for example, will add HCN and NH_3 under certain conditions.

1. Michael Condensation.   Malonic ester and acetoacetic ester add 1,4 to α,β-unsaturated esters in the presence of $NaOC_2H_5$—

examples of the Michael condensation (p. 517). Glutaric acid can thus be prepared from ethyl acrylate.

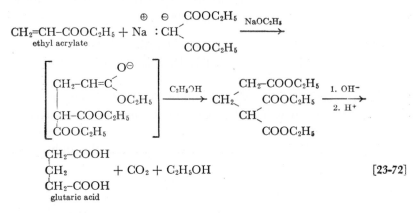

$$
CH_2=CH-COOC_2H_5 + \overset{\oplus}{Na} : \overset{\ominus}{CH} \overset{COOC_2H_5}{\underset{COOC_2H_5}{<}} \xrightarrow{NaOC_2H_5}
$$
ethyl acrylate

$$
\begin{bmatrix}
\overset{O^\ominus}{} \\
CH_2-CH=C \overset{}{\underset{OC_2H_5}{<}} \\
| \\
CH-COOC_2H_5 \\
COOC_2H_5
\end{bmatrix}
\xrightarrow{C_2H_5OH}
CH_2 \overset{CH_2-COOC_2H_5}{\underset{CH}{<}} \overset{COOC_2H_5}{\underset{COOC_2H_5}{<}}
\xrightarrow[2.\ H^+]{1.\ OH^-}
$$

$$
\begin{array}{l}
CH_2-COOH \\
CH_2 \qquad\qquad + CO_2 + C_2H_5OH \\
CH_2-COOH
\end{array}
\qquad\qquad [23\text{-}72]
$$
glutaric acid

Exercise 23-13. Write the synthesis in Expression 23-72 with acetoacetic ester instead of malonic ester.

Ethyl fumarate also adds malonic ester in the 1,4 manner, and a compound called tricarballylic acid is the final product.

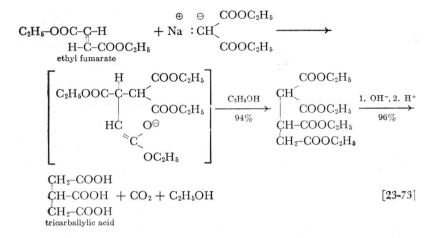

$$
\begin{array}{l}
C_2H_5-OOC-C-H \\
\qquad H-C-COOC_2H_5
\end{array}
\quad + \overset{\oplus}{Na} : \overset{\ominus}{CH} \overset{COOC_2H_5}{\underset{COOC_2H_5}{<}} \longrightarrow
$$
ethyl fumarate

$$
\begin{bmatrix}
C_2H_5OOC-\overset{H}{\underset{}{C}}-CH \overset{COOC_2H_5}{\underset{COOC_2H_5}{<}} \\
HC \quad O^\ominus \\
\overset{}{\underset{}{C}} \\
OC_2H_5
\end{bmatrix}
\xrightarrow[94\%]{C_2H_5OH}
\begin{array}{l}
CH \overset{COOC_2H_5}{\underset{COOC_2H_5}{<}} \\
| \\
CH-COOC_2H_5 \\
CH_2-COOC_2H_5
\end{array}
\xrightarrow[96\%]{1.\ OH^-,\ 2.\ H^+}
$$

$$
\begin{array}{l}
CH_2-COOH \\
CH-COOH \quad + CO_2 + C_2H_5OH \\
CH_2-COOH
\end{array}
\qquad\qquad [23\text{-}73]
$$
tricarballylic acid

2. Polymerization. The esters of acrylic and α-methylacrylic (methacrylic) acids (as well as some others) have been very important in the development of a polymer sold on the market under the names Plexiglas and Lucite.

The most useful one is the polymer of methyl methacrylate. Acetone cyanohydrin can be dehydrated, hydrolyzed, and esterified, all in a single step, by the proper choice of conditions. The

$$CH_3-\overset{O}{\overset{\|}{C}}-CH_3 + HCN \longrightarrow \underset{CH_3}{\overset{CH_3}{\diagdown}}\overset{OH}{\underset{CN}{C}} \xrightarrow[H_2SO_4]{CH_3OH}$$

$$CH_2=\underset{CH_3}{\overset{}{C}}-C\overset{O}{\underset{OCH_3}{\diagup}} + NH_4OSO_3H \qquad [23\text{-}74]$$

methyl methacrylate

monomer, methyl methacrylate, can be polymerized by benzoyl peroxide as a catalyst to a clear resin of high molecular weight, which can be molded into various shapes, cast into rods, or rolled into plates. The polymer has properties that give it many potential uses. The index of refraction is near that of crown glass; so it can be used in lenses and magnifying glasses. It transmits 96% of the light incident on a flat surface, whereas plate glass transmits 92%. This is not a large difference for a single surface; but, if many surfaces act on the light, it becomes very significant. Display lighting systems may be greatly enhanced in beauty and freed from glare by the use of the polymer. Rods of polymethyl methacrylate can be twisted into any desirable shape if softened for a time at about 200°C. The rod is said to have a perfect "memory" because it can be reheated and straightened out again. The polymer is so much easier to scratch than glass that it cannot be used in car windshields, but it has been used in airplanes and bomber noses.

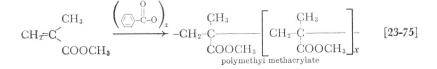

polymethyl methacrylate

Whether the structure is that indicated in Expression 23-75 is still uncertain.

An Unsaturated Nitrile

The only unsaturated nitrile made in commercial quantities is acrylonitrile, $CH_2=CH-CN$. It can be synthesized from acetylene (Exp. 5-15).

23-11. Reactions of Acrylonitrile

Acrylonitrile undergoes the usual hydrolysis reaction of a nitrile in basic solution and can be esterified directly in the presence of sulfuric acid.

$$CH_2=CH-C\equiv N + CH_3OH \xrightarrow[H_2O]{H_2SO_4} CH_2=CH-C\overset{O}{\underset{OCH_3}{<}} + NH_4OSO_3H \qquad [23\text{-}76]$$

The conjugated system of the double and the triple bond in acrylonitrile is endowed with considerably more activity than the other conjugated systems studied in this chapter. It takes part in Michael-type condensations with avidity, in which a strong base such as 30% KOH, 1% Na, a trace of alkali, or trimethylbenzyl ammonium hydroxide is the catalyst in various cases.

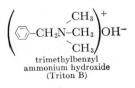

trimethylbenzyl
ammonium hydroxide
(Triton B)

A very large number of active hydrogens, such as hydrogens on O, S, and N, and an active methylene group take part in the Michael condensation with acrylonitrile. Several examples follow. Uses have not been found for all the compounds shown here, but much of this chemistry is new.

1. Phenol

$$\text{◯-OH} + \text{CH}_2\text{=CH-C≡N} \xrightarrow{\text{Triton B}} \left[\begin{array}{c} \text{CH}_2\text{-CH=C=NH} \\ \text{O} \\ \text{◯} \end{array} \right] \qquad [23\text{-}77]$$

$$\text{◯-O-CH}_2\text{-CH}_2\text{-C≡N} \xleftarrow[67\%]{\text{tautomerizes}}$$

2. Hydrogen Sulfide.

Both hydrogens in H_2S are active enough to add 1,4 to acrylonitrile when a trace of $NaOCH_3$ is present.

$$\text{CH}_2\text{=CH-C≡N} + \text{HSH} \xrightarrow{80°} \left[\begin{array}{c} \text{CH}_2\text{-CH=C} \\ \text{SH} \qquad \text{NH} \end{array} \right] \longrightarrow$$

$$\text{HS-CH}_2\text{-CH}_2\text{≡CN} \xrightarrow{\text{CH}_2\text{=CH-C≡N}} \left[\begin{array}{c} \text{CH}_2\text{-CH=C=NH} \\ \text{S-CH}_2\text{-CH}_2\text{-C≡N} \end{array} \right] \xrightarrow{93\%}$$

$$\text{N≡C-CH}_2\text{-CH}_2\text{-S-CH}_2\text{-CH}_2\text{-C≡N} \qquad [23\text{-}78]$$

3. Ethanolamine.

Ethanolamine, $H_2N\text{-CH}_2\text{-CH}_2\text{-OH}$, has three hydrogens attached to N and O, and each takes part in a Michael condensation with one molecule of acrylonitrile.

Exercise 23-14.   Write the Michael condensation for acrylonitrile and ethanolamine.

4. Indene.

Methylene groups situated between carbons carrying double bonds to oxygen (malonic ester, acetoacetic ester) have already been called active methylene groups, and the properties of such molecules have been studied. If the double bonds are carbon-to-carbon, the methylene group is still somewhat active and will take part in a Michael condensation with acrylonitrile. Indene

cyclopentadiene   indene   fluorene

reacts with three molecules of acrylonitrile to give a good insecticide. Not only do the two hydrogens on the methylene group react,

$$\text{(ring)} + 3CH_2\!=\!\underset{\underset{C\equiv N}{|}}{CH} \longrightarrow \text{(ring)} \overset{CH_2-CH_2-C\equiv N}{\underset{N\equiv C-CH_2-CH_2 \quad CH_2-CH_2-C\equiv N}{}} \qquad [23\text{-}79]$$

but so does the hydrogen on carbon 3. Cyclopentadiene adds six molecules of acrylonitrile.

Table 23-5 shows the number of acrylonitrile molecules that will react with each of a number of compounds containing active hydrogens. Some of these compounds can be hydrolyzed to useful polycarboxylic acids.

TABLE

23-5 | *Compounds Reacting with Acrylonitrile*

NAME	FORMULA	MOLECULES OF ACRYLONITRILE REACTING
Hydrogen sulfide	H_2S	2
Phenol	C_6H_5OH	1
Ethanolamine	$H_2N-CH_2-CH_2OH$	3
Indene	(see ¶ 4)	3
Cyclopentadiene	(see ¶ 4)	6
Fluorene	(see ¶ 4)	2
Acetophenone	$CH_3-CO-C_6H_5$	3
Cyclohexanone	$CH_2\begin{smallmatrix}CH_2-CH_2\\ \\CH_2-CH_2\end{smallmatrix}C\!=\!O$	4 (all α H's)
Chloroform	$CHCl_3$	1
2-Ethylhexanal	$CH_3(CH_2)_3CH(C_2H_5)CHO$	1 (α H)
Nitromethane	CH_3NO_2	3
Malonic ester	$CH_2(COOC_2H_5)_2$	2
Phenylacetonitrile	$C_6H_5-CH_2-CN$	2
Ethylene glycol	$HOCH_2-CH_2OH$	2

Exercise 23-15. Write structures for the addition compounds of acrylonitrile with chloroform, nitromethane, malonic ester, and phenylacetonitrile.

Exercise 23-16. Write the stepwise condensation of acrylonitrile with nitromethane.

5. Polymers and Copolymers. Acrylonitrile is produced commercially on a large scale for use in polymerization reactions, one of which has been mentioned (p. 110). Besides this use in an oil-resistant rubber (Buna-N), acrylonitrile can be copolymerized with styrene, the acrylates (p. 525), and vinyl chloride. The copolymers with vinyl chloride are useful in making fibers (Acrilan, Dynel, Orlon).

$$CH_2=CH + CH_2=CH \xrightarrow{\text{catalyst}} -\left[CH_2-CH-CH_2-CH\right]- \qquad [23\text{-}80]$$
$$CN Cl CN Cl _x$$

Another recent use of acrylonitrile is in the production of soil conditioners. Polyacrylonitrile can be hydrolyzed to a polyelectro-

$$CH_2=CH \longrightarrow -CH_2-CH-\left[CH_2-CH\right]-\xrightarrow{\text{NaOH}} -CH_2-CH\underline{}\left[CH_2-CH\underline{}\right]-$$
$$CN CN CN _n COONa COONa _n$$
$$\text{polyacrylonitrile} [23\text{-}81]$$

lyte that improves the water-holding power of some soils and still keeps the soil friable at high water content. It may be a help in reclaiming alkali soils and even in preventing soil erosion. One such soil conditioner is sold under the name Krilium.

SUMMARY

One cannot generalize on the effect of unsaturation in a molecule, for that effect depends not only on the position of the functional group with respect to the unsaturation but also on the nature of the functional group.

A double bond on the same carbon with an alcohol function (a vinyl alcohol) exists, with rare exceptions, only as the tautomeric aldehyde or ketone, but in the aromatic series is the predominant classical structure (phenol). When the double bond is one carbon removed from the alcohol function (an allyl alcohol), the special propensity to rearrange is latent in the displacement reactions of the OH group (allylic rearrangement). When the double bond is further removed from the OH group, the two groups are virtually independent of each other. Vinyl ethers do exist, and polyvinyl alcohol has been made.

The halogen atoms in unsaturated halides of the vinyl type (both aliphatic and aromatic) are quite inert, and in allyl halides the activity of the halogen is greatly enhanced if a normal alkyl halide is used as the standard in the two cases. Removal of the unsaturation beyond two carbons from the halogen results in a halogen of about normal activity.

Ketene, the only important compound bearing a carbonyl group and a double bond on the same carbon, is best described as having the properties of an acid anhydride. α,β-Unsaturated aldehydes and ketones contain a conjugated system of double bonds. There is always competition, in all their reactions, between 1,2-addition to the carbonyl and 1,4-addition to the system. Aldehydes more often add 1,2, but steric considerations at atoms 1 and 4 are of first importance in additions to the α,β-unsaturated ketones.

α,β-Unsaturated acids and esters also contain the conjugated system, and 1,4-additions predominate. Acrylonitrile, containing a conjugated double and triple bond, also reacts primarily by 1,4-addition.

The following name reactions are discussed: Meerwein-Ponndorf-Verley reduction, Claisen-Schmidt, Michael, Skraup, Knoevenagel, and Perkin.

REFERENCE

Young, "The Allylic Rearrangement," J. Chem. Education, 27, 357 (1950).

EXERCISES

(Exercises 1–16 will be found within the text.)

17. Arrange the following in the order of decreasing activity of the halogen atom:

   1. furfuryl bromide
   2. phenylacetyl chloride
   3. *o*-bromotoluene
   4. 2-chloro-1-phenylethane
   5. carbon tetrachloride
   6. sec-butyl bromide

18. Show the products of reactions, if there are any, between the following substances:

   1. vinyl bromide + HBr
   2. 1-chloro-1-propene + cold aqueous NaOH
   3. allyl bromide + Mg (ether), followed by CO_2 and hydrolysis
   4. allyl alcohol + Na, followed by allyl bromide
   5. 3-penten-2-ol + HBr

19. Indicate how to prepare

   1. α,β-dibromopropionic acid from allyl alcohol
   2. 1-buten-3-ol from acrolein
   3. crotonaldehyde from ethanol
   4. 1,3-dibromopropane from glycerol

20. Complete the reactions of the following compounds with $CH_3-CH=C=O$:

   1. C_2H_5OH
   2. HCl
   3. Br_2
   4. C_2H_5MgBr
   5. H_2O
   6. $C_6H_5NH_2$

21. Show the products of the following reactions:

1. crotonaldehyde + hydrogen (catalyst)
2. ketene + n-propyl amine
3. ketene + isopropyl magnesium bromide, then hydrolysis
4. cinnamaldehyde + aluminum isopropoxide
5. 4-methyl-3-penten-2-one + HCN

6. $\begin{matrix} CH_3 \\ \diagdown \\ \diagup \\ CH_3 \end{matrix}$ C=C=O + ethanol

7. acrolein + HBr

22. Synthesize the following compounds by the best available methods:

1. $CH_3CH=CH-CH_2COOH$
2. $CH_3-CH=CH-CH=CH-COOH$
3. $CH_3CH_2COCH(CH_3)COOC_2H_5$
4. $CH_3CH=C=O$
5. $C_6H_5CH=CHCOOC_2H_5$
6. $C_6H_5-C{\equiv}C-COOH$
7. $\begin{matrix} CH_2-CH-COOC_2H_5 \\ | \qquad\quad \diagdown \\ | \qquad\qquad C=O \\ | \qquad\quad \diagup \\ CH_2-CH_2 \end{matrix}$ (Dieckmann)
8. $\begin{matrix} CH_3 \\ \diagdown \\ \diagup \\ CH_3 \quad CN \end{matrix}$ C-CH$_2$COCH$_3$

9. $C_6H_5COCH=CHC_6H_5$
10. $HOOC(CH_2)_3COOH$
11. $\begin{matrix} CH_3-CH-COOH \\ | \\ C_2H_5 \end{matrix}$
12. $\begin{matrix} CH_3-CHCOCH_3 \\ | \\ CH_3 \end{matrix}$
13. $\begin{matrix} CH_2-COOH \\ | \\ CH-COOH \\ | \\ CH_2-COOH \end{matrix}$

PART FIVE

INSIDE THE TETRAHEDRON

24 Carbon-to-Metal Bonds

The making and breaking of carbon-oxygen, carbon-nitrogen, carbon-halogen, carbon-carbon, and carbon-sulfur bonds have been discussed in some detail for a few homologous series in the preceding chapters. The properties of bonds from carbon to some other elements will be the content of this chapter. Four of these bonds—carbon-magnesium, carbon-silicon, carbon-zinc, carbon-mercury—have at least some measure of covalent character.

In the following three chapters the profound changes in properties that come from the molecule's having a central atom other than carbon will be discussed. In Chapter 25 the effect of changing the central atom to silicon, a member of the same group as carbon, and in Chapters 26 and 27 the effect of changing the central atom to a member of Group V, VI, or VII, will be examined.

Grignard Syntheses

From the point of view of the synthetic chemist, the Grignard synthesis is one of the most versatile of all preparative methods. The purpose of this section will be to focus attention on all the Grignard reactions mentioned so far and to add a few useful new ones. You should review the use of the Grignard reagent in synthesizing primary, secondary (p. 211), and tertiary (pp. 227, 274) alcohols, carboxylic acids (p. 200), and ketones (p. 283).

24-1. Structure of the Grignard Reagent

The practical use of the Grignard reagent depends on its solubility in the inert solvent ether. Although the Grignard reagent has been written as RMgX, it is more correctly pictured as an equilibrium mixture of RMgX with R_2Mg and MgX_2. All the

$$2RMgX \overset{ether}{\rightleftarrows} R_2Mg + MgX_2 \qquad [24\text{-}1]$$

reactions may be represented, however, as taking place with the structure R–Mg–X.

24-2. Reactions of the Grignard Reagent

A. Addition Reactions

1. With Oxygen. Two reactions that need to be avoided in the use of the Grignard are those with oxygen and with water (Exp. 24-3). Oxygen adds quite rapidly to a Grignard to yield a peroxide. The peroxide formed reacts with a second molecule of

$$R\text{–}Mg\text{–}X + O_2 \longrightarrow R\text{–}O\text{–}O\text{–}Mg\text{–}X \overset{RMgX}{\longrightarrow} 2R\text{–}O\text{–}Mg\text{–}X \qquad [24\text{-}2]$$

Grignard to yield R–O–MgX. Hydrolysis of this product yields an alcohol, but not a useful one, in most cases, since the original Grignard is ordinarily made from the alcohol. How, then, is one able to run a Grignard reaction in air without extraordinary precautions? The answer is that the ether used as a solvent for the Grignard reaction is very volatile and keeps a vapor shield (not a perfect one, however) between the atmospheric oxygen and the Grignard reagent. Tertiary amines (which carry no hydrogen on the nitrogen) have been used in some syntheses from Grignard reagents.

2. With Active Hydrogens. Substances containing hydrogen attached to oxygen, nitrogen, a halogen, or sulfur lose hydrogen more or less readily to a Grignard reagent. One use made of such a reaction is the occasional preparation of a pure hydrocarbon.

$$RMgX + HOH \longrightarrow RH + Mg\overset{OH}{\diagup}X \qquad [24\text{-}3]$$

$$RMgX + HCl \longrightarrow RH + Mg \overset{Cl}{\underset{X}{\big\langle}} \qquad [24\text{-}4]$$

$$RMgX + R'OH \longrightarrow RH + Mg \overset{OR'}{\underset{X}{\big\langle}} \qquad [24\text{-}5]$$

$$RMgX + H_2S \longrightarrow RH + Mg \overset{X}{\underset{SH}{\big\langle}} \qquad [24\text{-}6]$$

$$RMgX + R'NH_2 \longrightarrow RH + Mg \overset{NHR'}{\underset{X}{\big\langle}} \qquad [24\text{-}7]$$

$$RMgX + R'COOH \longrightarrow RH + R'COOMgX \qquad [24\text{-}8]$$

An important use of one Grignard reagent, methyl magnesium iodide, is the determination of the presence of active hydrogens in unknown compounds. By measuring the amount of methane liberated from a weighed quantity of an unknown substance, we can determine the percentage of active hydrogen. Consequently, this

$$CH_3MgI + HY \longrightarrow CH_4 + Mg \overset{I}{\underset{Y}{\big\langle}} \qquad [24\text{-}9]$$

Zerewitinoff determination is frequently made in quantitative organic analysis. It appears likely, however, that the new reagent $LiAlH_4$ may replace the Zerewitinoff determination, for it is more convenient to use and is a superior reagent in many instances. It can be used satisfactorily for determining the active hydrogen in alcohols, glycols, phenols, and amines.

$$LiAlH_4 + 4HY \longrightarrow LiY + AlY_3 + 4H_2 \qquad [24\text{-}10]$$

3. With Double and Triple Bonds Bearing no Hydrogens

Exercise 24-1. Write equations for reactions between ethylmagnesium bromide and the following compounds: acetaldehyde, cyclohexanone,

isobutyronitrile, formaldehyde, ethyl benzoate, carbon dioxide, ethylene oxide.

Grignard reagents add to ketones rather readily, and some ketones can actually be synthesized through the use of a Grignard.
Two different circumstances have been used advantageously:

(1) When an acyl halide reacts with a Grignard, the addition
reaction is faster than with an ordinary carbonyl group; so nearly
all the acyl halide is used up to form a ketone before the product
(the ketone) reacts with any Grignard.

$$\text{R--C}\underset{\text{Cl}}{\overset{\text{O}}{\lessgtr}} + \text{R'MgX} \longrightarrow \left[\text{R--C}\underset{\text{R' Cl}}{\overset{\text{OMgX}}{\lessgtr}} \right] \longrightarrow \text{R--C=O} + \text{Mg}\underset{\text{Cl}}{\overset{\text{X}}{\lessgtr}} \qquad [24\text{-}11]$$

Exercise 24-2. Could an acid anhydride be used in place of an acyl
halide in Expression 24-11? If so, write the reaction. If not, why?

(2) Amides of the type R–CO–NR$_2'$ have been used in the
Grignard synthesis for making ketones. The intermediate addition product does not decompose spontaneously but is fairly stable
and may be hydrolyzed to a ketone, R–CO–R''.

$$\text{R--C}\underset{\text{NR}_2'}{\overset{\text{O}}{\lessgtr}} + \text{R''MgX} \longrightarrow \text{R--C}\underset{\text{R'' NR}_2'}{\overset{\text{OMgX}}{\diagup}} \xrightarrow{\text{H}_2\text{O}} \text{Mg}\underset{\text{OH}}{\overset{\text{X}}{\lessgtr}} + \text{R}_2'\text{NH} + \text{R--C--R''} \qquad [24\text{-}12]$$

Exercise 24-3. Could the reaction of Expression 24-12 be run on
N-ethylacetamide? If so, write the reaction. If not, why?

Exercise 24-4. Write an equation for the reaction between N,N-di-
ethylformamide and phenylmagnesium bromide.

4. 1,4-Addition.

Some compounds, such as nitroalkenes and
α,β-unsaturated ketones, add Grignard reagents in the 1,4 manner.
For example, 1-nitro-2-methylpropene will add ethyl magnesium bromide to give a complex that will hydrolyze to 2,2-dimethyl-

1-nitrobutane. These branched-chain nitroalkanes are not easily synthesized by any other method.

Exercise 24-5. Write a synthesis for 1-nitro-2-methylpropene.

$$CH_3-C\!\!=\!\!CH + C_2H_5MgBr \longrightarrow \underset{\underset{\displaystyle O}{C_2H_5}}{\overset{\displaystyle CH_3}{CH_3-C\!-CH\!=\!NOMgBr}} \xrightarrow[60\%]{HOH} \underset{C_2H_5}{\overset{CH_3}{CH_3-C\!-CH_2NO_2}}$$
$$\underset{CH_3\ NO_2}{}$$

[24-13]

Whether a Grignard reagent will add 1,2 or 1,4 to an α,β-unsaturated ketone is governed largely by the reactivity and size (the two factors are not unrelated) of the Grignard. An ethyl magnesium bromide, for example, adds 1,4 to a greater extent than the less reactive (and larger) phenyl magnesium bromide. A large group at the alkene end of the conjugated system will favor 1,2-addition while a bulky group at the carbonyl end will favor 1,4-addition. These statements are corroborated by the observations compiled in Table 24-1, taken from Fuson and Snyder, *Organic Chemistry*, 2nd ed. (John Wiley & Sons, New York, 1954), p. 287.

TABLE
24-1 | *Competitive Additions of Grignard Reagents*

	% 1,2-addition*			
	2-penten-4-one	benzal-acetone	benzal-acetophenone	2-methyl-2-penten-4-one
C_2H_5MgBr	25	40	1	100
C_6H_5MgBr	60	88	6	100

* The percentage remaining when each figure is subtracted from 100 gives the competitive product expected by 1,4-addition.

Exercise 24-6. Write the equation for addition of the Grignard reagent and subsequent hydrolysis of the major product in each of the eight cases of Table 24-1.

5. Barbier-Wieland Degradation. A Grignard reagent, phenyl magnesium bromide, may be used to find the number of methylene (–CH₂–) groups between a carboxyl group and any other point in a molecule. It is used commonly only on large and complex molecules. This Barbier-Wieland degradation may be illustrated on β-phenylpropionic acid. If the ethyl ester of this acid is treated with phenyl magnesium bromide, a tertiary alcohol

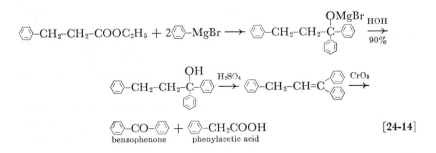

$$[\text{24-14}]$$

is obtained. Dehydration and oxidation then yield phenylacetic acid. The reaction can then be repeated on ethyl phenylacetate, and benzoic acid will be obtained. This result means that the original acid contained two –CH₂– groups between the benzene ring and the carboxyl group. In this particular example, of course, phenylacetic acid is a known compound that could be identified, and the second reaction would not have to be run.

B. *Substitution Reactions*

1. Mercury for Magnesium. Use of the mercury compound corresponding to the Grignard reagent has been a means of identifying many alkyl halides in a qualitative manner, for a large number of them are crystalline compounds having sharp melting points. The alkyl mercuric halides are much less reactive than the Grignard reagent: for example, they do not hydrolyze rapidly in alcohol containing water. One can readily prepare the alkyl

$$HgCl_2 + R\text{–}Mg\text{–}Cl \longrightarrow R\text{–}Hg\text{–}Cl + MgCl_2 \qquad [\text{24-15}]$$

$$HgBr_2 + R-Mg-Br \longrightarrow R-Hg-Br + MgBr_2 \qquad [24\text{-}16]$$

$$HgI_2 + R-Mg-I \longrightarrow R-Hg-I + MgI_2 \qquad [24\text{-}17]$$

mercuric halide simply by shaking an ether solution of the Grignard with solid mercury halide. The unreacted mercuric halide and the magnesium halide can be dissolved away from the product by dilute alcohol and water washes.

2. Silicon for Magnesium.   The Grignard reagent has become very useful in the preparation of alkyl silicon halides, the importance of which is discussed in Chapter 25. The reaction may be considered an exchange of the alkyl group from the reactive Mg atom to an atom more closely resembling carbon. The reaction may be used with $SiCl_4$ to produce four stages of substitution, the most felicitous one being the preparation of R_2SiCl_2. The following

$$SiCl_4 + RMgX \xrightarrow{\text{ether}} RSiCl_3 + MgXCl \qquad [24\text{-}18]$$

$$RSiCl_3 + RMgX \longrightarrow R_2SiCl_2 + MgXCl \qquad [24\text{-}19]$$

$$R_2SiCl_2 + RMgX \longrightarrow R_3SiCl + MgXCl \qquad [24\text{-}20]$$

$$R_3SiCl + RMgX \longrightarrow R_4Si + MgXCl \qquad [24\text{-}21]$$

compounds have been synthesized by this method: $C_2H_5SiCl_3$ in 80% yield, $(C_2H_5)_2SiCl_2$ (70%), and $(C_2H_5)_4Si$ (35%).

Alkyl groups from the Grignard reagent may be substituted in many other halides although the mercury and silicon compounds have found most use. In Group Ib, Cu, Ag, and Au halides will react with the Grignard reagent to form metal alkyls. In other groups the following elements are known to react: Group IIa, Be and Ca; Group IIb, Zn, Cd, and Hg; Group IIIa, B, Al, and Sc; Group IIIb, Ga, In, and Tl; Group IVb, Ge, Sn, and Pb; Group Vb, Sb and Bi; Group VIa, Cr; Group VII, Mn; Group VIII, Fe and Pt.

Reagents that are somewhat less reactive than Grignards occasionally find some use. For example, R_2Cd (a dialkyl cadmium) can be used in the synthesis of ketones from acyl halides or acid anhydrides, whereas the Grignard itself ordinarily reacts very readily with the product (see p. 538, however) as well as with the acyl halide. The syntheses of propiophenone and *p*-methoxy-acetophenone with alkyl cadmium compounds are shown in Expressions 24-22 and 24-23.

$$CH_3-CH_2-C \Big\langle \begin{matrix} O \\ O \end{matrix} + Cd-(C_6H_5)_2 \longrightarrow \left[\begin{matrix} C_6H_5 \\ C_2H_5-C-O-Cd-C_6H_5 \\ O-C-C_2H_5 \\ O \end{matrix} \right] \xrightarrow[76\%]{H^+} CH_3-CH_2-\overset{O}{\underset{}{C}}-C_6H_5$$

$$CH_3-CH_2-C \Big\langle \begin{matrix} O \end{matrix}$$

[24-22]

$$CH_3-O-C_6H_4-\overset{O}{\underset{}{C}}-Cl + Cd(CH_3)_2 \longrightarrow \left[\begin{matrix} O-Cd-CH_3 \\ CH_3-O-C_6H_4-C \Big\langle \\ Cl \\ CH_3 \end{matrix} \right] \xrightarrow[84\%]{H^+}$$

$$CH_3-O-C_6H_4-\overset{O}{\underset{}{C}}-CH_3 \qquad [24-23]$$

The organo-cadmium compound is prepared *in situ* from anhydrous cadmium chloride and the Grignard reagent.

$$2R-Mg-Br + CdCl_2 \longrightarrow CdR_2 + MgBr_2 + MgCl_2 \qquad [24-24]$$

Zinc, with an α bromoester (Reformatsky reaction, p. 464), will make an organo-zinc reagent that will add to ketones or aldehydes to give, after hydrolysis, β hydroxy esters. A Grignard can not be used here since it will react rapidly with esters as well as with aldehydes and ketones.

Exercise 24-7. Write the Reformatsky reaction to synthesize ethyl 1-hydroxycyclohexylacetate; to synthesize ethyl β-hydroxy-β-phenyl-propionate.

Metal Alkyls

24-3. Preparation of Metal Alkyls

1. Displacement of a Halogen. The first carbon-to-metal bond was recognized by Frankland in 1849 (p. 73). He prepared diethyl zinc by treating ethyl iodide with zinc. The product was spontaneously inflammable in air and poisonous. This method is

$$2Zn + 2CH_3CH_2I \longrightarrow Zn(CH_2CH_3)_2 + ZnI_2 \qquad [24-25]$$

applicable to the preparation of alkyls of metals in the first four groups and is especially convenient for lithium alkyls.

$$Li + nC_4H_9\text{-}Cl \xrightarrow[77\%]{ether} nC_4H_9\text{-}Li + LiCl \qquad [24-26]$$

$$NaPb + CH_3CH_2Cl \longrightarrow Pb(CH_2CH_3)_4 + NaCl + PbCl_2 \qquad [24-27]$$

Mercury alkyls and aryls are easily prepared from the corresponding halides and sodium amalgam (3%). The preparation of mercury aryls, which are useful in the making of some aryl carbon-silicon bonds (see p. 552), is aided by the presence of ethyl acetate.

$$CH_3CH_2I + Na(Hg) \longrightarrow NaI + HgI_2 + Hg(C_2H_5)_2 \qquad [24-28]$$

$$C_6H_5Br + Na(Hg) \xrightarrow{CH_3COOC_2H_5} NaBr + HgBr_2 + \underset{\text{m.p. }120°}{Hg(C_6H_5)_2} \qquad [24-29]$$

2. Replacement of Metals. Whereas zinc and lithium alkyls are easily prepared by the method just mentioned, sodium and the other metals of Group I form alkyls most conveniently by displacement of some other metal atom. Sodium and potassium alkyls are prepared from HgR_2 or ZnR_2.

$$Zn(CH_2CH_3)_2 + 2Na \longrightarrow Zn + 2Na\text{-}CH_2CH_3 \qquad [24-30]$$

3. From Grignard Reagents. Alkyls of less reactive metals can be made from Grignard reagents, as was suggested in Expression 24-24.

$$\text{\hexagon-MgI} + \text{CuI} \longrightarrow \text{Cu-\hexagon} + \text{MgI}_2 \qquad [24\text{-}31]$$

$$\text{CH}_3\text{MgBr} + \text{BeCl}_2 \xrightarrow[90\%]{} \text{Be(CH}_3)_2 + \text{MgBr}_2 + \text{MgCl}_2 \qquad [24\text{-}32]$$

$$\text{CH}_3\text{MgBr} + \text{CdCl}_2 \longrightarrow \text{Cd(CH}_3)_2 + \text{MgBr}_2 + \text{MgCl}_2 \qquad [24\text{-}33]$$

24-4. Properties of Metal Alkyls

The zinc alkyls are spontaneously inflammable. Other metal alkyls may be handled with less hazard, but all of them react with water more or less violently. The most important property (for this discussion) of the alkali metal alkyls is that they may often be used where a Grignard reagent fails. The lithium alkyls have frequently been useful in this regard. Recently (1944) some tertiary alcohols that had never before been made by any method have been made with lithium alkyls. The lithium alkyl is, for one thing, less sterically hindered than the corresponding Grignard reagent and much more reactive.

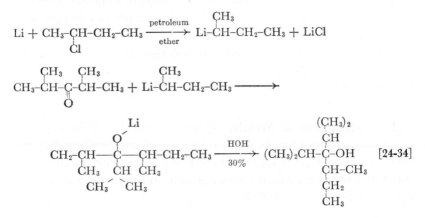

Exercise 24-8. Write out the preparation of 2,4-dimethyl-3-isopropyl-3-pentanol by the method of Expression 24-34.

SUMMARY

The very versatile Grignard reagent can be used to synthesize a wide variety of organic compounds by addition and to make just as varied a group of carbon-to-metal bonds by substitution. The use of metal alkyls instead of Grignards has advantages in particular cases.

EXERCISES

(Exercises 1–8 will be found within the text.)

9. Write reactions of phenyl magnesium bromide with the following substances and the subsequent hydrolysis, when appropriate:

    1. methyl ethyl ketone
    2. benzaldehyde
    3. carbon dioxide
    4. ethylene oxide
    5. oxygen
    6. ethanol

    7. acetonitrile
    8. formaldehyde
    9. ethyl acetate
    10. acetyl chloride
    11. 1-nitro-1-propene

10. How would you prepare the following compounds?

    1. lead tetraethyl
    2. sodium sec-butyl
    3. lithium phenyl
    4. ethyl mercuric iodide

    5. zinc diethyl
    6. cadmium dimethyl
    7. dimethyl silicon dichloride

11. Cite examples of using carbon-metal bonds in syntheses where they have some advantages over Grignard reagents.

12. What is the best method of preparing the following?

    1. 2,4-dimethyl-3-ethyl-3-pentanol
    2. 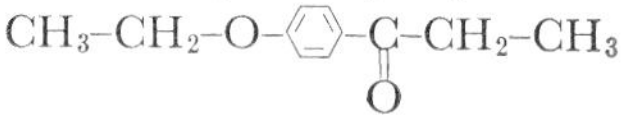 $CH_3-CH_2-O-\langle \rangle-\overset{\displaystyle O}{\underset{\displaystyle ||}{C}}-CH_2-CH_3$
    3. ethyl β-hydroxy-β-methylbutyrate

25 Covalent Silicon Compounds

As we seek to determine the influence of the central atom on the properties of covalent compounds, we must turn our inquiry toward what happens when another element of Group IV is substituted for carbon. Our attention in this direction will be centered on silicon because it has been studied more extensively, because it is of current interest, and because its chemistry—for the present, at least—is more important than that of any other member of the group.

Although silicon appears in the Periodic Table in the same group as carbon, the differences between them, in the chemistry of their covalent compounds, are more striking than the likenesses. Silicon does resemble germanium and tin in Group IV and boron in Group III. Carbon, silicon, and boron are the only elements that form atom-to-atom covalent bonds of the same element in even moderately stable open chains.* In contrast to carbon, however, only short chains are known. Six silicon atoms are joined in a chain in Si_6H_{14}, a compound bearing a formal resemblance to hexane, but the silicon chain is brittle when it contains more than four silicon atoms.

From the electronic structures of carbon and silicon, one might expect the two elements to form compounds of close resemblance. Unexpectedly, the only likenesses are due to the fact that both atoms form four bonds to other atoms and can be linked to atoms of their own kind.

Many of the differences between carbon and silicon can be accounted for by the following three generalizations about elements in the Periodic Table.

* Perhaps nitrogen, sulfur, and germanium should be added to this group. Nitrogen compounds containing chains of four nitrogen atoms have been reported.

1. The size of an atom increases in a group from top to bottom. The covalent radius of silicon is 1.17 A; that of carbon is 0.77 A.

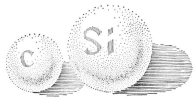

2. The electropositive character of an element in a group increases from top to bottom. Sulfur is more electropositive than oxygen; silicon is more electropositive than carbon. The effect of the relative electronegativities of carbon, hydrogen, and silicon in a hydrolysis reaction is discussed below (p. 549).
3. Elements in the first period (from Li to F) obey the Rule of Eight; those below may expand their shells to ten or twelve if the occasion demands it. The contrast between the ease of hydrolysis (and alcoholysis) of $SiCl_4$ and the difficulty of hydrolysis of CCl_4 can be accounted for on this basis (see pp. 570–571).

In this chapter we shall examine the properties of Si-Si, Si-H, Si-C, Si-O, and Si-Cl bonds in order to highlight these differences between carbon and silicon.

Silicon-Silicon and Silicon-Hydrogen Bonds (the Silanes)

25-1. Preparation of the Silanes

1. From Magnesium Silicide.   The silicon-hydrogen analogues of the hydrocarbons are called silanes: monosilane (SiH_4), disilane (Si_2H_6), etc. All the known silanes are prepared by treatment of magnesium silicide with HCl or another inorganic acid. The magnesium silicide is obtained by a high-temperature reaction between

$$4Mg + SiO_2 \longrightarrow Mg_2Si + 2MgO \qquad [25\text{-}1]$$

$$Mg_2Si + HCl \longrightarrow SiH_4 + Si_2H_6 + Si_3H_8 \ldots Si_6H_{14} \qquad [25\text{-}2]$$

sand (SiO_2) and magnesium. About 40% monosilane, 30% disilane, and smaller percentages of the higher silanes are obtained by the method. By fractionation, SiH_4 and Si_2H_6 can be separated as gases from the others, which are liquids.

2. From Silicon Tetrachloride. A much simpler method of making monosilane is available now through the use of lithium aluminum hydride. The reaction with $SiCl_4$ is run satisfactorily at room temperature in an ether solution.

$$SiCl_4 + LiAlH_4 \xrightarrow[99\%]{ether} LiCl + AlCl_3 + SiH_4 \qquad [25\text{-}3]$$

25-2. Properties of the Silanes

1. Thermal Decomposition. Monosilane is stable to high temperatures, but the higher homologues decompose at lower and lower temperatures, and hexasilane is unstable at room temperature upon long standing. This behavior is in sharp contrast to that of the hydrocarbons, which are completely stable at room temperature although they may be thermally cracked above 600°C. The cracking process is of the same nature for both types of compounds, however. The silanes break down into lower saturated and unsaturated compounds. Hexasilane, for example, decomposes into monosilane, disilane, and compounds of uncertain composition.

$$\underset{\text{hexasilane}}{SiH_3\text{-}SiH_2\text{-}SiH_2\text{-}SiH_2\text{-}SiH_2\text{-}SiH_3} \longrightarrow \underset{\text{monosilane}}{SiH_4} + \underset{\text{disilane}}{SiH_3\text{-}SiH_3} + ? \qquad [25\text{-}4]$$

2. Spontaneous Combustion. The silanes burn spontaneously in air, even below −180°C, with evolution of considerable heat from the formation of silica. The reaction goes with explosive vio-

$$SiH_3\text{-}SiH_2\text{-}SiH_3 + 5O_2 \longrightarrow 3SiO_2 + 4H_2O \qquad [25\text{-}5]$$

lence, which is thought to be due to preliminary dissociation of the hydrogen. Hydrogen and oxygen then combine rapidly. This property makes the study of silanes difficult, to say the least.

3. Hydrolysis. The silanes are such strong reducing agents that they will reduce water to hydrogen in alkaline solution. Every

$$H_3Si-SiH_3 + 4OH^- + 2H_2O \longrightarrow 2SiO_3^= + 7H_2 \qquad [25-6]$$

silicon-hydrogen bond and the silicon-silicon bond are broken, and hydrogen is evolved.

Exercise 25-1.   Indicate the changes in oxidation number of the various atoms involved in Expression 25-6.

A satisfactory explanation of this sharp contrast in behavior between the silanes and the alkanes is available in the differences in electronegativity among the atoms involved and in the character of the attacking agent, hydroxyl ion. Pauling has estimated the relative electronegativities of the atoms involved here as Si 1.8, H 2.1, and C 2.5; that is, silicon is more electropositive than hydrogen, and hydrogen is more electropositive than carbon. In disilane and ethane, then, the disposition of electrons is described by the exaggerated pictures below.

$$\begin{array}{ll} \delta^- \delta^+ \begin{array}{c} H \ \ H \\ | \ \ | \\ H: \ Si-Si-H \\ | \ \ | \\ H \ \ H \end{array} & \delta^+ \ \delta^- \begin{array}{c} H \ \ H \\ | \ \ | \\ H \ : C-C-H \\ | \ \ | \\ H \ \ H \end{array} \end{array}$$

The hydroxyl ion is attracted to the relatively electropositive silicon atom in disilane. In this case, a hydride ion $(H:^-)$ is expelled into the solvent. It picks up a proton from water, becoming a hydrogen molecule and regenerating the catalyst, OH^-. Eventually all Si-H bonds are broken by this process. The Si-Si bond is also easily broken, the final result being depicted in Expression 25-6.

In ethane, if the hydroxyl group attracted the hydrogen (proton) on carbon, water would be formed, but the fragment left would then pick up another proton from water and regenerate the OH⁻. No apparent reaction, then, would take place. Whether this exchange of ethane with the solvent does take place is not known, but evidence of similar exchanges is accumulating.

4. Halosilanes. A remarkable difference in behavior between silanes and alkanes is shown in the replacement of hydrogen by chlorine. Silanes react violently with chlorine and bromine, but dry HCl, in the presence of $AlCl_3$, will accomplish the replacement

$$SiH_4 + HCl \xrightarrow[100\%]{AlCl_3} SiH_3Cl + H_2 \qquad [25\text{-}7]$$

smoothly at 200°. This reaction also indicates the strong reducing properties of the silanes. The reaction goes progressively, if excess HCl is available, to SiH_2Cl_2, $SiHCl_3$, and $SiCl_4$.

Chloroform and carbon tetrachloride undergo a Friedel-Crafts type of reaction with trisilane. It is uncertain where the halogens are located in the product of the oxidation-reduction reaction.

$$SiH_3\text{-}SiH_2\text{-}SiH_3 + 4CHCl_3 \xrightarrow{AlCl_3} Si_3H_4Cl_4 + 4CH_2Cl_2 \qquad [25\text{-}8]$$

$$SiH_3\text{-}SiH_2\text{-}SiH_3 + 5CHCl_3 \xrightarrow{AlCl_3} Si_3H_3Cl_5 + 5CH_2Cl_2 \qquad [25\text{-}9]$$

Chlorosilane, a gas, is readily hydrolyzed to a compound called disiloxane and containing a Si-O-Si linkage. Disiloxane is not

$$2SiH_3Cl + HOH \xrightarrow[100\%]{} \underset{\substack{\text{disiloxane} \\ \text{b.p. } -15°}}{SiH_3\text{-}O\text{-}SiH_3} + 2HCl \qquad [25\text{-}10]$$

spontaneously inflammable; this and other evidence indicate that the Si-O bond has a powerful stabilizing influence on silicon compounds having Si-H bonds.

Exercise 25-2.   What is the analogue of disiloxane among carbon compounds? Can the carbon analogue be prepared in this manner?

Silicon-Carbon Bonds

25-3. Preparation of Silicon-Carbon Bonds

1. From Silicon and an Alkyl Chloride. A direct method of forming a Si-C bond has gained some importance commercially in the preparation of dimethyldichlorosilane. In the presence of copper sintered with silicon, methyl chloride will react with the silicon at 350°C to give good yields of $(CH_3)_2SiCl_2$ along with other substances. Hurd and Rochow have shown conclusively that the reaction takes steps shown in Expression 25-11. Free methyl rad-

$$2Cu + CH_3Cl \longrightarrow CuCl + Cu-CH_3 \qquad [25\text{-}11a]$$
$$\qquad\qquad\qquad\qquad \hookrightarrow Cu + CH_3 \cdot$$

$$Si + 2CuCl \longrightarrow 2Cu + \overset{\cdot}{\underset{Cl}{Si}}-Cl \qquad [25\text{-}11b]$$

$$\overset{\cdot}{\underset{Cl}{Si}}-Cl + 2CH_3 \cdot \xrightarrow[90\%]{} CH_3-\overset{CH_3}{\underset{Cl}{Si}}-Cl \qquad [25\text{-}11c]$$

icals (see pp. 657ff.) have been identified as having a short life during the process. This mechanism accounts for the fact that silicon compounds bearing 0–4 chlorine bonds and 4–0 carbon bonds are formed along with the indicated main product.

2. Modifications of the Wurtz Reaction. By using sodium metal on a mixture of silicon tetrachloride and alkyl halides, we can obtain the tetra-alkyl silicon compounds. The reaction cannot be controlled effectively enough to stop at less than complete alkylation. The large amount of heat evolved when the first unit

$$4C_2H_5Br + 8Na + SiCl_4 \xrightarrow[35\%]{} Si(C_2H_5)_4 + 4NaBr + 4NaCl \qquad [25\text{-}12]$$

of sodium halide is formed keeps the reaction going until all the halogens are removed. By dividing the reaction into two steps, so that the heat to be dissipated was evolved in two stages, Schumb was able to prepare hexaphenyldisilane by this modification of the

$$\text{⬡–Cl} + \text{Na} \xrightarrow{\text{toluene}} \text{Na–⬡} + \text{NaCl} \tag{25-13}$$

$$\text{Cl}_3\text{Si–SiCl}_3 + 6\text{Na–⬡} \xrightarrow[35\%]{} (\text{⬡})_3\text{–Si–Si–}(\text{⬡})_3 + 6\text{NaCl} \tag{25-14}$$

Wurtz reaction. The one-step process had always broken the Si-Si bond.

Other metals besides sodium, such as zinc, aluminum, and mercury, may be used to introduce Si-C bonds into molecules. A metal alkyl may be used as such, or, if the metal itself is used, an alkyl is undoubtedly formed as one step in the process of the reactions.

$$(\text{CH}_3)_2\text{SiCl}_2 + \text{CH}_3\text{Cl} + \text{Al} \xrightarrow[30\%]{350°} (\text{CH}_3)_3\text{SiCl} + \text{AlCl}_3 \tag{25-15}$$

$$\text{Hg–}(\text{⬡})_2 + \text{SiCl}_4 \xrightarrow[26\%]{} \text{⬡–SiCl}_3 + \text{⬡–HgCl} \tag{25-16}$$

$$\text{Zn}(\text{C}_2\text{H}_5)_2 + \text{SiCl}_4 \xrightarrow[\text{closed tube}]{160°} \text{Si}(\text{C}_2\text{H}_5)_4 + \text{ZnCl}_2 \tag{25-17}$$

Various examples are shown in Expressions 25-15–25-17. (For the preparation of diphenyl mercury in Exp. 25-16, see Exp. 24-29.)

The effect of the size of the silicon atom, in comparison with that of the carbon atom, is shown in the relative ease with which four phenyl groups can be put around silicon. By a Wurtz reaction it has been possible to synthesize tetraphenylsilane with bromobenzene and silicon tetrachloride. Tetraphenylmethane cannot be made in this way, and a Friedel-Crafts reaction between excess

$$\text{⬡–Cl} + \text{SiCl}_4 + 8\text{Na} \xrightarrow[95\%]{} (\text{C}_6\text{H}_5)_4\text{Si} + 8\text{NaCl} \tag{25-18}$$

benzene and carbon tetrachloride, which might be expected to yield the compound, gives triphenylchloromethane instead, presumably because of the steric factor.

$$3\text{⬡} + \text{CCl}_4 \xrightarrow[86\%]{\text{AlCl}_3} (\text{C}_6\text{H}_5)_3\text{CCl} + 3\text{HCl} \tag{25-19}$$

Silicon has its steric limitations, too. Four cyclohexyl groups cannot be introduced around silicon even with the more reactive lithium alkyl. The last chlorine in the product, tricyclohexylchloro-silane (Exp. 25-21), cannot be replaced even by a smaller lithium alkyl.

$$\begin{matrix} CH_2-CH_2 \\ CH_2{\Big\langle}\qquad{\Big\rangle}CHCl + 2Li \longrightarrow C_6H_{11}Li + LiCl \\ CH_2-CH_2 \end{matrix} \qquad\qquad [25\text{-}20]$$

$$3C_6H_{11}Li + SiCl_4 \xrightarrow[60\%]{} (C_6H_{11})_3SiCl + 3LiCl \qquad\qquad [25\text{-}21]$$
$$\text{tricyclohexyl-}$$
$$\text{chlorosilane}$$

3. The Grignard Reaction. Though the direct method of making new Si-C bonds has been adopted as an industrial process and the Wurtz reaction is useful in particular cases, the Grignard reaction is still the most versatile method of introducing Si-C bonds. One reason for its wide use is the ease of preparation of the alkyl magnesium halide. The reactions have already been outlined (p. 541).

25-4. Properties of Silicon-Carbon Bonds

1. Inertness. The silicon tetra-alkyls and tetra-aryls are very stable thermally. Silicon tetraphenyl, for example, can be distilled without decomposition at its boiling point, 425°C. It is not safe to say, however, that the presence of silicon in a molecule with carbon atoms stabilizes the compound. Silicon is more electropos-itive than carbon. It is possible to distort this difference in polarity by the introduction of electronegative atoms one, two, or even more carbon atoms away from the silicon atom. A phenyl group attached to silicon also makes a covalent silicon compound vulner-able to attack. Ethyl-n-propylbenzylphenylsilane, for example, hydrolyzes readily in sulfuric acid and splits the Si-C bond with the phenyl group. The silanol formed (see p. 555) spontaneously loses water to give a siloxane.

$$\begin{matrix} C_6H_5 & CH_2-C_6H_5 \\ & Si \\ C_2H_5 & C_3H_7 \end{matrix} \xrightarrow[H_2SO_4]{HOH} \bigcirc + \begin{bmatrix} C_6H_5-CH_2 \\ C_3H_7-Si-OH \\ C_2H_5 \end{bmatrix}$$

$$\begin{matrix} C_6H_5CH_2 & CH_2C_6H_5 \\ C_3H_7-Si-O-Si-C_3H_7 \\ C_2H_5 & C_2H_5 \end{matrix} \longleftarrow \qquad [25\text{-}22]$$

However, aryl groups on silicon can be nitrated and sulfonated without breaking a Si-C bond.

$$(C_2H_5)_3Si-\bigcirc \xrightarrow[\substack{\text{acetic anh.} \\ 98\%}]{HNO_3} (C_2H_5)_3Si-\bigcirc-NO_2 \qquad [25\text{-}23]$$

2. Effect of a Halogen on α, β, and γ Positions. A chlorine atom one, two, or three carbons away from a silicon atom in a molecule has a profound effect on the properties. The results are given here simply to indicate that the entire environment of the silicon atom must be taken into account before the behavior of a molecule containing silicon can be understood.

In the compounds of Expressions 25-24–25-28 the α halogen is slightly reactive, the β halogen very reactive, and the γ halogen moderately reactive, toward silver nitrate. The first compound is less reactive than n-hexyl chloride toward Ag^+. With bases the three compounds behave as shown.

$$\underset{\underset{Cl}{|}}{CH_3-CH-SiCl_3} \xrightarrow{OH^-} \underset{\underset{Cl}{|}}{CH_3-CH-Si(OH)_3} + \underset{100\%}{3Cl^-} \qquad [25\text{-}24]$$

$$Cl-CH_2-CH_2-SiCl_3 \xrightarrow{OH^-} CH_2{=}CH_2 + Si(OH)_4 + \underset{100\%}{4Cl^-} \qquad [25\text{-}25]$$

$$\xrightarrow[60\%]{\text{quinoline}} CH_2{=}CH-SiCl_3 + \text{quinoline hydrochloride} \qquad [25\text{-}26]$$

$$ClCH_2-CH_2-CH_2-SiCl_3 \xrightarrow{OH^-} ClCH_2-CH_2-CH_2-Si(OH)_3 + \underset{100\%}{3Cl^-} \qquad [25\text{-}27]$$

$$\xrightarrow[\substack{\text{alcohol}}]{OH^-\text{ in}} \underset{\underset{CH_2}{\diagdown}}{CH_2{-}\!\!-\!\!-\!\!-\!\!-CH_2} + Si(OH)_4 + 4Cl^- \qquad [25\text{-}28]$$

$$31\%$$

Silicon-Oxygen Bonds

25-5. Preparation of Silicon-Oxygen Bonds

A. Breaking Silicon-Carbon Bonds by Combustion

The chemistry of some of the silicon bonds already discussed involves the formation of Si-O bonds. Not only do the silanes, as we have seen, burn spontaneously in air to yield SiO_2 and H_2O, breaking Si-H and Si-Si bonds to form new Si-O bonds, but silicon tetramethyl, though stable and inert to most reagents, can also be burned.

$$CH_3-\underset{\underset{CH_3}{|}}{\overset{\overset{CH_3}{|}}{Si}}-CH_3 + 8O_2 \xrightarrow{\text{burning}} SiO_2 + 4CO_2 + 6H_2O \qquad [25\text{-}29]$$

B. Hydrolysis of Silicon-Chlorine Bonds

One type of reaction that can be controlled to give Si-O bonds is the hydrolysis of silicon-chlorine bonds.

1. Silanols and Siloxanes. Triethylchlorosilane can be titrated with a cold base to yield triethylsilanol, a liquid boiling at $77°/28$ mm. The silanols, in contrast to carbon alcohols, are extremely easy

$$(C_2H_5)_3SiCl + OH^- \xrightarrow[95\%]{} \underset{\text{triethylsilanol}}{(C_2H_5)_3Si-OH} + Cl^- \qquad [25\text{-}30]$$

to dehydrate to ether-type compounds, called siloxanes. The siloxane rather than the silanol, in fact, is frequently isolated in the process of working up the reaction mixture. Hexaethyldisiloxane is obtained from triethylsilanol by treatment with HBr, triethylchlorosilane by treatment with HCl. Triethylsilanol reacts with

$$\xrightarrow[48\%]{\text{HBr}} (C_2H_5)_3Si\text{-}O\text{-}Si(C_2H_5)_3 \qquad [25\text{-}31]$$
$$\underset{\text{hexaethyldisiloxane}}{}$$

$(C_2H_5)_3Si\text{-}OH$

$$\xrightarrow[77\%]{\text{HCl}} (C_2H_5)_3Si\text{-}Cl \qquad [25\text{-}32]$$
$$\underset{\text{triethylchlorosilane}}{}$$

sodium to give a quantitative yield of hydrogen and also dissolves in sodium hydroxide. Triethylchlorosilane will also react with

$$\xrightarrow{\text{Na}} (C_2H_5)_3Si\text{-}O\text{-}Na + \tfrac{1}{2}H_2 \qquad [25\text{-}33]$$

$(C_2H_5)_3Si\text{-}OH$

$$\xrightarrow[87\%]{\text{NaOH}} (C_2H_5)_3Si\text{-}O\text{-}Na + H_2O \qquad [25\text{-}34]$$

liquid ammonia to give, in 70% yield, the silicon analogue of an amine (called a silazane). The reaction can be reversed with HCl.

$$(C_2H_5)_3Si\text{-}Cl \left\{ \begin{array}{c} \text{liq. NH}_3 \\ \xrightarrow{\hspace{1cm}} \\ 70\% \\ \text{HCl} \\ \xleftarrow{\hspace{1cm}} \\ 80\% \end{array} \right\} (C_2H_5)_3Si\text{-}NH_2 \qquad [25\text{-}35]$$
$$\underset{\substack{\text{triethylsilazane} \\ \text{b.p. 134°C}}}{}$$

Exercise 25-3. Which of the reactions delineated in the preceding paragraph are similar in character to reactions in carbon compounds, and which are in sharp contrast? Using a comparable carbon compound, write, for each of the reactions given in Expressions 25-30–25-35, a reaction that will show a likeness or a difference.

2. Silanediols and Silicones. Dimethyldichlorosilane hydrolyzes readily to form a dihydroxy compound (presumably) that cannot be isolated because it condenses spontaneously to a polymeric substance. The intermediate dihydroxy compound might be

$$\begin{array}{c} CH_3 \\ \diagdown Si \diagup \\ CH_3 \end{array}\begin{array}{c} Cl \\ \\ Cl \end{array} + HOH \longrightarrow \left[\begin{array}{c} CH_3 \\ \diagdown Si \diagup \\ CH_3 \end{array}\begin{array}{c} OH \\ \\ OH \end{array}\right] \longrightarrow HO\text{-}\!\!\begin{array}{c} CH_3 \\ Si\text{-}O\text{-} \\ CH_3 \end{array}\!\!\left[\begin{array}{c} CH_3 \\ Si\text{-}O \\ CH_3 \end{array}\right]_x\!\!\begin{array}{c} CH_3 \\ Si\text{-}OH \\ CH_3 \end{array}$$

$$[25\text{-}36]$$

expected to lose one molecule of water intramolecularly, if we reason from the behavior of the corresponding carbon compound, to give a ketone-type compound. But silicon, in contrast to carbon, tries to maintain bonds to as many oxygens as possible. The great propensity of silicon to maintain single bonds to oxygen points up a big difference between it and carbon, which readily forms multiple bonds with oxygen. No silicon-to-oxygen double bond is known.

Only a few fairly stable substances containing the structure

$$\text{Si} \begin{matrix} \text{OH} \\ \text{OH} \end{matrix}$$

have been isolated. When the alkyl groups on silicon are large, such as tert-butyl groups, a stable silanediol and a diaminosilane

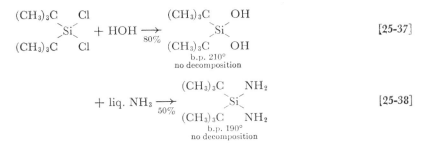

can both be isolated. Furthermore, it is not possible, even with sulfuric acid at 110°, to dehydrate the di-tert-butylsilanediol either to a polymer (silicone) or to a Si-O linkage.

When diphenyldichlorosilane is hydrolyzed, the dihydroxy compound does not lose appreciable amounts of water until a temperature of 100°C is reached. If it is boiled in water, a trimer (Exp. 25-39) and a tetramer are obtained, and chain polymers may be formed by heating at higher temperatures. If silicon has three bonds to chlorine, hydrolysis introduces the possibility of cross-linking long chains, and this would lead to three-dimensional polymers.

$$
CH_3\text{--}SiCl_3 + H_2O \longrightarrow \left[CH_3\text{--}Si\text{--}OH \atop \substack{OH \\ OH} \right] \longrightarrow H\text{--}\left[\text{--}O\text{--}\underset{\underset{|}{O}}{\overset{CH_3}{\underset{|}{Si}}}\text{--}O\text{--}\right]_x\text{--}\underset{\underset{|}{O}}{\overset{CH_3}{\underset{|}{Si}}}\text{--}O\text{--}\underset{\underset{|}{O}}{\overset{CH_3}{\underset{|}{Si}}}\text{--}OH
$$

[25-40]

25-6. Methyl Silicone Oils

Control of the end groups and of the length of the chain in these *silicone* polymers has been achieved to a remarkable degree. By hydrolyzing a mixture of pure trimethylchlorosilane and dimethyldichlorosilane, we might expect silicone compounds of definite composition. The products in which $n = 1\text{--}9$ are called methyl silicone oils. When $n = 1$ (Exp. 25-41), the oil has a boiling point

$$
2(CH_3)_3SiCl + n(CH_3)_2SiCl_2 + HOH \longrightarrow CH_3\text{--}\underset{\underset{CH_3}{|}}{\overset{CH_3}{\underset{|}{Si}}}\text{--}O\text{--}\left[\underset{\underset{CH_3}{|}}{\overset{CH_3}{\underset{|}{Si}}}\text{--}O\right]_n\underset{\underset{CH_3}{|}}{\overset{CH_3}{\underset{|}{Si}}}\text{--}CH_3
$$

[25-41]

of 153°C; when $n = 9$, the boiling point is 201°/4.7 mm. We synthesize silicone oils of specified viscosity by using the proper ratio of monochloro to dichloro compound.

The most important properties of these silicone oils—properties that make them useful as lubricants and hydraulic fluids—are that they are very stable to heat and that the change in viscosity is small over long temperature ranges. For example, though a certain hydrocarbon lubricating oil changed viscosity eighteen-fold when the temperature sank from 100°F to −35°F, a silicone oil increased in viscosity only seven-fold over the same temperature range.

25-7. Silicone Resin

If small amounts of CH_3SiCl_3 are introduced into the mixture just described for making an oil, some of the long chains may be cross-linked during hydrolysis, increasing the molecular weight very sharply. If the hydrolysis by water is helped along by heating, a resin that is an excellent electrical insulator is obtained.

25-8. Silicone Rubber

By a complete avoidance of cross-linking and by excluding $(CH_3)_3SiCl$, which allows termination of chains, a very long-chain silicone polymer may be obtained when $(CH_3)_2SiCl_2$ is hydrolyzed.

$$(CH_3)_2SiCl_2 \xrightarrow{\text{HOH}} -\underset{\underset{CH_3}{|}}{\overset{\overset{CH_3}{|}}{Si}}-O-\left[\underset{\underset{CH_3}{|}}{\overset{\overset{CH_3}{|}}{Si}}-O\right]_n- \qquad [25\text{-}42]$$

This silicone polymer, which has some of the properties of rubber and some of those of putty, has been given the name "bouncing putty."

Other R groups may be substituted for methyl in any of the monomers mentioned, with consequent modifications in properties. The methyl compounds described here have one unique feature: they contain no C-C bonds.

SUMMARY

The difference in properties between silicon compounds and carbon compounds is partially accounted for by the following differences between the silicon and carbon atoms: silicon is larger, silicon is more electropositive, silicon can expand its outer shell of electrons on demand. The following generalizations fail in enough cases to be disturbing but nevertheless are of some use in rationalizing the behavior of covalent silicon compounds: (1) Si-H and Si-Si bonds are extremely brittle but are stabilized by the presence of electronegative atoms (halogen or oxygen) on the same atom. (2) Si-C bonds are relatively inert unless the silicon also holds an electronegative atom.

The silicone oils, resins, and rubber are commercially valuable products. Heat stability and electrical insulating properties are their assets.

REFERENCES

1. E. G. Rochow, *Chemistry of the Silicones*, 2nd ed. (John W. Wiley & Sons, New York, 1951).
2. H. W. Post, *Silicones and Other Organic Silicon Compounds* (Reinhold Publishing Corporation, New York, 1949).

EXERCISES

(Exercises 1–3 will be found within the text.)

4. Select reactions that point up likenesses between carbon and silicon. Select other reactions that manifest differences between carbon and silicon.

5. Would it be possible to make a silicon analogue of benzaldehyde? of propionic acid? of the diethylacetal of acetone? Give a reason for each answer.

6. Could disilane be synthesized with lithium aluminum hydride as a reducing agent? How?

7. Synthesize the following compounds:

1. tetraphenylsilane
2. trimethylsilane
3. ethyltrichlorosilane
4. di-tert-butyldichlorosilane
5. diphenyldichlorosilane
6. bouncing putty

26 Covalent Compounds of Halogens

The elements in Groups IV, V, and **VI** form, with halogens, compounds that have considerable covalent character; that is, the compounds have low melting points or boiling points and are soluble in some organic solvents. No sharp division in character can be drawn, however, and the inclusion of a particular substance in this chapter is more or less arbitrary.

The Covalent Halides

Table 26-1 summarizes some of the properties of halogen-containing compounds that boil or sublime below 300°. This is a fair definition of a covalent halide.

26-1. Preparation of the Covalent Halides

1. Direct Combination. All of the covalent halides mentioned in Table 26-1 except those of carbon and nitrogen can be made most readily by direct combination of the two elements. Some of the elements require preheating to start the direct combination, but the heat of the reaction is usually enough to keep it going.

$$2P \text{ (red)} + 3Cl_2 \xrightarrow[94\%]{75°} 2PCl_3 \qquad\qquad [26\text{-}1]$$

$$PCl_3 + Cl_2 \xrightarrow[85\%]{25°} PCl_5 \qquad\qquad [26\text{-}2]$$

TABLE
26-1

Covalent Halides with One Central Atom

GROUP	FORMULA	M.P.	B.P.	COLOR, DENSITY, OTHER PROPERTIES
III	$GaCl_3$	78	201	colorless
	BCl_3	-107	13	colorless, fumes in air
	$AlCl_3$	85 (subl.)		colorless crystals, fumes in air
	$AlBr_3$	95	260	colorless crystals, fumes in air
IV	CCl_4	-23	77	colorless, $d = 1.5947$, solvent for oils and fats
	CF_4		-15	colorless
	CBr_4	92	190	colorless, $d = 3.42$
	CI_4	90 (subl.)		ruby-red, $d = 4.32$
	CF_2Cl_2		-30	colorless, odorless, nonpoisonous
	$SiCl_4$	-89	57	colorless, fumes in air, $d_0^4 = 1.5226$
	$SnCl_4$	-33	114	colorless, fumes in air, $d_0^4 = 2.2788$
	$TiCl_4$	-23	136	colorless, fumes in air, $d_0^4 = 1.7604$
	$GeCl_4$	-50	87	
	$ZrCl_4$	above 300 (subl.)		
V	NCl_3	explosive		yellow oil, soluble in benzene and $CHCl_3$ and ether
	NBr_3	explosive		dark-red oil
	NI_3	explosive		black powder
	PCl_3		76	colorless liquid, fumes in air
	PBr_3		173	colorless liquid, fumes in air
	PI_3	55		dark-red crystals
	PCl_5	148	160 (subl.)	white crystals, soluble in CS_2 and benzoyl chloride, fumes in air
	PBr_5		wide range	citron-yellow crystals
	$AsCl_3$	-18	130	colorless liquid, fumes in air
	$SbCl_3$	73	223	white crystals, fumes in air
	SbF_5		150	colorless, oily liquid, $d_0^{23} = 2.993$
	$SbCl_5$	-6	dec.	colorless, fumes in air, $d_0^{20} = 2.346$
VL	S_2Cl_2	-80	138	yellow-to-orange liquid, $d_0^4 = 1.7094$
	SCl_2	-78	59 (dec.)	red liquid, $d_{15}^4 = 1.622$
	SCl_4	-31	dec.	yellow solid at $-31°$
	SF_6	-51		colorless gas
	$SeCl_4$	191 (subl.)		
VII	$FeCl_3$	307 (subl.)		crystals, red by transmitted light, green by reflected
	$FeBr_3$			green crystals, metallic luster, soluble in ether and alcohol

$$2P \text{ (red)} + 3Br_2 \xrightarrow[94\%]{173°} 2PBr_3 \qquad\qquad [26\text{-}3]$$

$$2P \text{ (white)} + 5Br_2 \xrightarrow[0°]{CS_2} 2PBr_5 \qquad\qquad [26\text{-}4]$$

$$2Al + 3Br_2 \xrightarrow[85\%]{\Delta} 2AlBr_3 \qquad\qquad [26\text{-}5]$$

$$Sn + 2Cl_2 \xrightarrow{\Delta} SnCl_4 \qquad\qquad [26\text{-}6]$$

$$2S + Cl_2 \xrightarrow{25°} S_2Cl_2 \qquad\qquad [26\text{-}7]$$

$$S_2Cl_2 + Cl_2 \xrightarrow{25°} 2SCl_2 \qquad\qquad [26\text{-}8]$$

2. Special Methods. The nitrogen trihalides NCl_3, NBr_3, and NI_3 are explosive substances whose properties are not closely related to those of any other compounds in Table 26-1. Nitrogen trichloride, an aging agent for wheat flour (recently condemned), is made by treating a solution of an ammonium salt with chlorine gas. It may be collected in dilute benzene solution but still must be handled cautiously since it is explosive. The bromide and iodide may be prepared in an analogous manner.

$$(NH_4)_2SO_4 + 3Cl_2 \xrightarrow[59\%]{} NCl_3 + 3HCl + NH_4OSO_3H \qquad\qquad [26\text{-}9]$$

The carbon tetrahalides are prepared by high-temperature reactions. Carbon tetrachloride is made commercially by treating CS_2 with chlorine. The starting compound, CS_2, is synthesized by

$$C + 2S \xrightarrow[\text{furnace}]{\text{electric}} CS_2 \qquad\qquad [26\text{-}10]$$

direct combination of sulfur vapor with coke at the high temperature of the electric furnace. In the presence of antimony pentachloride, chlorine will react with CS_2 to give CCl_4 and sulfur monochloride, S_2Cl_2. This product also reacts with CS_2 to liberate sulfur, which reacts with chlorine to yield S_2Cl_2 again.

$$CS_2 + Cl_2 \xrightarrow{SbCl_5} CCl_4 + S_2Cl_2 \qquad\qquad [26\text{-}11]$$

$$CS_2 + 2S_2Cl_2 \longrightarrow CCl_4 + 6S \qquad [26\text{-}12]$$

$$2S + Cl_2 \longrightarrow S_2Cl_2 \qquad [26\text{-}13]$$

Both chlorine compounds are used in large quantities. Carbon tetrachloride is used as a fire extinguisher, being one of a few organic compounds that will not burn, as a solvent for fats, and in dry cleaning; sulfur monochloride is used in vulcanizing rubber.

A series of carbon halides known as Freons are useful as nontoxic, odorless, noninflammable refrigerants. These fluorine-containing compounds include CCl_3F, $CHCl_2F$, $CHClF_2$, $CCl_2F\text{-}CClF_2$, $CClF_2\text{-}CClF_2$, and CCl_2F_2, of which the last, Freon-12, is representative. It is made by fluorinating CCl_4 with SbF_3 in the presence of $SbCl_5$ as a catalyst (Swarts reaction). Recent patents reveal

$$3CCl_4 + 2SbF_3 \xoverset{SbCl_5}{\longrightarrow} 3CCl_2F_2 + 2SbCl_3 \qquad [26\text{-}14]$$

ways of making Freon-12 from chloroform or even methane. At 340°, a mixture of methane and chlorine treated with CrF_3 and HF gives nearly a quantitative yield of CCl_2F_2.

26-2. Chemical Properties of the Covalent Halides

1. Hydrolysis. The most important property of the covalent halides (of all the compounds listed in Table 26-1 except the carbon and nitrogen halides and SF_6) is the ease with which they are hydrolyzed. Most of them fume violently in moist air. The mechanism of the hydrolytic reactions is not known with certainty, but

$$PCl_3 + 3HOH \longrightarrow \underset{\overset{|}{OH}}{HO\text{-}P\text{-}OH} + 3HCl \qquad [26\text{-}15]$$

$$PCl_5 + HOH \longrightarrow O \leftarrow PCl_3 + 2HCl \qquad [26\text{-}16]$$

$$AsCl_3 + HOH \longrightarrow O{=}As\text{-}Cl + 2HCl \qquad [26\text{-}17]$$

$$SbCl_3 + HOH \longrightarrow O{=}Sb\text{-}Cl + 2HCl \qquad [26\text{-}18]$$

$$S_2Cl_2 + HOH \longrightarrow SO_2 + HCl + S + \text{several other products} \qquad [26\text{-}19]$$

$$2SCl_2 + 2HOH \longrightarrow SO_2 + 4HCl + S \qquad\qquad [26\text{-}20]$$

$$FeCl_3 + 3HOH \longrightarrow Fe(OH)_3 + 3HCl \qquad\qquad [26\text{-}21]$$

$$AlCl_3 + 3HOH \longrightarrow Al(OH)_3 + 3HCl \qquad\qquad [26\text{-}22]$$

$$TiCl_4 + 4HOH \longrightarrow Ti(OH)_4 + 4HCl \qquad\qquad [26\text{-}23]$$

one of the satisfying theories is that of Pfeiffer, which has been modified by others, especially Hückel. When PCl_3 is the halide, the first step is association of a proton from the water with the

$$[26\text{-}24]$$

unshared pair of phosphorus to give $HPCl_3^+$. This tetrahedral cation then reacts with water in at least three displacement steps to give phosphorous acid and three molecules of HCl.

Arsenic trichloride and antimony trichloride probably react in the same way, but the product, after two steps, is insoluble, and the reaction stops.

$$[26\text{-}25]$$

Exercise 26-1. The reactions of the sulfur chlorides are oxidation-reduction reactions as well as hydrolyses. What are the changes in oxidation number for sulfur in Expressions 26-19 and 26-20?

Hydrolysis of the halides of iron, aluminum, and titanium probably involves first an expansion of the octet surrounding the central atom by association of water molecules with the central atom and then loss of HCl. The geometry of an octahedron is involved and will not be discussed here.

Zirconium tetrachloride, which sublimes above 300°, may be considered as a borderline covalent-electrovalent compound. Hydrolysis takes place appreciably only in hot water and is accompanied by ionization.

$$ZrCl_4 + HOH \xrightarrow{\Delta} ZrOCl_2 + 2HCl \qquad [26\text{-}26]$$

$$ZrCl_4 \xrightarrow{HOH} Zr^{+4} + 4Cl^- \qquad [26\text{-}27]$$

In contrast to PCl_3, which has a free pair of electrons that can associate with a proton to start the hydrolytic reaction, carbon already has a completed octet in CCl_4. Consequently, carbon tetrachloride does not react with water at ordinary temperatures. However, it is dangerous to use CCl_4 to extinguish a flame burning at 2,000° if water is present. At this high temperature a reaction does take place, and phosgene, a poisonous gas, is formed.

$$CCl_4 + HOH \longrightarrow \underset{\text{phosgene}}{C{=}O{\overset{\diagup Cl}{\diagdown Cl}}} + 2HCl \qquad [26\text{-}28]$$

Analogously, SF_6 has about the sulfur atom a complete shell of twelve electrons, which gives a satisfied octahedron. Hence it also does not hydrolyze readily.

Since most of the practical uses of these covalent halides involve dry reactions, their hydrolysis is important only as something to be avoided. Reactions of the Friedel-Crafts type, for example, are run under anhydrous conditions because hydrolysis of the catalysts, $FeBr_3$, $AlCl_3$, $AlBr_3$, $ZnCl_2$, $SnCl_4$, $SbCl_5$, and others, must be avoided. The principal uses of the phosphorus halides are in the preparation of alkyl and acyl halides, and then the reaction with

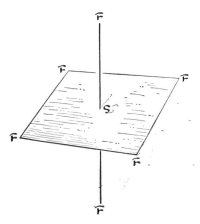

water is undesirable since it is faster than that of the halide with an alcohol or an acid.

2. Ammonolysis. The study of the ammonolysis reactions of the covalent halides, for a number of reasons (among them an accumulation of difficulties), is far from complete. The reactions result in more complex mixtures of products than the corresponding hydrolytic reactions; the products have found few practical uses; and, since the reactions are violent and many of them not readily controlled, clear-cut syntheses are difficult. The result is that the chemistry of these compounds has been neglected for what have appeared to be more promising fields.

The ammonolysis of the two phosphorus halides, however, has been studied extensively, and we shall discuss it here to indicate the paths that these reactions may take. The treatment of the subject will be, in part, a formal presentation that will be applied to the ammonolysis of other compounds later in this chapter.

If the reaction is assumed to be of the type written for the hydrolysis of phosphorus trichloride (p. 565), the end result is to replace chlorine by an NH_2 group.

$$\underset{\diagdown Cl}{\overset{\diagup Cl}{P-Cl}} + NH_3 \longrightarrow \underset{\diagdown Cl}{\overset{\diagup NH_2}{P-Cl}} \xrightarrow{NH_3} \xrightarrow{NH_3} \underset{\diagdown NH_2}{\overset{\diagup NH_2}{P-NH_2}} \qquad [26\text{-}29a]$$

$$PCl_3 + 6NH_3 \longrightarrow P(NH_2)_3 + 3NH_4Cl \qquad [26\text{-}29b]$$

It has been observed throughout this course that a carbon atom bearing two OH groups has an unstable configuration, which is usually stabilized by loss of water and formation of a multiple bond (a double bond in this case). The nitrogen analogue of an OH group—namely, an NH_2 group—exhibits the same behavior in carbon compounds. The Si-O bond (p. 556) shows the opposite tendency; that is, silicon forms two single bonds to oxygen (in the silicone polymers) in preference to one multiple bond. In the P-N system of compounds shown here, there is a case between these extremes. The P-N compound shown in Expression 26-29, $P(NH_2)_3$, exists, but it loses NH_3 (the nitrogen analogue of H_2O) readily to form the P-N series of compounds shown in Expression 26-30. In the P-N system of compounds, an $-NH_2$ group is called an amide; $=NH$, an imide; and $\equiv N$, a nitride or nitrile.

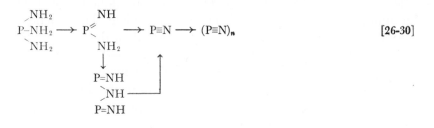

$$[26\text{-}30]$$

Phosphorus triamide, a yellow, unstable solid formed by the action of liquid ammonia on PBr_3 at $-70°C$, begins to decompose

$$PBr_3 + 6NH_3 \xrightarrow{\;-70°\;} P(NH_2)_3 + 3NH_4Br \qquad [26\text{-}31]$$

above $-25°C$ to yield diphosphorus tri-imide. The decomposition includes, as the first step, the formation of phosphorus amide imide, but this also is unstable and loses a molecule of ammonia (deammonation is rapid at $100°$) between each two molecules of the amide imide. The resulting product, the tri-imide, is a brown solid, insoluble in liquid ammonia when an ammonium salt is present.

$$2P\text{-NH}_2 \begin{matrix} ,\text{NH}_2 \\ \diagdown\text{NH}_2 \end{matrix} \xrightarrow{-25°C} 2P \begin{matrix} \text{NH} \\ \diagdown \\ \text{NH}_2 \end{matrix} \xrightarrow{100°} \begin{matrix} \text{P=NH} \\ \diagup \\ \diagdown\text{NH} \\ \text{P=NH} \end{matrix} \qquad [26\text{-}32]$$

When diphosphorus tri-imide is heated to 700°C at low pressure, it loses hydrogen rather than NH_3 to form tetraphosphorus hexanitride, P_4N_6, a white powder, nonvolatile, insoluble in water, but spontaneously inflammable in air. Continued heating yields a polymer, $(PN)_n$, which is stable in air.

Phosphonitrile, $P\equiv N$, which probably has only a transient existence before the stable polymer, $(P\equiv N)_n$, is formed, may also be made by combining phosphorus vapors and nitrogen at 1,000°C in an electric arc. This process has been patented for removing nitrogen from vacuum tubes.

The concept of ammonolysis and deammonation gives a qualitative picture of what to expect when PCl_3 is treated with NH_3. Since all the products shown in Expression 26-30 have actually been prepared, we might use the concept as a guide to the ammonolysis of other covalent halides, such as $AsCl_3$, $SbCl_3$, and $SiCl_4$. Not many of these have been reported in the literature. A cautionary note must be added, however, since one product not expected by the scheme, P_4N_6, was also obtained in the ammonolysis of PCl_3.

By a similar formal treatment, ammonolysis and deammonation of PCl_5 should lead to the products shown in Expression 26-33. (Five bonds are shown for P in every case and three bonds for N.

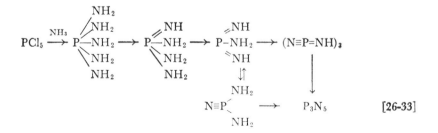

$$[26\text{-}33]$$

The actual structures are still not known with certainty.) All these compounds except phosphorus imide triamide and phosphorus pentamide have been reported in the literature.

3. Alcoholysis. Guided by the hydrolysis and ammonolysis reactions of the covalent halides, we may readily put the alcoholysis products to paper. In the laboratory, not all the possible esters have been characterized.

Borates can be obtained by treating BCl_3 with an alcohol; but, since this halide is a gas, they are more conveniently prepared by direct esterification of boric anhydride, B_2O_3.

$$BCl_3 + 3C_2H_5OH \longrightarrow B(OC_2H_5)_3 + 3HCl \qquad [26\text{-}34]$$

Alcoholysis of silicon tetrachloride may be controlled to give a

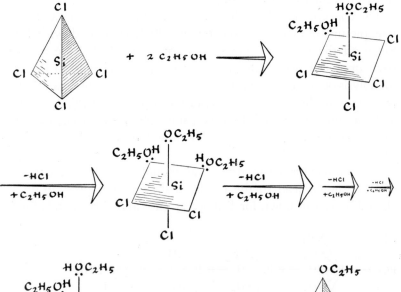

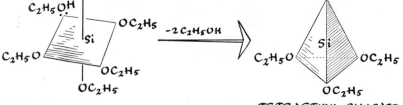

TETRAETHYL SILICATE

stepwise replacement of chlorine. (In contrast to the silicon halide, carbon tetrachloride does not undergo such alcoholysis.) This can be accounted for by the ability of silicon to expand its electron shell to ten or twelve electrons by coordinating one or two molecules of the alcohol in an octahedral configuration. Loss of hydrogen

$$SiCl_4 + C_2H_5OH \longrightarrow C_2H_5OSiCl_3 + HCl \qquad [26\text{-}35]$$

$$C_2H_5OSiCl_3 + C_2H_5OH \longrightarrow (C_2H_5O)_2SiCl_2 + HCl \qquad [26\text{-}36]$$

$$(C_2H_5O)_2SiCl_2 + C_2H_5OH \longrightarrow (C_2H_5O)_3SiCl + HCl \qquad [26\text{-}37]$$

$$(C_2H_5O)_3SiCl + C_2H_5OH \longrightarrow (C_2H_5O)_4Si + HCl \qquad [26\text{-}38]$$

chloride, coordination of more alcohol, and, finally, reversion to a tetrahedron, result in tetraethyl silicate. This compound has been used to harden stone, to arrest decay in wood, and in acid-proof cements.

Formation of an alkyl halide at the expense of the ester is an important competing reaction in the alcoholysis of phosphorus halides and indeed was once considered the main reaction (p. 145). By adding a tertiary amine to remove HCl, one may reduce the competing alkylation of hydrogen chloride to a minimum. Triethyl phosphite is formed.

$$PCl_3 + 3C_2H_5OH + 3(C_2H_5)_2NC_6H_5 \xrightarrow[92\%]{}$$
$$P(OC_2H_5)_3 + 3(C_2H_5)_2NC_6H_5 \cdot HCl \qquad [26\text{-}39]$$
$$\text{triethyl phosphite}$$

The halides of higher molecular weight in Group IV can be esterified if basic conditions are used. This suggests that these halides, as expected, are less reactive. Since the titanium and zirconium halides do not undergo ammonolysis, ammonia or amines may be used to remove HCl.

$$TiCl_4 + 4ROH + 4NH_3 \longrightarrow Ti(OR)_4 + 4NH_4Cl \qquad [26\text{-}40]$$

$$ZrCl_4 + 4ROH + 4C_5H_5N \longrightarrow Zr(OR)_4 + 4C_5H_5N \cdot HCl \qquad [26\text{-}41]$$
$$\text{pyridine}$$

4. Friedel-Crafts Reaction. Friedel-Crafts reactions have been run on PCl₃ to give an aryl-substituted phosphorus chloride in the presence of AlCl₃ as a catalyst. This product can be oxidized in

$$\text{⬡} + PCl_3 \xrightarrow[\substack{36 \text{ hours} \\ 9\%}]{AlCl_3} \text{⬡}-P\begin{smallmatrix}Cl\\ \\Cl\end{smallmatrix} + HCl \qquad [26\text{-}42]$$

the presence of chlorine and P₂O₅ to phenyl phosphonyl chloride, which undergoes the alcoholysis reaction of other oxychlorides (p. 581). The ester formed with allyl alcohol, diallylphenylphosphonate, can be polymerized to a hard, transparent, infusible resin or copolymerized with other unsaturated organic compounds.

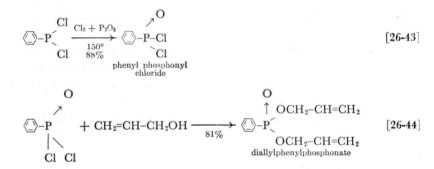

[26-43]

phenyl phosphonyl chloride

[26-44]

diallylphenylphosphonate

5. Grignard Reactions. The use of the Grignard reagent with covalent halides to form carbon-metal bonds has been discussed previously (pp. 541ff.).

The Oxychlorides

The oxychlorides of carbon, nitrogen, sulfur, and some other elements in Groups IV, V, and VI are foul-smelling substances the vapors of which have deleterious effects on the respiratory tract. Some, if not all, of the objectionable properties of these substances are due to their easy hydrolysis in the presence of moisture to yield

HCl. Some of the physical properties of these compounds are given in Table 26-2.

TABLE
26-2 | *Properties of Some Oxychlorides*

FORMULA	COMMON NAME	B.P.	DENSITY[t]	PROPERTIES
$COCl_2$	phosgene	8		colorless gas, penetrating odor, poisonous
$NOCl$	nitrosyl chloride	−6		orange-yellow gas, red liquid
NO_2Cl	nitryl chloride	−72		white crystalline solid, colorless gas
$POCl_3$	phosphorus oxychloride	105	1.673^{14}	colorless liquid
$VOCl_3$	vanadium oxychloride	127	1.865^{0}	yellow liquid
$SOCl_2$	thionyl chloride	79	1.677^{0}	colorless liquid
SO_2Cl_2	sulfuryl chloride	69	1.667^{20}	colorless liquid
$SeOCl_2$	selenium oxychloride	176	2.424^{22}	colorless, soluble in $CHCl_3$ and benzene
CrO_2Cl_2	chromyl chloride	117	1.96^{0}	red vapor, blood-red liquid, strong oxidizing agent
UO_2Cl_2	uranyl chloride			yellow needles
MoO_2Cl_2	molybdenyl chloride			light-yellow crystals
WO_2Cl_2	tungstenyl chloride			golden-yellow crystals

26-3. Preparation of Oxychlorides

1. Direct Combination of Chlorine and an Oxide. An oxychloride of at least one element in each of Groups IV, V, and VI can be prepared by direct combination of an oxide with chlorine in the presence of charcoal as a catalyst. Sunlight also is needed in the preparation of phosgene and nitrosyl chloride by this method.

$$CO + Cl_2 \xrightarrow{h\nu + \text{charcoal}} \underset{\text{phosgene}}{C{=}O\!\!\begin{matrix} \diagup Cl \\ \diagdown Cl \end{matrix}}$$

[26-45]

$$2NO + Cl_2 \xrightarrow[50°C]{h\nu + charcoal} 2Cl-N=O \qquad [26\text{-}46]$$
$$\text{nitrosyl}$$
$$\text{chloride}$$

Phosgene, also called carbonyl chloride, is a war gas with a penetrating odor and is very poisonous. Sulfur dioxide and chlorine combine readily in the presence of charcoal; we can also obtain sulfuryl chloride by passing chlorine through a solution of sulfur dioxide in camphor, an extraordinary solvent for this gas.

$$SO_2 + Cl_2 \xrightarrow[96\%]{camphor,\ 0°C} O{\leftarrow}\overset{\overset{Cl}{|}}{\underset{\underset{Cl}{|}}{S}}{\rightarrow}O \qquad [26\text{-}47]$$
$$\text{sulfuryl}$$
$$\text{chloride}$$

$$UO_2 + Cl_2 \xrightarrow{charcoal} UO_2Cl_2 \qquad [26\text{-}48]$$
$$\text{uranyl}$$
$$\text{chloride}$$

These direct combinations are oxidation-reduction reactions in which the central atom increases its oxidation state (by 1 or 2) while chlorine is changing from an oxidation state of 0 in elemental chlorine to -1 in each of the oxychlorides. Hence only compounds whose central atom forms a fairly stable oxide that can exist in the lower oxidation state will be available for this type of reaction. Of the compounds listed in Table 26-2, $POCl_3$, $VOCl_3$, $SOCl_2$, $SeOCl_2$, and NO_2Cl have not been prepared by direct combination.

2. Replacement of Oxygen by Chlorine.

Various chlorinating agents, such as the covalent chlorides and the oxychlorides discussed in this chapter, have been used to replace oxygen by two chlorines in a covalent oxide. For example, two oxychlorides can be prepared by the treatment of SO_2 with PCl_5. Other methods of preparing thionyl chloride illustrate the various chlorinating agents

$$SO_2 + PCl_5 \longrightarrow SOCl_2 + POCl_3 \qquad [26\text{-}49]$$
$$\qquad\qquad \text{thionyl} \quad \text{phosphorus}$$
$$\qquad\qquad \text{chloride} \quad \text{oxychloride}$$

that can be generally used to prepare other oxychlorides. Not every chlorinating agent can be used to make every oxychloride, however.

$$SO_2Cl_2 + PCl_3 \longrightarrow SOCl_2 + POCl_3 \qquad [26\text{-}50]$$

$$COCl_2 + SO_2 \xrightarrow[\text{charcoal}]{200°} SOCl_2 + CO_2 \qquad [26\text{-}51]$$

$$SO_3 + S_2Cl_2 + Cl_2 \xrightarrow{75°} 2SOCl_2 + SO_2 \qquad [26\text{-}52]$$

Strong heating is necessary in preparing the compounds of higher molecular weight. Phosgene can be used as a chlorinating agent on WO_3 at 350°, for example. Chromium VI oxide, CrO_3, can

$$WO_3 + COCl_2 \xrightarrow{350°} WO_2Cl_2 + CO_2 \qquad [26\text{-}53]$$

$$4WO_3 + S_2Cl_2 + 3Cl_2 \xrightarrow{\Delta} 4WO_2Cl_2 + 2SO_2 \qquad [26\text{-}54]$$

be chlorinated at elevated temperature by a number of reagents, such as $FeCl_3$, PCl_5, $CHCl_3 + O_2$, and CCl_4. If the oxide itself

$$3CrO_3 + 2FeCl_3 \xrightarrow{\Delta} 3CrO_2Cl_2 + Fe_2O_3 \qquad [26\text{-}55]$$

has little covalent character, all the oxygen may be replaced by chlorine to yield a covalent halide. If the second product is a volatile oxide such as carbon dioxide, this type of reaction will succeed. In Expression 26-56 the yield of $AlCl_3$ is 98%.

$$Al_2O_3 + 3COCl_2 \xrightarrow[\text{1,000°}]{\text{charcoal}} 2AlCl_3 + 3CO_2 \qquad [26\text{-}56]$$

26-4. Reactions of Oxychlorides

1. **Hydrolysis.** (See pp. 188–189 also.)
The oxychlorides, as might be expected from their structures,

resemble acyl chlorides in their properties. The central atom, which is electropositive with respect to oxygen and chlorine, is attached to one or two oxygens and one or more chlorines. Reactions analogous to the hydrolysis of acyl chlorides can be written for these oxychlorides. The acyl halides of high molecular weight react with water less rapidly than those of lower weight, and the same is true in this group of compounds. Here the increasing metallic character of the central atom is the most pronounced change as the molecular weight increases. In Group VI, as one goes from Cr to U, the covalent character of CrO_2Cl_2 changes to a predominantly electrovalent character in UO_2Cl_2, for which dissociation in water precedes hydrolysis. All the compounds shown in Expressions 26-57–26-62 fume in moist air, another way of saying that

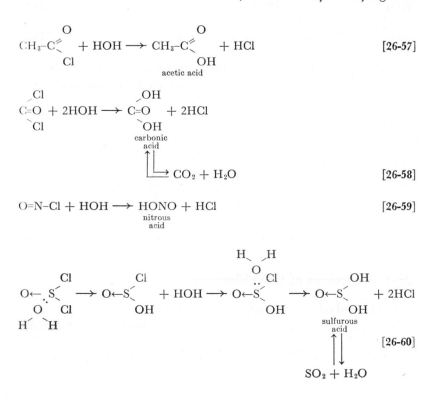

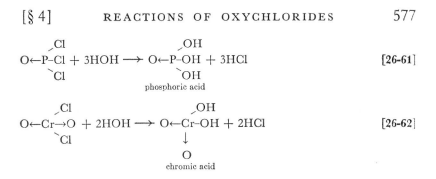

$$O \leftarrow P\begin{smallmatrix}\diagup Cl \\ \diagdown Cl\end{smallmatrix}\text{-Cl} + 3HOH \longrightarrow O \leftarrow P\begin{smallmatrix}\diagup OH \\ \diagdown OH\end{smallmatrix}\text{-OH} + 3HCl \qquad [26\text{-}61]$$

phosphoric acid

$$O \leftarrow Cr\begin{smallmatrix}\diagup Cl \\ \diagdown Cl\end{smallmatrix}\rightarrow O + 2HOH \longrightarrow O \leftarrow Cr\text{-OH} + 2HCl \qquad [26\text{-}62]$$

$$\downarrow$$
$$O$$

chromic acid

the hydrolytic reaction of each is rapid. Though the hydrolysis yields only products that are made less expensively by other methods, it is essential to know the reaction because it must be avoided in useful syntheses involving these substances. The probable mechanism for the hydrolysis of thionyl chloride is shown in Expression 26-60.

If a partial hydrolysis could be carried out on the oxychlorides containing more than one chlorine, one might expect to get chloro-substituted acids. The procedure is impractical because the reactions are rapid and violent, the violence being due to the great evolution of heat. POF_3, however, in which the halogen is more tightly held, can be hydrolyzed at a controllable rate in basic solution to give salts of $HO-POF_2$ and $(HO)_2POF$.

$$C\begin{smallmatrix}\diagup Cl \\ \diagdown OH\end{smallmatrix}=O$$
chlorocarbonic acid
(nonexistent)

$$O \leftarrow S\begin{smallmatrix}\diagup Cl \\ \diagdown OH\end{smallmatrix}\rightarrow O$$
chlorosulfonic acid
b.p. 152°C
$d_4^{20} = 1.753$

$$O \leftarrow P\begin{smallmatrix}\diagup F \\ \diagdown OH\end{smallmatrix}\text{-OH}$$
fluophosphoric acid
$d_4^{25} = 1.818$

Chlorophosphoric acid has not been prepared in the free state, but a number of esters (p. 578) and salts of the compound are known. Recently (1947) the preparation of pure fluophosphoric acid has been announced, and its salts had been known previously. It was made by treating anhydrous HPO_3 (metaphosphoric acid) with anhydrous liquid HF. The fluophosphoric acid does not attack glass as long as water is absent.

$$\underset{\underset{O}{\overset{\|}{}}}{HO-P}\to O + HF \longrightarrow O\leftarrow P\overset{\nearrow F}{\underset{\searrow OH}{-OH}} \qquad [26\text{-}63]$$

Most of the original interest in fluophosphoric acid came from the discovery of new chemical warfare agents in its derivatives. Some of the derivatives cause myopia in night vision and are highly toxic when inhaled or injected. One of them is di-isopropyl fluophosphate, which was synthesized from PCl_3 as in Expressions 26-64–26-66.

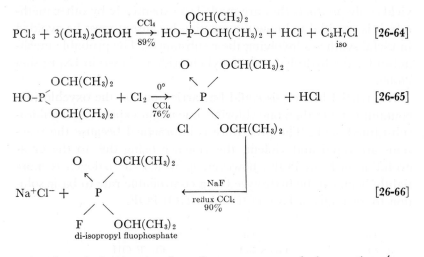

$$PCl_3 + 3(CH_3)_2CHOH \xrightarrow[89\%]{CCl_4} \underset{}{\overset{OCH(CH_3)_2}{HO-P-OCH(CH_3)_2}} + HCl + \underset{iso}{C_3H_7Cl} \qquad [26\text{-}64]$$

$$HO-P\overset{OCH(CH_3)_2}{\underset{OCH(CH_3)_2}{}} + Cl_2 \xrightarrow[\substack{CCl_4 \\ 76\%}]{0°} \overset{O\qquad OCH(CH_3)_2}{\underset{Cl\qquad OCH(CH_3)_2}{P}} + HCl \qquad [26\text{-}65]$$

$$Na^+Cl^- + \overset{O\qquad OCH(CH_3)_2}{\underset{F\qquad OCH(CH_3)_2}{P}} \xleftarrow[\substack{redux\ CCl_4 \\ 90\%}]{NaF} \qquad [26\text{-}66]$$
di-isopropyl fluophosphate

Another derivative involves first an ammonolysis reaction (see

$$POCl_3 + 4(CH_3)_2NH \xrightarrow[68\%]{} \overset{(CH_3)_2N\qquad O}{\underset{(CH_3)_2N\qquad Cl}{P}} + 2(CH_3)_2NH\cdot HCl$$

$$\overset{(CH_3)_2N\qquad F}{\underset{(CH_3)_2N\qquad O}{P}} \xleftarrow[67\%]{NaF} \qquad [26\text{-}67]$$

paragraph 2, below) of phosphorous oxychloride and then a reaction similar to the one just described. Two chlorines can be replaced by amine groups if four moles of the amine are used, after which the third chlorine can be replaced by fluorine with NaF in refluxing CCl_4.

Chlorosulfonic acid, or chlorosulfuric acid, is made commercially by the treatment of fuming sulfuric acid (70% SO_3) with dry HCl. The pure acid can be distilled from the mixture easily at 150°C.

$$O{\leftarrow}\underset{\underset{O}{\|}}{S}{\rightarrow}O + HCl \xrightarrow[100\%]{} O{\leftarrow}\underset{\underset{OH}{|}}{\overset{\overset{Cl}{|}}{S}}{\rightarrow}O \qquad\qquad [26\text{-}68]$$
chlorosulfonic
acid

Chlorosulfonic acid is widely used in organic chemistry—for example, for the introduction of the $-SO_2Cl$ group into an aromatic

$$\bigcirc + \underset{\underset{O}{\downarrow}}{\overset{\overset{O}{\uparrow}}{HO\text{-}S\text{-}Cl}} \longrightarrow \bigcirc\text{-}SO_2\text{-}Cl + H_2O \qquad\qquad [26\text{-}69]$$

hydrocarbon. A sulfonyl chloride (p. 583) is an acid chloride derived from a sulfonic acid rather than a carboxylic acid. Benzene sulfonyl chloride reacts rapidly with Ag^+ but is less readily hydrolyzed than benzoyl chloride.

2. Ammonolysis. The paths of the ammonolytic reactions of the oxychlorides are probably more completely worked out than those of the covalent halides, but many of the difficulties mentioned previously are encountered here also. Formally, the initial ammonolytic product may be considered as the one obtained by replacement of each chlorine with an amide (NH_2) group. Whether deammonation, dehydration, or a combination of these processes will follow depends on the particular compound and the conditions.

Ammonolysis of phosgene, or carbonyl chloride, yields urea, or carbamide, a diamide having very weak basic properties. Commer-

$$\underset{\text{phosgene}}{\overset{\displaystyle /Cl}{\underset{\displaystyle \backslash Cl}{C=O}}} + 4NH_3 \longrightarrow \underset{\substack{\text{urea, or}\\\text{carbamide}}}{\overset{\displaystyle /NH_2}{\underset{\displaystyle \backslash NH_2}{C=O}}} + 2NH_4Cl \qquad [26\text{-}70]$$

cially, urea is made from CO_2 and NH_3 at elevated temperature and pressure involving the equilibria of Expression 26-71.

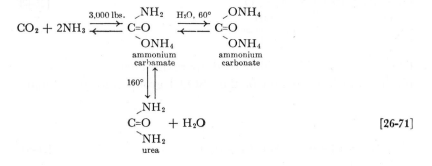

$$\qquad [26\text{-}71]$$

The technical grade of ammonium carbonate sold by chemical companies under that label is, in large part, ammonium carbamate —that is, an incompletely hydrolyzed product. Carbamic acid,

$$H_2N-\overset{\displaystyle O}{\underset{\displaystyle \backslash OH}{\overset{\|}{C}}}$$

, does not exist, but its salts and esters (p. 581) are

fairly stable. Reactions in which it should be formed yield CO_2 + NH_3 instead.

Sulfuryl chloride undergoes ammonolysis with liquid NH_3 to a white crystalline compound called sulfamide. The reaction is a

$$\underset{}{\overset{\displaystyle /Cl}{\underset{\displaystyle \backslash Cl}{O\leftarrow S\rightarrow O}}} + NH_3 \underset{86\%}{\overset{-40°}{\longrightarrow}} \underset{\text{sulfamide}}{\overset{\displaystyle NH_2}{\underset{\displaystyle NH_2}{O\leftarrow S\rightarrow O}}} + 2NH_4Cl \qquad [26\text{-}72]$$

violent one, and the mixture must be cooled and stirred continuously during the addition of the acid chloride to the liquid ammonia.

The reaction of POCl₃ with NH₃ will be discussed in Chapter 27. The reactions of the other oxychlorides result in products that have not been completely characterized or have not been found useful, and in some cases the chemists apparently have not run the anticipated reactions at all.

3. Alcoholysis.   In contrast to hydrolysis, which is a rapid reaction of phosgene and the other oxychlorides, alcoholysis can be controlled, and in some cases the products of a half-completed reaction can be isolated. Phosgene reacts with ethyl alcohol to yield ethyl chlorocarbonate, which in turn can be made to undergo further alcoholysis. The halogen left is still quite reactive and is

$$
\underset{Cl}{\overset{Cl}{>}}C{=}O \ + \ CH_3CH_2OH \ \xrightarrow{90\%} \ \underset{OCH_2CH_3}{\overset{Cl}{>}}C{=}O \ + \ HCl \qquad [26\text{-}73]
$$
ethyl
chlorocarbonate

$$
\underset{OC_2H_5}{\overset{Cl}{>}}C{=}O \ + \ C_2H_5OH \ \longrightarrow \ \underset{OC_2H_5}{\overset{OC_2H_5}{>}}C{=}O \ + \ HCl \qquad [26\text{-}74]
$$

displaced in reactions with water or ammonia as well as with the alcohol. Ethyl carbonate reacts with ammonia as other esters do, this one behaving as a diester.

$$
\underset{OC_2H_5}{\overset{Cl}{>}}C{=}O \ + \ HOH \ \longrightarrow \ \left[\underset{OC_2H_5}{\overset{OH}{>}}C{=}O\right] \ \longrightarrow \ C_2H_5OH + CO_2 \qquad [26\text{-}75]
$$

$$
\underset{OC_2H_5}{\overset{Cl}{>}}C{=}O \ + \ NH_3 \ \longrightarrow \ \underset{OC_2H_5}{\overset{NH_2}{>}}C{=}O \qquad [26\text{-}76]
$$
ethyl carbamate,
or urethane

$$
\underset{OC_2H_5}{\overset{OC_2H_5}{>}}C{=}O \ + \ 2NH_3 \ \longrightarrow \ \underset{NH_2}{\overset{NH_2}{>}}C{=}O \ + \ 2C_2H_5OH \qquad [26\text{-}77]
$$
urea

Alkyl nitrites are obtained in the alcoholysis of nitrosyl chloride. Removing the hydrogen chloride accompanying this reaction by means of pyridine reduces the violence of the reaction and allows good yields of the nitrites, especially of the long-chain alcohols.

Phosphorus oxychloride can be used to form esters of phosphoric and two chlorophosphoric acids if a tertiary amine is used to remove the hydrogen chloride. The principal use of this reaction has been in the study of the corresponding esters of the amidophosphoric acids. Such substances have received an increasing amount of

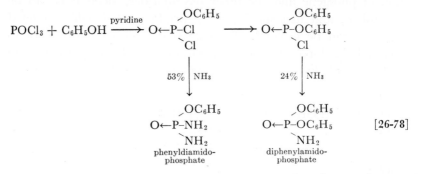

attention lately because of interest in the phosphorus-nitrogen system in biological substrates.

Tricresyl phosphate (TCP) is readily made by the same alcoholysis reaction.

$$3\langle\_\rangle\text{-OH} + POCl_3 \xrightarrow{R_3N} \text{[tricresyl phosphate]} \qquad [26\text{-}79]$$

tricresyl phosphate

Selenite, sulfite, and sulfate esters are obtainable by the alcoholysis of selenium oxychloride, thionyl chloride, and sulfuryl chloride, respectively. The replacement goes stepwise, and even with the short-chain alcohols the intermediate chloro esters, with care, can be successfully isolated. Adding pyridine favors a very rapid diester formation. One may make the sulfate esters more readily by simply dissolving alcohol in sulfuric acid.

Exercise 26-2.   Write equations for the reactions of $SeOCl_2$, $SOCl_2$, and SO_2Cl_2 with ethanol.

26-5. Reactions of Acid Chlorides

The vigorous reactivity of the oxychlorides is attenuated considerably if one of the halogens is replaced by a bond to carbon. Short-chain aliphatic acyl halides react very vigorously with water,

OXYCHLORIDE		ACID CHLORIDE	
$\begin{array}{c}\diagup Cl \\ C=O \\ \diagdown Cl\end{array}$	phosgene	$R-C\begin{array}{c}\diagup O \\ \diagdown Cl\end{array}$	an acyl chloride
$O=N-Cl$	nitrosyl chloride	————	
$\begin{array}{c}Cl \\ O=S\diagdown \\ \quad Cl\end{array}$	thionyl chloride	$\begin{array}{c}R-S-Cl \\ \parallel \\ O\end{array}$	a sulfenyl chloride
$\begin{array}{c}\diagup Cl \\ O\leftarrow S\rightarrow O \\ \diagdown Cl\end{array}$	sulfuryl chloride	$\begin{array}{c}O \\ \uparrow \\ R-S-Cl \\ \downarrow \\ O\end{array}$	a sulfonyl chloride
$\begin{array}{c}\diagup Cl \\ O\leftarrow P-Cl \\ \diagdown Cl\end{array}$	phosphorus oxychloride	$\begin{array}{c}Cl \\ R-P\diagdown \\ \downarrow\ Cl \\ O\end{array}$	a phosphonyl chloride

ammonia, and alcohols; but, as the chain is lengthened, the activity diminishes as in other homologous series.

1. Hydrolysis.   It has already been mentioned that benzoyl chloride hydrolyzes only slowly in water at room temperature. The same gradation in properties occurs in the other acid chlorides shown above. Benzenesulfonyl chloride reacts very slowly with water; phenylphosphonyl chloride reacts more rapidly than the sulfur analogue. The hydrolysis of p-toluenesulfonyl chloride is shown as an example of this reaction.

$$CH_3\text{-}\langle\bigcirc\rangle\text{-}SO_2Cl + HOH \longrightarrow CH_3\text{-}\langle\bigcirc\rangle\text{-}SO_3H + HCl \qquad [26\text{-}80]$$
p-toluenesulfonic acid

2. Ammonolysis. The reactions of benzoyl chloride and benzenesulfonyl chloride are frequently used as a means of identification of primary and secondary amines since the benzamides and benzenesulfonamides are generally solids, easily purified by recrystallization, and the melting points are widely separated for many compounds. Benzamide is prepared by the Schotten-Baumann method (p. 306).

The sulfonamides are made by a similar method called the Hinsburg reaction. In the case of a primary amine the SO_2 group

$$\langle\bigcirc\rangle\text{-}SO_2Cl + RNH_2 + 2OH^- \longrightarrow \langle\bigcirc\rangle\text{-}SO_2\text{-}\overset{\ominus}{\underset{\cdot\cdot}{N}}\text{-}R + Cl^- + 2H_2O$$

$$\langle\bigcirc\rangle\text{-}SO_2NHR \overset{H^+}{\longleftarrow} \qquad [26\text{-}81]$$

in the sulfonamide is sufficiently electron-attracting to allow the hydrogen on nitrogen to be replaced by a strong base. The soluble sulfonamide may be precipitated from solution by a strong acid such as HCl. Secondary amines give an insoluble precipitate when shaken with benzenesulfonyl chloride in dilute basic solution. Tertiary amines, having no hydrogen on nitrogen, cannot take part in the Hinsburg reaction.

$$\langle\bigcirc\rangle\text{-}SO_2Cl + R_2NH + OH^- \longrightarrow \langle\bigcirc\rangle\text{-}SO_2NR_2 + Cl^- + H_2O \qquad [26\text{-}82]$$

Phenylphosphonyl chloride will also react with ammonia and amines to yield phenylphosphonamides.

$$\langle\bigcirc\rangle\text{-}POCl_2 + RNH_2 \longrightarrow \langle\bigcirc\rangle\text{-}PO(NHR)_2 + 2RNH_2\cdot HCl \qquad [26\text{-}83]$$
a phenylphosphonamide

3. Alcoholysis. The acid chlorides react with alcohols and phenols in a completely analogous manner.

Exercise 26-3. Complete the following reactions:
1. benzoyl chloride + ethylene glycol + $OH^- \longrightarrow$

2. *p*-bromobenzenesulfonyl chloride + phenol + $R_3N \longrightarrow$
3. phenylphosphonyl chloride + allyl alcohol + $R_3N \longrightarrow$

SUMMARY

The covalent halides, except those of nitrogen and carbon, are prepared by direct combination of the elements. Covalent halides undergo hydrolysis, ammonolysis, and alcoholysis and either take part in or catalyze Friedel-Crafts reactions. The covalent oxyhalides can be prepared by direct combination of an oxide and a halogen or by replacement of an oxygen by a halogen. Their reactions are similar to those of the covalent halides. The oxyhalides are analogous in many ways to acyl halides.

REFERENCE

Clapp, "Some Chemistry of Covalent Compounds with a Single Central Atom," J. Chem. Education, 30, 584 (1953).

EXERCISES

(Exercises 1–3 will be found within the text.)

4. Synthesize the following:

1. CCl_4
2. $SiCl_4$
3. $SnCl_4$
4. $AlBr_3$
5. SO_2Cl_2
6. $SOCl_2$
7. NCl_3
8. CCl_2F_2
9. $P\begin{array}{l} \diagup\!\!\!\!=\ NH \\ \diagdown\ NH_2 \end{array}$
10. $O \leftarrow P\begin{array}{l} \diagup NH_2 \\ - NH_2 \\ \diagdown NH_2 \end{array}$
11. $C_{12}H_{25}ONO$
12. $O \leftarrow P(OC_2H_5)_3$
13. $[(CH_3)_2N]_2POF$
14. C_6H_5–SO_2NH_2
15. sulfanilamide, H_2N–C_6H_4–SO_2NH_2
16. C_6H_5–$SONHC_2H_5$

27 Mixed Aquo-Ammono Systems of Compounds

An American chemist, E. C. Franklin, introduced the idea of the *system* of compounds as a guide to the chemistry of new families of substances. Without calling it by name, we have used Franklin's method in the comparison of properties of substances derived from water and ammonia (p. 295). The *system* may be compared to the *functional group* in carbon compounds as a basis for guidance in remembering properties and predicting new reactions. Each covalent halide and oxyhalide mentioned in the last chapter is the starting point for an aquo, an ammono, and an aquo-ammono *system* of compounds.

27-1. Derivatives of Carbonic Acid

Hydrolysis of phosgene, $COCl_2$, yields carbonic acid, the basis for an aquo system of compounds of carbon. An aquo-ammono system may be obtained from carbonic acid by substitution of the ammonia analogues of $-OH$ and $=O$—namely, $-NH_2$ and $=NH$, respectively—in carbonic acid. This process, followed by deammonation, results in the derivatives shown in Expression 27-1. Strictly speaking, the last two compounds, guanidine and biguanid, belong only to an ammono system of compounds.

Some of the properties of each of these derivatives will be mentioned.

Groups possessing the same electronic configuration are defined

as isosters. Thus $: \overset{\cdot\cdot}{\underset{H}{N}} : H$ and $: \overset{\cdot\cdot}{\underset{\cdot\cdot}{O}} : H$ are isosters, as are the following pairs: : : Ö : and : : N̈ : H, H : C : : : and : : : N : , Na⁺ and Ne. Isosters have many properties in common.

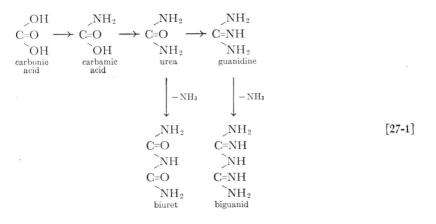

[27-1]

1. Carbamic Acid, Urea, and Biuret. Carbonic acid and carbamic acid are unstable substances and cannot be isolated though they may have some transient existence in solutions. Urea, or carbamide (p. 580), can be deammonated merely by heating above its melting point.

The deammonation product, isocyanic acid, trimerizes to cyanuric acid, but this reaction is in competition with the addition of urea to yield biuret (p. 440). Both these products are formed along with others. It has not been disproved, however, that the

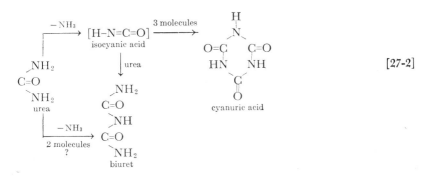

[27-2]

competition may be in the original deammonation—that is, between loss of NH_3 from one molecule of urea and loss of NH_3 from two molecules of urea.

2. Cyanamide and Guanidine. Dehydration of urea would yield $N\equiv C-NH_2$, but this reaction, which can be written on paper, has not been accomplished in the laboratory. Instead, cyanamide is made from calcium cyanamide $CaNCN$, by hydrolysis. Calcium cyanamide is used to defoliate cotton so that it can be harvested economically by machinery. Cyanamide, a white solid, is soluble

$$CaC_2 + N_2 \xrightarrow{1,100°} \underset{\substack{\text{calcium}\\\text{cyanamide}}}{CaN-C\equiv N} \qquad [27\text{-}3]$$

$$CaN-C\equiv N + 2H_2O \xrightarrow{65\%} Ca(OH)_2 + \underset{\text{cyanamide}}{H_2N-C\equiv N} \qquad [27\text{-}4]$$

in water and dimerizes upon heating. It will react with NH_4Cl in alcoholic solution, at 100° in a closed tube, to yield guanidinium

$$H_2N-C\equiv N + HNH-C\equiv N \xrightarrow[100\%]{\Delta} \underset{\text{dicyandiamide}}{H_2N-\overset{\overset{\displaystyle HN}{\|}}{C}-\overset{\overset{\displaystyle H}{|}}{N}-C\equiv N} \qquad [27\text{-}5]$$

chloride. Treatment of the salt with a stronger base liberates a white, crystalline, hygroscopic solid, guanidine. Heating guanidinium chloride to 185° yields biguanid and ammonium chloride.

$$H_2N-C\equiv N + NH_4Cl \longrightarrow \left(\underset{H_2N}{\overset{H_2N}{>}}C=NH_2\right)^{+} Cl^{-}$$

$$\underset{\text{guanidinium chloride}}{}$$

$$\underset{\substack{\\ \text{guanidine}}}{\overset{H_2N}{\underset{H_2N}{>}}C=NH} \overset{OH^-}{\longleftarrow} \Big|\overset{185°}{\longrightarrow} \underset{\text{biguanid}}{H_2N-\overset{\overset{\displaystyle HN}{\|}}{C}-\overset{\overset{\displaystyle H}{|}}{N}-\overset{\overset{\displaystyle NH}{\|}}{C}-NH_2} + 2NH_4Cl$$

$$[27\text{-}6]$$

3. Basic Strength of Carbonic Acid Derivatives. The ammonolysis of carbonic acid would be expected to lead to more basic substances since NH_3 is a weak base itself. Carbonic acid is a very weak acid; the strength of carbamic acid is not known; urea is nearly neutral although it forms a salt with a strong acid like nitric or oxalic; and the completely ammonolyzed product, guanidine, is as strong a base as KOH. Guanidine is a strong enough base to form a stable carbonate with the weak acid, carbonic. A comparison of the basic strengths in terms of dissociation constants is given in Table 27-1. The dissociation constant of a base is defined by Expression 27-8, in which the symbols have meanings comparable to those given in the definition of K_a (pp. 458–462).

$$\underset{\text{base}_1}{B} + \underset{\text{acid}_2}{H_2O} \rightleftarrows \underset{\text{acid}_1}{\overset{+}{BH}} + \underset{\text{base}_2}{OH^-} \qquad [27\text{-}7]$$

$$\frac{[BH^+][OH^-]}{[B]} = K_b \qquad [27\text{-}8]$$

TABLE
27-1 | *Comparison of Basic Strengths at 25°C*

COMPOUND	K_b	COMPOUND	K_b
Guanidine	too strong to measure	Pyridine	2.3×10^{-9}
		Aniline	4.6×10^{-10}
Ethyl amine	3.4×10^{-4}	Urea	1.5×10^{-14}
Cyclohexylamine	4.4×10^{-4}	Acetamide	3.1×10^{-15}
Ammonia	1.8×10^{-5}	Acetanilide	4.1×10^{-14} at 40°

27-2. Derivatives of Phosphoric Acid

When phosphoric acid is the starting compound, a formal treatment similar to that given for carbonic acid in the previous section results in the products shown in Expression 27-9. Phosphoric acid is first shown as the hypothetical substance having five OH groups attached to phosphorus. Subsequent ammonolysis, deammonation, and dehydration products are delineated.

$$
\begin{array}{ccccc}
\overset{\displaystyle\text{OH}}{\underset{\displaystyle\text{OH}}{|}}\!\!\!\begin{array}{c}\nearrow\text{OH}\\ \text{P-OH}\\ \searrow\text{OH}\end{array} \rightarrow &
\overset{}{\begin{array}{c}\nearrow\text{OH}\\ \text{O=P-OH}\\ *\,\searrow\text{OH}\end{array}} \rightarrow &
\begin{array}{c}\nearrow\text{NH}_2\\ \text{O=P-OH}\\ *\,\searrow\text{OH}\end{array} \rightarrow &
\begin{array}{c}\nearrow\text{NH}_2\\ \text{O=P-NH}_2\\ *\,\searrow\text{OH}\end{array} \rightarrow &
\begin{array}{c}\nearrow\text{NH}_2\\ \text{O=P-NH}_2\\ *\,\searrow\text{NH}_2\end{array}\\
\downarrow & \downarrow & \downarrow & \downarrow & \downarrow\\
\begin{array}{c}\text{O}\\ \text{O=P}\diagup\\ *\ \searrow\text{OH}\end{array} \rightarrow &
\begin{array}{c}\nearrow\text{OH}\\ \text{HN=P-OH}\\ \searrow\text{OH}\end{array} \rightarrow &
\begin{array}{c}\nearrow\text{NH}_2\\ \text{HN=P-OH}\\ \searrow\text{OH}\end{array} \rightarrow &
\begin{array}{c}\nearrow\text{NH}_2\\ \text{HN=P-NH}_2\\ \searrow\text{OH}\end{array} \rightarrow &
\begin{array}{c}\text{NH}_2\\ \text{HN=P}\diagup\\ *\ \searrow\text{O}\end{array}\\
\downarrow & \downarrow & \downarrow & \downarrow & \downarrow\\
\begin{array}{c}\text{O}\\ \text{O=P}\diagup\\ \searrow\text{NH}_2\end{array} \rightarrow &
\begin{array}{c}\text{O}\\ \text{HN=P}\diagup\\ *\ \searrow\text{OH}\end{array} \rightarrow &
\begin{array}{c}\text{OH}\\ \text{N=P}\diagup\\ \searrow\text{OH}\end{array} \rightarrow &
\begin{array}{c}\text{NH}\\ \text{HN=P}\diagup\\ \searrow\text{OH}\end{array} \rightarrow &
\begin{array}{c}\text{N=P=O}\\ *\end{array}
\end{array}
$$

$$[\textbf{27-9}]$$

1. Amido- and Diamido-phosphoric Acids.

Amidophosphoric acid, or phosphamic acid, $H_2N\text{-}P\diagup_{\searrow}^{OH}$ with O below, has been isolated, but it is unstable in air and hydrolyzes in water to the corresponding ammonium salt. Its esters have received more attention than the acid itself because of the importance of the sugar esters of nitrogen derivatives of phosphoric acid in biological systems. If phosphorus oxychloride is treated with phenol in pyridine, the two chlorine-containing compounds shown in Expressions 27-10 and 27-11 can be obtained, and these can be ammonolyzed to the corresponding amido- and diamido-phosphates.

$$
\begin{array}{c}
\begin{array}{c}\nearrow\text{Cl}\\ \text{O}\leftarrow\text{P-Cl}\\ \searrow\text{Cl}\end{array}
+ \langle\ \rangle\text{-OH} \xrightarrow{\text{pyridine}}
\begin{array}{c}\nearrow\text{O-}\langle\ \rangle\\ \text{O}\leftarrow\text{P-Cl}\\ \searrow\text{Cl}\end{array}
+ \text{pyridine}\cdot\text{HCl}\\[1.5em]
\begin{array}{c}\nearrow\text{O-}\langle\ \rangle\\ \text{O}\leftarrow\text{P-NH}_2\\ \searrow\text{NH}_2\end{array}
\xleftarrow[\ 53\%\ ]{\text{NH}_3}
\end{array}
$$

$$[\textbf{27-10}]$$

phenyl diamidophosphate

* Compounds marked with an asterisk are known, in the form of salts if not as free acids.

$$POCl_3 + 2C_6H_5OH + pyridine \longrightarrow O{\leftarrow}P\underset{Cl}{\overset{O-\bigcirc}{<}}_{O-\bigcirc} + 2\ pyridine\cdot HCl$$

$$\underset{\underset{\text{diphenyl amidophosphate}}{NH_2}}{O{\leftarrow}P}\overset{O-\bigcirc}{<}_{O-\bigcirc} \xleftarrow[24\%]{NH_3}$$ [27-11]

2. Phosphoryl Triamide.

The preparation of phosphoryl triamide has been reported, but later workers have not been able to repeat the preparation. However, substituted amines will react with $POCl_3$ to give well-characterized triamides in the presence of pyridine, which reduces the violence of the initial reaction.

$$O{\leftarrow}P\underset{Cl}{\overset{Cl}{<}}_{Cl} + 3\bigcirc{-}NH_2 \xrightarrow[\substack{CHCl_3 \\ 82\%}]{pyridine} O{\leftarrow}P\underset{NHC_6H_5}{\overset{NHC_6H_5}{<}}_{NHC_6H_5} + 3\ pyridine\cdot HCl$$ [27-12]

phosphoryl trianilide

3. Phosphoryl Imide Amide.

This compound, $O{\leftarrow}P\overset{\displaystyle{\parallel} NH}{\underset{\displaystyle NH_2}{}}$, can be extracted with ether from the reaction product of ammonia and PCl_5 in the presence of water at low temperatures. Heating it in the absence of air results in deammonation to phosphoryl nitride, $O{\leftarrow}P{\equiv}N$. Its inertness and high melting point have led to the idea that it is polymeric in character.

$$PCl_5 + 7NH_3 + H_2O \longrightarrow O{\leftarrow}P\underset{NH_2}{\overset{NH}{<}} + 5NH_4Cl$$

$$(O{\leftarrow}P{\equiv}N)_n + NH_3 \xleftarrow{\Delta}$$ [27-13]

4. Imidophosphoric Acid.

The only other compound in this series for which there is evidence of existence is imidophosphoric acid, $HN{=}P{\displaystyle{\nearrow O \atop \searrow OH}}$. It has been prepared, probably as a polymer,

by the action of NH_3 on P_2O_5 below $0°$. The compound is hydrolyzed in hot water.

$$H_2O + P_2O_5 + 2NH_3 \longrightarrow \left(HN{=}P{\nearrow O \atop \searrow OH} \right)_n + NH_4OPO_3H_2 \qquad [27\text{-}14]$$

An entire series of aquo-ammono-sulfo-phosphoric acids, in which sulfur is substituted for one or more of the oxygen atoms in phosphoric acid, can be written. A number of them are known but have not attained importance in the field.

27-3. Derivatives of Sulfuric Acid

Aquo-ammono-sulfuric acids (Exp. 27-15) have attained a more important position in the chemical laboratory than the corresponding phosphoric acids. One of them, indeed—sulfamic, or amidosulfuric, acid—is a well-known commercial product.

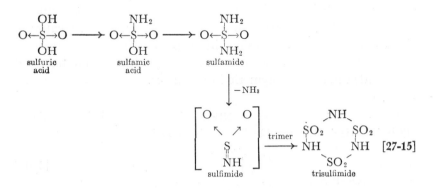

1. Sulfamic Acid. Sulfamic acid is a white, nonhygroscopic solid, of melting point $205°$, which may be grown into large crystals of flat plates. It is not infinitely soluble in water, as other inorganic acids are, its solubility being 24.3 g per 100 g of H_2O at 25°C. The ease of handling nonhygroscopic solids, in comparison with liquid acids, which are quite corrosive, is an attractive property of sulfamic acid. In contrast to sulfuric acid, it is a noncharring acid and can be used to advantage in the organic laboratory—for example, in catalyzing the polymerization of acetaldehyde to paraldehyde and the reverse reaction.

Sulfamic acid has been on the market since 1938 as an important chemical. It is made by dissolving urea in 70% fuming sulfuric acid (a solid) and then liberating the CO_2 from the urea disulfonic acid by warming it in 100% H_2SO_4, in which sulfamic acid is insoluble.

$$\underset{}{\overset{NH_2}{\underset{NH_2}{C=O}}} + 2SO_3 \longrightarrow \underset{\text{urea disulfonic acid}}{\overset{NHSO_3H}{\underset{NHSO_3H}{C=O}}} \xrightarrow{100\% \ H_2SO_4} CO_2 + \underset{\text{sulfamic acid}}{2H_2N-SO_3H}$$

$$[27\text{-}16]$$

The ammonium salt of sulfamic acid is a water-soluble fire retardant, which can be put on cloth merely by dipping and drying. Ammonium sulfamate is also a weed killer, especially effective on poison ivy and choke cherry. It does not permanently sterilize the soil since it slowly hydrolyzes to a fertilizer, ammonium sulfate.

$$\underset{\text{ammonium sulfamate}}{NH_4-O-SO_2-NH_2} + H_2O \xrightarrow{\text{slow}} \underset{\text{ammonium sulfate}}{NH_4-O-SO_2-ONH_4} \qquad [27\text{-}17]$$

2. Sulfamide.   The preparation of sulfamide has already been described (p. 580).

It should now be clear that the writing of formal relationships between oxygen acids and their nitrogen homologues is more than a mental exercise. For Franklin, the originator of the idea of a nitrogen system of compounds, it was a guide to the chemistry of new compounds. It cannot give all the answers, however. One final example will suffice to show its general usefulness.

Urea reacts with formaldehyde to give a high-molecular-weight resin, which may have the formula shown in Expression 27-18.

$$\begin{array}{c} \text{NH}_2 \\ \text{C=O} \\ \text{NH}_2 \end{array} + \text{H--}\overset{\text{H}}{\underset{}{\text{C}}}\text{=O} \longrightarrow \text{HOCH}_2\text{--N--C--N--CH}_2\text{OH}$$

$$-\left[\text{CH}_2\text{--N--C--N}\right]_x^- \overset{-\text{H}_2\text{O}}{\longleftarrow} \qquad\qquad [27\text{-}18]$$

Will sulfamide behave in like manner? A guess would be that it should:

$$\text{SO}_2\!\!\begin{array}{c}\text{NH}_2 \\ \\ \text{NH}_2\end{array} + \text{HCHO} \longrightarrow \text{HOCH}_2\text{--N--SO}_2\text{--N--CH}_2\text{OH} \overset{-\text{H}_2\text{O}}{\longrightarrow}$$

$$-\left[\text{CH}_2\text{--NH--SO}_2\text{--NH}\right]_y^-$$

SUMMARY

The derivatives of carbonic, phosphoric, sulfuric, and other acids are economically studied and profitably interrelated as systems of compounds. By replacing oxygen isosters with the corresponding nitrogen groups, we obtain the related aquo-ammono and ammono compounds. The method helps us to remember properties and predict new reactions.

EXERCISES

1. What are the sulfur analogues of the aquo-ammono-sulfuric acids?

2. Write the series of sulfo-aquo-ammono-phosphoric acids that can be obtained from $PSCl_3$ by suitable hydrolysis, ammonolysis, dehydration, and deammonation. Do you think any of these might lose H_2S instead of H_2O or NH_3 by suitable treatment?

3. How would you try to make the following?

   1. $As(NH_2)_3$ 3. $GeO(NH_2)_2$

   2. $SbO(NH_2)_3$ 4. ⬡–$SbOCl_2$

4. What is the nitrogen system analogue of the following oxygen system compounds?

   1. CH_3COOH

   2. $Si(OC_2H_5)_4$ 4. $\begin{array}{c} OC_2H_5 \\ \diagup \\ C\!=\!O \\ \diagdown \\ NH_2 \end{array}$

   3. $C_6H_5PO(NH_2)_2$

                       5. $CH_3\!-\!COCl$

Could any of the analogues be synthesized? How?

PART SIX

THE HEXAGON

28 Aromatic Character

The last part of this text will point to still other directions in which your interest in covalent chemistry may turn in the future if you pursue chemistry as a career. These chapters will be centered on a study of fused ring systems, rearrangements, colored substances, and more aromatic chemistry.

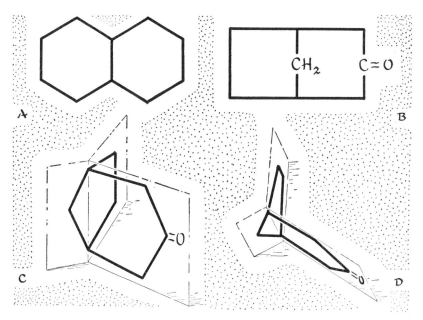

A fused ring system is one in which more than one atom is common to two rings. If the two rings are fused on adjacent atoms, two atoms will be common; if the two rings are fused on nonadjacent atoms, more than two atoms will be common.

This chapter will be devoted to the chemistry of polynuclear aromatic hydrocarbons and derivatives of benzene. Aromatic rings can be fused only on adjacent atoms; the rings are coplanar.

28-1. Definition of Aromatic Character

Carbon compounds were for many years called aromatic if they would undergo halogenation, sulfonation, nitration, and the Friedel-Crafts reaction. But, since all these reactions have been carried out on aliphatic compounds, there are actually only differences in degree and not in kind between the two classes. A better way to describe aromatic character is by the following attributes (emphasized by L. F. Fieser): (1) the low degree of unsaturation; (2) the ability to retain this peculiar saturation; (3) the ease of formation of the conjugated ring; (4) the behavior of some functional groups on the ring.

The low degree of unsaturation of aromatic compounds and the ability to retain this peculiar saturation are already familiar. Benzene does not add bromine readily, nor HX at all, and it does not decolorize neutral permanganate solution at room temperature, whereas these reagents add readily to an olefinic double bond.

The greater heat of hydrogenation (exothermic) of three aliphatic double bonds relative to that of benzene has been mentioned previously (p. 382) as evidence for the special stability of the benzene ring. It has been shown that the formation of benzene from

$$\bigcirc \longrightarrow \bigcirc + H_2 + 5.7 \text{ kcal} \tag{28-1}$$

dihydrobenzene (the reverse of addition) is actually an exothermic reaction. Besides these thermochemical data, the ease of formation of the aromatic ring can be demonstrated by a number of somewhat startling reactions.

1. Pinonic acid rearranges readily with the spontaneous loss of hydrogen and water to 2,4-dimethylphenylacetic acid.

$$3CH_3 \quad CH_3$$
$$CH_3\text{-}\overset{O}{\overset{\|}{C}}\text{-}CH \underset{C_2}{\overset{1}{\diagup}} CH\text{-}CH_2\text{-}COOH \longrightarrow$$
4   5   6CH₂

pinonic acid

$$3 \quad 2\,CH_3$$
$$CH_3\text{-}\langle \rangle\text{-}CH_2\text{-}COOH + H_2 + H_2O \qquad [28\text{-}2]$$
5   6
2,4-dimethylphenylacetic acid

2. Camphor is transformed into *p*-cymene in the presence of P_2O_5.

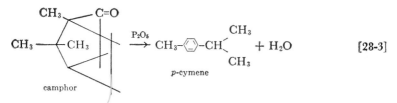

$$\text{camphor} \xrightarrow{P_2O_5} CH_3\text{-}\langle \rangle\text{-}CH \underset{CH_3}{\overset{CH_3}{\diagup}} + H_2O \qquad [28\text{-}3]$$

p-cymene

3. In the presence of a hydrogenation catalyst, Raney nickel, dimethylglyoxime is reduced in about 82% yield to tetra-methylpyrazine, an aromatic ring compound containing two nitrogens. This reaction takes place at 100°C, at 3,000 pounds pressure, and in the presence of hydrogen and ammonia!

$$\begin{matrix} 2CH_3\text{-}C\text{=}NOH \\ 2CH_3\text{-}C\text{=}NOH \end{matrix} + 5H_2 \xrightarrow[\text{liq. NH}_3]{\overset{100°C}{3,000\,\text{lbs.}}} \begin{matrix} CH_3\text{-}C \\ CH_3\text{-}C \end{matrix} \begin{matrix} N \\ \\ N \end{matrix} \begin{matrix} C\text{-}CH_3 \\ C\text{-}CH_3 \end{matrix} + 2NH_3 + 4H_2O$$

dimethylglyoxime

tetramethylpyrazine

[28-4]

About 18% of the glyoxime is reduced to 2,3-diaminobutane, but it is rather startling that an unreduced ring is obtained at all in the presence of liquid ammonia, a high hydrogen pressure, and a catalyst favorable to reductions.

4. One more example, aromatization of aliphatic hydrocarbons, does not illustrate ease of formation of the aromatic ring since it is a high-temperature reaction, but it does demonstrate dramatically that the barriers between aliphatic and aro-

matic chemistry are now completely broken down. Aromatization of an aliphatic hydrocarbon, of course, involves a combination of cyclization and dehydrogenation. Toluene can be produced very efficiently by passing n-heptane at 500° and atmospheric pressure over Cr_2O_3 on Al_2O_3. Much of the TNT

$$
\begin{array}{c}
CH_3 \\
| \\
CH_2 \\
CH_2 \quad CH_3 \\
| \quad\quad | \\
CH_2 \quad CH_2 \\
\diagdown CH_2 \diagup
\end{array}
\xrightarrow[500°,\ 1\ \text{atm.}]{Cr_2O_3\ +\ Al_2O_3}
\quad
\begin{array}{c} CH_3 \\ \bigcirc \end{array}
+\ 4H_2
\qquad\qquad [28\text{-}5]
$$

made from toluene during World War II came originally from a petroleum source rather than from coal tar. (See the reference at the end of this chapter.)

The difference in the behavior of functional groups between aliphatic and aromatic compounds is also largely a difference in degree and not in kind. In phenols the predominance of the enol form again demonstrates the stability of the conjugated ring. As we mentioned on page 494, the introduction of two and then three OH groups on a benzene ring makes the compound more aliphatic and increases the stability of the keto form at the expense of the enol form.

The formation of a diazonium salt of an aromatic primary amine, in contrast to the decomposition of, and loss of nitrogen from, an aliphatic primary amine, appears to be a difference in kind. Above 5°C, however, the diazonium salts in water solution also lose nitrogen readily to give a phenol. This apparent difference in kind may be due to lack of data on the diazotization of aliphatic amines at lower temperature, or, indeed, may merely indicate that the aliphatic diazonium salt has a very short life.

28-2. Polyenes

The unique low degree of unsaturation attained in benzene and its homologues makes it appropriate to say that no other hydrocarbons are as aromatic as benzene. Other aromatic hydrocarbons, however, are closely related to benzene in properties and show only differences in degree.

If two benzene rings substituted in an aliphatic compound are completely conjugated with aliphatic double bonds, these olefinic bonds do not exhibit the properties ordinarily associated with such bonds. None of the following four compounds adds HBr, and they are only very slowly attacked by bromine or alkaline permanganate solution. In tetraphenylethylene, bromine substitutes

⟨◯⟩–CH=CH–⟨◯⟩   ⟨◯⟩–CH=CH–CH=CH–⟨◯⟩
1,2-diphenylethylene   1,4-diphenylbutadiene

tetraphenylethylene

⟨◯⟩–CH=CH–CH=CH–CH=CH–⟨◯⟩
1,6-diphenylhexatriene

on the rings before it adds to the olefinic bond. It is perhaps significant that 1,6-diphenylhexatriene (containing six aliphatic carbons) is the most stable of the three diphenylpolyenes shown.

When the complete conjugation of these polyenes is destroyed, by loss of one of the double bonds, they become much more reactive. Hydrogen adds to 1,6-diphenylhexatriene in the 1,6 man-

$$\langle\bigcirc\rangle\text{–CH=CH–CH=CH–CH=CH–}\langle\bigcirc\rangle \xrightarrow[\substack{H_2O \\ 60\%}]{Al(Hg)} \langle\bigcirc\rangle\text{–CH}_2\text{–CH=CH–CH=CH–CH}_2\langle\bigcirc\rangle$$

[28-6]

ner; the resulting product is easily polymerized and, in contrast to the triene, adds bromine readily.

The chemistry of some polynuclear aromatic hydrocarbons will give a more nearly complete concept of what is meant by "aromatic character."

Polynuclear Aromatic Hydrocarbons

28-3. Naphthalene

Erlenmeyer suggested (1866) that the compound naphthalene, $C_{10}H_8$, which resembles benzene in many ways, should be assigned the structure numbered I below. Other structures, analogous to alternative proposed structures for benzene (Chap. 18), were also proposed, but at present the only modification of Erlenmeyer's proposal is the recognition of the contributing resonance structures, II and III.

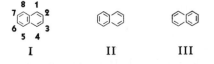

| I | II | III |

The ring (numbered as shown in I) has four equivalent positions, 1, 4, 5, and 8, which are referred to as α positions with respect to the carbons at which the two rings are fused. Positions 2, 3, 6, and 7 are, similarly, equivalent β positions.

One piece of evidence for the structure given above is that only two monosubstituted isomers, $C_{10}H_7Z$, have ever been prepared from naphthalene.

A. Reactions of Naphthalene

1. Nitration.   Nitration and halogenation reactions go exclusively to the α position. Naphthalene can be nitrated, under the same conditions as benzene, to give α-nitronaphthalene.

$$\text{HONO}_2, \text{H}_2\text{SO}_4, 50° \xrightarrow{\quad 92\% \quad} \text{NO}_2 \qquad\qquad [28\text{-}7]$$

The β isomer is made by diazotization of β-naphthylamine and replacement by the NO_2 group. This is best accomplished by use of diazonium borofluoride and treatment with $NaNO_2$ in the presence of copper powder.

$$\text{(naphthalene)}NH_2 + HONO \xrightarrow[97\%]{HBF_4} \left(\text{(naphthalene)}\overset{+}{N}\equiv N\right)BF_4^- \xrightarrow[\text{Cu powder}]{NaNO_2} \text{(naphthalene)}NO_2 \qquad [28\text{-}8]$$

2. Bromination. Bromination of naphthalene goes readily without a catalyst in CCl_4 at the boiling point of the solvent. Benzene is less reactive; that is, benzene is not brominated under the

$$\text{(naphthalene)} + Br_2 \xrightarrow[75\%]{CCl_4} \text{(naphthalene)}Br + HBr \qquad [28\text{-}9]$$

same conditions. β-Bromonaphthalene is obtained by the Sandmeyer reaction on the corresponding diazonium salt.

Exercise 28-1. Write a Sandmeyer reaction for the preparation of β-bromonaphthalene.

3. Sulfonation. Naphthalene will dissolve in concentrated sulfuric acid at about 80° to give α-naphthalenesulfonic acid. If the temperature of the reaction mixture is then elevated to 180°, or if naphthalene is sulfonated directly at this temperature, the β isomer is obtained. Only small amounts of the β isomer are formed at the lower temperature, and, conversely, only small amounts of the α isomer remain at the higher temperature. One can separate the isomers when they are present together by taking advantage of the greater solubility of calcium α-naphthalenesulfonate.

$$\text{(naphthalene)} \xrightarrow[H_2SO_4]{80°} \text{(naphthalene)}SO_3H \xrightarrow[70\%]{180°} \text{(naphthalene)}SO_3H \qquad [28\text{-}10]$$

4. Reduction. Reduction reactions also indicate that naphthalene is considerably more reactive than benzene. Whereas benzene cannot be reduced with sodium and alcohol, naphthalene is reduced with sodium in boiling alcohol to tetrahydronaphthalene (tetralin).

$$\text{(naphthalene)} \xrightarrow[80\%]{\text{Na} + \text{C}_2\text{H}_5\text{OH}} \text{(tetralin)}$$

[28-11]

The difference in ease of reduction of benzene and naphthalene is dramatized by this reaction, in which one of the benzene rings remains in the tetralin structure.

Tetralin, or naphthalene itself, can be reduced catalytically with platinum and hydrogen to decahydronaphthalene (generally called decalin).

$$\text{(tetralin)} \xrightarrow[91\%]{\text{Pt, CH}_3\text{COOH}} \text{(decalin)}$$

[28-12]

Another display of aromatic character is revealed in the reaction of brominated tetralin or decalin. Substitution with bromine takes place on the reduced ring at room temperature, but, surprisingly, HBr is lost upon mild heating, and the aromatic ring is reformed. This is a simple laboratory method of preparing dry HBr.

$$\text{(tetralin)} + \text{Br}_2 \longrightarrow \underset{\text{Br}}{\overset{\text{Br}}{\text{(dibromotetralin)}}} + 2\text{HBr}$$

$$\Big\downarrow \Delta$$

$$\underset{94\%}{\text{(naphthalene)}} + 2\text{HBr}$$

[28-13]

5. Oxidation. Evidence that naphthalene is easier to oxidize than benzene lies in the temperatures at which the two are catalytically oxidized by air in the presence of V_2O_5. Naphthalene is oxidized to phthalic anhydride at about 350°, but benzene must be heated to about 500° to give a good yield of maleic anhydride with the same catalyst, according to some recent patents in the literature.

$$[28\text{-}14]$$

When the naphthalene ring is negatively substituted (by chlorine or a nitro group, for example), the ring containing that group may be said to have electrons withdrawn from it, and hence the other ring is the more readily attacked by oxidizing agents (they are electron-seeking agents). Consequently, the oxidation of α-nitronaphthalene yields 3-nitrophthalic acid. If a group present on the

$$[28\text{-}15]$$

naphthalene ring supplies electrons to the ring (one such contributing resonance structure, in which Z may be O, N, or S, is shown alongside), as in α- or β-naphthol or α- or β-aminonaphthalene, the ring carrying the substituted groups is the one more readily oxidized. Hence oxidizing agents (which are seeking electrons) will react with β-naphthol and the other compounds just mentioned to give phthalic acid.

$$[28\text{-}16]$$

Exercise 28-2. Write electronic structures for the principal contributing resonance forms of β-naphthol.

B. Synthesis of Naphthalene

The proof of the structure of any organic molecule is considered incomplete until it has been synthesized by an unequivocal path (involving known reactions). If doubtful steps exist in a path (because of possible rearrangements, for example, or because a reaction is new and untried), another route must be chosen for the proof of structure. The synthesis of a substance may remain a problem long after its structure is known from its chemistry.

Naphthalene has been synthesized by a number of methods, one of which is reproduced in Expression 28-17. Succinic anhydride,

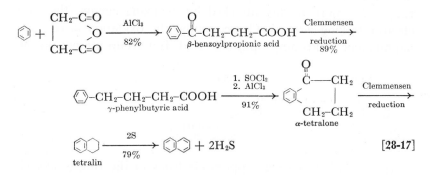

by a Friedel-Crafts reaction, can be condensed with benzene to β-benzoylpropionic acid. A Clemmensen reduction with zinc amalgam and HCl yields γ-phenylbutyric acid. One can close the ring by making the acid chloride with SOCl$_2$ and following this with a Friedel-Crafts reaction or, more propitiously, by direct condensation of the organic acid with H$_2$F$_2$. The product, α-tetralone, is reduced by the Clemmensen method, and the aromatic hydrocarbon is finally obtained by dehydrogenation with elemental sulfur at elevated temperature. This last reaction is a general method of aromatizing an alicyclic hydrocarbon.

28-4. Anthracene and Phenanthrene

The three polynuclear hydrocarbons mentioned in this chapter, naphthalene, anthracene, and phenanthrene, are obtained from coal tar along with benzene, toluene, the xylenes, and various phenols. Coal tar is the black viscous oil obtained by the destructive distillation of soft coal—that is, by heating soft coal in the absence of air. Fractional distillation will separate naphthalene both from the low-boiling aromatic hydrocarbons and from anthracene and phenanthrene, which boil near 340°C. The separation of anthracene and phenanthrene is based upon the greater stability of the picrate* of anthracene in alcohol solution relative to that of phenanthrene.

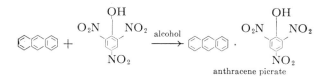

anthracene picrate

The structures generally written for anthracene and phenanthrene are given below. The contributing resonance structures are shown, but again, as in naphthalene, the greatest contribution is made by the structures that contain the greatest number of completely conjugated rings. The first and third structures for an-

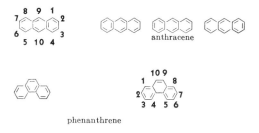

anthracene

phenanthrene

thracene (which are equivalent) and the first structure for phenanthrene make the greatest contribution to the total structures of these respective molecules.

The numbering systems for the two compounds are shown above. In anthracene there are three groups of equivalent positions: 1, 4, 5, and 8; 2, 3, 6, and 7; 9 and 10. In phenanthrene, however, equivalent positions occur only in pairs: 1 and 8; 2 and 7; 3 and 6;

* Picric acid, 2,4,6-trinitrophenol, forms, with aromatic hydrocarbons and ethers, a series of molecular compounds (see quinhydrone, p. 622) whose melting points can frequently be used to identify the hydrocarbons or ethers. They are highly colored, frequently not stable enough to be recrystallized, and of uncertain structure. These molecular compounds are called picrates though they are not salts or esters.

4 and 5; 9 and 10. The most reactive points in both molecules are the 9 and 10 positions.

A. *Reactions of Anthracene and Phenanthrene*

Since substitution reactions of anthracene and particularly of phenanthrene produce mixtures of monosubstituted products, they are not generally useful. Halogenation, however, will result in unique products from either compound.

1. Halogenation. The transition in character from an olefinic to an aromatic double bond is demonstrated very well by the reaction of bromine with phenanthrene. The double bond at 9,10 in phenanthrene is much more reactive than an aromatic double bond but less reactive than an aliphatic double bond. The reactivity may be accounted for, in part at least, by the great stability of the principal contributing resonance structure, the one in which there are three completely conjugated rings.

Phenanthrene will add Br_2 in CCl_4 solution or will substitute bromine in the presence of a catalyst such as $FeBr_3$. Here quite easily one may control the path of reaction to get either an aliphatic (9,10-dibromo) or an aromatic (9-bromo) product.

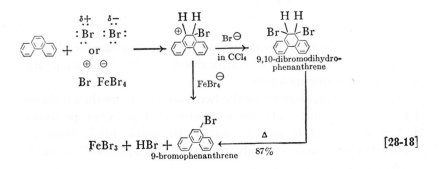

[28-18]

Anthracene adds a molecule of bromine readily to yield 9,10-dibromodihydroanthracene, which loses a molecule of HBr on heating to give the monosubstituted anthracene. This addition may be considered a 1,4-addition to the conjugated system (see also Exp. 28-20), resulting in a product that still has two isolated, conjugated rings. (Compare this reaction with the Diels-Alder reaction below.)

[28-19]

2. Reduction. Like naphthalene, anthracene and phenanthrene are easily reduced. Sodium amalgam and isoamyl alcohol yield the 9,10-dihydro compound with both compounds. Catalytic hydrogenation can be used to reduce the three rings completely.

$$\text{(anthracene)} \xrightarrow[\substack{C_5H_{11}OH \\ 50\%}]{Na(Hg)} \text{(9,10-dihydroanthracene)}$$

9,10-dihydroanthracene

[28-20]

3. Oxidation. Mild oxidation of naphthalene, anthracene, or phenanthrene yields a diketone. This type of diketone, conjugated in a ring system, is referred to as a quinone. Chromic anhydride will oxidize naphthalene to 1,4-naphthoquinone. Further oxidation gives phthalic acid (see p. 606). With phenanthrene and anthracene the vulnerable points are the 9 and 10 positions. Nitric acid acts as an oxidizing rather than nitrating agent on both of these compounds.

$$\text{(anthracene)} \xrightarrow[CH_3COOH]{CrO_3} \text{(anthraquinone)}$$

anthraquinone

[28-21]

Exercise 28-3. Assuming that CrO_3 is reduced to Cr^{+3}, balance the reaction in Expression 28-21.

4. Diels-Alder Reaction. Anthracene undergoes one reaction that distinguishes it from naphthalene and phenanthrene, the Diels-Alder reaction. Since this reaction is characteristic of compounds containing a conjugated diene system, it implies that anthracene is not altogether aromatic in its properties. The second reactant in the Diels-Alder reaction contains an olefinic bond that is activated by the presence of a functional group, commonly carboxyl, or of an acid anhydride.

Anthracene will react with maleic anhydride without any catalyst if a benzene solution of the two is refluxed. The reaction is a 1,4-addition of the alkene to the conjugated system terminating in

$$\text{[28-22]}$$

positions 9 and 10 on the anthracene molecule. In this process, as in all 1,4-additions, the diene system is replaced by one olefinic bond. In this case two isolated benzene rings appear in the product. The maleic anhydride forms a part of *two* new six-membered rings; such a bridged ring system is called an *endo* ring. The product in this reaction is named anthracene-9,10-endo-α,β-succinic anhydride; that is, the endo ring is formed in anthracene by the joining of carbons 9 and 10 to the carbons that are α and β to one carboxyl group.

Another Diels-Alder reaction between quinone (p. 621) and 1,3-butadiene is used in a synthesis of anthraquinone. After the addition of one molecule of butadiene to *p*-quinone, the olefinic bond in the quinone part is still activated by the two carbonyl groups. It therefore adds a second butadiene molecule. By dehydrogenating this molecule with selenium, we obtain anthraquinone.

$$\text{[28-23]}$$

anthraquinone

The wide use of the Diels-Alder reaction for synthetic purposes is scarcely suggested by the two examples cited here. The diene may be any of a wide variety of 1,3-alkadienes, including ring compounds like furane, cyclohexadiene, and cyclopentadiene, and the double or triple bond (dienophile) may be activated by almost any function with a multiple bond. A mechanism is suggested in Expression 28-24.

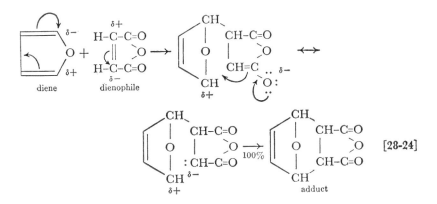

[28-24]

B. Synthesis of Anthracene and Phenanthrene

The synthesis of phenanthrene is analogous to that of naphthalene if we start with naphthalene rather than benzene.

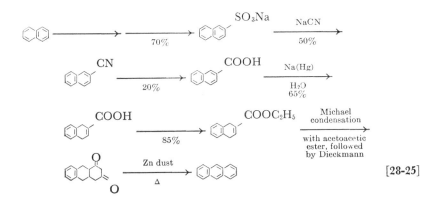

[28-25]

Exercise 28-4. Write a synthesis for phenanthrene by the method of Expression 28-17.

Anthracene also can be synthesized from naphthalene; see Expression 28-25. (bottom of p. 613.)

Exercise 28-5. Write the complete synthesis shown in Expression 28-25, supplying reagents and conditions for each reaction where they are not given and for the Michael condensation, supplying all the reactions necessary to get to the diketone.

28-5. Relative Reactivities of Some Polynuclear Hydrocarbons

Of the reactions discussed so far, the order of activity, with respect to ease of addition to double bonds, is the following: benzene < naphthalene < phenanthrene < anthracene < cyclohexene. For the first four the same order holds for ease of substitution and for oxidation reactions.

The heats of hydrogenation of these hydrocarbons are consistent with the order of reactivity. That of benzene (for the first molecule of H_2), the least reactive of the group, is negative, indicating an endothermic reaction.

TABLE

28- | *Heats of Hydrogenation of Some Cyclic Hydrocarbons*

benzene (first molecule of H_2)	−5.7 kcal/mole
naphthalene (first molecule of H_2)	9.4 kcal/mole (estimated)
phenanthrene (first molecule of H_2)	15. kcal/mole (estimated)
anthracene (first molecule of H_2)	18. kcal/mole (estimated)
cyclohexene	28.6 kcal/mole

Other Aromatic Compounds

Many reactions of functional groups are characteristic of both aliphatic and aromatic compounds, and such reactions have been

mentioned throughout this text. Some reactions of phenols, aldehydes, and ketones, characteristic only of aromatic compounds and
not previously discussed, will be given at this point to expand still
further the ideas to be conveyed by the phrase "aromatic character."

28-6. Reactions of Phenols

A. Esters and the Fries Reaction

Since phenols are themselves weak acids, they do not form esters
readily with carboxylic acids but will form them with acyl halides
or acid anhydrides. With these more reactive acylating agents the
phenol acts as an electron donor, according to the postulated behavior of these acid derivatives. (Read again pp. 284–288.) This
same gradation in character from alcohols to phenols is also shown
with $POCl_3$. The alcohol yields mainly an alkyl halide, but the
phenol reacts only to form a phosphate (p. 582) after two inter-

$$ArOH + POCl_3 \longrightarrow (ArO)_3P{\rightarrow}O + 3HCl \qquad [28\text{-}26]$$
$$\text{an aryl phosphate}$$

mediates are formed. A molecule of HCl is lost in each step of the
transformation (see Exp. 26-79). The aryl phosphate can be used
for making other products.

The organic esters of phenols undergo a reaction that has been
interpreted as an intramolecular rearrangement although the actual mechanism is still uncertain. In the presence of $AlCl_3$ the ester
is converted into a keto phenol. Whether the keto group is directed
to an o or a p position depends on the temperature in a number of

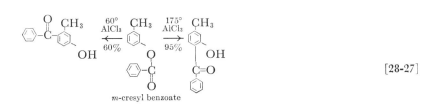

m-cresyl benzoate

[28-27]

the observed cases. In *m*-cresyl benzoate, below 100°, the keto group enters *p* to the hydroxy group; at 175° it enters *o* to the hydroxy group.

B. *Kolbe Reaction*

The Kolbe method of preparing salicylic acid, first reported in 1862, is still used with only slight modification. If sodium phenoxide is heated with CO_2 under pressure, sodium phenyl carbonate is formed and rearranges to sodium salicylate. A possible cyclic mechanism for the reaction is discussed on page 619.

$$\underset{}{\text{ONa}} + CO_2 \xrightarrow[\text{100 lbs.}]{160°} \underset{}{\text{O-C-ONa}} \xrightarrow[90\%]{} \underset{\text{COONa}}{\text{OH}} \qquad [\text{28-28}]$$

C. *Reimer-Tiemann Reaction*

Phenols that have no substituents in an *ortho* position react readily with chloroform in strong basic solution to yield an *o*-hydroxybenzaldehyde. Salicylaldehyde is obtained with little of the *para* isomer by this reaction. The formation of the *ortho* product may be

$$\underset{}{\text{OH}} + 3NaOH + CHCl_3 \xrightarrow[50\%]{} \underset{\text{salicylaldehyde}}{\overset{\text{OH}}{\underset{}{\text{CHO}}}} + 3NaCl + 2H_2O \qquad [\text{28-29}]$$

interpreted by a cyclic mechanism (p. 619), but the formation of the *p* isomer cannot be explained in that manner. The latter may

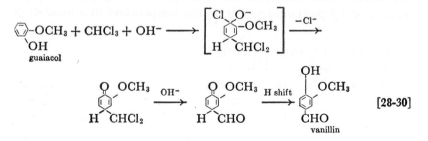

$$[\text{28-30}]$$

result from a 1,4-addition followed by hydrolysis. Vanillin, the principal constituent in extract of vanilla, is made synthetically by the Reimer-Tiemann reaction on guaiacol.

Salicylic acid may also be made by an application of this reaction using CCl_4, instead of $CHCl_3$, and a base.

$$\text{OH} \qquad\qquad \text{OH}$$

$$\bigcirc\text{ + } CCl_4 + 4KOH \longrightarrow \bigcirc\text{COOH} + 4KCl + 2H_2O \qquad [28\text{-}31]$$

25%
salicylic acid

D. Chelation and the Hydrogen Bridge

Ammonia readily accepts a proton from an acid to form an ammonium ion, but will release one only in the presence of a very strong base. HCl readily releases a proton to a base but will not accept one. Water, however, behaves with about equal reluctance either as a proton acceptor or as a proton donor. One may conclude from these observations and the strong electronegativity of oxygen that water molecules, by themselves, might dissociate to a small extent, but that they would be more likely to associate. This is known to be so. The hydrogen between the two

$$\text{HOH} + \text{HOH} \xrightleftharpoons{\qquad} \text{H}:\overset{..}{\underset{..}{\text{O}}}:\text{H}:\overset{..}{\underset{..}{\text{O}}}:\text{H} \xleftarrow{\qquad} \text{H}_3\text{O}^+ + \text{OH}^-$$
$$\text{H} \qquad\qquad\qquad [28\text{-}32]$$

oxygen atoms is said to form a "hydrogen bridge." Other atoms that have enough electrostatic attraction for hydrogen to form these weak hydrogen bridges are F, N, S, and Cl. The strength of this association is about 5% of the strength of the bonds between atoms within molecules. The hydrogen bridge may be considered a weak bond between hydrogen and an electronegative atom.

Hydrogen bridges have been suggested for a number of substances to account for certain of their properties and also for differences in the properties of some isomers. The student will recall

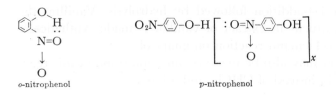

o-nitrophenol   *p*-nitrophenol

that *o*- and *p*-nitrophenol can be separated in the laboratory by steam distillation. At 100°, *o*-nitrophenol is volatile but *p*-nitrophenol is not. The intramolecular six-membered ring (shown above) formed by a hydrogen bridge from the phenol group to the nitro group makes intermolecular association less likely, and hence its volatility approaches that of naphthalene. In effect, the two functional groups are reduced to one with slightly modified properties. Any association that can take place by hydrogen bonds in *p*-nitrophenol will be intermolecular, will, in effect, increase its apparent molecular weight and decrease its volatility. The second ring formed in the *ortho* isomer is referred to as a chelate ring, from the Greek "chela," meaning claw.

Some other substances that form chelate rings are shown below. In acetoacetic ester and related compounds the electrostatic attrac-

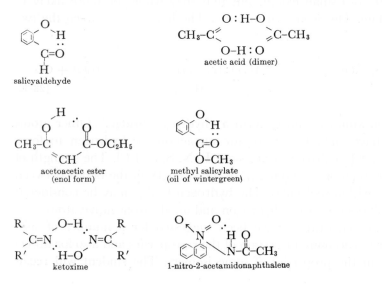

salicylaldehyde

acetic acid (dimer)

acetoacetic ester
(enol form)

methyl salicylate
(oil of wintergreen)

ketoxime

1-nitro-2-acetamidonaphthalene

tion of the ＼C=O group in the carbethoxy for hydrogen undoubtedly stabilizes the enol form to some extent. Carboxylic acids and
oximes frequently have molecular weights that we can account for
best by assuming some dimerization of the molecules, as shown in
the chelated rings above.

Chelation may occur in compounds without hydrogen-bridge
formation and may account for the predominance of *ortho* rather
than *para* substitution in such reactions as the Kolbe synthesis,
the Reimer-Tiemann reaction, and the Claisen rearrangement.
This does not necessarily imply that these reactions cannot go by
any other route.

In the chelation mechanism for the Kolbe synthesis (Exp. 28-33)
a double bond from the ring takes a part. The curved arrows indicate the electron shifts that are necessary if the transformation is
to proceed.

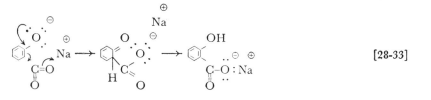

[28-33]

The Reimer-Tiemann reaction may be written in a comparable
fashion (Exp. 28-34).

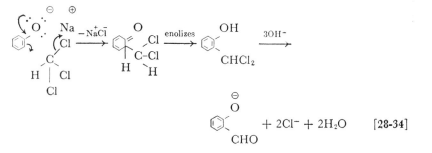

[28-34]

E. *The Claisen Rearrangement*

(See Chap. 29 also.) When allyl phenyl ether is kept at its boiling point for a long time, the boiling point slowly rises because of the formation of a new substance, *o*-allylphenol. The rearrangement is intramolecular and may be represented by an intermediate chelation-type process involving a double bond from the benzene ring.

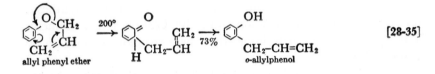

[28-35]

F. *Diazomethane on Enols and Phenols*

The close relation of enols and phenols has already been depicted in their reactions with Br_2 (Exp. 7-43 and 17-8) and with Fe^{+3} (Exp. 7-41 and p. 335). Another reagent that behaves similarly toward phenols and enols is diazomethane. An *O*-methyl ether of acetoacetic ester can be prepared with this unusually powerful methylating agent. It also reacts with phenols and carboxylic acids but not with alcohols or amines.

$$CH_3-\overset{OH}{\underset{}{C}}=CH-\overset{O}{\underset{}{C}}-OC_2H_5 + CH_2{=}\overset{+}{N}{=}\overset{-}{N} \longrightarrow CH_3-\overset{OCH_3}{\underset{CH-\overset{O}{\underset{}{C}}-OC_2H_5}{C}} + N_2 \qquad [28\text{-}36]$$

$$\langle\!\!\bigcirc\!\!\rangle\!\!-\!\!\overset{}{\underset{OH}{C}}H_3 + CH_2{=}\overset{+}{N}{=}\overset{-}{N} \longrightarrow \langle\!\!\bigcirc\!\!\rangle\!\!-\!\!\overset{}{\underset{OCH_3}{C}}H_3 + N_2 \qquad [28\text{-}37]$$

$$\langle\!\!\bigcirc\!\!\rangle\!\!-\!COOH + CH_2{=}\overset{+}{N}{=}\overset{-}{N} \longrightarrow \langle\!\!\bigcirc\!\!\rangle\!\!-\!COOCH_3 + N_2 \qquad [28\text{-}38]$$

Diazomethane is a poisonous gas, of boiling point $-24°C$, which is usually handled (with care) in ether or benzene solution. It is prepared at the time of use from methylurethane. Ethyl chlorocarbonate is treated with methyl amine, and the methylurethane is diazotized and hydrolyzed by KOH in the sequence shown in Expression 28-39.

$$ClCOOC_2H_5 + CH_3NH_2 \xrightarrow[90\%]{} \underset{\text{methyl urethane}}{CH_3NH\text{-}COOC_2H_5} \xrightarrow[H^+]{HONO}$$

$$\underset{\overset{|}{N=O}}{CH_3\text{-}N\text{-}COOC_2H_5} + KOH \longrightarrow [CH_3\text{-}N{=}N\text{-}OK] + C_2H_5OH + K_2CO_3$$

$$\downarrow$$

$$KOH + \underset{\text{diazomethane}}{CH_2{=}\overset{\oplus}{N}{=}\overset{\ominus}{N}} \qquad [28\text{-}39]$$

Exercise 28-6.   Write the electronic structure of diazomethane implied in Expression 28-39. Are there other likely structures for the molecule?

G. *Oxidation*

Phenols in general are easily oxidized compounds, and cresols, we have seen (p. 473), must be transformed into ethers before the side chain can be oxidized. Nitrations are carried out either on the ethers, which are then converted back to phenols, or else directly, at low temperature, on the phenols.

The dihydroxy benzenes are more readily oxidized than phenols, and the oxidation products have interesting properties. Mild oxidation of catechol and hydroquinone yields diketones that have lost much of their aromatic character. These diketones are called *o*-

OH
| OH   OH   OH

OH   OH
catechol   resorcinol   hydroquinone

and *p*-quinone, respectively, or the latter may simply be called quinone. Resorcinol is oxidized, but not so readily, and it cannot form a quinone.

$$[28\text{-}40]$$

$$[28\text{-}41]$$

Exercise 28-7. Balance the oxidation-reduction reaction implied in Expression 28-41, using $KBrO_3$ as the oxidizing agent in acid solution.

Aniline may be oxidized to quinone and NH_3 with sodium dichromate in acid solution.

Exercise 28-8. Write the equation for the reaction of aniline and dichromate ion in acid solution, and balance it.

1. Quinhydrone. If two alcohol solutions containing equimolecular quantities of quinone (yellow) and hydroquinone (white), respectively, are mixed, the solution immediately darkens, and soon dark-green crystals of a substance called quinhydrone separate. The original compounds have not undergone profound changes and may be recovered from the new product. Such loosely bound substances are called molecular compounds and are associated in some manner not entirely clear but certainly not involving ordinary covalent bonds. The dot in the formula

$$[28\text{-}42]$$

signifies that there is some bonding between the two parts. Many such compounds are known in which other quinones may be the first part and other phenols, phenolic ethers, etc., may be the second part. They are all highly colored substances, easily reduced or oxidized, and many are too unstable to be recrystallized. Picrates (footnote, p. 609) are also molecular compounds.

2. Photographic Developers. The ease of oxidation of the *o*- and *p*-dihydroxy compounds is paralleled by the corresponding amines and by *o*- and *p*-amino phenols. These last compounds are most frequently used as photographic developers. Metol, *p*-methyl-aminophenol sulfate, is one of the common ones. Metol and other photographic developers reduce silver bromide and silver iodide to metallic silver if these salts have first been exposed to light, but leave unexposed silver salts unaffected. The photographic film is an emulsion of these silver salts in gelatin bonded to a sheet of cellulose acetate. After the film is exposed to light (Exp. 28-43), it is developed (Exp. 28-44), and then the negative is fixed with "hypo," or sodium thiosulfate, which dissolves the unaffected silver salts (Exp. 28-45). The positive is made by exposing the negative to light, casting its shadow on printing paper (similar to the photographic film just described), and developing and fixing the paper. The asterisk placed by the silver in Expression 28-43 means that

$$AgBr + h\nu \longrightarrow Ag^*Br \qquad\qquad [28\text{-}43]$$

$$2Ag^*Br + CH_3\text{-}NH\text{-}\langle\bigcirc\rangle\text{-}OH + 2OH^- \xrightarrow[\text{base}]{\text{mild}}$$

$$2Ag + CH_3\text{-}N=\langle\bigcirc\rangle=O + 2Br^- + 2H_2O \qquad [28\text{-}44]$$

$$AgBr + S_2O_3^{=} \longrightarrow Br^- + (AgS_2O_3)^- \qquad\qquad [28\text{-}45]$$

this silver has been affected in some way by the light so that it is easily reduced to metallic silver. Without this action of light, the second reaction does not proceed.

3. Quinones. The most important property of *p*-quinone is the ease of reduction to hydroquinone. It behaves like a ketone in forming an oxime or a dioxime. Bromine will add to its double bonds as if they were olefinic. The conjugated nature of *p*-quinone

$$O=\langle\bigcirc\rangle=O + NH_2OH \longrightarrow O=\langle\bigcirc\rangle=NOH \xrightarrow{NH_2OH}$$

$$HON=\langle\bigcirc\rangle=NOH \qquad [28\text{-}46]$$

$$O=\langle \rangle=O + Br_2 \longrightarrow \underset{42\%}{O=\langle \rangle=O} \xrightarrow{Br_2} O=C\underset{CHBr-CHBr}{\overset{CHBr-CHBr}{\diagup}}C=O \qquad [28\text{-}47]$$

is demonstrated by the 1,4-addition of HCl. The chlorohydro-quinone formed may be oxidized to a chloroquinone and the addi-

$$O=\langle \rangle=O + HCl \longrightarrow \underset{93\%}{HO-\langle \rangle=O} \xrightarrow{} HO-\langle \rangle-OH \qquad [28\text{-}48]$$

tion of HCl repeated. The reaction of *p*-quinone in the Diels-Alder has been discussed already (p. 612).

H. Aminolysis

The hydroxyl group in phenol cannot be replaced directly by an amine group, but in a more reactive ring aminolysis and hydrolysis are reversible reactions. The *p*-methylaminophenol mentioned previously (p. 623) is made commercially from hydroquinone by heating with methyl amine.

$$\underset{\text{hydroquinone}}{HO-\langle \rangle-OH} + CH_3NH_2 \xrightarrow[73\%]{200°} \underset{p\text{-methylaminophenol}}{CH_3-NH-\langle \rangle-OH} + H_2O \qquad [28\text{-}49]$$

The naphthol ring being more reactive than the benzene ring, a second group is not needed to activate it. The –OH group may be replaced by –NH$_2$ simply by heating with ammonia and sodium hydrogen sulfite solution. The reverse reaction goes by boiling the resulting amine in aqueous sodium hydrogen sulfite. This is the Bucherer reaction.

$$\underset{\beta\text{-naphthol}}{\overset{OH}{\bigodot}} \underset{H_2O + NaHSO_3}{\overset{NH_3 + NaHSO_3}{\underset{\xleftarrow{\hspace{1cm}}}{\xrightarrow[96\%]{\hspace{1cm}}}}} \underset{\beta\text{-naphthylamine}}{\overset{NH_2}{\bigodot}} \qquad [28\text{-}50]$$

Exercise 28-9. Write equations for the synthesis of α-naphthol, β-naphthol, and α-naphthylamine by a method other than the Bucherer reaction.

28-7. Synthesis of Aromatic Carbonyl Compounds

Several methods of preparing aromatic aldehydes and ketones have been given previously. (Review Chaps. 11 and 12.) Besides the general methods that apply to both aliphatic and aromatic compounds, the Friedel-Crafts reaction is applicable to aromatic ketones, the Reimer-Tiemann reaction to phenolic aldehydes, and the Fries reaction to phenolic ketones.

1. Gattermann-Koch Reaction.   Adaptations of the Friedel-Crafts reaction have been made for the synthesis of aromatic aldehydes. In the presence of CuCl and $AlCl_3$, carbon monoxide and dry HCl will react with an aromatic hydrocarbon to yield an aldehyde. In effect, the CO + HCl corresponds to the unknown acid chloride that would be called formyl chloride, $H-C\overset{\displaystyle O}{\underset{\displaystyle Cl}{\big\backslash}}$.

$$CH_3-\hspace{-2pt}\langle\_\rangle + CO + HCl \xrightarrow[51\%]{AlCl_3 + CuCl} \underset{p\text{-tolualdehyde}}{CH_3-\hspace{-2pt}\langle\_\rangle-CHO} \qquad [28\text{-}51]$$

2. Gattermann Reaction.   A second modification of the Friedel-Crafts reaction may be extended to the preparation of phenolic aldehydes and aromatic ether aldehydes. Although the reaction appears to involve the formation of complexes with the metal halides, here it will be shown as a reaction analogous to the Gattermann-Koch. In the presence of $AlCl_3$, HCN and HCl will react to form an imino compound, which is readily hydrolyzed to an aldehyde.

$$H-C{\equiv}N + HCl \xrightarrow{AlCl_3} \left[H-\overset{\overset{\displaystyle NH}{\|}}{C}-Cl \right] \qquad [28\text{-}52]$$

$$CH_3-O-\hspace{-2pt}\langle\_\rangle + \left[H-C\overset{\displaystyle NH}{\underset{\displaystyle Cl}{\big\langle}} \right] \xrightarrow{AlCl_3} CH_3O-\hspace{-2pt}\langle\_\rangle-\overset{\overset{\displaystyle NH\cdot HCl}{\|}}{C}-H \xrightarrow[100\%]{H_2O} \underset{p\text{-anisaldehyde}}{CH_3O-\hspace{-2pt}\langle\_\rangle-CHO}$$
$$[28\text{-}53]$$

The HCN can be formed *in situ* with $Zn(CN)_2$ and HCl. The zinc chloride formed will act as catalyst on phenols that react in the presence of a milder Friedel-Crafts catalyst than $AlCl_3$. Not having to prepare anhydrous HCN is a further advantage since it is troublesome to handle. Resorcaldehyde is prepared by this method.

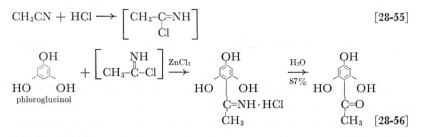

3. Hoesch Reaction. Phenolic ketones may be prepared by still a third modification of the Friedel-Crafts reaction. A nitrile will condense with a phenol in the presence of anhydrous $ZnCl_2$ and HCl in a manner analogous to that just shown for the preparation of resorcaldehyde. A popular antiseptic, n-hexylresorcinol, can be synthesized by the Hoesch reaction.

$$CH_3CN + HCl \longrightarrow \left[CH_3-\underset{Cl}{\overset{}{C}}=NH \right] \qquad [28\text{-}55]$$

Exercise 28-10. Show how to make n-hexylresorcinol, using the Hoesch reaction as part of the synthesis.

28-8. Reactions of Aromatic Carbonyl Compounds

Many of the reactions that are characteristic of both aliphatic and aromatic aldehydes have been discussed in Chapters 11, 12, and 23.

Exercise 28-11. Write an example of each of the following reactions: Perkin, Knoevenagel, Michael, Dieckmann, Meerwein-Ponndorf-Verley, Claisen-Schmidt, Cannizzaro, Strecker, and Clemmensen.

A few other important reactions, some of which apply only to aromatic aldehydes and ketones, will be given here to enhance our knowledge of carbonyl compounds.

1. Benzoin Condensation, Benzil, Benzilic Acid Rearrangement. Whereas strong alkalis (50%) will catalyze the reaction of an aromatic aldehyde to an intermolecular oxidation-reduction (the Cannizzaro reaction), milder alkalis—for example, aluminum

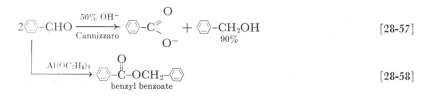

[28-57]

[28-58]

ethoxide—will yield an ester. The change in oxidation number is the same, however, in the two cases: one aldehyde group is oxidized to a carboxylate function, and the other is reduced to an alcohol.

Alkali cyanides, also basic reagents, cause a dimerization in which a new carbon-carbon bond is formed. The fact that these

$$\text{H} \quad \text{H} \quad \xrightarrow[\text{in alcohol}]{\text{NaCN}} \quad \text{H}$$

⬡-C=O + ⬡-C=O $\xrightarrow{83\%}$ ⬡-C—C-⬡
                                 OH O

benzoin   [28-59]

specific alkaline substances catalyze the benzoin condensation may be accounted for by the mechanism postulated in Expression 28-60. It may be that the cyanide ion has just the proper degree of

⬡-C=O + CN⁻ ⟶ [⬡-C-O⊖ ⇌ ⬡-C-OH + ⬡-C-O ⟶
        H   H   H   H
                       CN   CN   ⊕ ⊖

                       (a)   (b)

  CN H   −CN⁻   ⊕ H   H
⬡-C—C-⬡ ⟶ ⬡-C—C-⬡] ⟶ ⬡-C-C-⬡   [28-60]
  OH O⊖   OH O⊖   O OH

alkalinity to allow the proton in *a* to be transferred to the oxygen, leaving the charge on the carbon as shown in *b*. Structure *b* would then react with a new molecule of benzaldehyde, and the remaining reactions follow.

The keto alcohol, benzoin, is both easily oxidized and easily reduced. It can be reduced to the dialcohol with sodium amalgam in alcohol, and it can be oxidized to the diketone by potassium bromate or nitric acid.

$$[28\text{-}61]$$

Exercise 28-12. Write the three reactions indicated in the preceding paragraph, and balance the equations.

Benzil undergoes a base-catalyzed rearrangement to benzilic acid in alcoholic KOH. A possible mechanism for the rearrangement is discussed on page 638.

$$[28\text{-}62]$$

2. Auto-oxidation. Aldehydes are readily oxidized. Benzaldehyde, for example, is oxidized on standing in air to benzoic acid. This oxidation by the oxygen of the air has been given the name "auto-oxidation" (frequently shortened to "autoxidation"). It occurs on compounds other than aldehydes—for example, ethers, oils and fats, and many hydrocarbons.

The auto-oxidation of benzaldehyde occurs in a complex set of reactions, the first stable intermediate of which is perbenzoic acid. The reaction involves free radicals (pp. 657–665), probably both in the formation of perbenzoic acid and in the subsequent oxidation of the second molecule of benzaldehyde to benzoic acid. Sunlight catalyzes the addition of oxygen. Only the isolable intermediates are shown here.

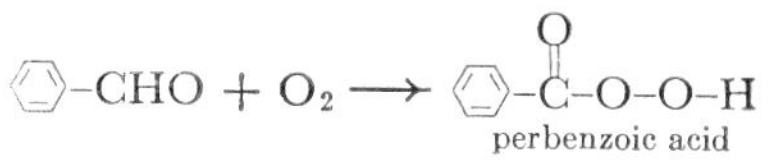

$$\bigcirc\text{-CHO} + O_2 \longrightarrow \bigcirc\text{-}\overset{O}{\overset{\|}{C}}\text{-O-O-H} \qquad\qquad [28\text{-}63]$$
perbenzoic acid

$$\bigcirc\text{-}\overset{O}{\underset{OOH}{C\big<}} + \bigcirc\text{-CHO} \longrightarrow 2\bigcirc\text{-COOH} \qquad\qquad [28\text{-}64]$$

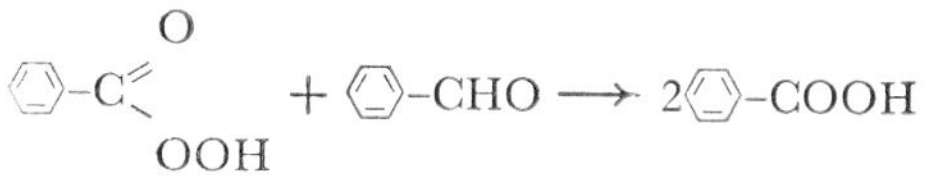

Even distilling benzaldehyde in air without special precautions to remove oxygen allows the formation of some benzoic acid. It will be recalled that a trace of hydroquinone was added to benzaldehyde in the laboratory to act as an antioxidant when it was allowed to stand overnight. Hydroquinone interrupts the formation of free radicals as intermediates in a chain reaction.

3. Mannich Reaction. The Mannich reaction is a general reaction of formaldehyde and NH_3, or of a primary or secondary amine, with a compound containing an active hydrogen. The active hydrogen may be on an active methylene group in malonic ester, on a methyl group in a methyl ketone, or on a carbon that is α to some aldehyde group. Acetophenone, for example, will react with dimethylamine hydrochloride and formaldehyde to give dimethylaminoethyl phenyl ketone.

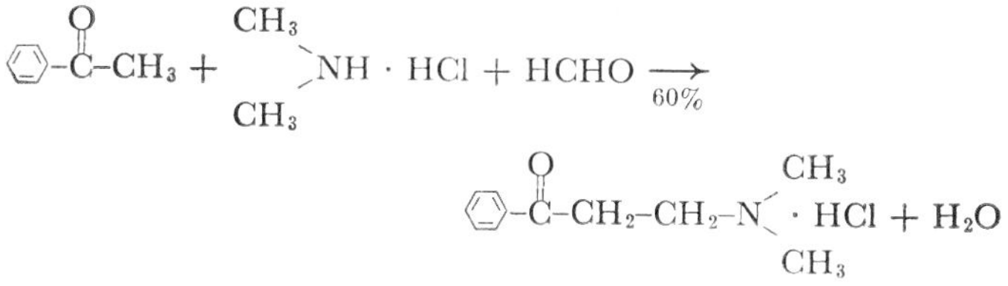

$$\bigcirc\text{-}\overset{O}{\overset{\|}{C}}\text{-CH}_3 + \overset{CH_3}{\underset{CH_3}{>}}\text{NH} \cdot \text{HCl} + \text{HCHO} \underset{60\%}{\longrightarrow}$$

$$\bigcirc\text{-}\overset{O}{\overset{\|}{C}}\text{-CH}_2\text{-CH}_2\text{-N}\overset{CH_3}{\underset{CH_3}{<}} \cdot \text{HCl} + \text{H}_2\text{O} \qquad [28\text{-}65]$$

The reaction is a useful synthetic tool for making unsaturated ketones, ketones with one more carbon than the original, β-amino alcohols, and many complex substances. For example, if the amine hydrochloride above is heated, it decomposes into dimethylamine hydrochloride and phenyl vinyl ketone. The unsaturated ketone is

$$\bigcirc\!\!-\overset{\overset{\displaystyle O}{\|}}{C}\!-CH_2\!-CH_2\!-N\!\!\begin{array}{c} CH_3 \\ \end{array}\!\cdot HCl \longrightarrow \bigcirc\!\!-\overset{\overset{\displaystyle O}{\|}}{C}\!-CH\!=\!CH_2 + \begin{array}{c} CH_3 \\ \end{array}\!\!NH \cdot HCl \quad [28\text{-}66]$$

phenyl vinyl ketone

a conjugated system and may be reduced to the saturated ketone. It will also undergo a Michael condensation. A better yield of the 1,4-addition product is obtained in many cases when the unsaturated ketone is not isolated.

Exercise 28-13. Write the Michael condensation for phenyl vinyl ketone and acetoacetic ester.

The β-amino ketones may be reduced readily, by zinc dust and HI or by hydrogen with a palladium-on-charcoal catalyst, to γ-amino alcohols, many of which are useful as local anesthetics. In methyl ethyl ketone the reactive hydrogen is the one on the

$$CH_3\!-\!\overset{\overset{\displaystyle O}{\|}}{C}\!-CH_2\!-CH_3 + HCHO + \begin{array}{c} CH_3 \\ \end{array}\!\!NH \cdot HCl \longrightarrow$$

$$CH_3\!-\!\overset{\overset{\displaystyle O}{\|}}{C}\!-\underset{\underset{\displaystyle CH_3}{|}}{C}H\!-CH_2\!-N\!\!\begin{array}{c} CH_3 \\ \end{array}\!\cdot HCl \overset{(H)}{\longrightarrow} CH_3\!-\!\underset{\underset{\displaystyle CH_3}{|}}{\overset{\overset{\displaystyle OH}{|}}{C}}H\!-\underset{}{C}H\!-CH_2\!-N\!\!\begin{array}{c} CH_3 \\ \end{array} \quad [28\text{-}67]$$

$-CH_2-$ group. When the product from this Mannich reaction is reduced to the amino alcohol, the latter can be esterified with p-aminobenzoic acid to yield the local anesthetic known as Tutocaine (as the hydrochloride).

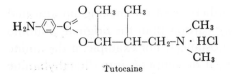

$$[28\text{-}68]$$

Tutocaine

SUMMARY

In defining aromatic character, we have emphasized the endowment of the ring itself rather than the four common substitution reactions of aromatic substances. Nearly every reaction mentioned in the chapter carries implicit significance with regard to the ease of formation of the aromatic ring or the behavior of some functional group on the ring. Various aromatic hydrocarbons are compared as either less or more aromatic than benzene itself, the standard.

The following named reactions are discussed: Diels-Alder, Fries, Kolbe, Reimer-Tiemann, Claisen rearrangement, Bucherer, Gattermann, Gattermann-Koch, Hoesch, and Mannich.

REFERENCE

Evans, "Recent Advances in Petrochemicals," J. Chem. Education, 32, 242 (1955).

EXERCISES

(Exercises 1–13 will be found within the text.)

Synthesize the following compounds by the best available methods:

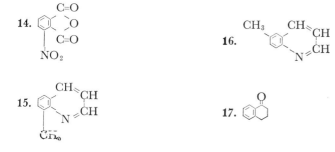

18.

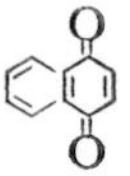

19.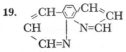

20. 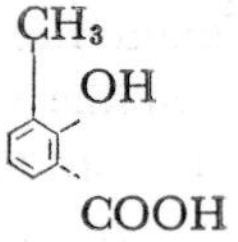
CH₃
OH
COOH

CH_3 / OH / $COOH$ substituted benzene

21. 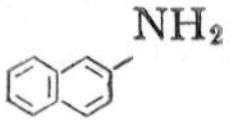
NH₂ — naphthalene

NH_2

22. 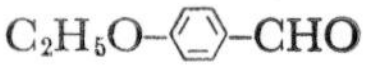 C_2H_5O—⟨⟩—CHO

23. HO—⟨⟩—
OH
$\underset{O}{\overset{OH}{C}}$—$C_5H_{11}$

24.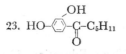

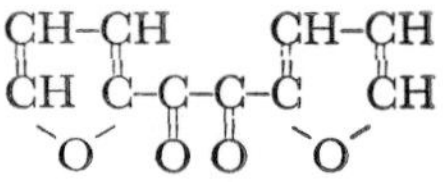

25.  ⟨⟩—CH—CH—⟨⟩
    OH   OH

26. 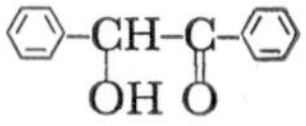 ⟨⟩—CH—C—⟨⟩
    OH   O

27. 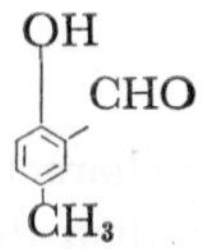
OH
CHO
CH₃

28. 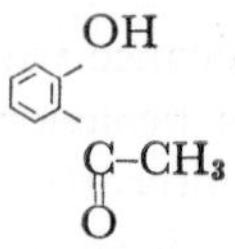
OH
C—CH₃
O

29. 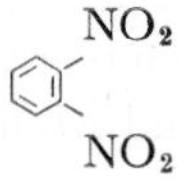
NO₂
NO₂

30. 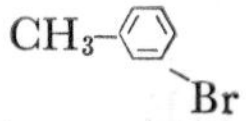 CH₃—⟨⟩
         Br
(from *p*-toluidine)

31. 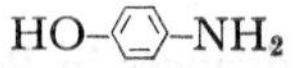 HO—⟨⟩—NH₂

32. 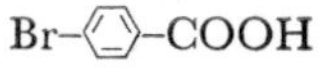 Br—⟨⟩—COOH

33. 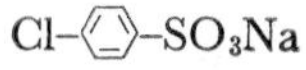 Cl—⟨⟩—SO₃Na

34. 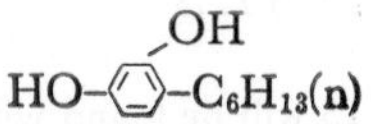 HO—⟨⟩—C₆H₁₃(n)
         OH

29 Rearrangements

The principle of *economy of explanation* has been implicit throughout the history of chemistry. In studying the chemistry of functional groups, for example, the chemist assumes that displacements of the OH group in alcohols take place at the site of the OH group and not at some other carbon in the chain. This invocation of the principle of economy of explanation is justified most of the time, but mistakes have been made because of it. For a while, after mistakes of this kind are observed, the behavior of such compounds is called exceptional. After enough exceptions to a rule are garnered by experience, a new *theory* may be forthcoming. In the chemistry of functional groups, when the replacing group does not appear at the site of the original function, the reaction is referred to as a rearrangement.

29-1. The Whitmore Theory of Rearrangements

In 1932, Whitmore proposed a theory of rearrangements that brought several known reactions under a single explanation. Whitmore's theory, a very general statement, is now popular with organic chemists because it relates many reactions that otherwise appear to be unrelated. More detailed pictures of the rearrangements to be discussed are available but do not warrant consideration here. The evidence for the mechanism of some rearrangements is better than for that of others, but the final word has probably not been said on any of the examples given. However, greater detail is not likely to invalidate Whitmore's general statement of the theory.

A number of substances that rearrange can be classed in one of

633

two categories represented by the following electronic structures:

$$:A:B:X: \quad \text{and} \quad :A::B:D:X:$$

Atoms A, B, and D are such atoms as carbon and nitrogen, neither of which is strongly electropositive or strongly electronegative; atom X is a strongly electronegative atom such as oxygen or a halogen.

The Case of $:A:B:X:$

If we postulate that the first step in a reaction involving molecule $:A:B:X:$ is the separation of X along with its eight accompanying electrons, leaving B with six electrons, then one of the four following events may take place.

$$:A:B:X: \longrightarrow :A:B + :X: \qquad \qquad [29\text{-}1]$$

1. Exchange. If X returns to B, an equilibrium will be established, and no net change will be apparent. Atom X may not return to the B to which it was originally attached, however, in which case the equilibrium includes an exchange. This reaction can be detected, not by chemical means, but by physical means using heavy isotopes or radioactive isotopes. Otherwise this event is not very exciting.

2. Displacement. If a second group, Y, with available pairs of electrons, is present in the reaction mixture, it may replace X in the original molecule, a process called a displacement reaction.

$$:A:B + :Y: \longrightarrow :A:B:Y: \qquad \qquad [29\text{-}2]$$

An example of this type of displacement is the reaction of an alcohol with HBr or PCl_5 or the reaction of an alkyl halide with OH^-

or NH_3. In fact, most of the reactions of this textbook are of the displacement type.

A B X   Y   A B Y

$$H : \overset{\overset{H}{\cdot\cdot}}{\underset{\underset{H}{\cdot\cdot}}{C}} : \overset{\overset{H}{\cdot\cdot}}{\underset{\underset{H}{\cdot\cdot}}{C}} : \overset{\cdot\cdot}{O} : H + \quad H \quad : \overset{\cdot\cdot}{\underset{\cdot\cdot}{Br}} : \quad \longrightarrow \quad H : \overset{\overset{H}{\cdot\cdot}}{\underset{\underset{H}{\cdot\cdot}}{C}} : \overset{\overset{H}{\cdot\cdot}}{\underset{\underset{H}{\cdot\cdot}}{C}} : \overset{\cdot\cdot}{\underset{\cdot\cdot}{Br}} : + H_2O \qquad [29\text{-}3]$$

$$H : \overset{\overset{H}{\cdot\cdot}}{\underset{\underset{H}{\cdot\cdot}}{C}} : \overset{\overset{H}{\cdot\cdot}}{\underset{\underset{H}{\cdot\cdot}}{C}} : \overset{\cdot\cdot}{\underset{\cdot\cdot}{O}} : H + Cl_4P \quad : \overset{\cdot\cdot}{\underset{\cdot\cdot}{Cl}} : \quad \longrightarrow \quad H : \overset{\overset{H}{\cdot\cdot}}{\underset{\underset{H}{\cdot\cdot}}{C}} : \overset{\overset{H}{\cdot\cdot}}{\underset{\underset{H}{\cdot\cdot}}{C}} : \overset{\cdot\cdot}{\underset{\cdot\cdot}{Cl}} : + POCl_3 + HCl \quad [29\text{-}4]$$

$$H : \overset{\overset{H}{\cdot\cdot}}{\underset{\underset{H}{\cdot\cdot}}{C}} : \overset{\overset{H}{\cdot\cdot}}{\underset{\underset{H}{\cdot\cdot}}{C}} : \overset{\cdot\cdot}{\underset{\cdot\cdot}{Br}} : + \quad : \overset{\cdot\cdot}{\underset{\cdot\cdot}{O}} : H^- \longrightarrow H : \overset{\overset{H}{\cdot\cdot}}{\underset{\underset{H}{\cdot\cdot}}{C}} : \overset{\overset{H}{\cdot\cdot}}{\underset{\underset{H}{\cdot\cdot}}{C}} : \overset{\cdot\cdot}{\underset{\cdot\cdot}{O}} : H + Br^- \qquad [29\text{-}5]$$

$$H : \overset{\overset{H}{\cdot\cdot}}{\underset{\underset{H}{\cdot\cdot}}{C}} : \overset{\overset{H}{\cdot\cdot}}{\underset{\underset{H}{\cdot\cdot}}{C}} : \overset{\cdot\cdot}{\underset{\cdot\cdot}{Br}} : + \quad H \quad : \overset{\cdot\cdot}{N} : H \longrightarrow H : \overset{\overset{H}{\cdot\cdot}}{\underset{\underset{H}{\cdot\cdot}}{C}} : \overset{\overset{H}{\cdot\cdot}}{\underset{\underset{H}{\cdot\cdot}}{C}} : \overset{\cdot\cdot}{\underset{H}{N}} : H + [HBr] \qquad [29\text{-}6]$$
$$\qquad\qquad\qquad\qquad\qquad\qquad\quad H \qquad\qquad\qquad\qquad H\ \ H\ \ H$$

3. β Elimination.   If a hydrogen atom is attached to A in the fragment $: \overset{\cdot\cdot}{A} : \overset{\cdot\cdot}{B}$, the loss of X from the original molecule may be followed by loss of a proton, leaving an alkene. This type of process probably occurs in the dehydration of an alcohol by sulfuric acid, aluminum oxide, or other dehydrating agent. The proton is kicked out from the carbon that is β to the original group X.

A B X   A B

$$H : \overset{\overset{H}{\cdot\cdot}}{\underset{\underset{H}{\cdot\cdot}}{C}} : \overset{\overset{H}{\cdot\cdot}}{\underset{\underset{H}{\cdot\cdot}}{C}} : \overset{\cdot\cdot}{\underset{\cdot\cdot}{O}} : SO_3H \xrightarrow{-OSO_3H^-} H : \overset{\overset{H}{\cdot\cdot}}{\underset{\underset{H}{\cdot\cdot}}{C}} : \overset{\overset{H}{\cdot\cdot}}{\underset{\underset{H}{\cdot\cdot}}{C}} \xrightarrow{-H^+} CH_2=CH_2 \qquad [29\text{-}7]$$

4. Rearrangement.   If, in the fragment $: \overset{\cdot\cdot}{A} : \overset{\cdot\cdot}{B}$, B has a greater attraction for electrons than A, a shift of an electron pair to B may occur, and this may be followed by events 2 and 3 mentioned above.

.

$$: \overset{..}{\underset{..}{A}} : \overset{..}{\underset{..}{B}} \longrightarrow \overset{..}{\underset{..}{A}} : \overset{..}{\underset{..}{B}} : \qquad [29\text{-}8]$$

If another atom or group is attached to A, the shift to B will include that atom or group along with its electron pair. The result is a rearrangement of the type exemplified in the pinacol-pinacolone rearrangement, the Hofmann hypobromite reaction (pp. 301 and 637), the benzilic acid rearrangement (pp. 628 and 638), and the Beckmann rearrangement (p. 639).

In the following examples of rearrangement of the ABX type, the forest can easily be obscured by the trees. The important thing is to watch for the atoms A, B, and X.

29-2. Pinacol-Pinacolone Rearrangement

Pinacol is prepared by the bimolecular reduction of acetone with a mild reducing agent such as magnesium amalgam and water (see p. 226). If the pinacol is then treated with a strong acid such as hydrochloric, the pinacol rearranges to pinacolone. The first step appears to be loss of water after association with the proton. The carbonium ion left at this point may be considered the fragment $: \overset{..}{\underset{..}{A}} : \overset{..}{\underset{..}{B}}$ in Whitmore's generalized picture of rearrangements.

$$[29\text{-}9]$$

Atom B is the carbon carrying the positive charge (in structure *a* of Expression 29-9), which has only six electrons in its outside shell. Atom A is the adjacent carbon, carrying two methyl groups and an OH group. Whitmore indicates the second step (to b) as a shift of a methyl group with its electron pair from atom A to atom B. The new fragment, $: A : B :$, is relieved of the deficiency of electrons on A by loss of a proton from the OH group, resulting in the product pinacolone.

Benzophenone undergoes this pinacol reduction by a very mild reducing agent, isopropyl alcohol in the presence of sunlight. The fact that this reaction goes implies that benzophenone is easier to reduce than acetone or, conversely, that acetone is a better oxidizing agent than benzophenone.

$$2\bigcirc\text{-}\underset{\underset{O}{\|}}{C}\text{-}\bigcirc + 2CH_3\text{-}\underset{\underset{OH}{|}}{C}H\text{-}CH_3 \xrightarrow[94\%]{h\nu} \underset{\underset{OH}{|}\;\;\underset{OH}{|}}{\bigcirc\text{-}\overset{\bigcirc}{C}\text{---}\overset{\bigcirc}{C}\text{-}\bigcirc} + 2CH_3\text{-}\underset{\underset{O}{\|}}{C}\text{-}CH_3 \quad [\textbf{29-10}]$$

benzopinacol

29-3. Hofmann Hypobromite Reaction

The Hofmann hypobromite reaction on an amide is depicted in the following sequence according to the Whitmore theory. First, a bromine replaces a hydrogen on the nitrogen in the amide.

$$R\text{-}\underset{\underset{O}{\|}}{C}\text{-}\underset{\overset{H}{..}}{N}:H + NaOBr \xrightarrow{OH^-} R\text{-}\underset{\underset{O}{\|}}{C}\text{-}\underset{\overset{H}{..}}{N}:Br \xrightarrow{OH^-} \left[R\text{-}\underset{\underset{O}{\|}}{C}\text{-}\underset{..}{\overset{\ominus}{N}}:Br\right] \xrightarrow{-Br^-}$$

$$\left[R\text{-}\underset{\underset{O}{\|}}{C}\text{-}\underset{..}{N}\right] \xrightarrow[\text{shifts}]{R:} \left[\underset{\underset{\oplus\;\;\ominus}{}}{\overset{\overset{O}{\|}}{C}}\text{-}\underset{..}{N}:R\right] \longrightarrow R\text{-}N=C=O \xrightarrow{HOH}$$

$$(a) \qquad\qquad\qquad (b)$$

$$\left[R:\underset{\overset{..}{H}}{\overset{..}{N}}:C\underset{OH}{\overset{\nwarrow}{\diagup^O}}\right] \xrightarrow{-CO_2} RNH_2 \qquad\qquad [\textbf{29-11}]$$

In the strong basic solution, the remaining hydrogen is lost to the OH^-. From this point we leave facts and account for the final results by hypothetical intermediates. The next step is shown as a loss of Br⁻, giving the neutral fragment *a*, in which nitrogen has six electrons. Nitrogen is more electronegative than carbon; so the Whitmore shift of a pair of electrons to nitrogen is reasonable. If the R group shifts with the electron pair, the Hofmann hypobromite reaction is accounted for. The product, R–N=C=O, resulting from

$$: \overset{::}{A} : \overset{..}{B} \longrightarrow \overset{::}{A} : \overset{..}{B} :$$

$$R : \overset{O}{\underset{..}{C}} : \overset{..}{N} \longrightarrow \overset{O}{\underset{..}{C}} : \overset{..}{N} : R \longrightarrow O = C = \overset{..}{N} : R \qquad [29\text{-}12]$$

the shift of the alkyl group with its electron pair, R :, is called an isocyanate. Isocyanates have been isolated from the hypobromite reaction and are stable compounds themselves. In strong basic solution, however, isocyanates yield amines and a carbonate (shown as [RNHCOOH] followed by loss of CO_2).

29-4. Benzilic Acid Rearrangement

The benzilic acid rearrangement also fits into the Whitmore explanation. If benzil is warmed with a strong base for a long time, it eventually goes into solution as benzilate ion. A reasonable se-

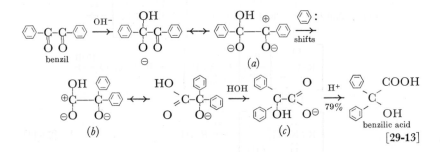

quence is presented in Expression 29-13. The first step is association of the hydroxyl ion with one carbonyl group, one resonating form of which is shown with an electropositive carbon (*a*). The Whitmore mechanism dictates a shift of a phenyl group with its pair of electrons to give structure *b*. Equilibration of *b* and its contributing resonance structures with water will yield the benzilate ion (*c*). Neutralization of the basic solution will then precipitate benzilic acid.

29-5. Beckmann Rearrangement

The double bond between carbon and nitrogen in an oxime, like other double bonds, restricts rotation. If the angle between the bonds C=N and N–OH were 180°, there would be only one form of any oxime. Since two different oximes of a number of carbonyl compounds are known, the angle must be less than 180°, producing *cis-trans* isomerism in oximes in which different groups are attached to the carbon carrying the oxime group.

$$
\begin{array}{cc}
\text{R–C–R}' & \text{R–C–R}' \\
\;\;\|\; & \;\;\| \\
\text{N–OH} & \text{HO–N}
\end{array}
$$

When oximes are treated with strong acids or (more often) with PCl$_5$ in ether, a substituted amide is formed. The reaction is known as the Beckmann rearrangement. In Expression 29-14 the curved

$$
\underset{\text{N–OH}}{\overset{\text{R–C–R}'}{C}} \xrightarrow[\text{ether}]{\text{PCl}_5} \left[\underset{\text{N–R}}{\overset{\text{HO–C–R}'}{\;}} \right] \xrightarrow[\text{shift}]{\text{enol}} \underset{\text{HN–R}}{\overset{\text{O=C–R}'}{\;}} \text{ or } \underset{\text{R}'-\text{C–NH–R}}{\overset{\text{O}}{\;}} \qquad [29\text{-}14]
$$

arrows indicate the groups that are known to move in the reaction —that is, the OH group and the group that is *trans* to it.

According to the Whitmore theory, the rearrangement takes place as in Expression 29-15. As in the other rearrangements

$$
\underset{\text{R}'\;\;\text{OH}}{\overset{\text{R}}{>}}\text{C=N} \xrightarrow[\text{lose H}_2\text{O}]{\text{add H}^+} \underset{\text{R}'}{\overset{\text{R}}{>}}\text{C=N}^{\oplus} \xrightarrow[\text{shifts}]{\text{R:}} \underset{\text{R}'\,\text{R}}{\text{C=N}}{\;}^{\oplus} \xrightarrow{\text{HOH}} \underset{\text{R}'\,\text{R}}{\overset{\text{OH}}{\text{C=N}}}{\;} \xrightarrow[\text{shift}]{\text{enol}} \underset{\text{R}'\,\text{R}}{\overset{\text{O}}{\text{C–NH}}} \qquad [29\text{-}15]
$$

$$
\qquad\qquad\qquad\qquad (a) \qquad\qquad\qquad (b)
$$

mentioned here, structures *a* and *b* show the Whitmore pattern,

$$: \overset{\cdot\cdot}{A} : : \overset{\cdot\cdot}{B} \qquad \overset{\cdot\cdot}{A} : : \overset{\cdot\cdot}{B} :$$

$$\begin{matrix} R \\ {\searrow} \\ R' \end{matrix} C{=\!=}N^{\oplus}_{\cdot\cdot} \qquad \overset{\oplus}{\underset{R'R}{C{=\!=}N:}} \qquad\qquad [29\text{-}16]$$

with the modification that the bond between A and B is a double
bond.

The Case of $: \overset{\cdot\cdot}{A} : : \overset{\cdot\cdot}{B} : \overset{\cdot\cdot}{\underset{\cdot\cdot}{D}} : \overset{\cdot\cdot}{\underset{\cdot\cdot}{X}} :$

A second group of rearrangements occur on molecules of the
type A=B–D–X (Exp. 29-17). Suppose that X separates from the
molecule A=BDX with its pair of electrons. Then the new positive
fragment may be stabilized by resonance between two canonical

$$: \overset{\cdot\cdot}{A}{=}\overset{\cdot\cdot}{B} : \overset{\cdot\cdot}{\underset{\cdot\cdot}{D}} : \overset{\cdot\cdot}{\underset{\cdot\cdot}{X}} : \longrightarrow \; : \overset{\cdot\cdot}{A}{=}\overset{\cdot\cdot}{B} : \overset{\cdot\cdot}{D}{}^{\oplus} \longleftrightarrow {}^{\oplus}: \overset{\cdot\cdot}{A} : \overset{\cdot\cdot}{B}{=}\overset{\cdot\cdot}{D} : \qquad [29\text{-}17]$$

$$(a) \qquad\qquad\qquad (b)$$

structures, *a* and *b*. If now a group Y attaches itself to fragment *a*,
a displacement reaction (event 2, p. 634) has occurred. If fragment *b* has greater stability than *a* (or even equal stability), Y may
join *b*, the result of which is a displacement with rearrangement.
The allylic rearrangement (p. 505) fits into this picture.

29-6. Claisen Rearrangement

Another quirk to liven up the picture still more is available in
the Claisen rearrangement. Suppose the fragment X, after detach-
ment from A=BDX, finds the canonical structure *b* (Exp. 29-17)
more to its liking than *a* but at a different electronic center in X.
Then a rearrangement involving two electronic shifts will have
occurred. Such an event takes place in the Claisen rearrangement.

The mechanism involving a chelate ring has already been applied to the Claisen rearrangement (p. 620), accounting for a preponderance of *ortho* isomer in the reaction. If a substituted allyl aryl ether is used, the R group will appear α to the ring, as shown in Expression 29-18, which gives added credence to the chelation mechanism.

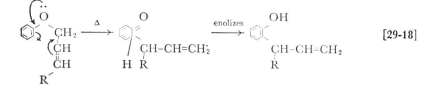

[29-18]

The Whitmore generalized picture of this chelation mechanism is shown electronically in Expression 29-19.

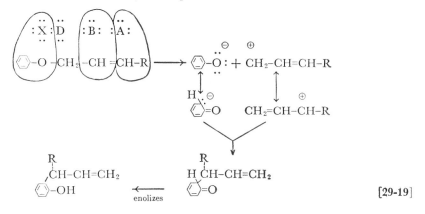

[29-19]

Both the negative fragment X, $\bigcirc\!-\!O:^{\ominus}$, and the positive fragment A=B–D, $\overset{\oplus}{CH_2}\!-\!CH\!=\!CH\!-\!R$, consist of at least two important resonance contributors. If the particular ones shown join, the formation of the new allylic phenol is accounted for (after enolization).

In the Claisen rearrangement, if both *ortho* positions are blocked, the rearrangement goes *para*, obviously not by a chelation mechanism.

SUMMARY

The generalized theory of rearrangements associated with the name of Frank C. Whitmore enables one to relate a number of reactions that appear, on superficial examination, to have little in common. The examples mentioned are the pinacol-pinacolone, Hofmann hypobromite, benzilic acid, Beckmann, and Claisen rearrangements.

EXERCISES

1. What product would be expected in the treatment of

$$(C_6H_5)_2\text{-C}\text{----}CH\text{-}CH_3$$
$$\quad\quad\quad |\quad\quad\quad |$$
$$\quad\quad\quad OH\quad NH_2$$

with nitrous acid? Account for it in terms of the Whitmore general theory of rearrangements.

2. Predict the product to be obtained by treatment of $(C_6H_5)_3C\text{-}N\text{-}OH$
$$\quad\quad\quad\quad\quad\quad\quad\quad\quad\quad\quad\quad\quad\quad\quad\quad\quad |$$
$$\quad\quad\quad\quad\quad\quad\quad\quad\quad\quad\quad\quad\quad\quad\quad\quad\quad CH_3$$

with strong acids.

3. When 2-methyl-2-phenyl-1-butanol is treated with thionyl chloride, the principal product is 2-chloro-2-methyl-1-phenylbutane. Show how this may come about.

4. The following reactions are given on page 109:

$$CH\equiv C\text{-}CH=CH_2 + HCl \longrightarrow CH_2=C=CH\text{-}CH_2Cl \xrightarrow[\text{HCl}]{\text{CuCl}} CH_2=\underset{\underset{Cl}{|}}{C}\text{-}CH=CH_2$$

Does Whitmore's theory throw any light on the formation of the 2-chloro-1,3-butadiene? If the copper ion could take part in forming a chelated intermediate, could you account for the product by this type of mechanism?

5. Action of silver acetate on 2,2-dimethyl-1-iodopropane gives

$$CH_3-\overset{\overset{O}{\|}}{C}-O-\overset{\overset{CH_3}{|}}{\underset{\underset{C_2H_5}{|}}{C}}-CH_3. \text{ How?}$$

6. If a Grignard is made from cinnamyl chloride, $C_6H_5-CH=CH-CH_2Cl$, and then treated with carbon dioxide, the main product is

$$\langle \bigcirc \rangle -\overset{|}{\underset{\underset{COOH}{|}}{C}}H-CH=CH_2.$$

Write a possible path for this reaction.

7. Write the pinacolone rearrangement for benzopinacol (Exp. 29-10) in terms of the Whitmore theory.

30 Colored Substances and Chemical Constitution

Colored things attract most people, whether the color appears in a sunset, a jewel, a bathing suit, a canyon, or a toy. The interpretation of color in terms of light phenomena has interested physicists for many years. Chemists have been interested in color because of its economic value in pigments and dyes and because of the scientific need to correlate it with chemical constitution and molecular and atomic structure. It will be assumed in this chapter that you know the elementary laws that govern the emission, transmission, and absorption of light.

If an object on which white light falls absorbs part of the wavelengths while it transmits or reflects others, it will be colored. In the case of yellow dye, the electronic structures of the chromophoric group (see p. 646) are such that the light having energy corresponding to wavelengths in the blue is absorbed by the group while the yellow is transmitted. Hence the person viewing the dye calls it yellow. The energy absorbed by the dye may be expended in small stages or re-emitted as scattered light in all directions.

30-1. Colored Elements and Small Covalent Molecules

Colored substances appear as elements, electrovalent compounds, covalent compounds, and coordinate covalent compounds. Only the constitution of covalent colored compounds will be discussed at any length here.

The elements that are colored (black, the absence of color, and white, the presence of all colors, will not be considered) are listed below. Sometimes one of the allotropic forms of an element is

544

colored while another is not—for example, red phosphorus and white phosphorus. Almost all the elements not listed here are described as gray, silvery, white, or silvery with a metallic luster.

Group Ib:

    Cu   a reddish-brown metal
    Au   a bright-yellow metal

Group Va:

    P   a red allotrope
    As   yellow and brown allotropes
    Sb   a yellow allotrope

Group VIa:

    S   yellow
    Se   red

Group VIIa:

    F   a greenish-yellow gas
    Cl   a greenish-yellow gas
    Br   a red liquid
    I   a blue-black solid, a violet vapor

Others:

    Si   a brown powder
    B   a brown or green powder

A few covalent compounds of the elements in Groups V, VI, and VII are colored. These are listed below. Two of them contain

Oxides:

Cl_2O	a yellow-red gas
ClO_2	a yellow-red gas
NO_2	a red-brown gas
N_2O_3	a red-brown gas

Sulfides:

P_4S_3	a yellow solid
P_4S_5	a yellow solid
P_4S_{10}	a yellow solid
NS_2	a red liquid
N_2S_5	a red liquid
N_4S_4	a red liquid

Halides:

PBr_5	a yellow solid
$NOCl$	an orange-yellow gas
$NOBr$	a red gas
S_2Cl_2	a light-yellow liquid
SCl_2	a red liquid
Se_2Cl_2	a red liquid
$SeCl_2$	a red liquid
Se_2Br_2	a red liquid
$SeBr_2$	a red liquid
$SeBr_4$	an orange-red solid

an odd number of electrons and are mentioned again in the section on free radicals (p. 657).

The transition elements in the Periodic Table, including Ni, Fe, Co, Cr, Mn, and Cu, form many salts that are highly colored. Most of the colored pigments used in paints are compounds of these elements. On the other hand, the coloring matter of flowers, other parts of plants, blood, furs, and hair, the dyes for cloth, leather, and synthetic materials, and indicators are all covalent compounds of carbon with other elements.

The organic colored compounds will be treated in three categories: dyes, indicators, and free radicals.

30-2. Colored Compounds: Dyes

A dye is a colored substance that will adhere to plant or animal fibers, animal tissue, or synthetic fibers. A good dye is one that will stick to one or more of these, will not be removed by many washings, will not fade upon continuous exposure to light, and is not harmed by mild heat or a mild oxidizing agent.

It has been observed that compounds containing certain functional groups are colored. These groups, called chromophores (color bearers), are usually present in an aromatic nucleus. Among the important chromophores are the following, all of which contain double bonds (the carbon-to-carbon double bond is a chromophore if multiple or combined with other groups):

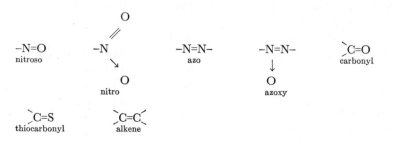

Combinations of these groups commonly appear in dyes. Though simple alkenes are not colored, conjugated systems involving alkene linkages frequently give color. For example, the *o*- and *p*-quinoid structures and other conjugated systems are colored.

In general, the presence of a chromophore in a molecule is not enough to make it a dye even though it is colored. Other groups, called auxochromes, must be present; these serve two functions, to intensify the color of the chromophore by stabilizing resonance structures, and to form bonds (probably hydrogen bridges) with the basic or acidic groups (not in cotton or linen) in the fiber. The principal auxochromes are $-NH_2$, $-NHCH_3$, $-N(CH_3)_2$, $-OH$, and $-OCH_3$, of which the first three are basic, the next one acidic, and the last neutral. The methoxy group, however, has unshared electron pairs on the oxygen and can form hydrogen bridges. In dye molecules, which are ordinarily of high molecular weight, acidic and basic groups help to make the dye soluble enough in water to be of practical use. The sulfonic acid group and the carboxyl group, for that reason, are often found in dyes, although they can scarcely be considered auxochromes.

Dyes are classified in two ways: by the chemist according to their chemical structure and by the dyer according to their method of application.

30-3. Methods of Application of Dyes

1. Direct Dyes.   Wool, silk, and nylon can be dyed directly in water solution or suspension if the dye molecule contains an acidic or basic group that will bind it to amphoteric protein substances. Very few dyes can be applied to cotton in this manner.

2. Ingrain, or Developed, Dyes.   All azo dyes are developed within the fiber when the cloth is dipped first into a diazonium salt solution and then into a solution of the coupling compound (p. 311). (The process may reverse this order.) Whereas a direct dye is coated on the outside of the fiber, the ingrain dye is developed within the fiber and hence stands a better chance of being fast to washing.

3. Mordant Dyes. Dyes containing the proper functional groups may be bound to the fiber by forming a chelated compound (p. 617) with a second substance that will itself adhere to cloth. The mordants are of two types: basic mordants (for acidic dyes), such as $Al(OH)_3$, $Cr(OH)_3$, and $Sn(OH)_4$; acidic mordants (for basic dyes), the principal one being tannic acid. The mordant is applied first, and then the chelated organo-metallic compound (with basic mordants), called a "lake," is formed by the application of the dye.

4. Vat Dyes. A reduced colorless form of a dye (a leuco compound) is applied to the cloth in a vat. Formerly the dye was reduced by a fermentation process in which the reducing action was provided by bacteria. Now sodium dithionite (hyposulfite), $Na_2S_2O_4$, is generally used as the reducing agent to produce the leuco compound. When the cloth is removed from the vat and exposed to the air, the dye is oxidized to its colored (and generally insoluble) form.

30-4. Chemical Constitution and Color

1. Azo Dyes. (See also pp. 311 and 654.) One of the few direct cotton dyes is Congo red, an azo dye obtained by coupling tetrazo-

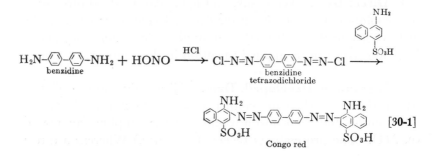

[30-1]

tized benzidine with naphthionic acid (1-amino-4-naphthalene-sulfonic acid). Congo red is also an acid-base indicator (see p. 654) that is red in basic solution and blue in acid solution.

Another interesting azo dye, known as toluidine red ("fire department red"), is made by coupling β-naphthol with diazotized 2-nitro-4-aminotoluene.

Exercise 30-1.   Write a synthesis of toluidine red.

Exercise 30-2.   Write a synthesis for the following azo dyes:

1. orange II: 2-hydroxy-1-naphthalenediazobenzene-4'-sulfonic acid
2. fast red A: 2-hydroxyazonaphthalene-4'-sulfonic acid
3. butter yellow: *p*-dimethylaminoazobenzene
4. alizarin yellow R: 4-nitro-3'-carboxy-4'-hydroxyazobenzene

2. Nitroso and Nitro Dyes.   Secondary and tertiary aromatic amines react with nitrous acid to give N-nitroso and C-nitroso compounds, respectively (see p. 308). In strong acid solution, N-nitroso-methylaniline rearranges to *p*-nitrosomethylaniline, a *dark-blue* solid, although not a dye.

$$\text{CH}_3\text{-N-}\underset{|}{\underset{\text{H}}{\bigcirc}} + \text{HONO} \xrightarrow[93\%]{} \text{CH}_3\text{-N-}\underset{|}{\underset{\text{NO}}{\bigcirc}} \xrightarrow[90\%]{\text{HCl}} \text{CH}_3\text{-}\overset{\text{H}}{\underset{|}{\text{N}}}\text{-}\bigcirc\text{-N=O} \qquad [30\text{-}2]$$

Phenol also can be nitrosated in the *para* position; β-naphthol reacts to give α-nitroso-β-naphthol, a green compound called gambine Y. The nitroso phenols are in tautomeric equilibrium with the corresponding quinone oxime. The nitrosation of β-naphthol and its tautomeric equilibrium are shown.

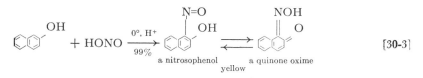

[30-3]

The use of a mordant for dyeing can be demonstrated with gambine Y. Aluminum hydroxide, for example, forms with it a green dye that may be represented as the chelated compound (Exp. 30-4). Two resonance forms are shown, and it will be recognized that the three gambine Y radicals attached to the aluminum will all have equivalent structures.

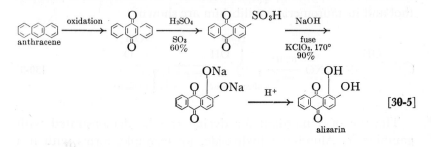

$$[30\text{-}4]$$

The student is already familiar with the yellow color in many nitro compounds, including *m*-dinitrobenzene, *p*-nitroaniline, and picric acid. The last will give a persistent yellow color to skin although it is too soluble to be used as a dye for cloth. It is also a strong acid, undesirable in a dye.

picric acid

3. Anthraquinone Dyes. A class of dyes containing the *p*-quinoid chromophore is derived from anthraquinone. Alizarin, a dihydroxy derivative of anthraquinone (Exp. 30-5), is the starting compound for a series of mordant dyes. It is prepared from anthraquinone by sulfonation and fusion of the β-sulfonic acid with alkali

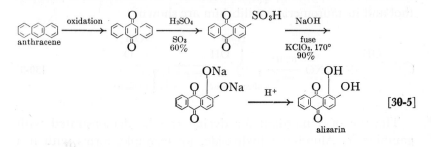

$$[30\text{-}5]$$

in the presence of an oxidizing agent. The β-sulfonic acid group is replaced by a hydroxyl group, but oxidation and fusion give a second hydroxyl group on the α position. When aluminum hydroxide is the mordant, alizarin gives a bright red dye known as Turkey red.

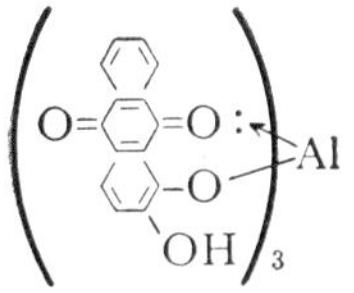

Turkey red

4. Indigo. Another conjugated system of double bonds, $O=C-C=C-C=O$, which may be considered a p-quinoid structure, is present in the most important of the vat dyes, indigo. Indigo is made from N-phenylglycine, which is obtained by heating aniline with chloroacetic acid. Sodium amide at the fusion temperature

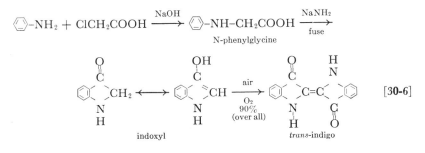

[30-6]

will effect a ring closure on N-phenylglycine, in which a molecule of water formed from the OH in the carboxyl group and an H from the *ortho* position on the ring is lost. By simply exposing indoxyl to air, we oxidize the methylene group and get indigo, thought to be the *trans* form.

Indigo is applied to cloth in the colorless, soluble, reduced form, obtained by treating indigo with sodium dithionite, $Na_2S_2O_4$. Exposure to air oxidizes the dye impregnated in the cloth to blue indigo.

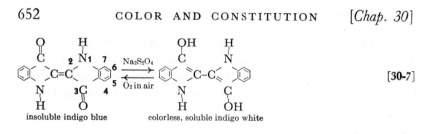

insoluble indigo blue   colorless, soluble indigo white

Exercise 30-3.   If the dithionite ion (Exp. 30-7) is oxidized to sulfite in basic solution, write the oxidation-reduction reaction indicated, and balance the equation.

Exercise 30-4.   Tyrian purple, the famous royal purple, is 6,6'-dibromoindigo. Write the structure for this molecule.

5. Triphenylmethane Dyes.   By oxidizing triphenylmethane and its derivatives, especially those having amine groups on one or more of the benzene rings in the original compound, we can prepare a series of colored substances, most of which are dyes. The effect of alkyl substitution and resonance on color changes is marked in these compounds. A red dye, pararosaniline, is obtained by the

$$3\langle\rangle + CHCl_3 \xrightarrow[16\%]{AlCl_3} (\langle\rangle)_3\text{-CH} \xrightarrow[\substack{H_2SO_4 \\ 65\%}]{HONO_2} (O_2N\text{-}\langle\rangle)_3\text{-CH} \xrightarrow[HCl]{Sn}$$

triphenylmethane

$$(H_2N\text{-}\langle\rangle)_3\text{-CH} \xrightarrow{PbO_2} H_2N\text{-}\langle\rangle\text{-}\underset{\underset{NH_2}{|}}{\overset{OH}{\underset{|}{C}}}\text{-}\langle\rangle\text{-}NH_2 \xrightarrow{H^+}$$

$$H_2N\text{-}\langle\rangle\text{-}\underset{\underset{NH_2}{|}}{C}\text{-}\langle\rangle\text{-}NH_2 \overset{\oplus}{\longleftrightarrow} H_2N\text{-}\langle\rangle\text{-}\underset{\underset{NH_2}{|}}{C}\text{=}\langle\rangle\text{=}\overset{\oplus}{NH_2} \qquad [30\text{-}8]$$

(*a*)   (*b*)

pararosaniline
(red)

series of reactions shown in Expression 30-8. There are three equivalent resonance structures (*b*) besides the somewhat less likely one (*a*). The resonance of the equivalent structures of *p*-

quinoid type is responsible for the color. (Structure *a* is less likely because nitrogen is more prone to accept a positive charge than carbon. Carbon with a $\oplus$ charge has only six electrons, and nitrogen has eight.)

Related triphenylmethane dyes, with their colors, are shown in Table 30-1. The first compound is not a dye, though it is colored.

TABLE
30-1 | *Triphenylmethane Dyes*

FORMULA	NAME	COLOR
1. $(\bigcirc)_2\text{-C}=\bigcirc=\overset{+}{\text{N}}\text{H}_2$		orange-red
2. $(\text{H}_2\text{N-}\bigcirc)_2\text{-C}=\bigcirc=\overset{+}{\text{N}}\text{H}_2$	pararosaniline	red
3. $\text{H}_2\text{N-}\bigcirc\text{-C}=\bigcirc=\overset{+}{\text{N}}\text{H}_2$ (with phenyl below)	Doebner's violet	violet
4. $(\text{CH}_3)_2\text{N-}\bigcirc\text{-C}=\bigcirc=\overset{+}{\text{N}}(\text{CH}_3)_2$ (with phenyl below)	malachite green	green
5. $(\text{CH}_3)_3\overset{+}{\text{N}}\text{-}\bigcirc\text{-C}=\bigcirc=\overset{+}{\text{N}}(\text{CH}_3)_2$ (with $\text{N}(\text{CH}_3)_2$ substituted phenyl below)	methyl green	green
6. $((\text{CH}_3)_2\text{N-}\bigcirc)_2\text{-C}=\bigcirc=\overset{+}{\text{N}}(\text{CH}_3)_2$	crystal violet	violet
7. $(\text{CH}_3)_2\overset{+}{\text{N}}\text{H-}\bigcirc\text{-C}=\bigcirc=\overset{+}{\text{N}}(\text{CH}_3)_2$ (with $\text{N}(\text{CH}_3)_2$ substituted phenyl below)	crystal violet (in acid sol.)	green
8. $((\text{CH}_3)_2\overset{+}{\text{N}}\text{H-}\bigcirc)_2\text{-C}=\bigcirc=\overset{+}{\text{N}}(\text{CH}_3)_2$	crystal violet (in strong acid)	yellow

It has been found necessary to have amine groups on two or more of the rings to get a triphenylmethane dye.

The effect of decreasing resonance contributions of certain forms to the total structure of a molecule is demonstrated graphically in the table. If we add acid to crystal violet, the color changes from violet through green to pale yellow as the acidity increases. Whereas there are three equivalent *p*-quinoid rings in structure 6, there are only two in 7; and in 8 only one structure can be written. Correspondingly, the color lightens considerably. These color changes, incidentally, are typical indicator reactions.

30-5. Colored Compounds: Indicators

An acid-base indicator is a weak acid or a weak base that changes color rapidly when the acidity of a solution containing the indicator increases or decreases slightly. The acidic groups are phenols, carboxylic acids, and sulfonic acids in various combinations; the basic groups are amines. Congo red, as we have seen, is an azo dye that can act as an indicator. In basic solution the

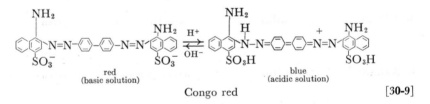

Congo red   [30-9]

anionic form probably predominates; in acid solution a shift from an azo to a *p*-quinoid structure (plus an azo group) is a likely explanation of the change in color.

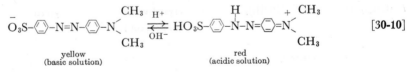

[30-10]

methyl orange

Methyl orange is also an azo dye, and it changes color for the same reason as Congo red: a shift from an azo to a *p*-quinoid structure, a change in chromophores.

A group of indicators known as phthaleins (derived from phthalic anhydride) also are colored because of a quinoid structure. The most important of these indicators is phenolphthalein, colorless in acid solution and red in basic solution, which is made by fusing

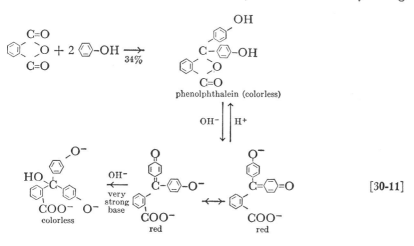

phenolphthalein (colorless)

[30-11]

colorless

red

red

phthalic anhydride with phenol. In very strong base, phenolphthalein is again colorless since the *p*-quinoid structure disappears.

Eosin (dye for red ink), fluorescein, and mercurochrome (the antiseptic) are derivatives of phthalic anhydride and resorcinol.

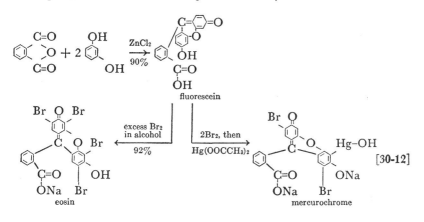

fluorescein

[30-12]

eosin

mercurochrome

The coloring matter in many flowers has the properties of an indicator. For example, the coloring matter in the blue cornflower, the red rose, and the red dahlia is the same dye at different acidities in the different flowers.

As indicated in the equilibrium reactions written for the indicators methyl orange, Congo red, and phenolphthalein, the addition of a base will shift an equilibrium in one direction and the addition of an acid will shift it in the other direction. The usefulness of an indicator depends on the sharpness of the color change and the narrowness of the range of acidity over which the change occurs.

The practical value of indicators is attested by the fact that every student of chemistry early learns the use of litmus paper. Litmus is an organic dye that happens to change color when the acidity is near that of pure water, pH 7, the neutral point (see Table 30-2). When one is neutralizing a strong base with a strong acid, litmus paper may be used as a test for the end point of the titration. When a strong base is used to titrate a weak acid—for example, sodium hydroxide and acetic acid—an

TABLE

30-2 | *pH Ranges of Some Indicators*

		Color in	
Indicator	*Range**	ACID	BASE
Methyl orange	pH 3.1–4.4	red	yellow
Phenolphthalein	8.3–10.0	colorless	red
Phenol red	6.8–8.4	yellow	red
Litmus	4.4–8.3	red	blue
Congo red	3.0–5.2	blue	red

* By "range" is meant the following: methyl orange is distinctly red at pH 3.1 and distinctly yellow at pH 4.4. Within the range 3.1–4.4 the color is changing. Litmus, though a common indicator, is not an especially good one, for its color change is over a long pH range.

indicator that changes on the basic side is necessary. In this case an indicator changing at about pH 9 is desirable. Phenolphthalein changes in that range (8.3–10.0). If a strong acid is titrated with a weak base, however, as HCl with carbonate ion, an indicator that changes color near pH 4 is necessary. Methyl orange can serve in that range. The pH ranges over which certain indicators may be used are gathered together in Table 30-2.

30-6. Free Atoms and Free Radicals

A free radical is a substance carrying one unpaired electron. This simple definition does not exclude atomic sodium or other atoms carrying single electrons, but in common usage elements are referred to as free atoms rather than free radicals. The term "free radical" is usually reserved for very short-lived alkyl free radicals and the longer-lived but reactive large organic molecules of the triarylmethyl type. There are a few simple molecules having an odd number of electrons (nitric oxide, chlorine dioxide, and nitrogen dioxide), and these have some of the properties associated with free radicals. A possible explanation of the relative stability of nitric oxide is the resonance contribution made by the canonical structures shown in Expression 30-15. There is evidence for the

$$\text{Na} \cdot \qquad\qquad\qquad\qquad\qquad\qquad\qquad\qquad\qquad \text{[30-13]}$$

free sodium atom

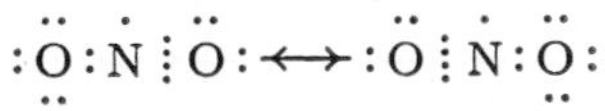

$$:\!\ddot{\text{O}}\!:\!\text{N}\vdots\ddot{\text{O}}\!: \longleftrightarrow :\!\ddot{\text{O}}\vdots\text{N}\!:\!\ddot{\text{O}}\!: \qquad\qquad\qquad \text{[30-14]}$$

nitrogen dioxide

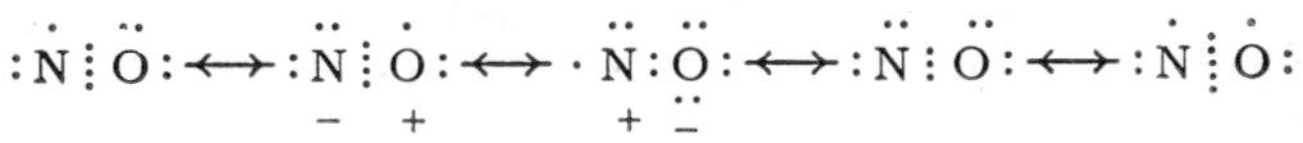

$$:\!\text{N}\vdots\ddot{\text{O}}\!: \longleftrightarrow :\!\ddot{\text{N}}\vdots\ddot{\text{O}}\!: \longleftrightarrow \cdot\,\ddot{\text{N}}\!:\!\ddot{\text{O}}\!: \longleftrightarrow :\!\ddot{\text{N}}\!:\!\ddot{\text{O}}\!: \longleftrightarrow :\!\text{N}\vdots\ddot{\text{O}}\!: \qquad \text{[30-15]}$$

$$\qquad\qquad - \quad + \qquad\quad + \quad -$$

nitric oxide

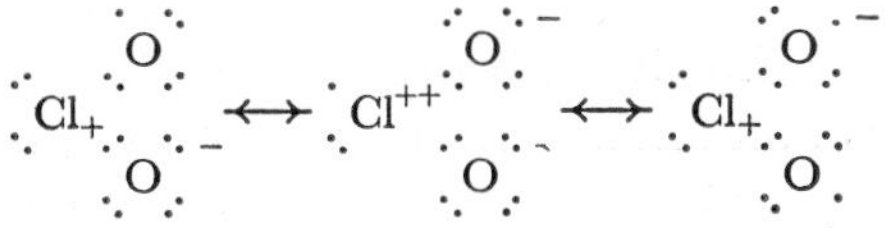

$$\text{[30-16]}$$

chlorine dioxide

existence of the three-electron bond in NO shown in the last two structures.

30-7. Reactions of Free Atoms

1. Chlorine with Hydrogen. The combination of chlorine and hydrogen does not take place in the dark if both substances are completely dry. Sunlight, however, will initiate the reaction, which then proceeds with explosive violence by a chain mechanism, the first step of which is the photochemical dissociation of molecular chlorine into free atoms. Since a chlorine atom is regenerated in

$$Cl_2 + h\nu \longrightarrow 2Cl \cdot \qquad [\textbf{30-17a}]$$

$$Cl \cdot + H_2 \longrightarrow HCl + H \cdot \qquad [\textbf{30-17b}]$$

$$H \cdot + Cl_2 \longrightarrow HCl + Cl \cdot \qquad [\textbf{30-17c}]$$

Expression 30-17c, the last two reactions could, on the face of it, continue indefinitely—that is, until all the H_2 and Cl_2 were exhausted. Occasionally, however, two chlorine or two hydrogen atoms may collide to give a chlorine or a hydrogen molecule, interrupting the chain reaction.

2. Chlorine with Alkanes. The stepwise halogenation of methane was one of the first reactions studied in this course (p. 52). It proceeds by a mechanism involving free chlorine atoms in a chain in which alkyl and substituted alkyl free radicals are postulated as intermediates in the reaction (Exp. 2-3). Few chlorine atoms are needed to initiate the reaction since they are continuously regenerated as in Expression 30-17.

3. Sodium with Halogens and Alkyl Halides (Wurtz). Sodium vapor at 275°C, when passed into chlorine gas, burns with a bright-yellow flame, forming sodium chloride and liberating considerable heat. The reaction is a transfer of an electron from the free sodium

$$Na \cdot + Cl_2 \longrightarrow \overset{+}{Na}\overset{-}{Cl} + Cl \cdot + 35 \text{ kcal} \qquad [\textbf{30-18}]$$

atom to the chlorine molecule, which splits into a chloride ion and a free chlorine atom.

The Wurtz reaction may proceed by this same type of free-radical mechanism. (See p. 72.)

30-8. Preparation of Free Radicals

1. Gomberg's Method.   The preparation in 1900 of the first free radical involving an unpaired electron attached to a carbon atom was viewed with much skepticism by chemists, since the idea that carbon always carried four bonds in compounds had become firmly entrenched. Moses Gomberg, working at the University of Michigan, reported an attempt to prepare hexaphenylethane, which ended in the formation of a free radical instead. Gomberg had treated triphenylchloromethane with metallic silver, as shown in Expression 30-19. Hexaphenylethane was formed, as Gomberg

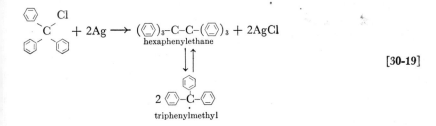

[30-19]

expected; but, if the reaction was run in benzene in the absence of oxygen, dissociation into triphenylmethyl ensued. Gomberg had first run the reaction in air, but the hexaphenylethane always contained oxygen upon analysis. In the next section, on reactions of free radicals, this reaction with oxygen is discussed (p. 661). Zinc, mercury, sodium, and potassium, in addition to silver, have been used to remove the halide.

2. Conant's Method.   A second general method of preparing free radicals of the triarylmethyl type, due to Conant, is the reduction of triarylmethyl halides. A triarylmethylcarbinol is treated with concentrated HCl, and the resulting triarylmethyl halide is treated with strong reducing agents in an inert atmosphere to give the free radical. The reaction is shown (Exp. 30-20) for triphenylmethyl radicals. The triphenylmethyl chloride is partially ionized in acetone, glacial acetic acid, or absolute ethyl alcohol. Strong reducing agents that may be used here are vanadium II and titanium III chlorides.

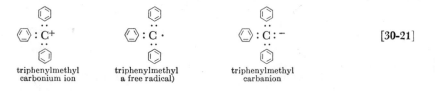

$$[30\text{-}20]$$

It may be well at this point to draw a clear distinction between a carbonium ion, a carbanion, and a free radical. A carbonium ion has a positive charge, due to the presence of only six electrons in the outside shell of a carbon (shared with three combining groups). A carbanion has a negative charge, due to the presence of eight electrons in the outside shell of the carbon that shares only six with three combining groups. A free radical, with carbon as the seat of an unpaired electron, has a total of seven electrons in its outside shell, six of which are shared with three combining groups.

$$[30\text{-}21]$$

triphenylmethyl   triphenylmethyl   triphenylmethyl
carbonium ion   a free radical)   carbanion

30-9. Reactions of Triarylmethyl Free Radicals

The great reactivity of a carbon lacking an electron to complete its octet is indicated in a number of reactions with $Ar_3C\cdot$, a triarylmethyl free radical. Most of the reactions are additions involving the formation of a complete octet. A few reactions, however, involve the loss of an electron to yield a carbonium ion. Carbonium ions are known only in solution and have not been isolated.

A. Addition Reactions

1. With Halogens

$$2Ar_3C \cdot + I_2 \longrightarrow 2Ar_3C : I \qquad\qquad\qquad [30\text{-}22]$$

2. With Oxygen. A peroxide is formed by the addition of oxygen to a free radical. Such a peroxide was formed in Gomberg's

$$2Ar_3C \cdot + O_2 \longrightarrow Ar_3C : \overset{..}{\underset{..}{O}} : \overset{..}{\underset{..}{O}} : CAr_3 \qquad\qquad [30\text{-}23]$$

attempted synthesis of hexaphenylethane (p. 659) when he carried out the reaction without excluding air. The analyses for carbon and hydrogen were always too low for the expected hydrocarbon,

$$(C_6H_5)CCl + Ag \longrightarrow \underset{\text{not isolated}}{(C_6H_5)_3C \cdot} + AgCl$$

$$(C_6H_5)_3C\text{-}O\text{-}O\text{-}C(C_6H_5)_3 \overset{O_2}{\longleftarrow}\!\!\bigg| \qquad\qquad [30\text{-}24]$$

$C_{38}H_{30}$. It was cognizance of the low values in the carbon-hydrogen analyses that led Gomberg to his far-reaching discovery.

3. With Nitric Oxide. By treating a free radical with nitric oxide we get a blue-green nitroso compound that very rapidly loses its color by dimerization.

$$Ar_3C \cdot + NO \longrightarrow \underset{\text{blue-green}}{Ar_3C : \overset{..}{N} \overset{..}{\underset{.}{\cdot}} \overset{..}{O} :} \longrightarrow \underset{\text{colorless}}{(Ar_3C\text{-}NO)_2} \qquad [30\text{-}25]$$

4. Reduction. In the presence of platinum as a catalyst, hydrogen will add to a free radical.

$$2Ar_3C \cdot + H_2 \overset{Pt}{\longrightarrow} 2Ar_3C : H \qquad\qquad\qquad [30\text{-}26]$$

5. With Active Hydrogens. A hydrogen atom attached to an oxygen, nitrogen, or sulfur readily reacts with a triarylmethyl free

radical. The fragment left after the loss of the hydrogen atom may
then combine with a second free radical.

$$2Ar_3C \cdot + HOH \xrightarrow{\text{trace of } I_2} Ar_3C : H + Ar_3C : OH \qquad [30-27]$$

$$2Ar_3C \cdot + \text{\textcircled{}}-NH-NH_2 \longrightarrow Ar_3C : H + Ar_3C : \overset{H}{\underset{|}{N}}-\overset{H}{\underset{|}{N}}-\text{\textcircled{}} \qquad [30-28]$$

$$2Ar_3C \cdot + HCl \longrightarrow Ar_3C : H + Ar_3C : Cl \qquad [30-29]$$

B. *Replacement Reaction*

The free radical triphenylmethyl will dissolve in liquid sulfur
dioxide to give a solution that will conduct an electric current.
Conductivity, of necessity, is due to the presence of charged particles. One explanation of this conductivity assumes the ability of
the solvent to solvate the electron—that is, to separate it from the
triphenylmethyl group. The idea came from the fact that the same
theory has been accepted as an explanation of the conductivity of
a solution of sodium in liquid ammonia.

$$Ar_3C \cdot + SO_2 \longrightarrow Ar_3C^+ + e^-(SO_2)_x \qquad [30-30]$$

$$Na \cdot + NH_3 \xrightarrow{\text{Fe cat.}} Na^+ + e^-(NH_3)_x \qquad [30-31]$$

It is also possible to form the triphenylmethyl carbanion in liquid
ammonia by the dissociation of its sodium salt.

$$Na : C(C_6H_5)_3 \xrightarrow{\text{liq. } NH_3} Na^+ + {}^-: C(C_6H_5)_3 \qquad [30-32]$$
$$\text{deep-red color}$$

30-10. Color in Free Radicals

Free radicals are, in general, highly colored substances. Triphenylmethyl is yellow in benzene solution. If other aryl radicals
are formed in benzene solution, the color may be still deeper;
phenyl-α-naphthyl-p-biphenylmethyl, for example, is deep-brown
in benzene solution.

The color in these triarylmethyl free radicals, as we saw earlier in this chapter, has been correlated with the fact that substances containing quinoid structures are colored. The quinoid chromophore is present in many of the resonance structures that can be written for this type of free radical. The deepening of the color as one goes to free radicals with more possible resonance structures has been noted by a number of investigators (Table 30-3).

TABLE 30-3 | *Properties of Some Triarylmethyl Free Radicals*

FORMULA	% DISSOCIATED IN 2–3% BENZENE	COLOR
$(\bigcirc)_3 C \cdot$	3	yellow
$(\bigcirc-\bigcirc)_3-C \cdot$	100	deep-violet
$(\bigcirc)_2-C-\bigcirc-\bigcirc$	15	orange-red
$\bigcirc-C-(\bigcirc-\bigcirc)_2$	79	red
$\bigcirc-C-\bigcirc-\bigcirc$ (with fused ring)	>90	brown
$(\bigcirc)_3-C-C-(\bigcirc)_2$	100	bright-red
$C(\bigcirc)_2$ (with fused ring)	——	bright-red

Triphenylmethyl, in addition to the structure commonly written for it, has nine resonance structures nearly equivalent in energy.

five other equivalent structures   two other equivalent structures

[30-33]

There are six equivalent *o*-quinoid structures and three equivalent *p*-quinoid structures (Exp. 30-33) without the Kekule resonance contributions from each benzene ring. The effect of replacing a phenyl group by a biphenyl is to increase the number of resonance forms, as indicated by two representative structures in diphenyl-biphenylmethyl (Exp. 30-34).

[30-34]

Exercise 30-5. How many other canonical resonance forms can you write for diphenylbiphenylmethyl?

30-11. Polymerization by Free Radicals

A large number of commercial polymerizations are catalyzed by substances that have been shown to produce free radicals. Perhaps the most widely used of these catalysts is benzoyl peroxide,

$$\text{○-}\overset{\overset{O}{\|}}{C}\text{-O-O-}\overset{\overset{O}{\|}}{C}\text{-○}$$

When it decomposes, a free benzoate radical is formed and initiates the polymerization of the monomer. Benzoate free radicals may

$$\text{○-}\overset{\overset{O}{\|}}{C}\text{-O-O-}\overset{\overset{O}{\|}}{C}\text{-○} \longrightarrow 2\,\text{○-}\overset{\overset{O}{\|}}{C}\text{-O} \cdot \longrightarrow \text{○} \cdot + CO_2 \qquad [30\text{-}35]$$

themselves decompose into phenyl free radicals, which are also very reactive, as might be expected. The polymerization is then initiated by the free radical from the peroxide (probably either one) and propagates itself in the way shown for polyvinyl chloride in Expression 30-36. The polymerization may be terminated by removal of a hydrogen from another monomer or another chain or from water if traces of moisture happen to be present, or in other ways.

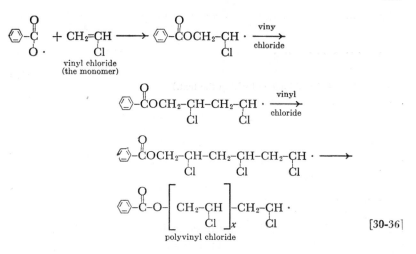

[30-36]

Benzoyl peroxide may be made by the action of sodium peroxide on benzoyl chloride.

[30-37]

SUMMARY

The interplay of resonance effects in chromophores and auxochromes determines, in large part, what substances will be dyes, indicators, or colored free radicals. Dyes are classified by the chemist with regard to functional groups but by the dyer according to the procedure of application. Indicators are weak acids or weak bases generally containing some solubilizing group. Free radicals play a determining role in a large number of organic reactions, their color being an interesting facet of their behavior.

REFERENCE

De LaMare and Vaughan, J. Chem. Education, 34, 10 (1957).

EXERCISES

(Exercises 1–5 will be found within the text.)

6. Would you expect any of the following substances to be colored?

1. 4-nitro-3'-carboxy-4'-hydroxyazobenzene
2. 4-dimethylamino-3'-carboxy-4'-hydroxyazobenzene
3. 4-nitroso-1-naphthol

Explain.

7. What would be the structure of eosin in acid solution? Would you expect it to be colored in acid solution?

8. Malachite green changes from green to yellow when the acidity is increased. Explain.

9. An iron mordant is used with naphthol green B, sodium 1-nitroso-2-naphthol-6-sulfonate, to make a green dye. Write a structure for the mordant dye.

REVIEW PROBLEMS

1. Identify the following terms and named reactions:

1. Perkin reaction	10. Zerewitinoff determination
2. Diels-Alder reaction	11. Van Slyke titration
3. Bucherer reaction	12. Reimer-Tiemann reaction
4. Beckmann rearrangement	13. Skraup synthesis
5. Clemmensen reduction	14. Kolbe synthesis
6. Friedel-Crafts reaction	15. Gabriel synthesis
7. Hell-Volhard-Zelinsky reaction	16. Strecker synthesis
8. Claisen condensation	17. Fries rearrangement
9. Sandmeyer reaction	18. chain reaction
	19. benzoin condensation

20. benzilic acid rearrangement
21. coupling of diazonium salt
22. zwitterion
23. phthaleins
24. lactone
25. lactam
26. polypeptide
27. free radical

28. inductive effect
29. chromophore
30. asymmetric molecule
31. carbonium ion
32. carbanion
33. isoelectric point
34. chelation

2. How does the theory of resonance and electronic structure apply to or explain the following?

1. structure of benzene
2. acidity of various halogenated acids
3. basicity of various nitrogen compounds
4. relative reactivity of halides (alkyl, aryl, allyl, and vinyl)
5. orientation of benzene derivatives
6. color of various compounds
7. Claisen rearrangement

3. Indicate (a) a general synthetic method and (b) a typical reaction for each of the following classes of compounds:

1. dicarboxylic acids
2. halogen-substituted acids
3. α,β-unsaturated acids
4. hydroxy acids
5. amino acids
6. keto acids
7. unsaturated alcohols and halides
8. sulfonic acids
9. anthracene
10. triphenylmethane
11. naphthalene

4. Outline the various products of reduction of nitrobenzene.

5. Diagram the interrelationships of urea, guanidine, phosgene, diethyl carbonate, and carbon dioxide.

6. Indicate the use of each of the following in the preparation of longer-chain acids:

1. malonic ester
2. acetoacetic ester
3. aldol condensation
4. Reformatsky reaction

7. Indicate the products of reaction of a Grignard reagent with

1. ethanol
2. aniline
3. ethyl acetate
4. ethylene oxide
5. acetyl chloride
6. acetonitrile

7. benzophenone
8. allyl bromide
9. $CH_3-CH=CH-\overset{\overset{\displaystyle O}{\displaystyle \|}}{C}-C_6H_5$

8. Classify dyestuffs (a) by method of application and (b) by skeleton structure.

9. Indicate a preparation and use for organic compounds of

1. mercury
2. zinc
3. lead

4. lithium
5. sodium

Index

An italicized page number refers to a table on the page.

669

13370

Clapp, Leallyn

Chemistry of the covalent
  bond.

INTERNATIONAL ATOMIC WEIGHTS 1956

NAME	SYMBOL	ATOMIC NUMBER	ATOMIC WEIGHT*
Actinium	Ac	89	227
Aluminum	Al	13	26.98
Americium	Am	95	[243]
Antimony	Sb	51	121.76
Argon	A	18	39.944
Arsenic	As	33	74.91
Astatine	At	85	[210]
Barium	Ba	56	137.36
Berkelium	Bk	97	[249]
Beryllium	Be	4	9.013
Bismuth	Bi	83	209.00
Boron	B	5	10.82
Bromine	Br	35	79.916
Cadmium	Cd	48	112.41
Calcium	Ca	20	40.08
Californium	Cf	98	[249]
Carbon	C	6	12.011
Cerium	Ce	58	140.13
Cesium	Cs	55	132.91
Chlorine	Cl	17	35.457
Chromium	Cr	24	52.01
Cobalt	Co	27	58.94
Copper	Cu	29	63.54
Curium	Cm	96	[245]
Dysprosium	Dy	66	162.51
Erbium	Er	68	167.27
Europium	Eu	63	152.0
Fluorine	F	9	19.00
Francium	Fr	87	[223]
Gadolinium	Gd	64	157.26
Gallium	Ga	31	69.72
Germanium	Ge	32	72.60
Gold	Au	79	197.0
Hafnium	Hf	72	178.50
Helium	He	2	4.003
Holmium	Ho	67	164.94
Hydrogen	H	1	1.0080
Indium	In	49	114.82
Iodine	I	53	126.91
Iridium	Ir	77	192.2
Iron	Fe	26	55.85
Krypton	Kr	36	83.80
Lanthanum	La	57	138.92
Lead	Pb	82	207.21
Lithium	Li	3	6.940
Lutetium	Lu	71	174.99
Magnesium	Mg	12	24.32
Manganese	Mn	25	54.94
Mendelevium	Mv	101	[256]
Mercury	Hg	80	200.61
Molybdenum	Mo	42	95.95
Neodymium	Nd	60	144.27
Neon	Ne	10	20.183
Neptunium	Np	93	[237]
Nickel	Ni	28	58.71
Niobium	Nb	41	92.91
Nitrogen	N	7	14.008
Osmium	Os	76	190.2
Oxygen	O	8	16.0000
Palladium	Pd	46	106.4
Phosphorus	P	15	30.975
Platinum	Pt	78	195.09
Plutonium	Pu	94	[242]
Polonium	Po	84	210
Potassium	K	19	39.100
Praseodymium	Pr	59	140.92
Promethium	Pm	61	[145]
Protactinium	Pa	91	231
Radium	Ra	88	226.05
Radon	Rn	86	222
Rhenium	Re	75	186.22
Rhodium	Rh	45	102.91
Rubidium	Rb	37	85.48
Ruthenium	Ru	44	101.1
Samarium	Sm	62	150.35
Scandium	Sc	21	44.96
Selenium	Se	34	78.96
Silicon	Si	14	28.09
Silver	Ag	47	107.880
Sodium	Na	11	22.991
Strontium	Sr	38	87.63
Sulfur	S	16	32.066
Tantalum	Ta	73	180.95
Technetium	Tc	43	[99]
Tellurium	Te	52	127.61
Terbium	Tb	65	158.93
Thallium	Tl	81	204.39
Thorium	Th	90	232.05
Thulium	Tm	69	168.
Tin	Sn	50	118.
Titanium	Ti	22	
Tungsten	W	74	
Uranium	U	92	
Vanadium	V	23	
Xenon	Xe	54	
Ytterbium	Yb	70	
Yttrium	Y	39	
Zinc	Zn	30	
Zirconium	Zr	40	

* A value given in brackets is the mass number of the most stable known isotope.

WITHDRAWN